Semiconducting and Insulating Materials 1998

Proceedings of the 10th Conference
on Semiconducting and Insulating Materials (SIMC-X)

1–5 June 1998
Berkeley, California, USA

1998 IEEE Semiconducting and Insulating Materials Conference

Copyright and Reprint Permission

IEEE Catalog Number   98CH36159 (softbound)
98CB36159 (casebound)

ISBN   0-7803-4354-9 (softbound)
0-7803-4355-7 (casebound)
0-7803-4356-5 (microfiche)

Library of Congress   97-80479

Additional copies of this publication are available from

IEEE Operations Center
P.O. Box 1331
445 Hoes Lane
Piscataway, NJ 08855-1331 USA

1-800-678-IEEE
1-732-981-1393
1-732-981-9667 (FAX)
email: customer.services@ieee.org

The SIMC-X Conference

Organized by

Lawrence Berkeley National Laboratory

This volume has been prepared by

Zuzanna Liliental-Weber and Carla Miner
SIMC-X Editors

Committees

Acknowledgments

The Chair of the conference wants to thank our generous sponsors.

For sponsoring the conference:

IEEE Electron Device Society

For financial support of the conference and conference scholarships:

Air Force Office of Scientific Research

Lawrence Berkeley National Laboratory

Wright Patterson Air Force Base

For gracious contribution to the opening reception:

AIXTRON

Editorial

The tenth Conference on Semiconducting and Insulating Materials (SIMC-X) was held at the Radisson Hotel located at the beautiful waterfront in Berkeley, CA, USA, from June 1 to June 5, 1998. The SIMC is a series of biannual conferences initiated in 1980 in Nottingham, UK. This conference has traditionally been the leading meeting for fundamental materials problems of compound semiconductors of relevance for electronic and optoelectronic devices. The tenth, jubilee, conference attracted 120 participants from 27 countries. More than 100 abstracts were selected. The papers were delivered orally or during poster presentations. The conference brought together specialists from various fields, including material scientists, solid state physicists, device and process engineers. During the meeting growth, characterization, theory, device applications, and materials problems related to semiconducting and insulating compound semiconductors such as GaAs (including low-temperature grown off-stoichiometric materials), InP, III-nitrides, SiC, and II-VI compounds were discussed.

This conference was sponsored by the IEEE Electron Device Society, and supported by the Air Force Office of Scientific Research, Lawrence Berkeley National Laboratory, and Wright Patterson Air Force Base with gracious contribution to the opening reception from AIXTRON. These financial contributions helped the organizers to support 16 participants (some invited speakers and some participants from East European and Asian countries) who otherwise would have had financial difficulties to participate in this truly international conference, and accept 24 students with almost half reduced rates. I would like to give wholehearted thanks for this generous help.

The Conference consisted of ten oral sessions with ten invited and 55 contributed presentations as well as 46 poster presentations distributed into two evening sessions. The International Advisory Committee and Program Committee took active part in shaping the program, nominating invited speakers and rating the contributed abstracts.

Despite this active program, the atmosphere always remained relaxed and informal. Coffee breaks and lunches, the possibility to walk at the waterfront, gave participants time to relate to colleagues scientifically and socially. The social program included a free afternoon with a tour for wine tasting to a local California winery. The highlight of the evening banquet in the Radisson Hotel was a Chopin concert played by Leesa Dahl from the SF Conservatory of Music.

Finally, I want to thank the conference secretary for the preparation and updating of the conference web page and abstract collection. I want to express my gratitude to all colleagues from my laboratory for their help with the preparation of the conference abstract book, in order to save money to allow a larger number of students to participate in this exciting meeting for reduced rates, as well as all colleagues who helped in the manuscript room with paper collection and computer work, during and after the conference, and with the preparation of manuscripts for publication. I also thank all distinguished colleagues for reviewing all papers for publication.

I want to wish Dr. C. Jagadish (Australian National University), the organizer of SIMC-XI, a lot of success in the organizing of the next conference starting the new Millennium on the southern hemisphere, in Canberra, Australia.

Zuzanna Liliental-Weber
Conference Chair

Contents

I. Crystal Growth

II. Semi-Insulating III-V Compounds

III. Non-Stoichiometric III-V Compounds

IV. Defects in III-V Compounds

V. Low-Dimensional Structures

VI. Group III-V Oxides

VII. Group III-Nitrides

VIII. Silicon Carbide

IX. Wide Gap II-IV Compounds

X. Devices

I.

Crystal Growth

High Pressure Solution Growth and Physical Properties of GaN crystals

I. Grzegory

*High Pressure Research Center, Polish Academy of Sciences,
Sokolowska 29/37, 01-142 Warsaw, Poland*

The recent results in high pressure crystallization of GaN from the solutions of N in pure Ga and in its alloys with II group metals (Mg, Ca, Zn) are discussed. It is shown that high resistivity (10^4-$10^6\Omega$cm) GaN crystals of improved structural quality can be grown from solutions containing Mg. The structural, electrical and optical properties properties of GaN crystals grown without and with the intentional doping are compared.

It is also shown that atomically flat, thermally stable surfaces of GaN substrates are possible to obtain by mechanical and mechano-chemical polishing .

Some most interesting results concerning homoepitaxial growth by MOCVD and MBE are reviewed.

A. Introduction

GaN is the most important material for blue and ultraviolet optoelectronics. The device structures [1] are usually grown on foreign substrates which results in a high density of dislocations above 10^8cm^{-2}·.The application of high N_2 pressure gives a unique possibility of growing of GaN single crystalline substrates which allows to lower the dislocation density in epitaxial layers by 3-4 orders of magnitude. During the last two years the surface area of pressure grown GaN single crystalline platelets has been increased by more than one order of magnitude (from 4-6mm^2 to 60-100mm^2).

B. Crystallization experiment

B. 1.   *Thermodynamical conditions*

Gallium nitride is a strongly bonded compound (bonding energy of 8.9 eV/atom pair) in comparison with typical III-V semiconductors like GaAs (bonding energy of 6.5 eV/atom pair [2]). Consequently, the free energy G(T) of the crystal is low in relation to the reference state of free N and Ga atoms. On the other hand, the N_2 molecule is also strongly bonded (4.9eV/atom). Therefore, the free energy of GaN constituents at their normal states, Ga and N_2, becomes quite close to that of the crystal. It is illustrated in Fig. 1 where the free energy of GaN (1 mole) and the free energy of the system of its constituents (Ga+1/2N_2) is shown as a function of temperature and N_2 pressure. With increasing temperature, G(T) of the constituents decreases faster than G(T) of the crystal and at higher temperatures, the nitride becomes thermodynamically unstable. The crossing of G(T) curves determines the equilibrium temperature where GaN coexists with its constituents at given N_2 pressure. The application of pressure increases the free energy of the constituents in much higher degree than G(T) of the crystal. As a consequence the equilibrium point shifts to higher temperatures and GaN stability range extends. The equilibrium p_{N2} - T conditions for GaN were determined experimentally by Karpinski et al. [3] and the curve following from these data is shown in Fig. 2.

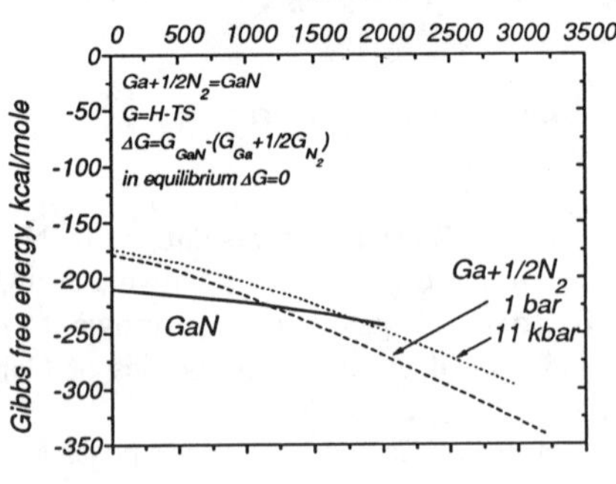

Fig.1 Gibbs free energy of GaN and its constituents

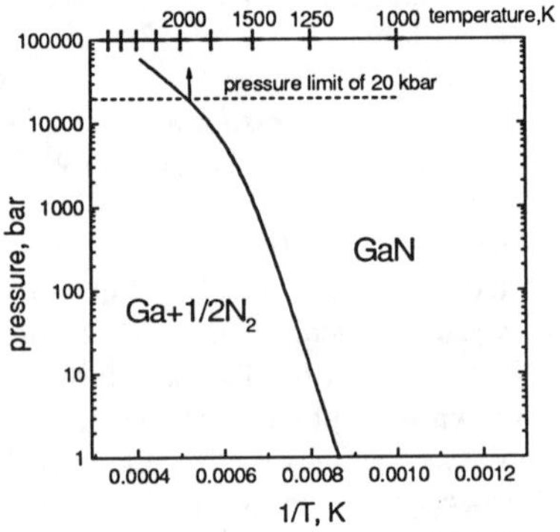

Fig. 2 Equilibrium curve for GaN [3]

Crystallization processes discussed in this paper have been carried out at N_2 pressure up to 20 kbar which corresponds to a GaN stability limit of 1960K. These conditions are marked in Fig.2. The extension of the GaN stability range allows to grow GaN crystals from a solution in liquid Ga. However even the highest available temperature of 1960 K is quite far from the melting temperature of GaN (close to 3000K as calculated by Van Vechten [4]). Therefore the N concentrations are not high (about 1 at.%) and the growth experiments have to be long to get high quality crystals with dimensions appropriate for research and applications.

B.2. Apparatus and procedure

At present GaN is crystallized in pressure chambers with volume up to 1500 cm^3 allowing crucibles with a working volume of 50-100 cm^3. The high pressure - high temperature reactor consisting of the pressure chamber and the multizone furnace is equipped with additional systems necessary for: annealing in vacuum, electronic stabilization and programming of pressure and temperature, and cooling of of the pressure chamber. Pressure in the chamber is stabilized with a precision better than 10 bar . The temperature is measured by a set of thermocouples arranged along the furnace and coupled with the standart input power controllers. This allows stabilization of temperature to ±0.2 deg. and programmable changes of temperature distribution in the crucible.

GaN crystals presented in this paper were grown from the solutions in pure liquid gallium and in Ga alloyed with 0.2 - 0.5 at.% of Mg, Ca or Zn which are the most common acceptors in GaN.

The supersaturation in the growth solution has been created by the application of a temperature gradient of 2-20 °C/cm along the axis of the crucible. The slow cooling method was not applied due to the small concentrations of nitrogen in the liquid gallium .

The crystallization experiments were performed without intentional seeding and the crystals nucleated spontaneously on the crucible walls at the cooler zone of the solution. Typical duration of the process was 120 - 150 hours.

B.3 Crystals

The GaN crystals grown by this method are wurzite, mainly in the form of hexagonal platelets or hexagonal needles. High supersaturations favor growth into the c-direction which leads to the needle like forms.

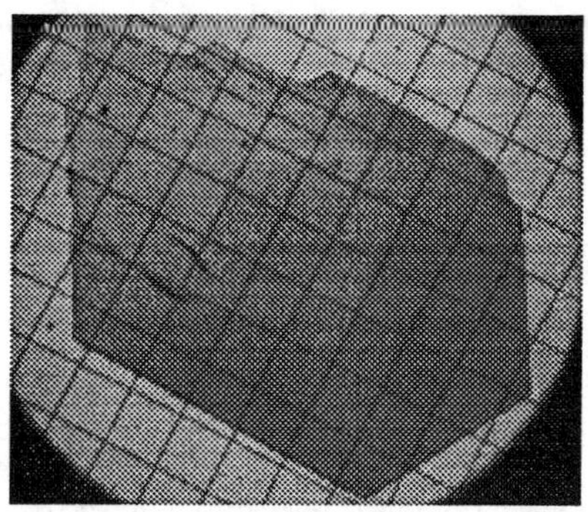

Fig.3 GaN crystal grown from the solution in liquid Ga at high N_2 pressure. The distance between two grid lines corresponds to 1mm.

The crystals, in the form of hexagonal platelets grown slowly, with a rate <0.1mm/h into $\{10\underline{1}0\}$ directions (perpendicular to the c-axis), are usually single crystals of perfect morphology suggesting stable layer-by-layer growth. They are transparent, with flat mirror like faces. Three dimensional, hexagonal growth figures growing during cooling of the system after crystallization in the temperature gradient, are observed on the crystal surfaces. A GaN platelet grown without intentional doping is shown in Fig. 3. The growth is strongly anisotropic being much faster in directions perpendicular to the c-axis.

This relation is valid at supersaturations corresponding to the average growth rate in $\{10\underline{1}0\}$ directions of 0.05 - 0.1mm/h. If the supersaturation is too high, the edge nucleation on the hexagonal faces of GaN platelets is often observed, which is the first step to unstable growth. This tendency for unstable growth is stronger for one of the polar $\{0001\}$ faces of the platelets. On this side, morphological features such as macrosteps, periodic inclusions of solute or cellular growth can be observed. The opposite surface is mirror like and often atomically flat. Only this side of the crystals grown from solutions in pure Ga, can be etched by alkaline water solutions [5]. It was identified by CBED [6] and XPD measurements [7] as the $(000\underline{1})_N$ surface.

Crystals grown from solutions containing Mg, Ca or Zn also form hexagonal platelets similar to the undoped samples but crystals containing Mg grow slightly slower in directions perpendicular to the c-axis and faster into c-directions resulting in relatively smaller but thicker platelets.

For crystals doped with Mg the etching behavior is different which means that the flat face is always inert to etching and only the opposite side can be etched. However the polarity of the Mg doped crystals was not yet determined and it is still unclear if Mg doping changes the growth mechanism or etching mechanism.

C. Physical properties of pressure grown crystals

GaN crystals grown from the pure gallium solutions are highly conductive showing metallic behaviour (Table1) in the whole temperature range of 4.2 - 300K. High free electron concentrations of 3-6 x 10^{19} cm^{-3} with mobilities of 30-90 cm^2/Vs [8] have been found. The main residual impurity detected in the crystals by SIMS [9] was oxygen at an estimated concentration in the range of 10^{18} - 10^{19} cm^{-3}. It is well established that oxygen is a single donor in GaN. However since the concentration of free electrons in the crystals is higher than the estimated concentration of the impurity, the N-vacancy as an additional source of free electrons is often proposed [10,11].

Following theory [10,12], the most probable native defects in GaN crystals which are highly n-type, are Ga-vacancies (V_{Ga}^{3-}). This is due to their low formation energy even in strongly Ga-rich conditions of crystallization. The presence of negatively charged Ga vacancies at concentrations of 10^{18} cm^{-3}, in n-type pressure grown GaN crystals, was detected by positron annihilation experiments [13].

The introduction of 0.2-0.5 at% of Ca or Zn into the growth solution does not change the character of the electrical properties of the crystals. Crystals still show the metallic type of conductivity though the free electron concentration decreases by factor of a few. In contrast, the addition of Mg into the growth solution drastically changes the electrical properties of GaN crystals increasing their resistivity by orders of magnitude (Table 1).

Table 1 Electrical properties of GaN crystals

crystal	conductivity type	ρ Ωcm, 300K	carrier concentration,cm^{-3}
GaN	metallic	$10^{-3}10^{-2}$	3-6 x 10^{19}, n-type
GaN : Mg	hopping	10^{4}-10^{6}	----

This increase of electrical resistance in GaN:Mg crystals is related to a drastic decrease of free electron concentration. The temperature dependence of the resistivity for these samples is typical for hopping conductivity [14] (Fig.4), which suggests that the Fermi level lies within the gap. The optical absorption data [12] also indicates that the free carrier concentration in the Mg-doped GaN is very low (Fig. 5). The free carrier absorption which dominates the low energy part of the absorption spectra for the undoped GaN disappears completely for crystals grown from Mg containing solutions.

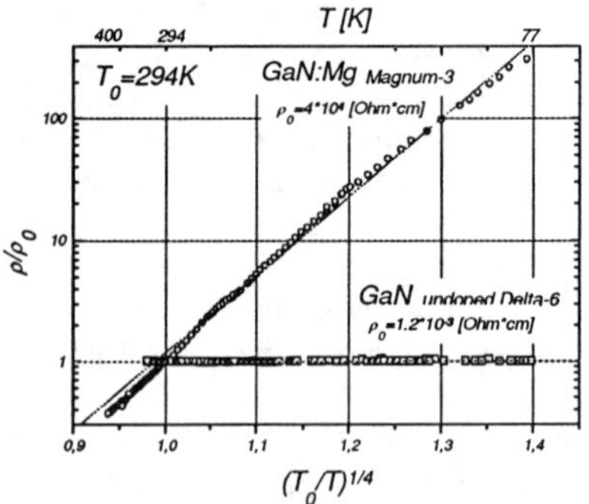

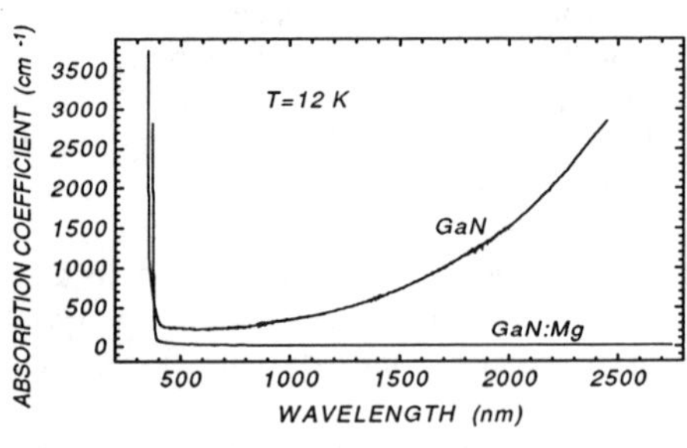

Fig.4 Temperature dependence of resistivity for GaN crystals grown from solutions of both pure liquid Ga and of Ga alloyed with Mg

Fig. 5 Optical absorption for GaN crystals grown from solutions of both pure liquid Ga and of Ga alloyed with Mg

In contrast to the highly conductive n-type crystals, no Ga-vacancies were detected by positron anihilation [15] in the highly resistive GaN:Mg crystals. This shows that doping of GaN substantially modifies its stoichiometry which is in qualitative agreement with the results of *ab initio* calculations [10,12] which predict the decrease of V_{Ga} concentration with decreasing Fermi energy.

Crystals grown from solutions in pure Ga show strong yellow photoluminescence (PL) and also relatively weak PL close to the energy gap. Both signals disappear in crystals obtained from solutions doped with Mg, Zn and Ca. Instead a strong broad PL occurs at energies of 2.8 - 3.2 eV (at 80K). The mechanism of this luminescence is not well understood.

The structure of the GaN crystals was investigated by X-ray diffraction methods [16] (Table 2). For the conductive crystals, the shape of the X-ray rocking curves ((0002)CuKα reflection) often depends on the size of the crystal. The full widths at half maximum (FWHM) were 20-30 arcsec for 1mm crystals and 30-40 arcsec for 1-3mm ones. For larger platelets the

6

rocking curves often split into a few ~30 40arcsec peaks showing a presence of low angle (1-3 arcmin) boundaries separating grains of 0.5-2mm in size.

Table 2 Lattice parameters for GaN crystals and homoepitaxial layers

Sample	a, Å	c, Å	FWHM X-ray rocking curve (0002) refl.
-GaN conductive, bulk	3.1881-3.1890	5.1856-5.1864	30"-40"
-undoped homoepitaxial GaN on undoped GaN bulk	3.1881*	5.1844	30"-40"
-GaN:Mg, bulk	3.1876	5.1846	20"-30"
-undoped homoepitaxial GaN on GaN:Mg bulk	3.1876	5.1846	20"-30"

*lattice parameter *a* for the substrate and layer is always the same

TEM examination by Z. Liliental - Weber et al. [6], determined that the $(000\underline{1})_N$ surface of conductive GaN crystals is often atomically flat (2-3 monolayer steps present) and that the crystal under this surface is practically free of extended defects.

On the opposite side a number of extended defects like stacking faults, dislocation loops and Ga microprecipitates were observed. The relative thickness of this part usually consists of 10% of the entire thickness of the platelet. Such pecularities result from different growth modes into the two polar directions of the crystal in the Ga-rich conditions of the crystallization process.

The lattice constants [16] for undoped GaN and the highly conductive Ga:Zn and GaN:Ca bulk crystals vary slightly for different crystals as well as for different areas of the individual crystals. This is caused by the differences in free electron concentration and can lead to certain strains in the crystals.

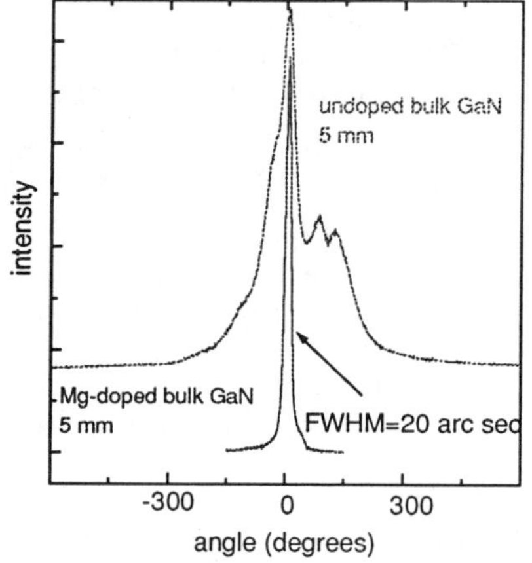

Fig 6 X-ray rocking curves for GaN crystals

For the Mg-doped crystals there are no free electrons and therefore the lattice parameters are uniform for each individual crystal and the same for different samples [17]. The rocking curves are narrower (Fig.6) and no low angle boundaries are observed even for 8mm single crystalline GaN:Mg platelets. The strains in the undoped GaN crystals are extremely small (<0.02%) but their presence can adequately explain the wider rocking curves than for the Mg-doped samples.

D. Wet etching and surface preparation

The surfaces of the pressure grown GaN crystals are usually covered by the growth figures on surface layer resulting from the cooling of the system. Therefore to obtain epi-ready surfaces it is necessary to subject them to mechanical and mechano-chemical polishing.

Mechanical polishing with diamond micropowders leads to the formation of highly damaged surfaces as was shown by RBS [18] (Fig.7) and recently by TEM [19].

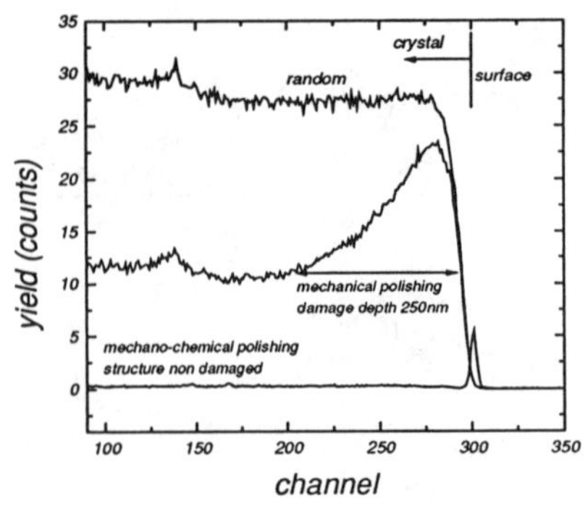

Fig. 7 RBS signals for polished GaN surfaces, Ref. 18

It was shown that bulk GaN crystals (highly conductive, undoped) can be etched in aqueous solutions (10N - 1N) of KOH and NaOH [5]. However only one of the polar {0001} surfaces of bulk crystals is attacked by the applied etchants. The free etching of this surface ((000$\underline{1}$)$_N$ as it was identified by XPD [7]) is strongly anisotropic and results in the formation of numerous stable pyramids 100 - 200nm in height.

As was mentioned before for Mg-doped crystals also only one {0001} surface etches; however it is not yet determined which polarity it is.

The same aqueous solutions (10N - 1N) of KOH and NaOH have been used for mechano-chemical polishing of the chemically active GaN surface [5] giving atomically flat surfaces.

The RBS measurements [18] (Fig.7) and recently TEM [19], indicate that the applied procedure allows removal of the subsurface damage resulting from the mechanical polishing.

E. Homoepitaxy

There is no systematic data on homoepitaxy because only recently the effective method of surface preparation has been developed. Nevertheless many important results on homoepitaxy reported up to now indicate new possibilities for GaN physics and technology. The homoepitaxial growth of GaN has been studied by both MOCVD [20,21,22] and MBE [23, 24]. The experiments have been performed on both polar surfaces of GaN crystals. The "as grown" [20,21] as well as the mechanically [25] and recently the mechano-chemically [24] polished surfaces were used.

The X-ray diffraction from MOCVD homoepitaxial layers and from GaN substrates ("as grown", highly conductive substrates have been used in that work) have been measured [17]. The curves of the layers were narrower than for the substrates indicating improvement of crystal quality during the epitaxial growth. The lattice constant parallel to the interface was the same for both substrate and layer whereas the c constant (perpendicular to the interface) of the layer was smaller by 0.005 0.02%.This effect is related to the fact that the lattice constants for undoped GaN crystals are slightly enlarged due to the presence of a high density of free electrons. It was observed [17] (Table 2) that for Mg-doped substrates the lattice constants of the substrates and the undoped layers are perfectly identical.

The high quality of homoepitaxial layers grown by MOCVD was confirmed by photoluminescence measurements [20]. Very narrow (~0.5meV) lines corresponding to bound exciton and close D-A pairs transitions have been reported. Very narrow excitonic lines have been also observed for homoepitaxial GaN grown by MBE on (000$\underline{1}$) "as grown" surfaces of GaN substrates [23]. A lower energy line related to acceptor bound excitons (ABE), lying at 3.4663 eV had a half width of 0.5 meV. Also the two next lines at 3.4709 eV and 3.4718 eV had similar widths. These lines, attributed to excitons bound to two different donors (DBE), were distinctly separated despite a small (1meV) distance between them. In the measurements [26] of PL at 2K, for MOCVD layers a relative chemical shift has been observed also for the

ABE. Two acceptor bound exciton lines with a relative chemical shift of 0.8meV were resolved. The lines were the narrowest ever reported for GaN (FWHM 0.3meV).

Reflectance and photo-luminescence spectra of exciton-polaritons in GaN homoepitaxial layers grown by MOCVD have been recently reported [27]. The energies of the transverse excitons were found to be: $E_{TA} = 3.4767\pm0.0003eV$, $E_{TB}= 3.4815\pm0.0003eV$, $E_{TC} = 3.4986\pm0.0008eV$. Using the obtained exciton energies the parameters describing the valence band splitting $\Delta_{so}=17.9\pm1.2$ meV and $\Delta_{cr}=8.8\pm0.8$ meV have been estimated.

Recently, a near ideal atomic step flow growth mode has been achieved by Cohen et al. [24] using MBE with an NH_3 nitrogen source. The growth experiments were performed on GaN pressure grown crystals mechano-chemically polished with an intentional misorientation of 0.5 - 2°. For growth on the substrates without an intentional misorientation ideal RHEED behaviour was observed by the same group [24]. The period of the RHEED oscillations corresponded to the growth rate in contrast to the growth of GaN on sapphire where defects reduce the effective Ga flux. No pinholes were observed in contrast to GaN on sapphire.

The growth of $Al_xGa_{1-x}N$ ternaries have been also studied by MBE with an RF plasma source of nitrogen [28]. The $Al_xGa_{1-x}N$ layers grown on bulk GaN crystals exhibited a very low roughness (3-5 Å RMS), as was measured using X-ray reflectivity. The layers of x=0.2 were not relaxed even to a thickness of 3000Å. For x=1, a layer of about 1500 Å was partially relaxed, whereas one of about 100 Å was fully strained.

A comprehensive study of homoepitaxial homojunction GaN LEDs grown on bulk pressure grown crystals has been performed [24]. Single peak blue emission at 420 nm with a line width of 60nm was obtained. A high quality of the homoepitaxial diodes was revealed by a comparison with diodes heteroepitaxially grown on sapphire. The homoepitaxial devices were twice as bright as LEDs grown on sapphire.

F. Conclusions

High quality GaN single crystalline platelets with surface area of 60 - 100mm^2 have been grown by the high pressure method in a reproducible way.

Full compensation of residual donors in the crystals was achieved by doping with Mg during crystallization.

Atomically flat surfaces of GaN substrates have been obtained by mechano-chemical polishing with alkaline water solutions.

It was demonstrated that near ideal growth of homoepitaxial layers is possible by both MOCVD and MBE methods.

Acknowledgements

The work on GaN crystallization is supported by the Polish Committee for Scientific Research Grants no 7 7834 95C/2399 and 7T08A 007 13.

References

[1] S. Nakamura and G. Fasol, "The Blue Laser Diodes" (Springer-Verlag, Berlin/Heidelberg, 1997).
[2] W. A. Harrison, "Electronic Structure and Properties of Solids" (Freeman, San Francisco, 1980).
[3] J. Karpinski., J. Jun and S. Porowski , J. Cryst. Growth, 66 (1984) 1.
[4] J.A. Van Vechten, Phys. Rev. B 7 (1973) 1479.

[5] J.L. Weyher, S. Müller, I. Grzegory and S. Porowski, J. Cryst. Growth 182 (1997) 17.
[6] Z. Liliental-Weber, S. Ruvimov, Ch. Kisielowski, Y. Chen, W. Swider, J. Washborn, N. Newman, A. Gassmann, X. Liu, L. Schloss, E. R. Weber, I. Grzegory, M. Bockowski, J. Jun, T. Suski, K. Pakula, J. Baranowski, S. Porowski, H. Amano, I. Akasaki, Mat. Res. Soc. Symp. Proc. 395 (1996) 351.
[7] M. Seelmann-Eggebert, J. L. Weyher, H. Obloh, H. Zimmermann, A. Rar and S. Porowski, Appl. Phys. Lett. 71 (1997) 2635.
[8] P. Perlin, J. Camasel, W. Knap, T. Talercio, J.C. Chervin, T. Suski, I. Grzegory and S. Porowski, Appl Phys Lett. 67 (1995) 2524.
[9] A. Barcz and T. Suski, unpublished
[10] P. Boguslawski, E. Briggs and J. Bernholz, Phys. Rev. B 51 (1995) 17255.
[11] Wook Kim, A. E. Botchkarev, A. Salvador, G. Popovici, H. Tang and H. Morkoc, J. Appl. Phys. 82 (1997) 219.
[12] J. Neugebauer and C. G. Van de Walle, Phys. Rev. B 50 (1994) 8067.
[13] K. Saarinen, T. Laine, S. Kuisma, P. Hautojarvi, L. Dobrzyński, J. M. Baranowski, K. Pakuła, R. Stępniewski, M. Wojdak, A. Wysmołek, T. Suski, M. Leszczyński, I. Grzegory and S. Porowski, Phys. Rev Lett. 79 (1997) 3030.
[14] S. Porowski, M. Bockowski, B .Lucznik, I. Grzegory, M. Wroblewski, H. Teisseyre, M. Leszczynski, E. Litwin-Staszewska, T .Suski, P. Trautman, K. Pakula and J.M. Baranowski, Acta Phys. Pol.A 92 (1997) 958.
[15] K. Saarinen et al., submitted to MRS Fall Meeting, Boston, 1997.
[16] M. Leszczynski, I. Grzegory, H. Teisseyre, T. Suski, M. Bockowski, J. Jun, J. M. Baranowski, S. Porowski and J. Domagala, J. of Cryst. Growth 169 (1996) 235.
[17] J. Domagała, P. Prystawko, M. Leszczyński, T. Suski and S. Porowski, unpublished.
[18] M. Conway, J. S. Williams and C. Jagadish, private communication.
[19] Z. Liliental-Weber, private communication.
[20] J. Baranowski, Mat. Res. Soc. Symp. Proc. 449 (1997) 393.
[21] K. Pakula, A. Wysmolek, K. P. Korona, J.M. Baranowski, R. Stepniewski, I. Grzegory, M. Bockowski, J. Jun, S. Krukowski, M. Wroblewski, and S. Porowski, Solid State Commun.97 (1996) 919.
[22] A. Pelzmann, C. Kirchner, M. Mayer, M. Schauler, M. Kamp, K. J. Ebeling, I. Grzegory, M. Leszczyński, G. Nowak and S. Porowski, presented on ICNS'97, Tokushima, Japan.
[23] H. Teisseyre, G. Nowak, M. Leszczyński, I. Grzegory, M. Boćkowski, S. Krukowski, S. Porowski, M. Mayer, A. Pelzmann, M. Kamp, K. J. Ebeling and G. Karczewski, MRS Internet J. Nitride Semicond. Res. 1 (1996) 13.
[24] R. Held, S. M. Seutter, B. E. Ishaug, A. Parhamovsky, A. M. Dabiran, P. I. Cohen, C. J. Palmstroem, G. Nowak, I. Grzegory and S. Porowski, submitted to MRS Fall Meeting, Boston, 1997.
[25] F. A. Ponce, D. P. Bour, W. T. Young, M. Saunders and J. W. Steeds, Appl. Phys. Lett. 69 (1996) 337.
[26] I. Ivanov, A. Henry, B. Monemar and J. M. Baranowski, unpublished result.
[27] R. Stępniewski, K. P. Korona, A. Wysmołek, J. M. Baranowski, K. Pakuła, M. Potemski, G. Martinez, I. Grzegory and S. Porowski, Phys. Rev. B 56 (1997) 15151.
[28] A. Barski, M. Leszczynski, I. Grzegory, S. Porowski, unpublished.

Growth of GaN and AlGaN by High Temperature Vapor Phase Epitaxy

S. Fischer, F. Anders, M. Theis, G. Steude, T. Christmann, D.M. Hofmann, B.K. Meyer

I. Physikalisches Institut, Justus-Liebig-Universität Giessen, D-35392 Giessen, Germany

We investigated the influence of the growth temperature on High Temperature Vapor Phase Epitaxy of GaN. An almost direct proportionality between the growth rate and the Ga vapor pressure is observed. At optimum conditions growth rates as high as 210 μm/h (T=1150°C) are achieved. The maximum growth rate is believed to be limited by supply of ammonia and starting decomposition of GaN. Under optimum GaN growth conditions AlGaN layers were grown starting from previously alloyed Al-Ga as well as from co-evaporation of Ga and Al. Adding Al leads to a significant reduction of growth rate and increased the homogeneity of the layers. However, in almost all cases phase separation is found. Besides the binary GaN and AlN phases an intermediate AlGaN phase appears.

A. Introduction

Considerable progress in the growth of group III nitrides, especially GaN, has been achieved over the past couple of years. Still being a crucial point is the question of the ideal substrate material. The highest quality nitrides are currently grown on epitaxial lateral overgrown substrates, finally utilizing homoepitaxy [1]. The hydride vapor phase epitaxy technique is already widely explored for thick film GaN growth [1].

However, considerable less is known about High Temperature Vapor Phase Epitaxy (HTVPE), another technique suitable for GaN thick film growth. This method originates from the Sublimation Sandwich Method used for SiC growth [2] and growth rates as high as 300 μm/h have been reported so far [3]. Here we present a study of the influence of the growth temperature on the growth rates, proposing a simple dependency on the Ga vapor pressure. Furthermore we employed the HTVPE technique the first time for the growth of AlGaN.

B. Experimental

GaN and AlGaN layers were grown by HTVPE on c-oriented Al_2O_3 and 6H-SiC substrates. Precursors are NH_3 as group V source and direct evaporation of the respective group III element. The growth took place in a horizontal reactor setup as described previously in Ref. [4]. For GaN growth the NH_3 flow was set to 25 sccm and the growth pressure to 20 mbar. 500 mg Ga were loaded each run into the crucible. The influence of the growth temperature on the layer properties was investigated in the range from 800 to 1300°C. Growth temperature refers to the temperature of the group III crucible, whereas the substrate temperature is 25°C lower. For comparison each growth lasted for 20 minutes. No buffer layer was grown prior deposition.

AlGaN growth was conducted from pre-alloyed Al-Ga (at 600°C for 7 min) as well as from co-evaporation of Ga and Al. Al atomic fraction was varied from 0 to 95%. The growth conditions were set again to NH_3 flow of 25 sccm, growth temperature of 1150°C and pressure of 20 mbar.

The structural properties of the layers were characterized by x-ray diffraction. The film thickness and hence the growth rate was determined in cross section optical microscopy equipped with micro photoluminescence (PL). The optical properties were investigated by means of PL and absorption measurements.

C. Results and Discussion

C1. GaN growth

The growth rate, determined as film thickness divided by growth time, is plotted over the growth temperature in Fig.1 GaN was grown on Al_2O_3 (circles) and 6H-SiC (triangles) substrates. The films were up to 70 μm thick.

Starting with 800°C (on Al_2O_3) a steep increase in the growth rate with temperature (note the semi logarithmic plot) is observed up to 1150°C. The growth rate remains constant then and starts to drop rapidly over 1200°C. On SiC GaN was grown starting at 1000°C because at lower temperatures only poor material in terms of optical properties was obtained. Up to 1100°C the growth rate increased also rapidly but started to level off and decreased already at lower temperatures. In addition the evolution of Ga vapor pressure (Ref. [5]) with temperature is plotted in Fig.1.

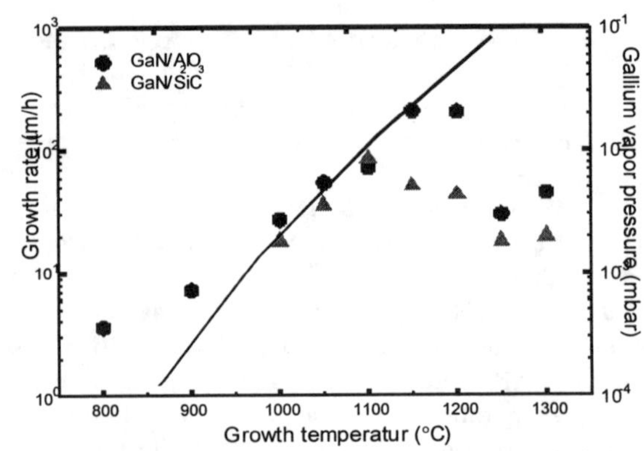

Fig.1: Growth rates of GaN deposited on Al_2O_3 (circles) and 6H-SiC (triangles) are plotted vs. growth temperature. In addition the evolution of Ga vapor pressure with temperature is shown (solid line, right axis).

Assuming that up to a certain temperature, depending on the growth pressure, the residual NH_3 pressure provides enough N species, the growth should be Ga limited. As one can see the steep increase in growth rate matches with the increase in Ga vapor pressure supporting our assumption. Once the growth process switches from Ga limited to N limited the growth rate should stay at least constant as found for temperatures above 1100°C. With a further increase in temperature the decomposition of GaN will limit the growth since GaN starts to decompose above T=850°C in vacuum [6].

The reason why the growth rate of GaN on 6H-SiC above 1100°C differs significantly is not understood so far. In addition it has to be noted that the films on 6H-SiC exhibit lower structural quality over the entire temperature range in terms of x-ray diffraction line width.

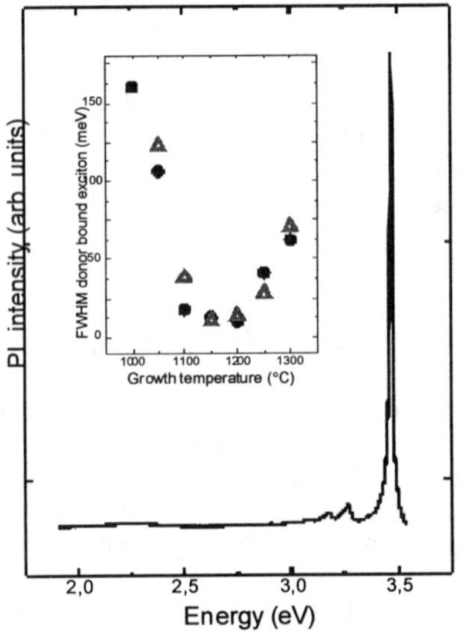

Fig.2: PL of GaN grown at 1150°C. The inset shows the dependence of FWHM of the donor bound exciton line on the growth temperature.

In Fig.2 PL spectrum of GaN grown at T=1150°C is shown. Besides the dominant donor bound exciton emission (3.472 eV) contributions of donor acceptor pair recombinations (zero phonon line at 3.27 eV) and the yellow luminescence are observed. In the inset of Fig.2 the optical properties of the grown layers (on Al_2O_3 (circles) and 6H-SiC (triangles)) are composed in terms of donor bound exciton line width recorded at T=4K. Two features can be seen. There is an optimum growth temperature in the range of 1100 - 1200°C. Contrary to the growth rates and the structural properties no difference can be seen between the two substrate materials. This indicates that the optical properties depend mainly on the growth temperature.

C1. AlGaN growth

The AlGaN was grown under optimum GaN growth conditions on Al_2O_3. The achieved growth rate was 2-5 µm/h. This is considerably lower than what one would expect for pure AlN (12 µm/h) calculating from of Ga and Al vapor pressure at 1150°C [5]. Alloying effects could be the reason for the low growth rates. Due to the low growth rate the overall homogeneity in terms of surface roughness of the layers increased.

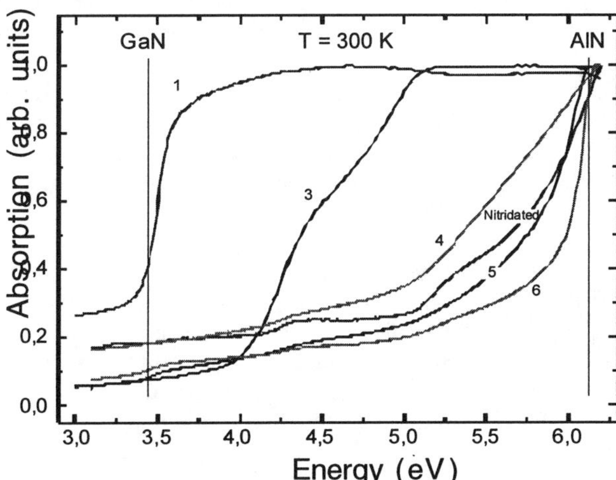

Fig.3: Absorption spectra of different $Al_xGa_{1-x}N$ samples when 9.5 - 95% atomic Al was added to the Ga melt. In addition the spectrum of a nitridated Al_2O_3 substrate is shown. Sample #2 could not be measured due to an opaque film.

In Fig.3 absorption spectra recorded at 300K of films grown with a atomic fraction of 9.5 to 95% Al in the crucible. In addition the absorption spectrum of a Al_2O_3 substrate, nitridated in a flow of 25 sccm NH_3 at 1150°C and 20 mbar is plotted.

In general the absorption edge shifts to higher energy with increasing Al content in the crucible. Besides sample 1 (low Al content) all spectra exhibit a broad tailing and additional absorption steps around 3.6, 4.3 and 5.25 eV. An absorption step around 4.3 eV in AlN single crystals was attributed to O impurities in Ref. [7]. The strong appearance of this step in the nitridated sample supports this assignment.

The samples were also investigated by x-ray diffraction in the Θ-2Θ mode. The diffraction patterns are displayed in Fig.4. The peak positions for binary GaN and AlN phases are marked with arrows. Slight shifts of the whole pattern might be due to residual strain. Again sample 1 is the only one exhibiting a single diffraction peak. The other samples contain an almost binary GaN and AlN phase as well as an intermediate AlGaN phase.

All samples containing the GaN-like phase show also an absorption around 3.6 eV, except sample 3. This indicates that the 3.6 eV absorption corresponds to a GaN-like phase in the samples. Assuming the validity of Vegard's law the intermediate phase peaking around $2\Theta=35.15°$ could correspond to the 4.3 eV absorption as well as to assignment of O impurities in a AlN phase, which again is observed in all samples. Furthermore a peak corresponding to the 5.25 eV absorption could not be found.

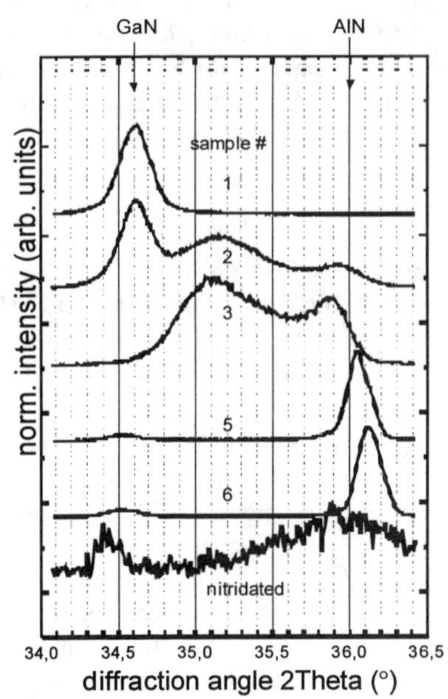

Fig.4: Θ-2Θ x-ray diffraction pattern of samples shown in Fig.3. From sample #4 no diffraction pattern could be obtained. At the bottom the pattern of the nitridated Al_2O_3 substrate is shown. The position of GaN and AlN (0002) diffraction peaks are marked with arrows. From Sample 4 no signal could be obtained.

The GaN-like phase in the nitridated sample is believed to be a memory effect of our crucible. The observed phase separation occurs independent of pre alloying or co-evaporation of Ga and Al. Therefore it must be due to processes in the gas phase or on the substrate.

D. Summary and conclusion

The influence of the growth temperature on the growth rate during HTVPE growth of GaN was studied from 800 to 1300°C. We find that the growth rate depends in a wide temperature range up to the maximum value (210 μm/h) on the Ga vapor pressure (Ga limited). It is limited by the supply of ammonia and the starting decomposition of GaN.

In addition we employed the HTVPE technique the first time for the growth of AlGaN. Within the layers we observe strong phase separation in a GaN, AlN, and an intermediate AlGaN phase.

Acknowledgments

The authors thank the Volkswagen Foundation for financial support within the "Photonik" program. S.F. gratefully acknowledges the Hanns-Seidel-Foundation for a scholarship funded by the BMBF.

References

[1]   see e.g. Proceedings of the ICNS'97, Tokoshima, Japan, October 27-31 1997.

[2]   Y.A. Vodakov, E.N. Mohkov, M.G. Ramm, A.D. Roenkov, Krist. Tech. 14, (1979) 729.

[3]   C. Wetzel, D. Volm, B.K. Meyer, K. Pressel, S. Nilsson, E.N. Mokhov, P.G. Baranov, Appl. Phys. Lett. 65, (1994) 1033.

[4]   S. Fischer, C. Wetzel, W.L. Hansen, E.D. Bourret-Courchesne, B.K. Meyer, E.E. Haller, Appl. Phys. Lett. 69 (1996) 2716.

[5]   G.L: Weissler, R.W. Carlson, eds: "Vacuum Physics and Technology", Methods of Experimental Physics Vol. 14 (Academic Press, New York, 1979).

[6]   O. Ambacher, M.S. Brandt, R. Dimitrov, T. Metzger, M. Stutzmann, R.A. Fischer, A. Miehr, A. Bergmaier, G. Dollinger, J. Vac. Sci. Technol. B 14, (1996) 3532.

[7]   G.A. Slack, T.F. McNelly, J. Crystal Growth 34, (1976) 263.

Phosphorus and arsenic incorporation during chemical beam epitaxial growth of strained GaAs$_{1-x}$P$_x$ layers on GaAs (100) substrates

D. Wildt, B. J. García, J. L. Castaño, and J. Piqueras

Laboratorio de Microelectrónica, Departamento de Física Aplicada C-XII, Universidad Autónoma de Madrid, Cantoblanco 28049 Madrid, Spain

Phosphorus and arsenic incorporation during chemical beam epitaxial growth (CBE) of GaAs$_{1-x}$P$_x$ using triethylgallium (TEGa), tertiarybutylarsine (TBAs) and tertiarybutylphosphine (TBP) as precursors has been studied. Reflection high-energy electron diffraction intensity oscillations are used to measure the arsenic and phosphorus incorporation during group V controlled growth on a Ga-rich surface for different TBAs and TBAs+TBP fluxes. The so obtained phosphorus mole fraction is compared with the phosphorus composition measured by x-ray rocking curves on the strained GaAs$_{1-x}$P$_x$ layers grown by conventional CBE with simultaneous supply of the group III and group V precursors and by atomic layer epitaxy (ALE) alternating the group III and group V fluxes. The phosphorus incorporation rate during CBE growth is lower than that measured during ALE and group V controlled growth but is still much higher than the incorporation rate reported for molecular beam epitaxial growth using elemental sources. Photoluminescence spectra show clearly strain effects in the heavy and light hole excitonic transitions.

A. INTRODUCTION

In$_{1-x}$Ga$_x$As$_{1-y}$P$_y$ based optoelectronic devices are interesting for applications in telecommunications. The range of bandgaps covered with this material includes the optimum wavelengths for minimum dispersion (1.3 μm) and minimum absorption loss (1.55 μm) in several kinds of optical fibers. However, the reproducible growth of III-V compounds containing both arsenic and phosphorus was not among the many achievements of molecular beam epitaxy (MBE) due to problems related with phosphorus mole fraction control [1-3]. Gas source MBE (GSMBE) or chemical beam epitaxy (CBE) using arsine and phosphine allow to overcome the problem of the phosphorus mole fraction control, but the use of these highly toxic group V hydrides presents severe handling risks.

CBE growth of GaAs$_{1-x}$P$_x$ using triethylgallium (TEGa), tertiarybutylarsine (TBAs) and tertiarybutylphosphine (TBP) as precursors is an alternative and less dangerous growth method, because these precursors are liquids with sub-atmospheric vapor pressures [4].

B. EXPERIMENT

The uncracked TEGa flux is introduced into the CBE growth system through a low temperature cell at 70 °C, whereas TBAs and TBP fluxes are introduced and cracked through the same high temperature cell at 700 °C.

The beam equivalent pressure (BEP) as measured by an ionization gauge placed in the position of the substrate holder has been taken as an indirect measurement of the flux for each precursor. The BEP reading depends on the temperature of the cracking cell, not only because of the kinetic energy of the thermalized molecules, but also because of the decomposition of the precursor gases. To avoid this dependence, the calibration has been done at low temperature (80 °C).

The growth was performed on semi-insulating GaAs (100) substrates. After thermal oxide desorption in the growth chamber, a GaAs buffer layer was grown on the substrate surface at T=550 °C, using simultaneous supply of TEGa and TBAs. After the buffer layer growth, an intense specular spot in the reflection high-energy electron diffraction (RHEED) screen was obtained, while a clear 2x4 surface reconstruction was observed.

To study the phosphorus and arsenic incorporation into the grown layers, we have grown $GaAs_{1-x}P_x$ layers with fixed TBAs flux while varying the TBP flux using three different growth techniques at the same growth temperature: (a) the growth by CBE, with simultaneous supply of TEGa, TBAs and TBP, adjusting the V/III flux ratio to obtain intense and long lasting RHEED intensity oscillations; (b) atomic layer epitaxy (ALE), where the TEGa and TBAs+TBP fluxes were alternated; (c) the group V-controlled growth, in which the growth of $GaAs_{1-x}P_x$ is performed when As and P atoms arrive on a Ga-rich sample surface, the amount of pre-deposited Ga exceeding the equivalent of several monolayers. This method is of high interest as it gives direct insight on the incorporation of the phosphorus into the layer.

C. RESULTS AND DISCUSSION

The RHEED intensity oscillations observed during the CBE growth are similar to those observed in the case of standard MBE growth, where the growth rate is controlled by the arrival of Ga atoms on a group V rich surface. The growth rate during the CBE growth, as deduced from the frequency of the RHEED oscillations, is proportional to the TEGa flux, being approximately independent of the group V population and composition of the surface if an excess of group V atoms is present on the sample surface, as it is confirmed by the observed 2x4 surface reconstruction during the growth [5].

The RHEED signal during the growth by ALE is shown in figure 1. Each curve has been obtained during the growth with the same TEGa and TBAs fluxes but with different TBP flux, as indicated in the insets of this figure. With increasing TBP flux, the group V deposition time was decreased, as indicated on the right side of each set of RHEED oscillations. The lowest oscillation was obtained during the growth of GaAs, when no TBP flux was used. The amplitude of the oscillations remains constant during the growth, indicating a growth rate of 1 ml per cycle and no surface roughening during the growth. The abrupt changes in the RHEED intensity are related to changes in the surface reconstruction, being 4x6 during the group III deposition cycle, and 2x4 during the group V deposition cycle. Although RHEED intensity oscillations are very interesting to obtain information about growth rates or composition during the growth of III-V compound semiconductors or even ternary alloys containing two different column III atoms, they do not provide information about the composition of III-V alloys containing two different column V atoms, as in our case.

In order to obtain such information about the composition of the $GaAs_{1-x}P_x$ layers grown by both methods, high resolution x ray diffraction (HRXRD) rocking curves were used. Figure 2 represents the HRXRD rocking curves obtained near the (004) GaAs reflection using the $CuK\alpha$ line, the lowest curve corresponding to the lowest TBP flux. It can be seen from this figure that for increasing TBP flux the corresponding $GaAs_{1-x}P_x$ satellite peak shifts to higher angles, indicating a higher P content in the epitaxial layer. Assuming that the grown thickness (between 200 and 50 nm, depending on the P content of the sample) is always lower than the critical thickness and therefore no strain relaxation can be expected, the P content can be calculated for each layer from the satellite peak shift and the elastic theory.

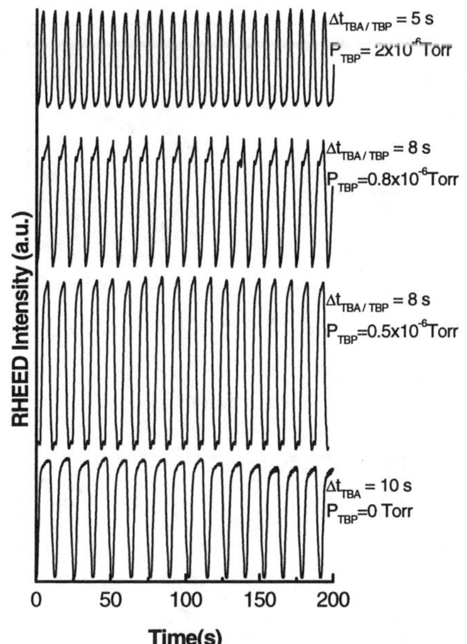

Fig. 1: RHEED intensity oscillations during GaAs$_{1-x}$P$_x$ growth by ALE.

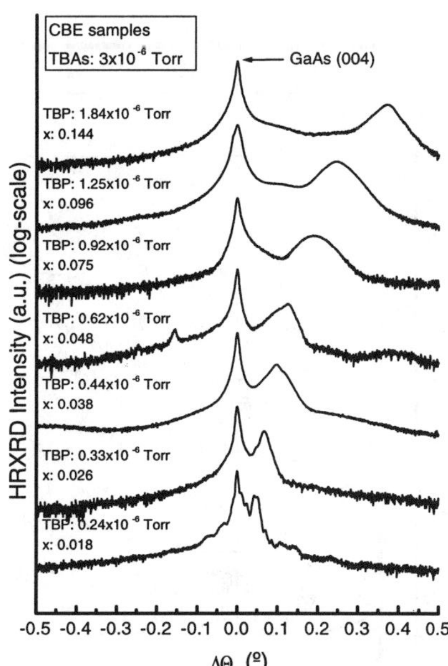

Fig. 2: HRXRD rocking curves of GaAs$_{1-x}$P$_x$ layers grown by CBE.

Figure 3 represents the RHEED intensity oscillations during the group V controlled growth at the same substrate temperature used during the previous growth techniques. After stopping the TBAs flux (the same TBAs flux that was used for layers grown by CBE or ALE), a gallium rich surface is intentionally formed by deposition of TEGa during 35 s.. After a short annealing time, the TBAs and TBP fluxes are simultaneously turned on. The growth proceeds while consuming the predeposited Ga, giving rise to RHEED intensity oscillations. The observed oscillation frequency depends only on the group V flux, while the number of oscillations depends on the amount of Ga previously deposited.

The lowest curve in fig. 3 belongs to the only TBAs controlled growth, while the remaining curves have been obtained for increasing TBP fluxes. The same experiment shown in fig. 3 has been previously done without using TBP, changing only the TBAs; a linear dependence of the growth rate on the TBAs flux was observed (not shown here), indicating that the growth is controlled by the arrival of As atoms to the sample surface. It can be clearly seen from Fig. 3 that the growth rate increases with increasing TBP flux, indicating now that the growth is controlled by the arrival of both As and P atoms to the Ga rich surface.

The experiment illustrated in fig. 3 has been repeated for three different TBAs fluxes, the results of them being summarized in fig. 4. The measured growth rate during the group V controlled growth has been plotted in this figure as a function of the TBP BEP, the growth rate without TBP flux corresponding to the growth rate of GaAs for each of the three TBAs fluxes. A linear increase of the growth rate for increasing TBP and TBAs fluxes can be observed, being the slope almost the same for the three TBAs flux values.

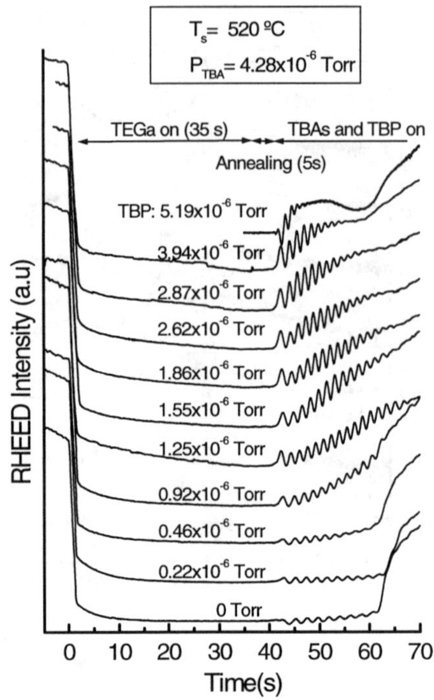

Fig. 3: RHEED intensity oscillations during the As+P controlled growth for increasing TBP flux and fixed TBP flux.

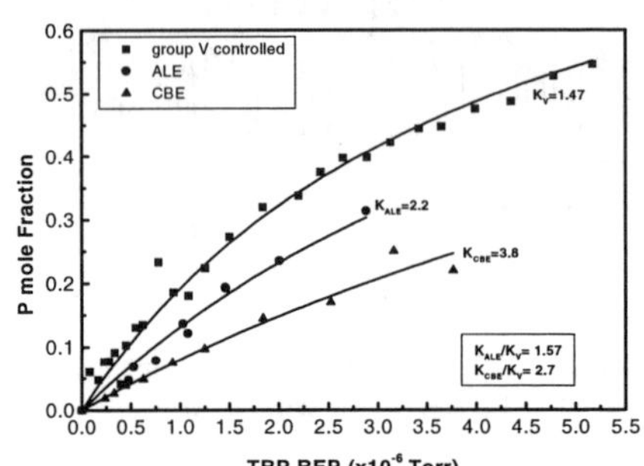

Fig. 4: GaAsP growth rate as a function of the TBP BEP for different TBAs BEP.

Fig. 5: P mole fraction as a function of the TBP BEP for different growth methods.

The observed linear dependence of the growth rate on both TBAs and TBP fluxes indicates that there is no competition between As and P to incorporate to the layer during growth. Under these conditions, the incorporation of each element, As and P, is expected to be proportional to their respective flux. Assuming that the sticking coefficient for both As and P is close to unity at the growth temperature during the group V controlled growth, the P mole fraction can be calculated either from flux values or from growth rates as:

$$x = \frac{F_{TBP}}{F_{TBP} + F_{TBAs}} \quad (1) \qquad x = \frac{v_{GaAsP} - v_{GaAs}}{v_{GaAsP}} \quad (2)$$

being F_{TBP} and F_{TBAs} the TBP and TBAs fluxes, respectively, while v_{GaAsP} is the growth rate of GaAsP measured during the growth with the fluxes F_{TBP} and F_{TBAs}, and v_{GaAs} is the growth rate of GaAs during the growth with only F_{TBAs}. Similar equations are used to calculate, for example, the Al mole fraction during the growth of GaAlAs and GaAs in standard MBE.

In our experiments, the TBAs and TBP fluxes are not known, although they can be estimated from the BEP measurements BEP$_{TBAs}$ and BEP$_{TBP}$, because both magnitudes are proportional:

$$F_{TBP} = I_{TBP} \cdot BEP_{TBP}$$
$$F_{TBAs} = I_{TBAs} \cdot BEP_{TBAs} \quad (3)$$

the proportionality constant being different for both chemical species because ion gauge measurements are sensitive to the mass of each molecule and to the mass of their decomposition products during the ionization process.

Taking into account (3), equation (1) can be written as :

$$x_F = \frac{BEP_{TBP}}{BEP_{TBP} + k_V \cdot BEP_{TBAs}} \qquad (4)$$

being

$$k_V = \frac{I_{TBAs}}{I_{TBP}} \qquad (5)$$

the relative sensitivity of the ion gauge to TBAs and TBP. Combining eqs.(2) and (4):

$$\frac{v_{GaAsP} - v_{GaAs}}{v_{GaAsP}} = \frac{BEP_{TBP}}{BEP_{TBP} + k_V \cdot BEP_{TBAs}} \qquad (6)$$

The results plotted in figure 4 have been fitted according to this equation, obtaining the value k_V=1.47.

A similar analysis can be done with the data obtained for the ALE and CBE grown layers. Because RHEED data do not contain information about P mole fraction, these data have been calculated from the HRXRD spectra. The results are plotted in figure 5 together with the results of the group V-controlled growth. The values of the fitting to the equation (4) are k_{ALE}=2.2 for ALE and k_{CBE}=3.8 for CBE grown layers.

This difference can be understood on the basis that sticking coefficients for As and P containing species are lower than unity, so the equation (4) remains valid if the k_V parameter is changed by:

$$k = \frac{I_{TBAs} \cdot S_{TBAs}}{I_{TBP} \cdot S_{TBP}} \qquad (7)$$

where S_{TBAs} and S_{TBP} are the effective sticking coefficients for the As and P atoms arriving to the substrate in the fluxes of each respective cracked precursors.

From the ratio of the k values obtained for ALE and CBE growth and the k value for the group V controlled growth, the dependence on the relative sensitivity of the ion gauge can be eliminated, obtaining the ratio of the sticking coefficients of both species:

$$\frac{S_{TBAs}}{S_{TBP}} = \frac{k_{ALE}}{k_V} = 1.5 \qquad \frac{S_{TBAs}}{S_{TBP}} = \frac{k_{CBE}}{k_V} = 2.6 \qquad (8\ a,b)$$

These results show that the sticking coefficient of As containing species during ALE growth is about 1.5 times higher than the sticking coefficient of the P species while this ratio is 2.6 for CBE growth.

The 4 K photoluminescence spectra of the grown layers, either by CBE or ALE, show interesting features. Exciton peaks are seen beside a main peak corresponding to the C acceptor. Both peaks shift to higher energies for increasing P mole fraction, indicating an increasing energy gap.

As a consequence of the tensile biaxial strain of the epitaxial layers, the valence band is splitted into the heavy and the light hole bands, so that two exciton peaks can be observed.

Figure 6 represents the position of both exciton lines as a function of the P mole fraction obtained from HRXRD. Lines represents the calculated excitonic gaps with and without strain, using published data for the deformation potentials of GaAs and GaP and the Vegard´s law. Because there is a good agreement between PL and HRXRD data it can be assumed that the strain of grown layers is not relaxed.

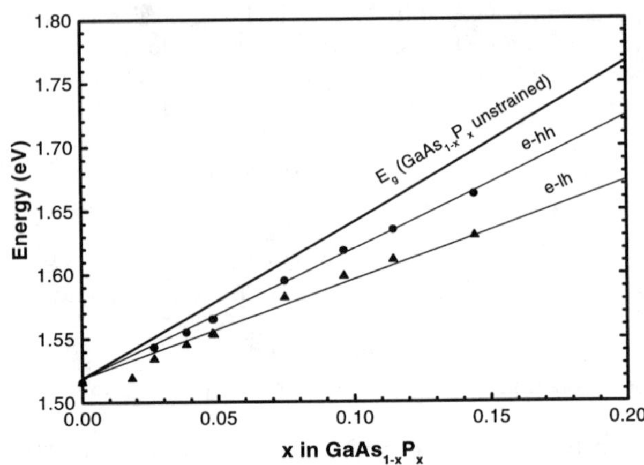

Fig.6: 4K PL exciton peak energies as a function of the P mole fraction. Lines represent theoretical calculations for the unstrained material, light hole and heavy hole excitons.

D. CONCLUSIONS

The growth of strained GaAs$_{1-x}$P$_x$ layers in a CBE system using TBAs, TBP and TEGa has been demonstrated.

Epitaxial strained layers have been grown by three different growth methods, CBE, ALE and group V controlled growth. The growth rate depends linearly on the TBAs and TBP fluxes, showing the possibility to control and reproduce the phosphorus mole fraction.

The dependence of the P mole fraction on the TBAs and TBP fluxes is different for the three growth methods. This is probably due to the different sticking coefficient of As and P because of the different Ga population on the sample surface during the different growth procedures.

Grown layers show good optical properties, as deduced from the exciton peaks on the PL spectra, showing the splitting of the valence band as a consequence of the lattice mismatch between the substrate and the epitaxial layer.

This work has been supported by the Spanish CYCIT under contract MAT97-0291.

REFERENCES
[1] M. B. Panish, Prog. Cryst. Growth Charact. **12**, 1 (1986).
[2] C. T. Foxon, B. A. Joyce, and M. T. Norris, J. Cryst. Growth **49**, 132 (1980).
[3] L. Samuelson, P. Omling, and H. G. Grimmeiss, J. Cryst. Growth **61**, 425 (1983).
[4] M. Aït-Lhouss, J. L. Castaño, B. J. García, and J. Piqueras, J. Appl. Phys. **78**, 5834 (1995).
[5] M. Aït-Lhouss, J. L. Castaño, and J. Piqueras, Mater. Sci. Eng. B **28**, 155 (1994).

High Resolution EL2 and Resistivity Topography of SI GaAs Wafers

M. Wickert, R. Stibal, P. Hiesinger, W. Jantz and J. Wagner

Fraunhofer Institut Angewandte Festkörperphysik, Tullastr. 72, D-79108 Freiburg, Germany

M. Jurisch, U. Kretzer and B. Weinert

Freiberger Compound Materials, Am Junger Löwe Schacht 5, D-09599 Freiberg, Germany

The mesoscopic inhomogeneity of LEC grown semi-insulating (SI) GaAs wafers has been investigated with $EL2^{\circ}$ absorption topography (EAT), photoluminescence topography (PLT), point contact topography (PCT) and contactless resistivity mapping (COREMA). Significant progress with respect to sensitivity of EAT and lateral resolution of COREMA has been achieved. High resolution topograms of wafers cut from ingots subject to standard and modified annealing procedures are presented. Direct comparison of $EL2^{\circ}$ and resistivity topograms reveals significant differences in the mesoscopic contrast and a contrast reversal for modified annealing. These observations can be explained very satisfactorily by assuming mesoscopic inhomogeneity of the intrinsic acceptor concentration which is modified during annealing. A model involving generation of Ga vacancies by dissolution of As_{Ga} antisites and gettering of the interstitial As at precipitates is presented and discussed.

A. Introduction

The macro- and mesoscopic homogeneity of liquid-encapsulated Czochralski (LEC) SI GaAs wafers has been improved in recent years by systematic optimization of ingot annealing processes. But mesoscopic fluctuations of resistivity, related to the cellular distribution of dislocations, continue to stimulate research- and application-oriented investigations, requiring topographic optical and electrical characterization techniques with high sensitivity and lateral resolution. PLT has been widely used for qualitative homogeneity assessment, although the correlation between PL intensity and resistivity appears to be incidental and often is not observed at all. Hence, high resolution, high sensitivity topograms of the resistivity itself (obtained with PCT and COREMA) and of the EL2 concentration (obtained with EAT) are needed. Evaluation of these topograms not only qualifies the material, but also allows elucidating details in the defect redistributions and the resulting changes of the compensation processes.

B. Experimental Details

The $EL2^{0}$ absorption topography [1] has been enhanced in sensitivity to obtain precise mapping of concentration variations down to 1% with a lateral resolution of 30 μm. Mesoscopic resistivity mappings with comparable lateral resolution can be achieved by PCT [2]. We have implemented this technique and corroborated its reliability by direct comparison with measurements of the Bergakademie Freiberg group [3] and with COREMA measurements. The non-contacting resistivity technique COREMA [4] thus far offered a lateral resolution of about 2.5 mm, insufficient for mesoscopic investigations. We have achieved an improvement of three orders of magnitude by designing a 80 μm diameter capacitive probe with upgraded measurement circuitry. Using the same step size of 30 μm as PCT, the topograms can be directly compared. For large cell structures in the mm range, the relative variation of conductivity is comparable for PCT and μ-COREMA. As will be shown below, mesosocopic variations on a smaller scale encountered in LEC grown material are somewhat underrated by μ-COREMA due to still insufficient lateral resolution. The PLT measurements were performed with a commercial Philips SPM 200 system.

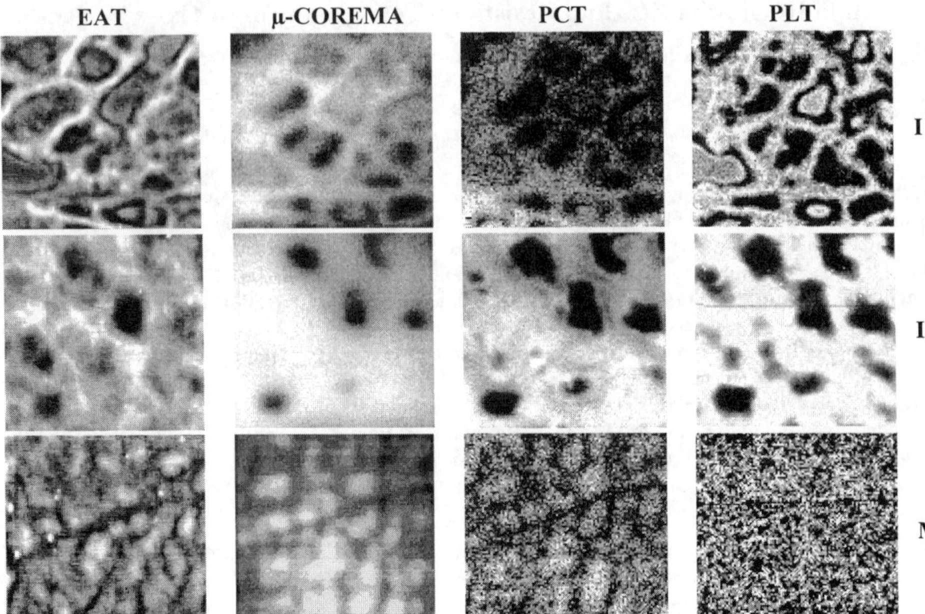

Fig.1: 3x3 mm² topograms of EL2⁰ concentration, conductivity - measured by μ-COREMA and PCT - and photoluminescence (from the left to the right). In bright areas the respective values are high. In each topogram the contrast is optimized. Samples are taken from 100 mm GaAs wafers homogenized by different ingot annealing processes. Sample I was treated with a standard singlestep process at about 1100°C, sample II with a standard twostep process including a second phase at about 800°C, sample M with a modified multistep process.

C. Results

For each annealing process we have selected a sample showing distinct mesoscopic variation. The arbitrary choice of samples, containing different carbon concentrations, implies that the topograms presented in Fig. 1 and the mesoscopic variation data summarized in Table 1 must not be used to compare the general quality of the respective annealing processes. Rather, the examples are chosen to clearly demonstrate the characteristic features. For the quantitative evaluation and argumentation in the next section, only *relative* values pertaining to a given sample will be compared.

Sample	[C]	EAT	μ-COREMA	PCT	PLT	σ-contrast	EL2⁰-contrast
	$[10^{14}\,\text{cm}^{-3}]$	Stdv [%]	Stdv [%]	Stdv [%]	Stdv [%]	$(\sigma_{cw}-\sigma_{ci})/\sigma_{mean}$	$(\text{EL2}^{0}_{cw}-\text{EL2}^{0}_{ci})/\text{EL2}^{0}_{mean}$
I	39	5.3	4.6	10.1	28.4		
II	4.1	3.8	9.1	26.5	29.9	+0.88	+0.25
M	5.7	1.8	1.9	8.7	5.4	-0.33	-0.07

Table 1: Standard deviation of topograms in Fig. 1 and contrast between cell wall (cw) and cell interior (ci) for sample II and sample M. Contrast in conductivity σ is stronger than contrast in [EL2⁰]. Sample II shows normal contrast with higher [EL2⁰] in cell walls. Contrast is reversed for sample M with lower [EL2⁰] in cell walls.

Sample I shows a clearly delineated structure with narrow cell walls. We find good correlation of all four topograms and a "normal" contrast exhibiting, as previously observed, high conductivity, high [EL2^0] and high PL intensity at the cell walls. A relative maximum of [EL2^0] is observed in the cell interior. In sample II the cell walls appear much broader, covering most of the sample area homogeneously. [EL2^0] monotoneously drops towards the centers of the remaining cells. Some cells revealed by EAT are not detected by PCT and μ-COREMA. These bubble-type cells are lying entirely below the surface, indicating that the probing depth of both resistivity techniques must be considerably smaller than the sample thickness. The comparatively small low conductivitiy areas, embedded in wide areas of higher conductivity, appear at a reduced size when measured with μ-COREMA. We attribute this to the smaller lateral resolution of this technique, which also explains the generally lower contrast recorded by μ-COREMA (see Table 1). This effect is even more pronounced for sample M which exhibits significantly smaller cells. Thus PCT, in spite of the destructive and slow measurement procedure, remains an indispensible technique for quantitative mapping of mesoscopic conductivity variations due to its better resolution. The PL topogram of sample M is very homogeneous, while EAT and the conductivity topograms show *reversed* contrasts, i.e. lower [EL2^0] and conductivity in the cell walls. To demonstrate this behaviour clearly, two large area [EL2^0] topograms are presented in Fig. 2. Such reversed contrasts are observed and topographically recorded here for the first time in ingot annealed material.

The carbon concentration of sample II and sample M are comparable, whereas for sample I the carbon concentration differs significantly. Hence we restrict the comparison of the relative change in [EL2^0] and the relative change in conductivity σ between cell wall and cell interior to sample II and sample M (see Table 1). As justified above, PCT data have been used for the evaluation of conductivity contrast .

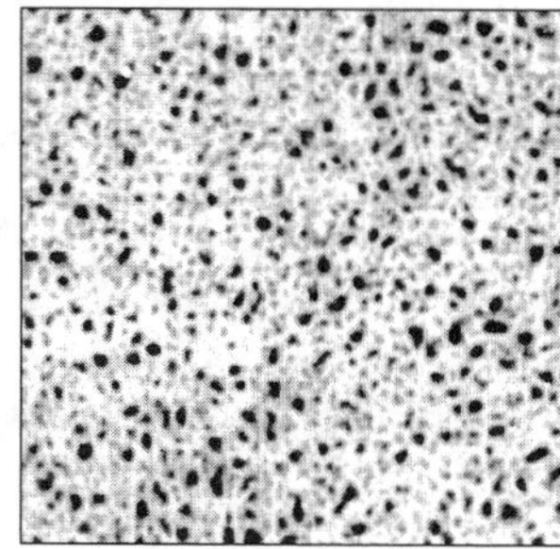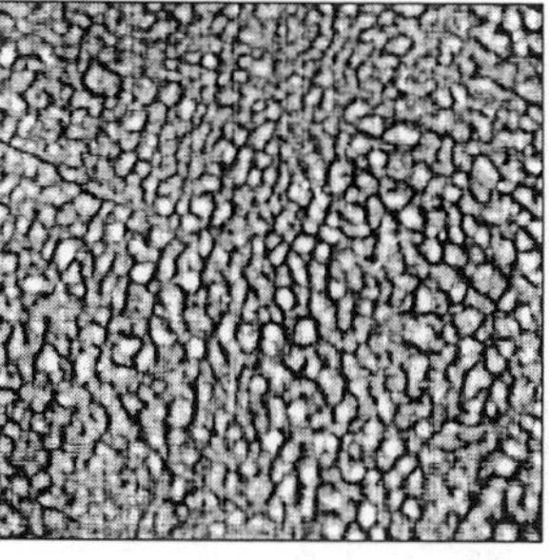

Fig. 2: 15x15 mm² topograms of EL2^0 concentration. Bright areas mean high EL2^0 concentration. The left sample was treated with a standard twostep process and shows normal contrast with brighter cell walls (Stdv 3.8%). The right sample was treated with a modified multistep process and clearly demonstrates EL2^0 contrast reversal on a high level of EL2^0 homogenization (Stdv 1.8%).

D. Discussion

According to the compensation model [5] the conductivity is given by the ratio of [EL2^0] $\cong$ 1.3x10^{16} cm^{-3} over the net acceptor concentration, which is determined by the carbon concentration $N_C \cong 10^{15}$ cm^{-3} plus the concentration of residual acceptors N_{RA} minus the concentration of donors $N_{SD} \cong 10^{14}$ cm^{-3} shallower than EL2. Since [EL2^0] $> N_C + N_{RA} > N_{SD}$, the Fermi level is pinned at the first EL2 ionization energy level. We assume that N_{SD} is laterally homogeneous and does not change significantly upon annealing. This assumption, although not proven experimentally, appears permissible in the present context because, for SI material with resitivitiy $\rho \geq 10^7$ Ωcm, shallow acceptors dominate over the influence of shallow donors. If, as is evident from Table 1 for sample II and sample M, conductivity varies more strongly than [EL2^0], then - assuming N_C to be spatially homogeneous -

necessarily the residual acceptor concentration N_{RA} must vary on a mesoscopic scale. We find that, after standard twostep annealing (sample II), N_{RA} is lower in the cell walls as compared to the cell interior. In sample M, exhibiting contrast reversal, the same consideration implies that N_{RA} is higher in the cell walls. Hence we obtain contrast reversal not only for $EL2^0$ and conductivity, but also for N_{RA}. Since the modified tempering involves only a small change in temperature on the order of 100°C as compared to the standard twostep annealing process, it is unlikely that correspondingly small differences in an assumed acceptor diffusion can account for the observed N_{RA} contrast reversal. We, therefore, suggest that annealing *generates* residual acceptors, i.e. changes N_{RA} with an efficiency that depends sensitively on temperature and is different in the cell wall and cell interior. For the modified multistep annealing the postulated N_{RA} generation rate must be higher in the cell walls. Note that these arguments remain valid even if N_C-N_{SD} were spatially inhomogeneous.

We suggest the following model for laterally inhomogeneous acceptor generation: First, EL2 dissociates into an interstitial As and a Ga vacancy, the latter forming a triple acceptor [6]. Subsequently, As_i segregates at arsenic precipitates decorating dislocations. Due to the reduction of $[As_i]$ preferentially near dislocations the re-generation of EL2 is selectively reduced in the cell walls. This model explains an increased N_{RA} generation in the cell walls, resulting in higher local resistivity. The concurrent $EL2^0$ contrast reversal is due to the reduction of total EL2 concentration as well as the higher EL2 ionization resulting from increased RA compensation. The redistribution of excess As to larger size precipitates for an annealing process comparable to the modified multistep annealing process has been reported [7, 8]. The proposed processes are known to be critically dependent on temperature in the range $T \geq 900$°C [9], qualitatively accounting for the observed contrast reversal within a narrow interval of annealing temperatures.

E. Conclusion

The dependence of mesoscopic optical and electrical inhomogeneities in SI GaAs wafers on tempering has been studied and is interpreted with a model suggesting laterally inhomogeneous generation of residual acceptors. This model is in accordance with previously discussed redistribution processes of excess As and accounts very satisfactorily for the observed contrast reversal. However, the model itself and the microscopic nature of the intrinsic acceptor remain highly speculative and need further experimental support. The conductivity contrast reversal implies that proper adjustment of annealing parameters should generate material with perfect electrical homogeneity.

References

[1] J. Windscheif, M. Baeumler and U. Kaufmann, Appl. Phys. Lett. **46** (1985) 661.

[2] W. Siegel, G. Kühnel, J. M. Niklas, M. Jurisch, B. Hoffmann, Semicond. Sci. Technol. **11** (1996) 851.

[3] C. Reichel, S. Schulte and W. Siegel, TU Bergakademie Freiberg, private communications.

[4] R. Stibal, J. Windscheif and W. Jantz, Semicond. Sci. Technol. **6** (1991) 995.

[5] G. M. Martin, J. P. Farges, G. Jacob and J. P. Hallais, J. Appl. Phys **51** (1980) 2840.

[6] G. A. Baraff and M. Schlüter, Phys. Rev. Lett. **55** (1985) 1327.

[7] B. Hoffmann, Ph.D. Thesis, Technische Universität Erlangen-Nürnberg (1996).

[8] J. L. Weyher, P. Gall, Le Si Dang, J. P. Fillard, J. Bonnafé, H. Rüfer, M. Baumgartner and K. Löhnert, Semicond. Sci. Technol. **7** (1992) A45-A52.

[9] H. Wenzl, W. A. Oates and K. Mika, in Handbook of Crystal Growth I, D. T. H. Hurle, Ed., Elsevier Science Publishers, Amsterdam (1993) 128.

Electrical and optical properties of annealed semi-insulating GaAs grown by vertical zone melt technique

Z-Q. Fang[a], D. C. Reynolds[a], D. C. Look[a], M. G. Mier[b], R. L. Jones[b], and R. L. Henry[c]
[a] University Research Center, Wright State University, Dayton, Ohio 45435
[b] Air Force Research Laboratory, Wright-Patterson Air Force Base, Ohio 45433
[c] Naval Research Laboratory, Washington, D.C. 20375-5000

Abstract - Electrical and optical properties of undoped semi-insulating GaAs grown by the vertical zone melt (VZM) technique and annealed at 950 °C under As overpressure have been characterized. The 950-°C annealing significantly improves the uniformity and increases both EL2 concentration and mobility. Cu incorporation into the VZM materials has been observed.

A. Introduction

The vertical zone melt (VZM) growth technique has been used to produce high quality, low dislocation density, semi-insulating (SI) bulk GaAs. This technique also allows for zone refinement (ZR) and Ga zone leveling (ZL) of the GaAs ingots, resulting in lower concentrations of residual impurities and the native defect EL2, respectively [1,2]. The VZM GaAs material without post-growth anneal often shows lower mobility and less uniformity as is typical for unannealed GaAs grown by the liquid-encapsulated Czochralski (LEC) method. Both mobility and EL2 concentration in VZM SI GaAs can be significantly increased by 950-°C annealing [3]; such improvements have also been reported for whole boule annealing of LEC GaAs. In this paper, we present a comprehensive comparison of electrical and optical properties of VZM materials with and without anneal. The characterization techniques include Hall-effect, near-infrared absorption, temperature dependent photocurrent (TDPC), thermally stimulated current (TSC) (including NTSC, i.e. TSC signals normalized by TDPC using 1.13 eV light), photoluminescence (PL), and selective pair PL (SPL) (see Ref. [4] for details of the techniques). In contrast to the observations in LEC-grown SI GaAs, we find evidence for Cu incorporation in VZM materials. The Cu-related defects before and after annealing and the role of As overpressure in the enhancement of EL2 are discussed below.

B. Experimental and results

As reported in Ref. [3], a 14-cm-long crystal of ZR VZM GaAs with a diameter of 3 cm was whole-boule annealed by lowering the temperature of the sealed growth ampoule containing the crystal and 2-atm As_2 (at 1200 °C) to 600 °C. After a 15-min. soak at 600 °C, the temperature was raised to 950 °C and held there for 4 hr, after which it was lowered to room temperature over 7 hr. Three wafers were cut from the seed, the middle and the tail of the crystal for near-infrared absorption measurements. Square samples (6 x 6 mm^2 in size) were then taken from the central portion of each wafer for electrical and optical characterizations. For comparison, the same characterizations were also performed on wafers and samples cut from a ZR VZM GaAs crystal without anneal.

The 300-K electrical properties and the neutral EL2 concentrations for the annealed and non-annealed VZM GaAs are shown in Table I, which indicates a significant increase of mobility and EL2 concentration by the 950-°C annealing. Typical TSC and TDPC spectra for annealed and non-annealed samples are pictured in Fig. 1. The TSC and TDPC spectra for the

Table I 300-K electrical properties and EL2^0 concentration of VZM SI GaAs

	Unannealed	ZR VZM	Annealed	ZR VZM	
	Seed	Tail	Seed	Middle	Tail
ρ (10^7 Ωcm)	3.99	9.83	4.06	1.51	0.72
n (10^7 cm^{-3})	3.13	1.17	2.06	5.45	11.36
μ (cm^2/V s)	5010	5430	7500	7590	7630
EL2^0 (10^{15}cm^{-3})	7.3	7.5	10-20[a]	10-20[a]	20-30[a]

[a] Accurate value cannot be obtained due to non-uniform thickness of the polished wafer.

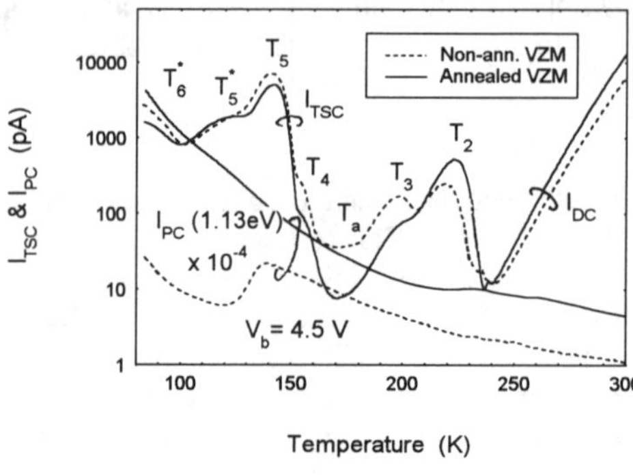

Fig. 1 Typical TSC and TDPC spectra

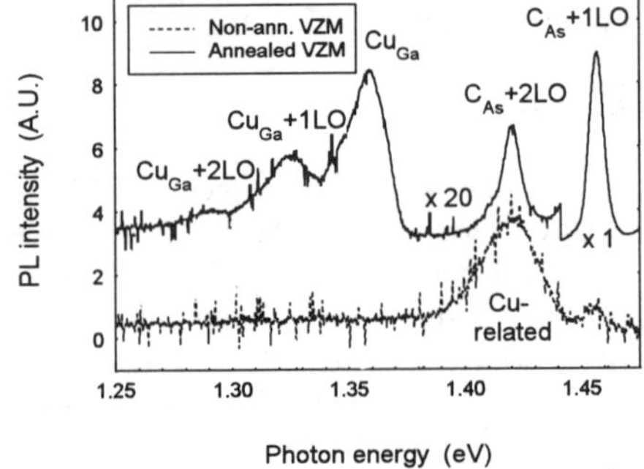

Fig. 2 Typical 4.2-K PL spectra

annealed VZM GaAs show six traps as well as a higher ratio of T_2/T_3 and a higher PC at 82 K, and are very similar to those for LEC SI GaAs, indicating a higher EL2 concentration and a lower compensation [5,6]. However, the non-annealed VZM GaAs behaves very differently; i.e. 1) the overall PC is about one order of magnitude lower than that of annealed VZM material; 2) there is a serious IR quenching of the PC at low T<140 K; 3) the ratio of T_2/T_3 is close to unity; and 4) there exists a Cu-related hole trap T_a at ~ 170 K , with E_T= 0.43 eV [7]. The NTSC of T_5 is found to be obviously reduced by the 950-°C annealing (not shwon).

Typical PL spectra in the deep level region for annealed and non-annealed VZM samples are shown in Fig.2. In the annealed sample, we can observe: 1) the phonon replicas of the C-related transitions at 1.456 eV (C$_{As}$+1LO) and 1.420 eV (C$_{As}$+2LO); and 2) the Cu-related transition at 1.359 eV (Cu$_{Ga}$) and its two phonon replicas [8]. In the non-annealed sample, on the other hand, we find that in addition to the transition C$_{As}$+1LO at 1.456 eV, there exist a broad peak at 1.418 eV and nearly no Cu$_{Ga}$-related features. The overall PL intensity in the near-band-edge and deep-level regions for the non-annealed sample was found to be about two orders of magnitude lower than that for the annealed sample.

SPL enables a definite identification of the residual shallow acceptors as well as a quantitative estimate of their relative concentrations [9]. With 8211-Å pumping light, different excited states corresponding to C:2P$_{3/2}$, C:2S$_{3/2}$, C:2P$_{5/2}$(Γ_8) and Zn:2S$_{3/2}$ can be clearly identified in the samples, taken from the seed, the middle, and the tail of the annealed ZR VZM crystal, as shown in Fig. 3. As compared to the emission C:2S$_{3/2}$, the relative intensity of the emission Zn:2S$_{3/2}$ is found to be largely increased from the seed to the tail due to the Zn segregation, which is identical with the results reported by Moore and Henry for two, non-annealed ZR VZM ingots [10]. They used a method that consists of moving the Fermi level to the ground state of the shallowest acceptor present (usually carbon) by diffusing copper into the material (which passivates the EL2) and then measuring the acceptors by means of 1s-2p infrared absorption spectra. In contrast to the increase of Zn towards the tail, the relative intensity of the Cu_{Ga}-related PL emission at 1.359 eV in the annealed ZR VZM crystal, if normalized by the intensity of the C_{As}-related PL emission at 1.492 eV (not shown), shows a decrease from 3.8 x 10^{-3} at the seed, to 1.2 x 10^{-3} at the middle, then to 7.5 x 10^{-4} at the tail. To discuss the behavior of Cu-related defects during 950-°C annealing, we also present a NTSC comparison for three samples taken from the seed, the middle, and the tail wafers, as shown in Fig. 4. The main NTSC traps, like T_2, T_3, and T_5, have been extensively studied (see Ref. [6] and related literature) and are believed to be related to point defects, such as As_{Ga} and V_{As}. In Fig. 4, the NTSC peak heights for T_2 (related to As_{Ga}) and T_5 (related to both As_{Ga} and V_{As}) show an increase towards the tail. Unlike the case in non-annealed VZM material, T_3, which might be a V_{As}-related defect complex, is not a prominent trap in annealed VZM GaAs.

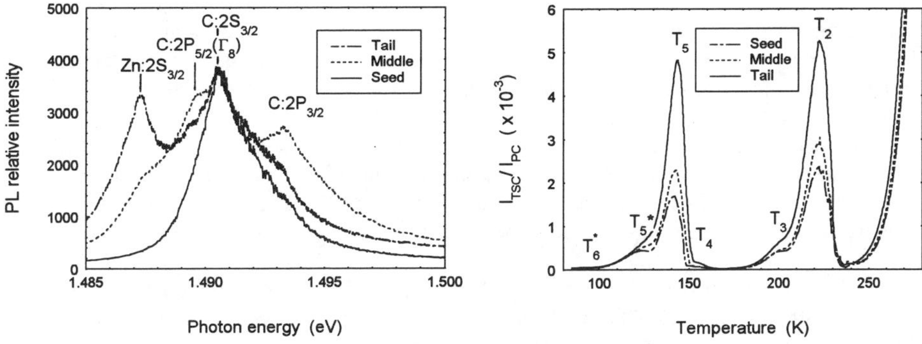

Fig. 3 2-K SPL spectra for annealed samples   Fig. 4 NTSC spectra for annealed samples

C. Discussion

There is no doubt that Cu is incorporated into the annealed VZM GaAs because of the presence of well-known Cu_{Ga}-related PL features. The broad peak at 1.418 eV found in non-annealed VZM GaAs may seem to be due to the transition C_{As}+2LO at 1.420 eV, but it is not, because the relative intensities of the 1.418-eV peaks found in these samples are comparable to or even higher than that of the transition C_{As}+1LO. As a secondary phonon replica of the C-related transition, the relative intensity of the transition C_{As}+2LO at 1.420 eV found in the annealed samples is only a few per cent of that of the transition C_{As}+1LO at 1.456 eV. At present, we do not know the exact origin of the emission at 1.418 eV, although it might be also Cu-related, like Cu_i-V_{Ga}. Copper is a fast interstitial diffuser in GaAs and is widely believed to complex with many impurities and defects in GaAs to produce a variety of deep centers in the band gap. Annealing can cause a reconstruction of Cu-related impurity-defect complexes. For example, in an early PL study of defects and impurities in GaAs grown by the horizontal Bridgman (HB) technique and annealed with and without adding Cu [11], many Cu-related centers were identified and discussed, such as Cu_{Ga} (1.356 eV), Cu_{Ga}-V_{As} (1.347 eV), Cu_{Ga}-$(V_{As})_2$ (1.31 eV), Cu_{Ga}-Si_{Ga} (1.27 eV), and another Cu-related feature (1.467 eV).

As observed in annealed LEC GaAs, the increase of mobility and EL2 concentration in annealed VZM GaAs is probably related to the re-distribution of excess As in the lattice and As precipitation at dislocations. However, the following impurity-defect reactions might happen during 950-°C annealing. First, a Cu-related defect, such as Cu_i-V_{Ga}, in conjunction with an As overpressure annealing condition, may play some role in enhancing EL2 by the reaction: $As_i + Cu_i$-$V_{Ga} \rightarrow As_{Ga} + Cu_i$ or $Cu_{Ga} + As_i$. Both As_i and Cu_i will attempt to occupy the V_{Ga} site, which explains the anti-correlation between As_{Ga} and Cu_{Ga} in the three samples, from the seed to the tail, as mentioned above. Secondly, another important native defect EL6, which is characterized as T_5 in the TSC spectrum [12], can be converted to EL2 by the defect reaction V_{Ga}-As_i-V_{As} (EL6) + $As_i \rightarrow As_{Ga}$ (core of EL2), as reported by Fang et al. on undoped HB-grown GaAs after 800-°C annealing [13]. Lower mobilities sometimes observed in SI GaAs are often linked with non-uniformity in the crystal. As the EL2 concentration is increased and homogenized, the VZM materials become more uniform and have higher mobility.

D. Conclusions

Electrical and optical properties of undoped SI GaAs materials grown by the VZM technique and annealed at 950 °C under As overpressure have been characterized by different methods, including Hall-effect, near-infrared absorption, TSC, TDPC (using 1.13 eV light), 4.2-K PL, and 2-K selective pair PL. 950-°C annealing of the VZM GaAs under 2-atm As vapor pressure significantly improves the material uniformity and increases both the EL2 concentration and the mobility. Cu incorporation into the VZM materials has been confirmed by both PL and TSC measurements. In addition to C and Cu, Zn, which can be segregated towards the tail of the crystal, is found to be another residual impurity in VZM materials.

We wish to thank T. Cooper and W. Rice for Hall-effect and photoluminescence measurements, respectively. The work of Z-QF, DCR, and DCL was supported by U.S. Air Force Contract No. F33615-95-C-1619. Part of the work was performed at the Air Force Research Laboratory, Wright-Patterson Air Force Base, OH, and partial support was received from the Air Force Office of Scientific Research.

References

1. R. L. Henry, P. E. R. Nordquist, R. J. Gorman, S. B. Qadri, J. Crystal Growth <u>109</u>, 228 (1991).
2. R. L. Henry, P. E. R. Nordquist, R. J. Gorman, J. S. Blakemore, and W. J. Moore, Nuclear Instruments & Methods in Physics Research A <u>380</u>, 30 (1996).
3. P. E. R. Nordquist, R. L. Henry, J. S. Blakemore, S. B. Saban, and R. J. Gorman, in <u>Semi-insulating III-V Materials, Warsaw</u>, edited by M. Godlewski (World Scientific, 1994) p. 47.
4. Z-Q. Fang, D. C. Reynolds, D. C. Look, N. G. Paraskevopoulos, T. E. Anderson, and R. L. Jones, J. Appl. Phys. <u>83</u>, 260 (1998).
5. H. Yoshida, M. Kiyama, T. Takebe, K. Fujita, and S.-I. Akai, Jpn. J. Appl. Phys. <u>36</u>, 19 (1997).
6. Z-Q. Fang, D. C. Look, and M. G. Mier, J. Electron. Mater. <u>27</u>, 62 (1998).
7. Z-Q. Fang, D. C. Look, and R. L. Jones, J. Electron. Mater. <u>26</u>, L29 (1997).
8. Z. G. Wang, H. P. Gislason, and B. Monemar, J. Appl. Phys. <u>58</u>, 230 (1985).
9. E. S. Koteles, J. Kafalas, S. Zemon, and P. Norris, Inst. Phys. Conf. Ser. <u>83</u>, 63 (1987)
10. W. J. Moore and R. L. Henry, Appl. Phys. Lett. <u>70</u>, 738 (1997).
11. J. van de Ven, W. J. A. M. Hartmann, and L. J. Giling, J. Appl. Phys. <u>60</u>, 3735 (1986).
12. Z-Q. Fang and D. C. Look, Mater. Res. Soc. Symp. Proc. <u>442</u>, 405 (1997).
13. Z-Q. Fang, T. E. Schlesinger, and A. G. Milnes, J. Appl. Phys. <u>61</u>, 5047 (1987).

Implant Isolation Study of $In_{0.53}Ga_{0.47}As$

M.Almonte [1,2], K.M.Yu [2], E.E.Haller [1,2] M.C.Ridgway [3], H.Hou [4] and J.Mirecki-Millunchick [4*]

[1] *Materials Science and Mineral Engineering, University of California, Berkeley, CA 94720*
[2] *Ernest Orlando Lawrence Berkeley National Laboratory, Berkeley, CA 94720*
[3] *Department of Electronic Materials Engineering, Australian National University, Australia*
[4] *Sandia National Laboratories, Alburquerque, NM 87185*

Abstract

Effects of implantation in $In_{0.53}Ga_{0.47}As$ due to damage by implantation of Ne^+ ions and to compensation by implantation of Fe^+ ions are reported. The former only involves damage induced effects while the latter leads to damage and dopant induced compensation. In addition, the changes in the electrical properties of the layers are correlated to the lattice damage (damage induced effects) and/or the diffusion of the compensating dopants (dopant induced compensation). Structural characterization of the layers is performed with channeling Rutherford Backscattering Spectrometry (RBS). The distribution of the compensating dopants in the as-implanted and annealed layers is examined by Secondary Ion Mass Spectrometry (SIMS). The thermal stability of these damage and compensation induced effects producing implant isolation is discussed.

A. Introduction

$In_{0.53}Ga_{0.47}As$, lattice-matched to InP, is a semiconductor material of technological importance for numerous electronic and optoelectronic device applications which include photodiodes to be developed for the 1.3-1.55 μm wavelength range. For many applications of $In_{0.53}Ga_{0.47}As$ there exists an important need to understand the formation of highly resistive layers by ion implantation. Semi-insulating layers are used in the active region of high-speed metal-semiconductor-metal photodetectors [1] and perhaps most importantly for device isolation in integrated circuits. Due to the small bandgap (0.75 eV at 300K) of $In_{0.53}Ga_{0.47}As$, this semiconductor alloy has a room temperature intrinsic carrier concentration $n_i = 6.7x10^{11}cm^{-3}$. The calculated intrinsic resistivity is approximately 900 Ohm-cm.

There are two methods to achieve electrical isolation using implantation. Lattice damage in the epilayers is created by bombardment with energetic inert or isoelectronic ions (damage induced effects). Dopant induced compensation involves the implantation and activation of impurities which form deep compensating donor or acceptor levels in the bandgap of the semiconductor pinning the Fermi level near the center of the gap. Transition metal impurities are typically used. The position of the Fe^{2+}-Fe^{3+} acceptor level in $In_{0.53}Ga_{0.47}As$ was determined to lie 0.39 ± 0.02 eV above the valence band edge [2]. There have been several reports on Fe^+ implantation of $In_{0.53}Ga_{0.47}As$ [3-7]. The electrical behavior as a function of a wide range of annealing temperatures has not been reported since no Hall effect measurements have been performed. A correlation of the electrical and diffusive properties of fully isolated Fe^+

---

*Currently at the Department of Materials Science and Engineering, University of Michigan, Ann Arbor, MI 48109 implanted $In_{0.53}Ga_{0.47}As$ layers (with uniform Fe concentration) would be of particular interest for forming layers of high resistivity.

B. Experiment

The $In_{0.53}Ga_{0.47}As$ layers were grown by Metal Organic Vapor Phase Epitaxy (MOVPE) on <100> semi-insulating InP (Fe doped) substrates and were unintentionally doped n-type. The thicknesses of the layers were 0.4 μm (Ne^+ implantation) and 0.72 μm (Fe^+ implantation). The implantations were performed at room temperature (RT) and for Fe also at liquid nitrogen (LN_2) temperatures. Implantation schedules consisting of multiple energy implants were used to obtain uniform concentrations over the entire $In_{0.53}Ga_{0.47}As$ layer. The Ne^+ ion implant energies ranged from 50 to 200 keV. Doses were chosen to obtain an average volume concentration of $5x10^{15}cm^{-3}$, $5x10^{17}cm^{-3}$, $5x10^{19}cm^{-3}$. The Fe doped samples were implanted with energies between 90 and 1250 keV to an average concentration of $2x10^{18}cm^{-3}$. The implanted samples were annealed in an AG Associates Heatpulse 210T rapid thermal annealer in a N_2 atmosphere. Annealing of the implanted samples was carried out with a proximity cap for either 5s (Fe^+ implanted samples) or 30 s (Ne^+ implanted samples) in the temperature range from 300 to 900°C . The $In_{0.53}Ga_{0.47}As$ epilayers were characterized electrically by room temperature resistivity and Hall effect measurements. The ion implantation damage in the layers was examined using channeling Rutherford Backscattering Spectrometry (RBS) with a 1.95 MeV He^+ ion beam aligned along the <110>. The Secondary Ion Mass Spectrometry (SIMS) analysis was performed using a Cameca double-focusing magnetic sector mass spectrometer configured with a O_2^+ primary ion beam (8 keV) and positive secondary ion mass spectrometry. The As SIMS signal was monitored to locate the interface between the $In_{0.53}Ga_{0.47}As$ epilayer and the InP substrate and was not quantified.

C. Results

The channeled RBS sprectra for the as-implanted samples are shown in Fig. 1. None of the implanted samples were amorphous. A maximum resistivity of 5 Ω-cm was measured for the

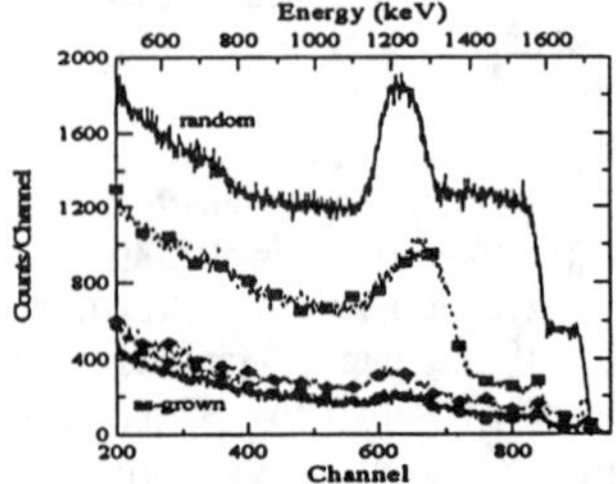

Fig. 1. 1.95 MeV He^+ <110> aligned backscatter spectra for as-implanted samples. $5x10^{15}cm^{-3}$ (●), $5x10^{17}cm^{-3}$ (◆), and $5x10^{17}cm^{-3}$ (■).

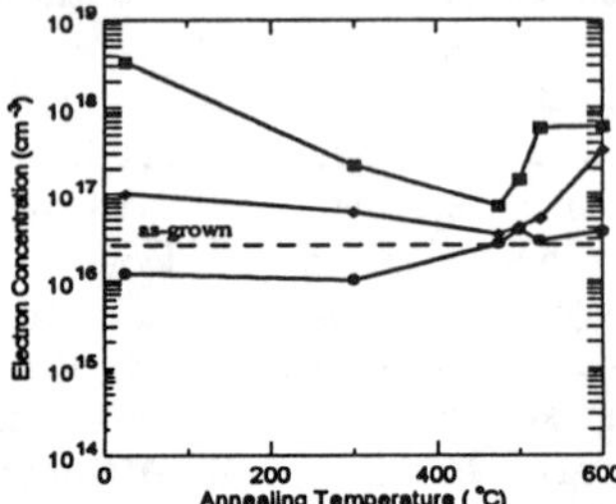

Fig 2. Hall Effect measurements for samples annealed for 30s. $5x10^{15}cm^{-3}$ (●), $5x10^{17}cm^{-3}$ (◆), and $5x10^{17}cm^{-3}$ (■).

layer implanted with $5x10^{17}$ atoms cm^{-3} and annealed at 300°C (30s). From the Hall Effect measurements an increase in electron concentration higher than the original value is observed

(Fig. 2). In the as-implanted samples the concentration of these carriers increases with implant ion concentration. The damage induced carriers observed in Ne^+ implanted and annealed layers suggests the formation of defect related levels which are close to the conduction band edge (E_{CB}) which is consistent with a Fermi stabilization energy (E_{FS}) close to the E_{CB} [8]. These results suggests the formation of shallow donors which are formed by implantation and are activated upon annealing. These donors may be attributed to implantation induced defect complexes which are not annihilated, but are converted into a new stable state after annealing at 600°C.

The electrical measurements for the Fe as-implanted (LN_2 temperature) and annealed samples are shown in Fig. 3. Similar electrical behavior was observed for the sample implanted at RT. The optimal annealing temperature for maximum resistivities observed for both of the samples implanted at LN_2 and RT was 800°C. The highest resistivities for the samples implanted at LN_2 and RT were approximately 320 Ω-cm and 770 Ω-cm, respectively. The concentration of deep Fe acceptors was greater than that of the donor concentration in the as-grown material. Since the concentration of carriers is still higher than the intrinsic value for the layers implanted at RT and annealed, we believe this may be attributed to damage related effects. From the results of the Ne^+ implanted layers, damage induced carriers were observed upon annealing of the $In_{0.53}Ga_{0.47}As$ layers. The Fermi energy for the sample with the highest compensation of carriers ($n \sim 2.7 \times 10^{13} cm^{-3}$) was calculated to be approximately E_{CB}-0.23 eV which correlates with the E_{FS} for irradiated $In_{0.53}Ga_{0.47}As$ [8]. It appears that the damage may not be completely annealed before electrical activation occurs. Ideally for maximum free carrier compensation we would like to anneal out all of the damage related effects before the dopants are electrically activated. Annealing at a higher temperature of 875°C for 5s resulted in the rapid out-diffusion of the Fe (RT implantation) which is evident from the SIMS profile (Fig.4) and is consistent with the increase in carrier concentration observed from the Hall effect measurements.

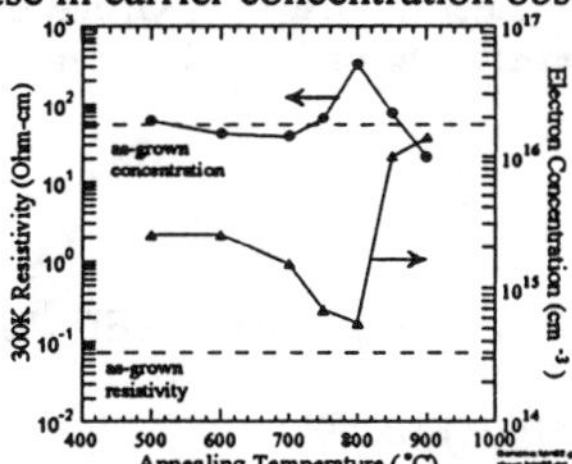

Fig. 3. Hall Effect measurements for Fe^+ doped samples implanted at LN_2 temperature.

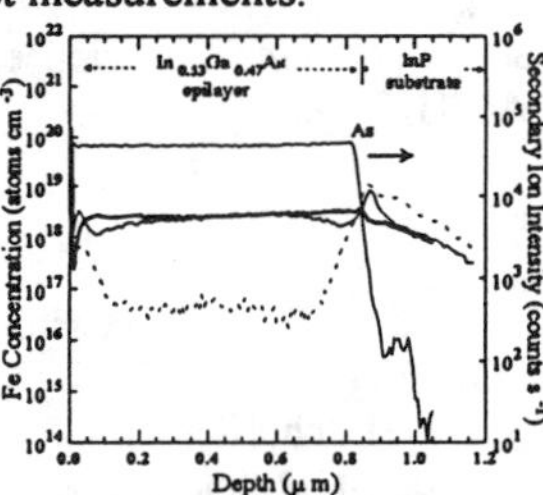

Fig. 4. SIMS depth profile for layers implanted (bold solid line) with Fe (RT) and annealed at 800°C (solid line) and 875°C (dashed line) for 5 s.

A simple, rough estimate of the diffusion coefficient (D_{Fe}) was obtained using the relation
$$x = (Dt)^{1/2} \qquad (1)$$
for the annealing conditions of 800°C for 5s. The D_{Fe} is approximately 10^{-11} cm²/s. To our knowledge there exist no accurate measurement of the D_{Fe} for Fe in $In_{0.53}Ga_{0.47}As$. We plan to measure D_{Fe} at a number of temperatures to further our understanding of Fe diffusion in $In_{0.53}Ga_{0.47}As$ layers.

D. Conclusions

The electrical properties of Ne^+ and Fe^+ implanted $In_{0.53}Ga_{0.47}As$ have been investigated. In addition, the thermal stability of the implant isolation in the layers was also examined as a function of implant dose and annealing temperature. The Ne^+ implanted layers exhibited a maximum resistivity $\rho_{max} = 5$ Ω-cm. Damage induced carriers were observed in the Ne^+ implanted $In_{0.53}Ga_{0.47}As$. The stability of the resistive state of these layers extended up to a temperature of approximately 300°C. A higher resistivity of $\rho_{max} = 770$ Ω-cm can be reached with Fe^+ implantation. The free carrier concentration measured for the highest compensated layer was ~2.7×10^{13} cm^{-3}. Since this concentration of free carriers is still higher than the intrinsic value we expect that this electrical resistivity can be further increased. Further electrical studies of the Fe^+ implanted layers will help to determine the role that damage induced carriers play in dopant induced compensation. The semi-insulating properties of these layers were stable up to 800°C. No significant out-diffusion of Fe was observed for the samples annealed up to this temperature. Through a thorough understanding of the mechanisms of activation and redistribution of the Fe^+ in the annealed layers, it should become possible to optimize the implantation and annealing conditions to produce $In_{0.53}Ga_{0.47}As$ layers of near uniform resistivities sufficiently high for device isolation purposes.

Acknowledgements

The authors would like to thank Andreas Stonas for his assistance in implantation and Robert Clark Phelps (Charles Evans & Associates) for performing the SIMS analysis on the samples. This work was supported by the Director, Office of Energy Research, Office of Basic Energy Sciences, Materials Sciences Division of the U.S. Department of Energy under Contract No. DE-AC03-76SC00098.

References

[1] E.H. Böttcher, D. Kuhl, F. Hieronymi, E. Dröge, T. Wolf, and D. Bimberg, IEEE Quantum. Electron. **QE-28** (1992) 2343
[2] B. Srocka, H Scheffler, and D. Bimberg, Phy. Rev. B **49** (1994) 10259
[3] S.J. Pearton, C.R.Abernathy, M.B.Panish, R.A.Hamm, and L.M. Lunardi, J. Appl. Phys. **66** (1989) 656
[4] S.M.Gulwaldi and M.V. Rao, J. Appl. Phys. **69** (1991) 162
[5] B. Gruska, H. Ullrich, R. K. Bauer, D. Bimberg, K. Wendel, J. Appl. Phys. **73** (1993) 4825
[6] M.V.Rao, N.R. Keshavarz-Nia, D.S. Simons, P.M. Amirtharaj, P.E. Thompson, T.Y. Chang, and J.M. Kuo, J. Appl. Phys. **65** (1989) 481
[7] M.V.Rao, S.M. Gulwaldi, S.Mulpuri, D.S. Simons, P.H. Chi, C. Caneau, W-P. Hong, O.W. Holland, H.B. Dietrich, J. of Elect. Mat., **21** (1992) 923
[8] W. Walukiewicz, in "Mat. Res. Soc. Symp. Proc." **300**, (Materials Research Society, Pennsylvania, 1993) 421

Growth and orientation of GaN epilayers on NdGaO$_3$ by hydride vapor phase epitaxy

Akihiro Wakahara[*], Teruaki Nishida, Kenji Kawano, and Akira Yoshida
Dept. of Electrical & Electronic Eng., Toyohashi University of Technology
Toyohashi 441-8580, Japan
Youji Seki and Osamu Oda
Central Research Lab., Japan Energy Corp., Toda 335, Japan

Abstract

Growth characteristics of GaN on NdGaO$_3$ (011) and (101) substrates by hydride vapor phase epitaxy are investigated. The epitaxial relationship is found to be GaN (0001) / NdGaO$_3$ (011) with GaN [10-10] // NdGaO$_3$ [100] for NdGaO$_3$ (011) and GaN(11-24)/ NdGaO$_3$(101) with GaN[11-2-4]// NdGaO$_3$[10-1] for the NdGaO$_3$(101). When the GaN is directly grown on NdGaO$_3$ substrates around 1000°C, no deposits of GaN are obtained due to the reduction of the NdGaO$_3$ substrate by the presence of NH$_3$ at high temperature. In order to avoid the substrate reduction, low-temperature grown GaN is effective as the protective layer.

A. Introduction

GaInN is a direct band-gap semiconductor which can emit from red (1.9eV for InN) to ultra-violet (UV, 3.4eV for GaN) region. It has, therefore, received much attention as a promising material for optoelectronic devices. Although, recent progress of heteroepitaxial growth technique for group-III nitrides has achieved bright blue/green light-emitting-diodes and CW operation of laser diodes (LDs) [1], GaN epilayers indicate high defect density due to the lack of a lattice matched substrate. For growing GaN related materials, sapphire is commonly used as a substrate. However, there exists large lattice mismatching between GaN and sapphire and it may hinder the realization of LDs. In order to improve the structure perfection, various substrates such as SiC[2,3], ZnO[4], Si[5], GaAs[6], spinel[7], and epitaxial lateral growth (ELO)[8] have been examined. Although, the sapphire substrates have made a great achievement for GaN related devices, such as light emitting diodes, laser diodes, and power electronic devices, there exist a large mismatching of the lattice constants and thermal expansion coefficient between the substrate and the GaN. In order to find a quasi-lattice matched substrate, various substrates such as LiGaO$_2$ and SiC have been examined. NdGaO$_3$ is one of the Ga perovskites and has the possibility of a lattice matched substrate for GaN. However, a few reports are published on GaN growth on NdGaO$_3$[9]. In this work, we grow GaN on both (011) and (101) oriented NdGaO$_3$ and investigate its possibility as a substrate for GaN epitaxial growth.

B. Experiments

GaN was grown on both (011) and (101) oriented NdGaO$_3$ substrates by hydride vapor phase epitaxy (HVPE) using Ga, HCl, and NH$_3$ as the sources. The growth temperature was varied from 600 to 1000°C. The GaN growth was begun by simultaneously supplying HCl and NH$_3$. In order to see the effect of surface nitridation, the substrates were pretreated

---

Corresponding e-mail: wakahara@eee.tut.ac.jp

with NH$_3$ at 1000°C for 0-10min. The crystalline quality and the epitaxial orientation were characterized by reflection high electron energy diffraction (RHEED), x-ray diffraction (XRD), and back Laue measurements. The surface morphology was observed by atomic force microscope (AFM).

C. Results and discussion

C.1 Direct growth

Figure 1 shows RHEED pattern of GaN layer on (a) NdGaO$_3$ (011) and (b) (101) substrates grown at 800°C. For the NdGaO$_3$ (011) substrate, the c-axis of GaN is perpendicular to the substrate surface. On the other hand, the c-axis of GaN layer tilts about 39° from the normal direction. From the back Laue analysis, the epitaxial relationship of GaN on NdGaO$_3$ (011) is found to be GaN (0001) / NdGaO$_3$ (011) with GaN [10-10] // NdGaO$_3$ [100]. In this case, Ga atom spacing at the substrate surface is 5.428A for which lattice mismatching is 1.7% if the hexagonal unit cell of GaN is put on the Ga atoms of the substrate (Fig.2(a)). On the other hand, for the NdGaO$_3$(101) substrate, the epitaxial relationship is GaN(11-24)/ NdGaO$_3$(101) with GaN[11-2-4]// NdGaO$_3$[10-1]. In this case, all lattice site of GaN are corresponding to that of the substrate, just like a GaAs/Si (Fig.2(b)). After this, we focus to the GaN/NdGaO$_3$ (011) because of its small lattice mismatch.

Figure 3 shows the RHEED pattern of GaN/NdGaO$_3$ (011) at various growth temperatures. The RHEED pattern changes from a spotty to a ring pattern by increasing the temperature. When the growth is carried out at 1000°C, the sample surface is hazy and GaN is not deposited on the substrate. From the lattice constant calculated by the radius of the ring in the RHEED pattern, the sample surface consists of GaN, NdN, and NdGaO$_3$. The results suggest that the NdGaO$_3$ substrate is resolved and nitrided by NH$_3$. In order to verify this

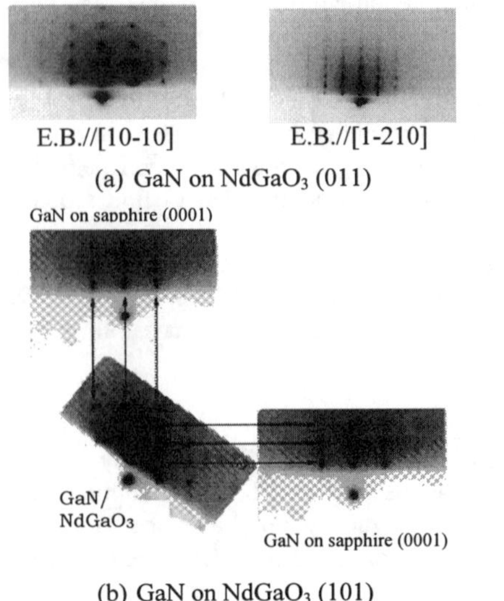

Fig.1: RHEED pattern of GaN layer on (a) NdGaO$_3$ (011) and (b) (101) substrates. For GaN on NdGaO$_3$ (101), the pattern is compared with that of GaN on sapphire.

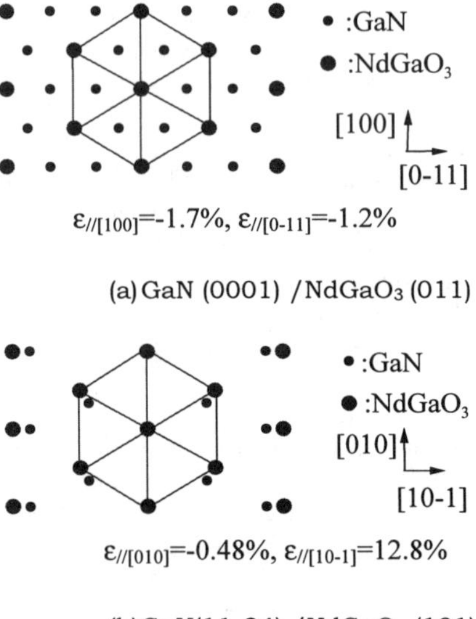

Fig.2: Epitaxial alignment and lattice mismatch of GaN on NdGaO$_3$ substrates.

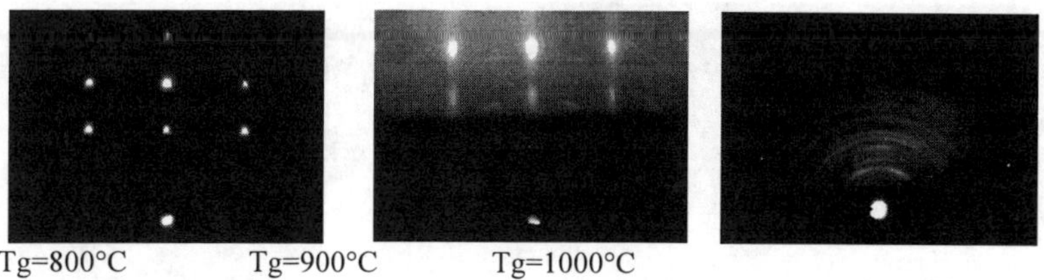

Fig.3: RHEED pattern of GaN on NdGaO₃ (011) at various growth temperature.

hypothesis, intentional nitridation at 1000°C for 10min is applied to the NdGaO$_3$ substrate. As a result, the RHEED pattern of the nitrided surface is almost the same as that of the sample grown at 1000°C. The substrate nitridation occurs when the growth temperature exceeds 900°C, because the NdN (222) diffraction is clearly observed in the XRD profile. In our growth system, GaCl reaches the substrate about 2min later than the NH$_3$, even though both gases start to flow at the same time. Therefore, the degradation of the GaN layer is due to the dissolution and nitridation of the NdGaO$_3$ substrate by NH$_3$.

C.2 Two step growth

Since the NdGaO$_3$ substrate is dissolved by the presence of NH$_3$ at high temperature and it is difficult to control the arrival time of the source gases in HVPE growth, no GaN films are obtained as mentioned in the previous section. In order to avoid the substrate dissolution before the growth, a two-step growth procedure, in which low-temperature grown GaN was used as a protective layer, was tried. The growth temperature of the GaN protective layer was chosen as 600°C, at which temperature the NdGaO$_3$ substrate does not react with the NH$_3$. Figure 4 shows the RHEED pattern and AFM image of as grown GaN protective layers. When the layer thickness is 20nm, RHEED indicates clear GaN-related spots with some extra-spots. These extra-spots are due to the NdGaO$_3$ (011) substrate. From the AFM image, the grown surface is covered with many small islands. The results mean that a part of the substrate is not covered with the GaN layer. By increasing the GaN thickness to 100nm, the extra-spot in the RHEED disappears and the GaN islands become large. On the contrary, when the layer thickness exceeds 300nm, the RHEED pattern becomes ring shaped, which means polycrystallinity. By

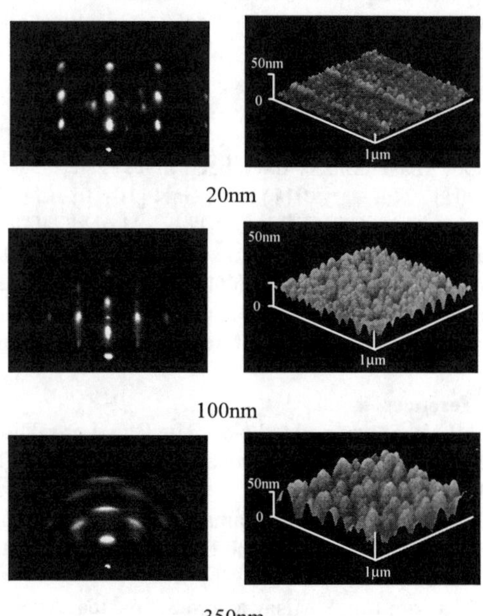

Fig.4: Thickness dependence of RHEED pattern and AFM image of low-temperature grown GaN protective layer. Electron beam was along [10-10].

Fig.5: RHEED pattern and surface morphology
of GaN film. Electron beam was along [10-10].

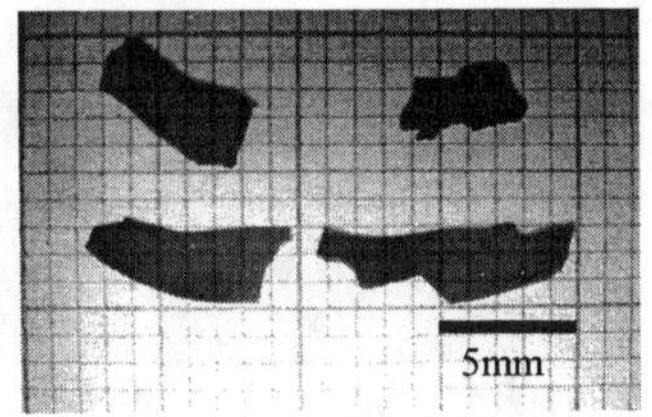

Fig.6:Free standing GaN. The
film thickness is 20μm.

applying thermal annealing at 1000°C, a single crystalline GaN protective layer cannot be obtained for a protective layer thickness of 300nm. Therefore, a GaN layer thickness around 100nm seems to be the optimum value for the protective layer from the standpoint of both surface coverage and crystalline quality.

Single crystalline GaN films can be obtained at 1000°C on $NdGaO_3$ substrates by using a low-temperature grown GaN protective layer, as can be seen in Fig.5. When the GaN thickness exceeds several micrometers, the GaN films separate from the $NdGaO_3$ substrate. Fig. 6 shows the free-standing GaN wafers. The maximum size of 10mm×7mm is achieved in the preliminary results. This result suggests that a freestanding GaN substrate can be achieved by using a $NdGaO_3$ substrate and the HVPE method. However, it was difficult to measure x-ray diffraction and electrical properties, because the separated GaN films warp strongly. Further investigations of the defects, electrical and optical properties, and also the mechanism of separation are necessary to realize high quality GaN wafers.

D. Conclusions

Growth characteristics of GaN on $NdGaO_3$ (011) and (101) substrates by hydride vapor phase epitaxy have been investigated. The epitaxial relationship was found to be GaN (0001) / $NdGaO_3$ (011) with GaN [10-10] // $NdGaO_3$ [100] for $NdGaO_3$ (011) and GaN (11-24)/ $NdGaO_3$(101) with GaN[11-2-4]// $NdGaO_3$[10-1] for the $NdGaO_3$(101). When GaN was directly grown on $NdGaO_3$ substrates at about 1000°C, no deposits of GaN were obtained due to the reduction of the $NdGaO_3$ substrate by the presence of NH_3 at high temperature. The reduction of the substrate can be avoided by using a low-temperature grown GaN protective layer. A freestanding GaN wafer with a maximum size of 10mm × 7mm was achieved.

References
[1] S.Nakamura and G.Falso, The Blue Laser Diode, (Springer-Verlag, Berlin, 1997).
[2] T.Sasaki and T.Matsuoka, J. Appl. Phys. 64 (1988) 4531.
[3] M.E.Lin, B.Sverdlov, G.L.Zhou, and H.Morkoc, Appl. Phys. Lett. 62 (1993) 3479.
[4] T.Detchprohm, K.Hiramatsu, H.Amano, and I.Akasaki, Appl. Phys. Lett. 61 (1992) 2688.
[5] T.Lei, M.Fanciulli, R.J.Molnar, T.D.Moustakas, R.J.Graham, and J.Scanlon, J. Appl. Phys. 59 (1991) 383.
[6]H.Okumura, S.Misawa, and S.Yoshida, Appl. Phys. Lett. 59 (1991) 1058.
[7]A.Kuramata, K.Horino, K.Domen, K.Shinohara, and T.Tanahashi, Topical Workshop on III-V Nitrides (Nagoya, Japan, 1994) SP-4.
[8] A.Usui, H.Sunakawa, A.Sakai, and A.A.Yamaguchi, Jpn. J. Appl. Phys. 36 (1997) L899.
[9] H.Okazaki, A.Arakawa, T.Asahi, O.Oda, and K.Aiki, Solid-State Electron. 41 (1997) 263.

II.

Semi-Insulating III-V Compounds

Improvement of uniformity and electrical properties of Fe-doped InP by wafer annealing

J. Jiménez, M. Avella, E. de la Puente

Física de la Materia Condensada, ETS Ingenieros Industriales,
47011 Valladolid, Spain. e-mail: jimenez@hp9000.uva.es

Abstract

Thermal treatments are crucial steps for improving the quality of semi-insulating InP substrates. The result of these treatments depends on the starting material and the specific annealing conditions. The best results in terms of the electrical properties and homogeneity are obtained by annealing Fe-doped InP wafers. The influence of the specific annealing conditions is discussed. The best results are obtained for long annealing times (50 hours) and slow cooling rates, which give good homogeneity, high carrier mobility and high enough electrical compensation.

A. Introduction

Semi-insulating InP is a basic material for high speed electronics and OEIC's (Optoelectronic Integrated Circuits) for fibre optic communication systems. The growth and preparation of substrates with precise requirements is a crucial step in the development of these technologies. Among the major requirements of these substrates are high resistivity, high mobility and whole wafer uniform properties [1]. Commercially available LEC (Liquid Encapsulated Czochralski) substrates are prepared by adding iron to the polycrystalline charge [1,2]. Iron enters the In site in the InP lattice giving a midgap acceptor (E_a=0.64 eV) [3], that compensates the residual shallow donors. However, the presence of iron raises several challenging problems. These problems are normally related to its small distribution coefficient (< 0.001) [1], its high diffusivity [4] and its incomplete electrical activation [1,4]. The consequence of the low segregation coefficient is a pronounced axial distribution profile [5]. The low electrical activation demands that high amounts of iron be added to the charge, above that strictly necessary for compensating the shallow donors. The high diffusivity together with the high doping amounts are responsible for the iron contamination of the epilayers grown on these substrates [6], which is detrimental for the device performance.

From the point of view of the radial uniformity the problems are not smaller, since typical macroscopic and microscopic lateral fluctuations appear. Doping growth striations are commonly observed [7,8]. These inhomogeneities arise from non convective transport in the melt, the curvature of the growth interfaces and other growth features.

In addition to these macroscopic and mesoscopic non uniformities, a number of microdefects are present, e.g. grown-in dislocations, microprecipitates and grain boundaries [9]. These microdefects play an important role in the properties of the substrates and in the thermal treatments as we will see later [10-12].

Following these considerations, a great effort is necessary in order to improve the quality of the wafers actually available. As it was shown for other substrates, e.g.GaAs [13], adequate thermal treatments seem to be the best way for achieving advances in the optimization of the semi-insulating InP substrates. Strong research has been carried out in the last years about thermal treatments to obtain high quality semi-insulating InP substrates. In spite of this, there is not still consensus about the best choice of the annealing parameters.

B. Thermal Treatments and Electrical Compensation

The aim of annealing is to produce semi-insulating substrates with resistivities above 10^7 Ω•cm, high carrier mobility, improved purity and good homogeneity (better than 10%). As it was already mentioned most of the problems are related to the presence of iron. According to such purposes several annealing procedures have been proposed. The main variables to be considered for the annealing are: i.-the as-grown material properties, ii.-the annealing temperature, iii.-the annealing atmosphere, iv.-the annealing time, v.-the post annealing cooling rate.

The choice of the as-grown material is one of the main issues discussed in this work. Different types of as-grown materials are considered for the thermal treatment: 1.-nominally undoped InP, 2.-lightly doped Fe-doped ($n>10^{15}$ cm^{-3}), 3.-semi-insulating Fe-doped, 4.-nominally undoped material annealed in the presence of an iron source. As well differences between ingot and wafer annealing have to be considered.

Since Klein et al. [14] showed the possibility of enhancing the resistivity of nominally undoped InP by annealing under phosphorous pressure, a great deal of attention was addressed to this procedure [15-17]. There are two ways to account for the resistivity increase, i.e. the generation of native deep defects that compensate the residual shallow donors or the massive loss of shallow donors, until their concentration falls down below a threshold value and they become compensated by deep acceptors.

It was soon agreed that the main mechanism producing the high resistivity of undoped InP after annealing is the compensation of the remaining shallow donors by residual iron [16,17]. In fact strong donor outdiffusion occurs during the annnealing step. Resistivities above 10^7 $\Omega\bullet$cm and mobilities as high as 4000 cm^{2}V^{-1}s^{-1} have been produced by this procedure [15-17]. However, the reproducibility was very poor, which has been associated with the lack of control over the iron concentration and the competition with other contaminants coming from the red phosphorous sources [18]. Anyway, it is admitted that in the absence of a native deep level, as EL2 in GaAs [19], a minimum iron concentration is required to obtain semi-insulating substrates; below such a concentration, contaminants play an important role in the compensation [18,20].

Electrically active iron can be in two charge states, either neutral, Fe^{3+}, or singly ionized, Fe^{2+}. Basically [Fe^{2+}] equals the net donor concentration, N_D-N_A. Iron is not uniformly distributed. However, at the mesoscopic scale the net donor concentration is rather uniform, which is in agreement with its distribution coefficient (~0.8) [1]. The main inhomogeneities are expected at the microscopic scale in the neighbourhood of the microdefects, where the strain field influences both iron and donor distributions.

The resistivity is controlled by the compensation ratio, [Fe$_{In}$]/(N_D-N_A). It is important to note that the influence of the compensation ratio fluctuations on the resistivity fluctuations is critical for low compensation ratios [21]. This suggests that two of the objectives of the treatments are low Fe content and high compensation ratio, which should ensure high resistivity, high mobility and uniformity.

In order to understand the annealing mechanisms it is interesting to know the levels participating in the electrical compensation in as-grown material. Typical residual shallow donors are Si and S, while shallow acceptors are Zn and Mg. It was demonstrated that [Fe^{2+}] is larger than the net concentration of these residual donors; which should imply the existence of missing donors [22]. This missing donor was identified as the defect giving rise to a line at 2315 cm^{-1} in the infrared absorption spectrum [5,22-25]. It was associated with a hydrogen related complex, which has been identified as an Indium vacancy, V_{In}, fully surrounded by 4 hydrogen atoms (V_{In}(PH)$_4$). The presence of hydrogen in high enough concentration is due to the wet boron oxide used as an encapsulant in LEC pullers. This defect was demonstrated to behave as a shallow donor [23,24] and the role it plays in the thermal treatments has to be carefully considered.

C. Thermal Treatments of Fe-Doped InP

Both wafers and thick blocks were annealed. The specific conditions of these treatments were as follows: Several semi-insulating wafers were selected from the same region of a LEC crystal. They were about 0.7 mm thick and had resistivities in the range from 5 to 7.2 x10^6 $\Omega\bullet$cm. The wafers were cleaved along a diameter: half wafer was used for reference and the other half was etched and sealed in a quartz ampoule together with an amount of red phosphorus sufficient to give ~1 Bar pressure (this pressure was selected to avoid surface degradation, once it was known that P in-diffusion was almost negligible).

Table I

Treatment (sample)	ELECTRICAL PROPERTIES						ANNEALING CONDITIONS			
	BEFORE ANNEALING			AFTER ANNEALING						
	resist. (Ωcm)	mobility (cm^2/Vs)	Hall conc. (cm^{-3})	resist. (Ωcm)	mobility (cm^2/Vs)	Hall conc. (cm^{-3})	temp. (°C)	durata (h)	press. (atm)	cool. (°C/min)
NF1 (159-53)	7.27×10^6	1500	5.72×10^8	7.3×10^7	1900	4.5×10^7	900	50	< 1	6
NF2 (159-60)	5.77×10^6	1112	9.72×10^8	4.4×10^7	3266	4.3×10^7	900	10	1	6
NF3 (159-59)	6.95×10^6	1390	6.47×10^8	3.51×10^7	2659	6.68×10^7	900	5	1	6
NF4 (159-57)	5.1×10^6	1002	1.2×10^9	3.37×10^7	3114	5.95×10^7	900	5	1	1
NF6 (159-55)	6.94×10^6	1462	6.16×10^8	2.78×10^7	3635	6.16×10^7	900	50	1	1

The annealing temperature was 900 °C. The annealing times were 5, 10 or 50 hours, while the cooling rate was either 1°C/min or 6°C/min. The specific treatments are given in Table I together with the electrical transport data. Also two blocks of a semi-insulating InP crystal were annealed. They were cut from the central part of a LEC ingot and had thickness of about 2 cm. These blocks were annealed separately at the same temperature of 900 °C for 60 hours under a P pressure of 1 Bar, while the cooling rates were 5 and 1°C/min, respectively. After the treatment the blocks were sliced in 0.7 mm thick wafers and were characterized for electrical properties, and both local and macroscopic uniformity.

D. Wafer Uniformity

Since the uniformity is a crucial issue for the application of these substrates in integrated optoelectronics, experimental techniques allowing spatially resolved characterization are most interesting in order to understand the origin of the inhomogeneities and their evolution with the thermal treatments. Studies about the homogeneity of semi-insulating InP have been carried out by resistivity [21], infrared transmission [8] and PL [7]. Resistivity and IR mapping give a comprehensive view of the whole wafer uniformity, however, they do not resolve microscopic inhomogeneities, since the probe size is at best tenths of µms. In contrast PL mapping is able to give information about both long and short scale inhomogeneities, since it combines scanning with micrometer size probe beams.

Recently we have set up a scanning photocurrent experimental system that is an excellent complement to scanning PL. Both techniques have been described elsewhere [10-12,26]. Here, we summarize the main features that can be analysed by these procedures in semi-insulating InP. SPL is measured at room temperature, the intensity of the intrinsic PL peak at λ_{max} is mapped. The efficiency of the band to band luminescence emission is governed by the minority carrier lifetime, which is controlled in semi-insulating InP by recombination at iron levels. It is well known that the intensity of the band to band luminescence is inversely proportional to the electrically active iron concentration [7]. Therefore, in a first approximation the fluctuations of the luminescence intensity reflect the distribution of the electrically active iron. Since luminescence is excited by above band gap light, the penetration depth lies below 200 nm The spot size is ~2 µm diameter. However, the probed volume is determined by the minority carrier diffusion length, which can extend further away in semi-insulating InP.

In contrast to PL, scanning PC is excited by extrinsic light, in particular 1.06 and 1.32 µm lines from a cw:YAG laser. Thus, the excitation is done throughout the sample thickness. The beam diameter is about two µms. Since the sample is transparent, the beam spreads out throughout the sample resulting in a lateral resolution of a few µms. Photocurrent is a complex function of carrier photogeneration, recombination and mobility. The interpretation

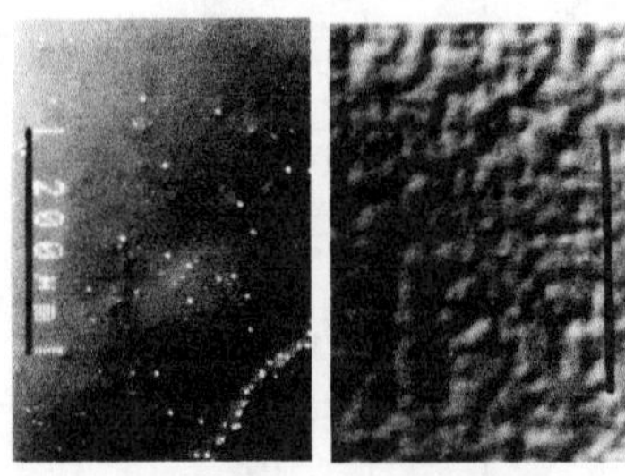

Figure 1.- Nomarski micrographs of as-grown (left) and wafer annealed (right) InP:Fe. Scale bar is 200 μm

interpretation of the PC maps requires modelling the corresponding electronic transitions involved for the excitation wavelengths used which are related to iron. A full discussion about the meaning of the photocurrent contrast was presented in ref.26. The results presented therein were referred to fully compensated material, with compensation ratio above 3. The main conclusions obtained were as follows: i.-photocurrent increases with $[Fe^{3+}]$, ii.-photocurrent decreases with $[Fe^{2+}]$, iii.-the photocurrent intensity increases with the compensation ratio $[Fe_{In}]/(N_D-N_A)$.

The transport data measured by the Hall effect at room temperature are reported for the different annealing procedures in table I. A general observation is the increase of the resistivity and mobility by the thermal treatment. This should mean that the compensation ratio, $([Fe^{3+}]+[Fe^{2+}])/[Fe^{2+}]$ is increased after annealing. This is the consequence of two combined processes, the outdiffusion of iron and shallow donors. However, the rate of donor removal seems to be faster than the rate of iron outdiffusion. The massive removal of donors is associated with the outdiffusion of H in the $V_{In}(PH)_4$ complex, which was found to anneal out [5, 23-25].

As-grown wafers show several inhomogeneities revealed by DSL (Diluted Sirtl with Light) etching [9], i.e. growth striations, microdefects embedded in the matrix and dislocations decorated by microdefects. Besides, the photocurrent map exhibits a pronounced gradient of current between the centre and the rim of the wafer with a variation that can reach about one order of magnitude. This suggests a higher iron content at the periphery. Superimposed are some local PC fluctuations. After annealing there are significant changes in the morphology revealed by DSL etching. First of all, no growth striations were observed even for the shortest annealing time (5 hours). The microdefects along dislocations were progressively dissolved by the annealing (the longer the treatments the smaller the defects). The general surface aspect is quite different according to the annealing duration. The surface presents an orange peel aspect for the longer annealing times, Fig.1. Differences are also observed when the cooling rate changes from fast (6°C/min) to slow (1°C/min), which is surely related to the interstitial mechanism of iron diffusion [4]. Step profiling shows that the dislocation atmospheres are revealed as depressions by DSL etching. Taking into account that the etching rate is controlled by the lifetime of photoholes at the surface, we have to conclude that the Cottrell atmosphere spreading is produced by the reduction of $[Fe^{2+}]$, which is the main capture centre for holes. Regarding PL and PC maps, the main observations are as follows: PL shows the existence of growth striations for short annealing times, however they fully disappear for annealing time over 40 h. The PC maps show a rather flat PC distribution, specially for the long annealing times, Fig.2. All that represents an improvement of the Fe homogeneity at both long and short range. Some iron gettering is observed around scratches [10], which is reason to handle wafers for annealing carefully.

Wafer annealing significantly changes the distribution of the impurities and native defects in

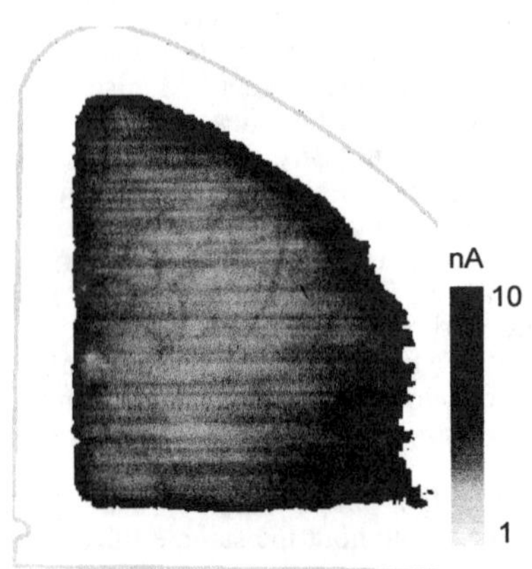

Figure 2.- PC map of sample NF6 Image size 10 x 12 mm^2.

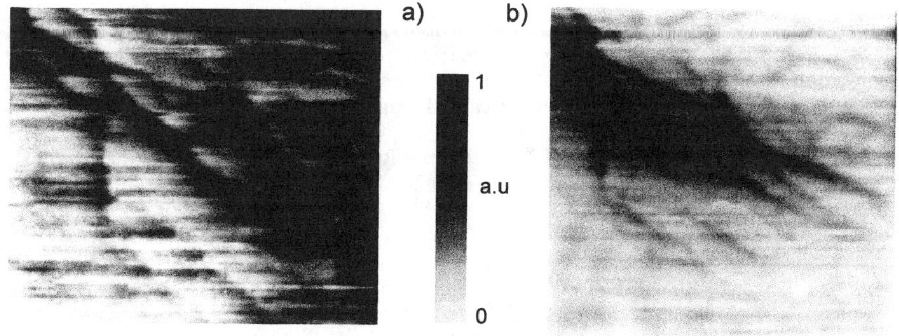

Figure 3.- a) SPL (1 x 0.9 mm^2) and b) SPC map (1x1 mm^2) of a microdefect in the ingot annealed sample. Grey levels are normalised

Fe-doped InP, smoothing the Cottrell atmosphere shape and the microdefect nature and distribution.

The resistivity is enhanced for all the wafers cut from the blocks but it is particularly increased in the wafers from the block extremes with respect to central wafers[10]. This is consistent with out diffusion of shallow donors. The ingot-annealed material is significantly different from the wafer-annealed one. Matrix microdefects are also found after annealing. This probably means that microdefects are related to species with low diffusion rate, thus, they need a free surface next to them in order to be removed. In the case of wafer-annealed material the closest surface (sink) is a maximum of 300 μm away.

The Cottrell atmospheres around dislocations were seen to be more extended in annealed material, in a similar way to what was observed in annealed wafers. On the other hand, decoration microdefects do not dissolve, but rather seem to grow. This effect is probably related to the indiffusion of iron; since it cannot migrate to the surface as was the case for annealed wafers, it is gettered by the strain field of the dislocation, preventing the dissolution of the microprecipitates. A depression is revealed by DSL etching around the dislocation. This should mean that [Fe^{2+}] is strongly reduced, which demonstrates active donor gettering by the dislocation core. In fully compensated material anticorrelated PL and PC patterns are expected, since in a rough approach PL is inversely proportional to [Fe$_{In}$] while PC is directly proportional. However, this is not so simple when the compensation ratio is considered [11]. In fact, iron depressed regions give high PL intensity and can give high PC intensity when the ratio [Fe^{3+}]/[Fe^{2+}] is high. Around crystallographic defects there are three concurrent observations: high PL, high PC and fast etching rate, which can be explained on the bases of reduced [Fe$_{In}$], and high compensation ratio, Fig.3. This shows that crystallographic defects play an important role as donor and iron sinks.

Regarding growth striations, PL maps show that they were not removed by crystal annealing, Fig.4, contrary to what happened for annealed wafers. This fact demonstrates once again the role played by the free surfaces during the annealing step in order to homogenize the iron distribution. With regard to the long range homogeneity, both PL and PC maps do not show significant improvements in wafers cut from the annealed blocks. All these observations show that crystal annealing does not provide optimal results.

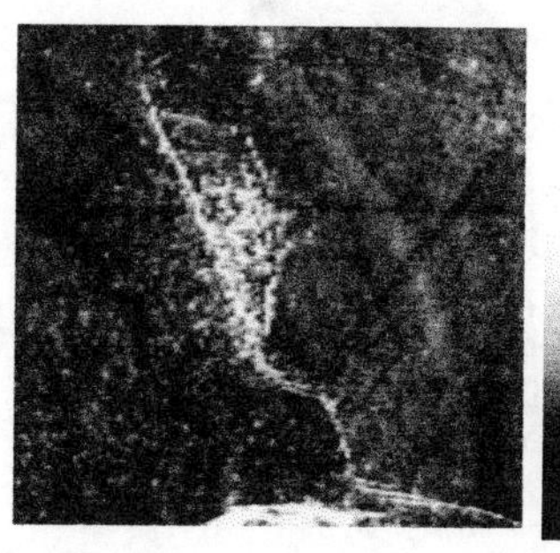

Figure 4.- SPL map of sample NF7. Map size 7.5x7.5 mm^2. Scale is

In conclusion, long annealing time (~50 hours) and slow cooling rate are the most advantageous conditions for obtaining good electrical properties (table I) and homogeneity. Wafer annealing gives better homogeneity over a wafer and from wafer to wafer than what can be obtained from ingot annealing, which exhibits axial gradients and microdefects.

Authors wish to thank Dr. Fornari (MASPEC) for supplying samples and fruitful discussions. This work was done under CE contract (SMT-CT95-2024).

References

[1] G. Müller, Phys. Scripta 35 (1991) 201
[2] C.R. Zeisse, G.A. Antypas, C. Hopkins, J. Cryst. Growth 64 (1983) 217
[3] S.G.Bishop, "Deep Centers in Semiconductors", ed. by S.T. Pantelides (Gordon and Breach, New York 1986) ch.8
[4] M. Brozel, E.J. Foulkes, B. Tuck, Electron. Lett. 17 (1982) 532
[5] G.Wittmann, F. Mosel, G.Müller, A.Seidl, A. Winnacker, Mater. Sci. Eng. B 20 (1993) 91
[6] J. Chevrier, M. Armand, A.M. Huber, N.T. Linh, J. Electron. Mater. 9 (1980) 745
[7] J.Y. Longere, K. Schohe, S. Krawczyk, R. Coquille, H. L´Haridon and P.N. Favennec, J. Appl. Phys. 68 (1990) 755
[8] W.Meier, H.Ch.Alt, T.Vetter, J.Völkl and A.Winnacker, Semicon. Sci. Tech. 6 (1991) 297
[9] R.Fornari, C.Frigeri, J.L. Weyher, S.K. Krawczyk, F. Krafft and G. Mignoni, Proc. of 7th Semi-insulating III-V Materials Conference, Ed. C. Miner, W. Ford and E. Weber, IOP Bristol 1993, p. 39
[10] M. Avella, J. Jiménez, A. Alvarez, R. Fornari, E. Gilioli, A. Sentiri, J.Appl. Phys. 82 (1997) 3832
[11] J.Jiménez, R.Fornari, M.Curti, E. de la Puente, M.Avella, L.F.Sanz, M.A. González, A. Alvarez, Mater. Res. Soc. Symp. Proc (MRS Fall Meeting, Boston Dec.1997)(to be published)
[12] J.Jiménez, R. Fornari, Defect Diff. Forum (Pt.A) 157-159 (1998) 103
[13] M.Yokogawa, S. Nishima, K.Matsumoto, H.Morishita, K.Fujita, S.Akai, Inst. Phys. Conf. Ser. 74 (1984) 29
[14] P.B.Klein, R.L.Henry, T.A.Kennedy, N.D.Wilsey, Mater. Sci. Forum 10-12 (1986) 1259
[15] D. Hofmann, G. Müller, N. Streckfuss, Appl. Phys. A 48 (1989) 315
[16] K.Kainosho, H.Shimakura, H.Yamamoto and O.Oda, Appl. Phys. Lett 59 (1991) 932
[17] G.Hirt, D. Wolf and G.Müller, J. Appl. Phys. 74 (1993) 5538
[18] R. Fornari, A. Zappettini, E. Gombia, R. Mosca, K.Cherkaoui, G.Marrachki, J.Appl. Phys. 81 (1997) 7604
[19] G. Baraff, Proc. 7th conference on Semi-insulating III-V Materials, edited by C.J. Miner, W. Ford and E.R. Weber (Adam Hilger, Bristol 1993) p.11
[20] A. Kalboussi, G.Marrachki, G. Guillot, K.Kainosho, O.Oda, Appl. Phys. Lett. 61 (1992) 2583
[21] D. Wolf, G. Hirt. F. Mosel, G. Müller and J. Völkl, Mater. Sci. Eng. B28 (1994) 115
[22] F.X.Zach, J.Appl. Phys. 75 (1994) 7854
[23] R.Darwich, B. Pajot, B. Rose, D. Robein, D. Theys, R. Rahbi, C. Porte, F. Gendron, Phys. Rev. B 48 (1993) 17776
[24] A. Zappettini, R.Fornari, R.Capelletti, Mater.Sci.Eng. B 45 (1997) 147
[25] D.F. Bliss, G. Bryant, D. Gabbe, G. Iseler, E.E. Haller and F.X.Zach, Proc. of 7th Int. Conf. on InP and Rel. Mater., Sapporo 1995, IEEE Catalog No. 95CH35720, p. 678
[26]A.Alvarez, M.Avella, J.Jiménez, M.A.González and R.Fornari, Semicond. Sci. Technol. 11 (1996) 941.

Semi-insulating InP crystal wafers characterized by different nondestructive techniques

Masayoshi Yamada[a], Masayuki Fukuzawa[a], Masanobu Akita[a], Martin Herms[a][†],
Masayuki Uchida[b], and Osamu Oda[b]

[a]*Dept. of Electronics and Information Science, Kyoto Institute of Technology,
Matsugasaki, Sakyoku, Kyoto 606-8585, Japan*
[b]*Materials and Components Laboratories, Japan Energy Corporation,
3-17-35 Niizo-Minami, Toda-shi, Saitama 335-8502, Japan*

Nondestructive techniques of mapping of residual strain, photoluminescence and resistivity were utilized to optimize the multiple wafer annealing (MWA) procedure of undoped or slightly Fe doped InP crystal wafers under phosphorous atomosphere. The annealing procedure optimized did not additionally produce unwanted residual strain but reduced and homogenized it. In conclusion, the MWA has proved to be a promising method to obtain semi-insulating InP crystals without undesired high Fe doping.

A. Introduction

Semi-insulating (SI) InP is a key material not only for electronic devices such as high electron mobility transistors and hetero-bipolar transistors but also for optical devices such as laser diodes and photo diodes used in high-speed multimedia communication systems. SI InP crystals are industrially produced by doping with a high concentration ($> 10^{16}$ cm^{-3}) of Fe, which behaves unfavourably in the fabrication of devices, because of the low activation efficiency of ion implantation[1], the out-diffusion of Fe into epitaxial layers[2,3], as well as the slip-like defect formation in epitaxial layers[4]. On the other hand, the desired change to semi-insulating behavior of undoped or extremely low Fe-doped InP crystals has been also achieved by annealing in phosphorus atomosphere, as reported by several authors[5-11]. The multiple wafer annealing (MWA)[9-11] seems to be the most promising technique to achieve SI InP substrates without high Fe doping. The MWA process consists of two steps: a) the first step serves the conversion of the wafers from conductive to the SI state, and b) the second step is performed to improve the uniformity of resistivity. However, there is the risk that the chosen MWA procedure does not homogenize wafer properties but causes additional inhomogeneities; in particular, residual strain due to thermal stress during MWA.

In this paper, we present the experimental results obtained on SI InP single crystal wafers treated by single-step wafer annealing (SWA) and MWA procedures[9-11]. In order to check both the reproducibility and the wafer uniformity, we have employed nondestructive techniques to measure residual strain, photoluminescence and resistivity, for which we have developed by ourselves a scanning infrared polariscope (SIRP)[12], a scanning photoluminescence spectrometer (SPL)[13] and a contactless resistivity profiler[14].

B. Experimental procedures and results

We performed SWA as well as MWA on slightly-iron-doped 2-inch-diameter InP (100) wafers, keeping the as-grown adjacent wafers for reference. The SWA was done at 950°C for 40h under phosphorous pressure of 1atm. For MWA, subsequently the second-step annealing was done at 807°C for 40h under phosphorous pressure ranging from 30atm to 50atm[11]. The annealing procedures resulted in a slight decrease of Fe concentration compared with the as-grown level. The Fe concentration was 2.5-3$\times 10^{15}$cm^{-3} after SWA and 3.5-5$\times 10^{15}$cm^{-3} after MWA. It is noted here that the as-grown wafers having these amount of Fe concentration are usually conductive but become semi-insulating after SWA and MWA[9-11].

[†]On leave from Institute of Experimental Physics, Freiberg University of Mining and Technology, D-09596 Freiberg, Germany

B1. Residual strain

In order to measure the distribution of residual strain, we used an improved version of SIRP. Figures 1 (a) and (b) are the two-dimensional maps of residual strain component $|S_r - S_t|$ measured in the SWA and as-grown adjacent wafers, respectively.

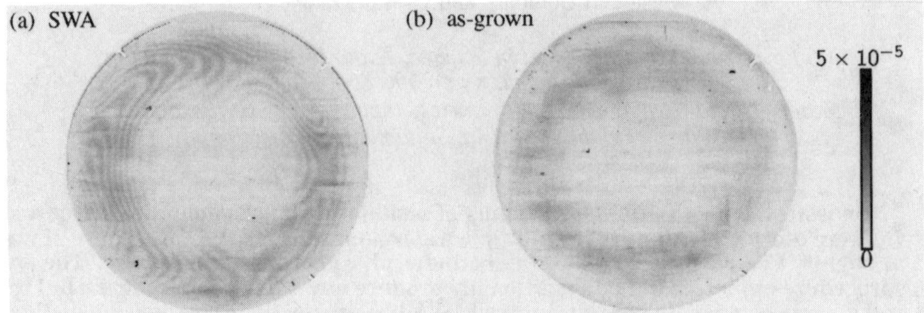

Figs. 1 (a) and (b). Two-dimensional maps of residual strain component $|S_r - S_t|$ measured in the (a) SWA and (b) as-grown adjacent wafers.

The value of $|S_r - S_t|$ averaged over the whole wafer is 7.5×10^{-6} for the SWA wafer and 6.4×10^{-6} for the as-grown wafer. As reported in some cases[15-18], excess thermal stresses during wafer annealing produce an abnormal increase of residual strain. However, the SWA procedure employed here does not seem to confirm such a case, although there is found a slight increase in the present case only. The streak patterns observed additionally are probably caused by multiple reflection. The levels of residual strain in the SWA and as-grown wafers examined here are much lower than those observed in conventional LEC-grown InP wafers[12,19] and almost comparable with those observed in VCZ-grown InP wafers[19].

Figures 2 (a) and (b) are the two-dimensional maps of the residual strain component $|S_r - S_t|$ measured in the MWA and as-grown adjacent wafers, respectively. The value of $|S_r - S_t|$ averaged over the whole wafer is 7.4×10^{-6} for the MWA wafer and 9.3×10^{-6} for the as-grown one. It is found that the MWA procedure employed here homogenizes and decreases the residual strain rather than increases it.

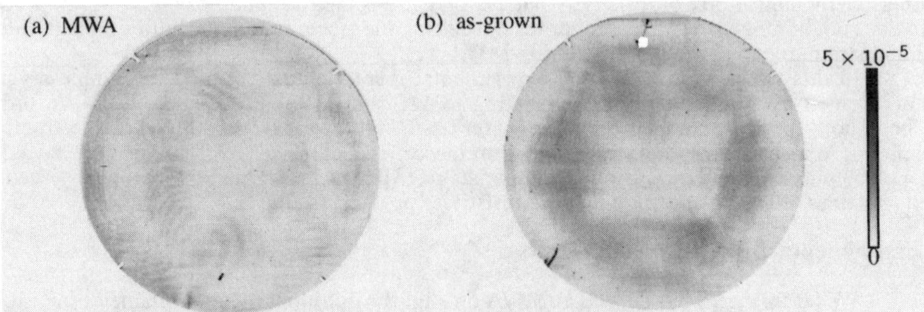

Figs. 2 (a) and (b). Two-dimensional maps of residual strain component $|S_r - S_t|$ measured in the (a) MWA and (b) as-grown adjacent wafers.

B2. Photoluminescence

Spectrally resolved SPL was measured at room temperature on 200×200 mesh points by using the SPL spectrometer developed by ourselves[13]. The spectral data obtained by successive scans were stored as an array which includes the x-position, the y-position, the intensity of incident laser light, the selected wavelength and the PL intensity corresponding to that wavelength.

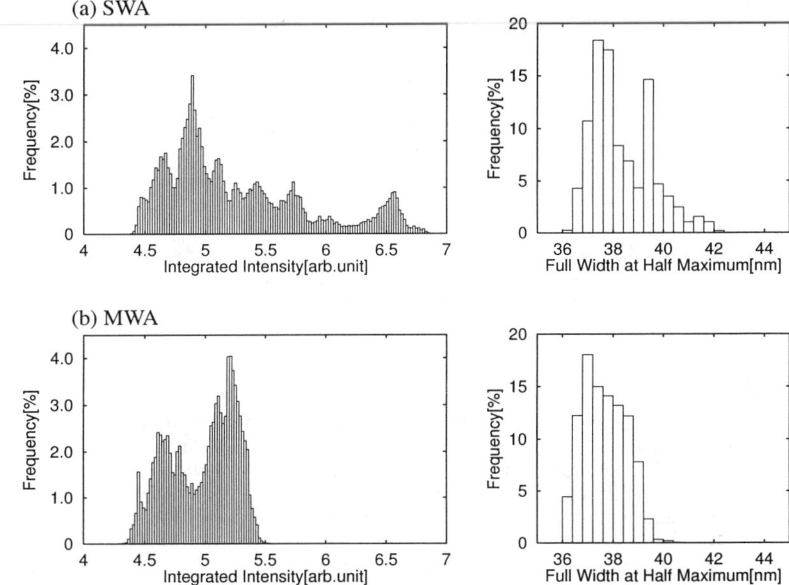

Figs. 3 (a) and (b). Histograms of integrated intensity and FWHM evaluated from the PL spectra measured at each point in the (a) SWA and (b) MWA wafers.

In order to check the homogeneity of SWA and MWA, we calculated the full width at the half maximum (FWHM) and the intensity integrated over the wavelengths near the bandgap from the PL spectra measured at each point under a relatively high excitation condition because the integrated intensity depends mainly on the lifetime related to nonradiative recombination processes[20]. The two-dimensional maps of the intergrated intensity and FWHM in the SWA, MWA, and corresponding as-grown adjacent wafers are not related to those of the residual strain shown in Figs. 1 and 2 but proved to be fairly flat. Figures 3 (a) and (b) are the histograms of the integrated intensity and of the FWHM in the SWA and MWA wafers. It is found that the integrated intensity and FWHM in the SWA wafers are more dispersive than those in the MWA one. This is an indication that the MWA procedure plays an important rule in homogenizing wafer properties, in comparison with the SWA only.

B3. Resistivity

Resistivity was measured in the SWA and MWA wafers by using a contactless resistivity profiler[14]. The measurement method was essentially based on the time-dependent charge measurement (TDCM)[21] but much improved so that the transient response might not be affected both by the bias and leak currents in the charge amplifier and by the variation of the air gap between the probing electrode and the wafer examined. The diameter of the probing electrode used was 0.9mm.

Figures 4 (a) and (b) show the resistivity profiles measured along the [011] direction in the SWA and MWA wafers. The resistivity value averaged over the measured points is 6.9×10^7 ohm·cm in the SWA and 8.0×10^7 ohm·cm in the MWA. The standard deviation is 35% in the SWA and 4.1% in the MWA. It is clear that the variation of resistivity is much smaller in the MWA wafer than in the SWA one. This result is in accord with that of SPL.

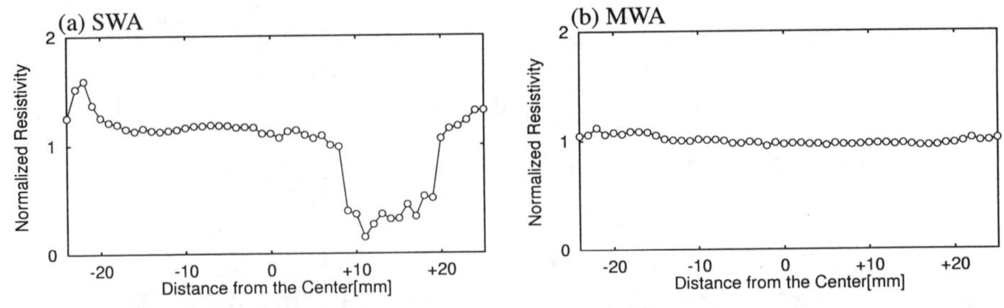

Figs. 4 (a) and (b). Resistivity profiles measured along the [011] direction in the (a) SWA and (b) MWA wafers.

C. Concluding remarks

We have demonstrated that nondestructive techniques such as SIRP, SPL and TDCM are very useful to characterize semi-insulating InP crystal wafers. In particular, SIRP is the most important for one to find adequate conditions in various thermal treatments. It may be concluded that the MWA procedure optimized by using the nondestructive characterization techniques presented here is a promising method to achieve SI InP substrates without undesired high Fe doping.

References

[1] K. Kainosho, H. Shimakura, H. Yamamoto, T. Inoue and O. Oda, in *"Proc. First Int. Conf. on InP and Related Mater. Okulahoma, 1989"* 312.
[2] D. E. Holmes, R. G. Wilson and P. W. Yu, Appl. Phys. 52 (1981) 3396.
[3] E. W. A. Young and G. M. Fontijn, Appl. Phys. Lett. 56 (1990) 146.
[4] C. J. Miner, D. G. Knight, J. M. Zorzi and M. Ikisawa, Inst. Phys. Conf. Ser. 135 (1994) 181.
[5] D. Hofmann, G. Müller and Streckfuss, Appl. Phys. A48 (1989) 315.
[6] G. Hirt, D. Hofmann, F. Mosel N. Schäfer and G. Müller, J. Electron. Mater. 25 (1991) 1065.
[7] K. Kainosho, H. Okazaki and O. Oda, in *"Proc. 5th Int. Conf. on InP and Related Mater. Paris, 1993"*, (Post-Deadline Papers) 33.
[8] D. Wolf, G. Hirt, F. Mosel, G. Müller and J. Völkl, Mater. Sci. Eng. B28 (1994) 115.
[9] M. Uchida, K. Kainosho, M. Ohta and O. Oda, in *"Proc. 8th Int. Conf. on InP and Related Mater. Schwäbisch Gmünd, 1996"* 43.
[10] O. Oda, M. Uchida, K. Kainosho, M. Ohta, M. Warashina and M. Tajima, in *"Proc. 9th Int. Conf. on InP and Related Mater. Hyannis, 1997"* 404.
[11] M. Uchida, K. Kainosho, M. Ohta and O. Oda, J. Electron. Mater. 27 (1998) 8.
[12] M. Yamada, M. Ito and M. Fukuzawa, in *"Proc. 9th Int. Conf. on InP and Related Mater. Hyannis, 1997"* 209.
[13] M. Fukuzawa, Y. Okada and M. Yamada, unpublished.
[14] M. Fukuzawa, M. Akita and M. Yamada, unpublished.
[15] M. Yamada, T. Shibuya and M. Fukuzawa, Inst. Phys. Conf. Ser. 136 (1994) 505.
[16] M. Yamada, M. Fukuzawa, T. Kawase, M. Tatsumi and K. Fujita, Inst. Phys. Conf. Ser. 145 (1996) 447.
[17] M. Yamada, M. Fukuzawa and K. Ito, Inst. Phys. Conf. Ser. 155 (1997) 901.
[18] M. Herms, M. Fukuzawa, M. Yamada, G. Zychowitz and J. R. Nicklas, to be presented in *"4th Int. Workshop on Expert and Control of Comp. Semicond. Mater. and Technol. Cardiff, 1998"*.
[19] M. Fukuzawa and M. Yamada, J. Electron. Mater. 25 (1996) 337.
[20] S. K. Krawczyk, in *"The Encyclopedia of Advanced Materials"* (Pergamon, 1994) 2318.
[21] R. Stibal, J. Windscheif and W. Jantz, Semicond. Sci. Technol. 6 (1991) 995.

The electrical properties of irradiated silicon: semi-insulating silicon

B.K. Jones, M. McPherson and J. Santana*
Department of Physics, Lancaster University, Lancaster, LA1 4YB, UK.
**Departmento de Electronica Sistemas e Informatica, Instituto Technologico y de Estudios*
Superiores de Occidente (ITESO), Tlaquepaque, Mexico.
b.jones@lancaster.ac.uk

Abstract

Clear experimental evidence is presented that irradiated silicon behaves electrically like semi-insulating GaAs and hence should be considered semi-insulating. The analysis follows that for a relaxation semiconductor.

A, Introduction

The description of materials as 'semi-insulating' is usually applied very generally. High resistivity compound semiconductors are the most common such materials with highly defected semiconductors or insulators sometimes also considered. In all cases the material is very far from a simple ideal lattice of pure material. With such complex materials it is very difficult to separate the intrinsic properties of the lattice from the extrinsic properties of the impurities and defects. Consequently there is not a good description of their electrical properties and much effort is spent on characterising the extrinsic properties such as trap levels and densities. The presence of an exemplar material with nearly ideal intrinsic properties would enable the best description of a semi-insulator to be determined. We propose that irradiated silicon is such a material and it has electrical properties which can be measured and clearly described by existing theory for a relaxation semiconductor.

B. Samples and Measurements

The samples were silicon P-I-N photodiodes (Hamamatsu S3590-04) irradiated or reirradiated at fluences of 3.39, 8.29, 25.0 and 39.4 x 10^{13}ncm^{-2}. The irradiations were carried out in ISIS using 1MeV neutrons. There is some variability in the results since two different devices were used in the sequence. The measurements made were of I-V, C-V(f) over a wide range of temperatures and characterisation of the traps by various transient techniques such as PICTS. It should be noted that, as well as generating generation-recombination (g-r) centres, the irradiation of N-type material produces an excess of acceptors so that it type-inverts to P-type.

C. Relaxation Semiconductor Theory

The active regions of most commercial devices need material with long minority carrier lifetime. This type of material has dominated the literature of semiconductors. This conventional 'lifetime' material has $\tau_D \ll \tau_0$ while a 'relaxation' material has $\tau_D \gg \tau_0$. Here $\tau_D = \rho\varepsilon\varepsilon_0$ (ρ is the resistivity) is the dielectric relaxation time which is the time in which a space charge is neutralised by the flow of the free carriers. The minority carrier generation or recombination lifetime is τ_0 .

A relaxation material thus needs a high resistivity and a high density of active g-r centres. That is they need to be in a biased structure so that the Fermi energy is at their mid-gap position. This type of material has a large literature [1-5]. These conditions are readily available in doped or defected compound semiconductors. However in those materials there are normally so many deep levels, other than g-r centres, that the properties are dominated by the defects rather than the intrinsic material. Such deep levels only interact significantly with one band are in contrast to g-r centres which interact approximately equally with each band. Irradiated silicon is almost ideal with readily recognisable relaxation behaviour which can be distinguished from the deep level effects.

If a reverse bias is applied to a relaxation semiconductor P^+-N junction the free carriers are not completely removed to cause a depletion region since new e-h pairs are created, on demand, to fill the space. This results in this region having a single quasi-Fermi level pulled to near mid-gap where the g-r process is most efficient. To maintain a steady state $n_0\mu_n = p_0\mu_p$ and with the equilibrium condition for g-r centres, $n_0p_0 = n_i^2$ we get $n_0 = p_0(\mu_p/\mu_n) = n_i(\mu_p/\mu_n)^{1/2}$. The carrier densities n_0 and p_0 are independent of the specific material, contacts or traps. Since $\mu_p << \mu_n$ the material is 'P-type'. This is not to be confused with the type inversion mentioned earlier. Also the resistivity is then at its maximum value given by $\rho^{-1}_{max} = (2\ e\ n_i\ (\mu_n\mu_p)^{1/2}) \sim (8 \times 10^8\ \Omega\ cm)^{-1}$ at room temperature for GaAs and $(3 \times 10^5\ \Omega\ cm)^{-1}$ for silicon. As well as this dynamic relaxation material, intrinsic material also has high resistivity with $n = p = n_i$ so that $\rho^{-1} = (2\ e\ n_i\ (\mu_n + \mu_p)/2)$ and compensated material has a high resistivity which can reach intrinsic values. It is created by the presence of an excess of deep levels to absorb the excess carriers. Such high resistivity material is necessary to enable a semiconductor to show relaxation properties when suitable dynamic conditions are present. The similarity of the properties can make the experimental differentiation between these materials difficult. However for silicon the properties follow relaxation theory very well so that it can be considered relaxation-like and semi-insulating.   Since the properties are insensitive to the contact details the analysis also applies to Schottky-Ohmic structures.

D. Results

There are specific properties of a P-I-N diode which are found experimentally to agree with theory and cannot be explained by the normal lifetime analysis. These demonstrate that the intrinsic material is relaxation-like in the nearly trapless case where there is no space-charge. They are:
* Ohmic characteristics in forward and reverse bias with a material resistivity of ρ_{max} and with a temperature dependence of the current corresponding to n_i (i.e. $E_g/2$) [6].
* Forward bias breakdown by controlled avalanche.
* A constant internal field region.
* Negative capacitance for forward bias[7].
* Peak in capacitance at low reverse bias.
* Modifications to the C-$V^{1/2}$ variation.
* Suppression of shot noise.

If there is a space charge then effects specific to deep levels occur:

- There is a $V^{1/2}$ dependence of the current at voltages before the depletion region reaches the far contact.
- There is a frequency and temperature dependence of the capacitance.

These experimental properties have been observed and methods of analysis have been developed to determine those parameters which may be used as measures of the g-r centre density, the space charge density and the 'relaxation-like-ness' [8]. Figs.1 and 2 show the I-V and C-V plots obtained and are described in the captions.

E. Conclusions

Silicon which has been damaged by radiation has properties very similar to SI-GaAs. It behaves as an almost ideal relaxation material [9,12]. Also Au or Pt doping, which is used to 'kill' the minority carrier lifetime, gives similar effects and such material can also be analysed as relaxation-like [13]. There are applications for this semi-insulating silicon which include: device substrates, well defined resistivity resistors and zero-capacitance structures.

The use and understanding of this analysis can be used to separate the intrinsic properties of a semi-insulating material from its intrinsic properties.

Acknowledgements

The authors appreciate the interactions within the members of RD8 and RD48 who are developing high energy particle detectors. Financial help has been obtained from PPARC, the National Council for Science and Technology (CONACYT) Mexico, the Government of Lesotho and CERN.

References

[1] W. van Roosbroeck and H.C. Casey Jr., *"Transport in relaxation semiconductors"*, Phys.Rev. **B5**, 2154 (1972).
[2] J.C. Manifacier and R.Ardebili, *"Bulk and contact effects in p+-SI-n+ semi-insulating GaAs structures"*, J. Appl. Phys. 77 (1995) 3174-85.
[3] H.J. Queisser. *"Semiconductors in the relaxation regime"*, Proc. ESSDERC 1972 145-88 IOP Conf Proc No 15 (1972)
[4] N.M. Haegel. *"Relaxation semiconductors: in theory and in practice"*, Appl.Phys. **A53** (1991) 1-7.
[5] G.H. Dohler and H. Heyszenau, *"Conduction in the relaxation regime"*, Phys. Rev. **B 12**, (1975), 641.
[6] B.K. Jones, J. Santana and M. McPherson, *"Ohmic I-V characteristics in semi-insulating semiconductor diodes"*, Solid State Comm. **105** (1998) 547-9.
[7] B.K. Jones, J. Santana and M. McPherson, *"Negative capacitance effects in semi-insulating semiconductor diodes"*, Solid State Comm. (1998) to be published.
[8] B.K. Jones, *"Irradiated silicon detectors"*, ROSE report, http://http1.brunel.ac.uk:8080/research/rose/rose.html
[9] B.K. Jones, J. Santana and M. McPherson, *"The electrical properties of GaAs analysed as a relaxation semiconductor"*, this conference.

[10] B.K. Jones, J. Santana and M. McPherson, *"Semiconductor detectors for use in high radiation damage environments-Semi-insulating GaAs or Silicon?,* Nucl. Instrum. and Methods. **A395** (1997) 81-7.

[11] J. Santana and B.K. Jones, *"Semi-insulating GaAs as a relaxation semiconductor"*, J. Appl. Phys. **83**,(1998) in press.

[12] J.M. Santana-Corte, *"GaAs diodes in the relaxation regime used for radiation detection"* PhD Thesis, Lancaster University, 1997.

[13] M. McPherson, *"Irradiated silicon detectors as relaxation devices"*, PhD Thesis, Lancaster University, 1998.

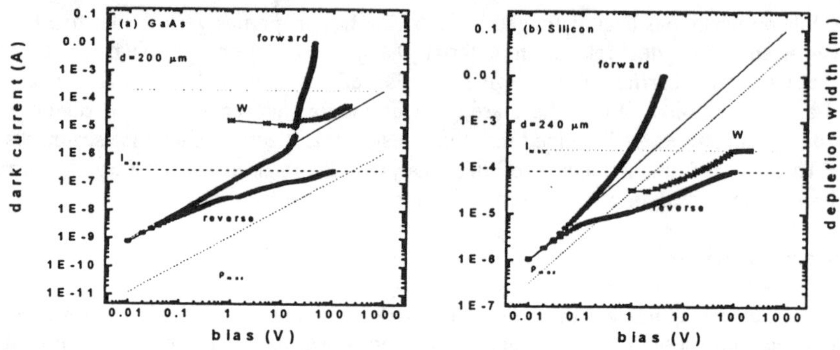

Fig. 1 The I-V characteristics of (a) a non-irradiated semi-insulating GaAs diode (SQ-20) and (b) an irradiated Si diode (CSD1-1i). Also shown is the calculated maximum resistance (broken line) and the depletion width calculated from the C-V data to show that the current derives from the depleted and relaxation region.

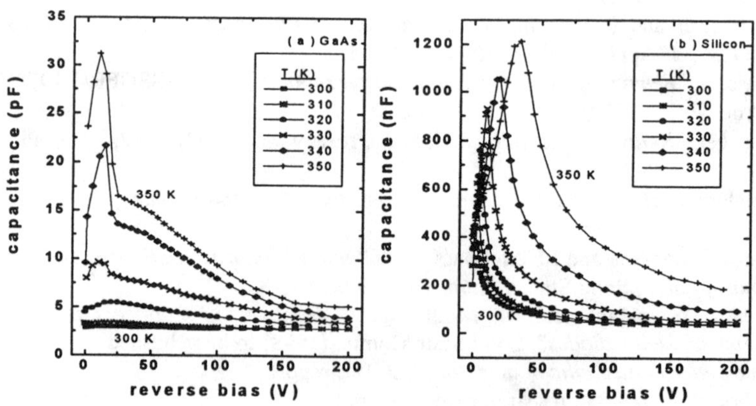

Fig. 2 The C-V, with temperature as parameter for (a) a GaAs diode (SQ-20) and (b) an irradiated silicon diode (CSD1-1i) to show that both have a peak in the capacitance due to relaxation effects and a temperature dependence due to the particular active deep level.

Carbon in semi-insulating gallium arsenide: A comparative study between FTIR, SSMS and CPAA

H. Ch. Alt*, B. Wiedemann**, J. D. Meyer**, R. W. Michelmann** and K. Bethge**
* *Physikalische Technik, Fachhochschule München, D-80335 München, GERMANY*
***Institut für Kernphysik, Univ. Frankfurt, D-60486 Frankfurt, GERMANY*

Local vibrational mode absorption was used to investigate the electrically active fraction of C in semi-insulating GaAs. A strictly linear relationship to the total chemical C concentration measured by spark source mass spectrometry is found. Using charged particle activation analysis as a reference method, a new calibration factor $f_{77} = (7.2 \pm 0.2) \times 10^{15}$ cm^{-1} for the absorption integral at 77 K is derived. Based on the temperature dependence of the absorption a calibration factor $f_{300} = (7.5 \pm 0.5) \times 10^{15}$ cm^{-1} is suggested for the room temperature measurement.

A. Introduction

Carbon is the dominating acceptor in state-of-the-art semi-insulating (s.i.) GaAs grown by the liquid-encapsulated Czochralski (LEC), vertical Bridgman (VB), and vertical-gradient-freeze (VGF) technique. The development of low- and high-carbon doped material with precisely defined C content is necessary for the different applications in electronic industry. Therefore, improved analytical methods for the quantitative determination of the electrically active as well as the total chemical carbon concentration are required.

Fourier-transform infrared (FTIR) absorption spectroscopy is used as a routine method for the evaluation of the C concentration in many laboratories. Substitutional carbon (C_{As}) is giving rise to a local vibrational mode (LVM) in the IR absorption spectrum at 582 cm^{-1} (77 K) and at 580 cm^{-1} (300 K), respectively. As the oscillator strength of the band is not known theoretically, a calibration with the help of some reference method is necessary. This has been done in the past using temperature-dependent Hall effect [1], secondary ion mass spectrometry [2], charged particle activation analysis (CPAA) [3,5], radio-tracing [4], and photon activation analysis [6]. The published calibration factors f = [C]/(integrated infrared absorption I_α) scatter over a wide range (see Table I). The purpose of this work was to study the relation between the electrically active [C_{As}] and the total chemical carbon concentration [C], and to improve the precision of the IR calibration factor.

B. Experimental

IR absorption measurements were performed in the temperature range between 77 and 300 K on a high-resolution FTIR instrument (Bruker IFS 113v). A highly sensitive mercury cadmium telluride detector was used in combination with long pass optical and electronic filtering to avoid signal distortions. The spectral resolution was selected according to the line width [7]. Absorption spectra were calculated taking into account multiple internal reflections. The detection limit at 77 K for the C_{As} band is well below 1×10^{13} cm^{-3}.

Spark source mass spectrometry (SSMS) was used as the analytical method to measure the total chemical carbon concentration. A SSMS detection limit of 4×10^{13} cm^{-3} in

Table I.   Calibration factors for the C_{As} LVM absorption given in literature.

Reference	Exp. ref. method	[C] (10^{16} cm^{-3}]	FTIR res. (cm^{-1})	FTIR method	f_{77} (10^{15} cm^{-1})	f_{300} (10^{15} cm^{-1})
Hunter et al.[1]	TDH[a]	?	?	$\alpha_{max}\times\Delta$	11	-
Homma et al.[2]	SIMS[b]	2.7 - 3.6	0.5	$\alpha_{max}\times\Delta$	-	9.5 ± 2.9
Kadota et al.[3]	CPAA[c]	0.03 - 3	0.063	I_α	7 ± 1	-
Brozel et al.[4]	RTT+H[d]	1.6	?	$\alpha_{max}\times\Delta$	-	8 ± 2
Arai et al.[5]	CPAA	<0.1 - 4.4	1.0	$\alpha_{max}\times\Delta$	9.2 ± 2	11.8 ± 2
Yoshioka et al.[6]	PAA[e]	0.1 - 5.1	?	$\alpha_{max}\times\Delta$	-	14
this work	SSMS, CPAA	0.03 - 1.6	0.06/0.5	I_α	7.2 ± 0.2	7.5 ± 0.5

[a]Temperature-dependent Hall effect
[b]Secondary ion mass spectrometry
[c]Charged particle activation analysis
[d]Radio-tin tracing + Hall effect
[e]Photon activation analysis

favorable cases could be achieved by an improved sample preparation, ultrahigh vacuum conditions (below 5×10^{-11} mbar), and computerized microphotometric data acquisition. The ^{12}C(^{3}He,α)^{11}C and ^{12}C(d,n)^{13}N reactions are used for the determination of C in GaAs. The detection limit in optimized samples is about 4×10^{14} cm^{-3} .

A large number of polycrystalline samples (from synthesis under nitrogen or argon atmosphere) and monocrystalline samples (grown by low- and high-pressure LEC, VB, and VGF) with C concentrations between 2×10^{14} and 2×10^{16} cm^{-3} were investigated by both FTIR and SSMS. Some of the monocrystalline LEC samples (see below) were used for CPAA.

C. Results and discussion

The determination of the SSMS relative sensitivity coefficient (RSC) was performed using monocrystalline LEC samples with high carbon concentrations $> 10^{15}$ cm^{-3}. The samples were pre-characterized by SSMS with respect to homogeneity and possible cross-contamination with other impurities. The results are depicted in Fig. 1. An excellent linearity is found between the SSMS concentration (with RSC=1) and the carbon concentration determined by CPAA. The true RSC (RSC=[C]$_{SSMS}$/[C]$_{CPAA}$) calculated from these data is 3.1±0.1.

A linear relation is also found between the LVM integral at 77 K and the SSMS signal (Fig. 2). Taking into account the large variation of the carbon concentration (3×10^{14} cm^{-3} < [C] < 1.7×10^{16} cm^{-3}) and the different crystal growing conditions, it must be concluded that carbon in s.i. GaAs is predominantly incorporated as the electrically active C_{As} acceptor. The results give an improved calibration factor f_{77} of $(7.2\pm0.2)\times10^{15}$ cm^{-1} for the LVM of C_{As} at 77 K.

The temperature dependence of the LVM band was re-investigated. Excited state

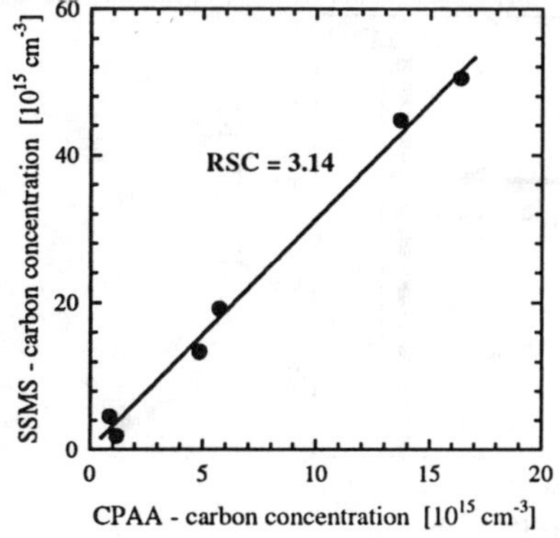

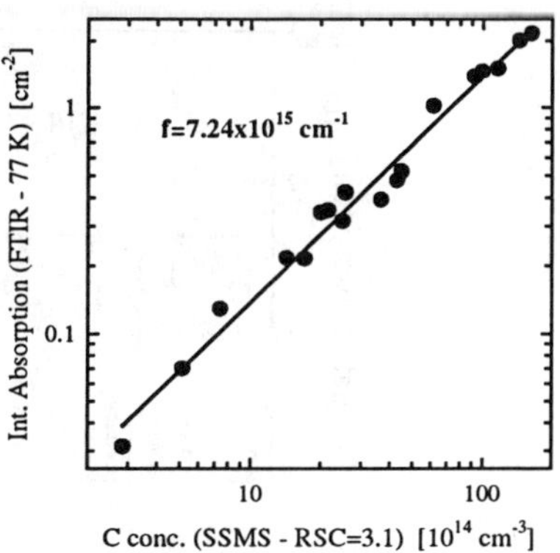

Fig 1: SSMS concentration (RSC=1) as a function of the carbon concentration determined by CPAA.

Fig. 2: Correlation between the integrated absorption of the C_{As} line at 77 K and the SSMS concentration using RSC=3.1.

absorption must be included for the accurate evaluation of the LVM integral at room temperature (Fig. 3). The side band at 576.5 cm^{-1} was already reported [8] and must be attributed to the anharmonically shifted $\Gamma_5(v=1) \rightarrow \Gamma_5(v=2)$ transition. The relative intensity at RT is ≈ 12 % of the main line. The band at 566.2 cm^{-1} was predicted theoretically [9]. Due to the $\Gamma_5(v=1) \rightarrow \Gamma_1(v=2)$ transition, its intensity should be 1/3 of the 576.5-cm^{-1} band. Taking the area of the main band and the 576.5-cm^{-1} side band, the average ratio of the RT band to the 77-K band is 0.972 (Fig. 4). This is close to the value expected theoretically, assuming that the oscillator strength itself is independent of temperature. Therefore, for the RT measurement, the calibra-

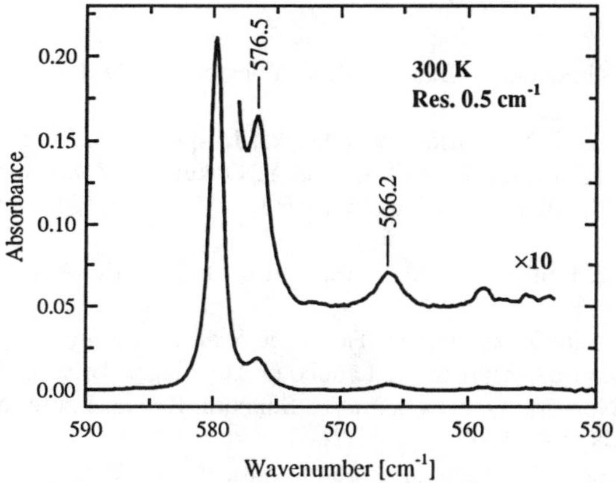

Fig. 3: RT IR absorption spectrum of a sample with $[C_{As}]=1.9\times10^{16}$ cm^{-3} after subtraction of the reference. Low-energy side bands at 576.5 and 566.2 cm^{-1} are due to excited-state transitions.

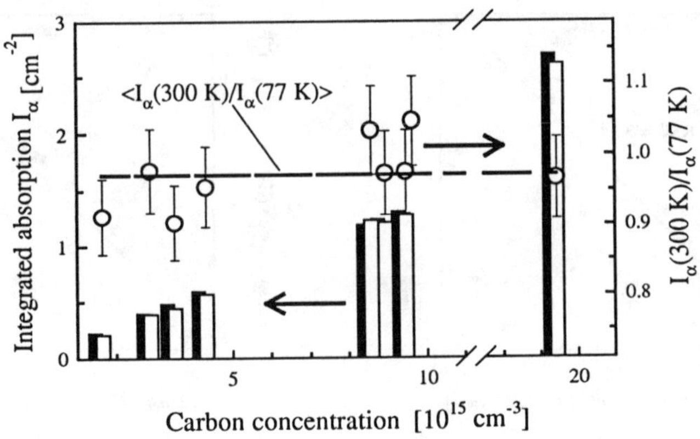

Fig. 4: Integrated absorption of the C_{As} band: 77 K (black bars), 300 K (white bars), and ratio (open circles). The dashed line is the average of the ratio (0.972).

tion factor $f_{300} = (7.5 \pm 0.5) \times 10^{15}$ cm^{-1} is proposed.

The errors given for f_{77} and f_{300} do not include systematic errors, which are caused mainly by an uncertainty of $\pm 5\ \%$ in the energy loss functions used by the CPAA method.

D. Conclusion

Carbon in s.i. GaAs is predominantly incorporated as substitutional C_{As}. Improved calibration factors are given for the FTIR method. The 77-K factor is close to the value given by Kadota et al. [3] (see Table I), however, the error margins could be reduced.

References

[1] A. T. Hunter, H. Kimura, J. P. Baukus, H. V. Winston, and O. J. Marsh, Appl. Phys. Lett. 44, 74 (1984).
[2] Y. Homma, Y. Ishii, T. Kobayashi, and J. Osaka, J. Appl. Phys. 57, 2931 (1985).
[3] Y. Kadota, K. Sakai, T. Nozaki, Y. Itoh, and Y. Ohkubo, in *Semi-Insulating III-V Materials: Hakone 1986*, edited by H. Kukimoto and S. Miyazawa (Ohmsha, Tokyo, 1986), p. 201.
[4] M. R. Brozel, E. J. Foulkes, R. W. Series, and D. T. J. Hurle, Appl. Phys. Lett. 49, 337 (1985).
[5] T. Arai, T. Nozaki, J. Osaka, and M. Tajima, in *Semi-Insulating III-V Materials: Malmö 1988*, edited by G. Grossmann and L. Ledebo (Adam Hilger, Bristol, 1988), p. 201.
[6] A. Yoshioka, K. Nomura, O. Kawakami, K. Shimura, K. Masumoto, and M. Yagi, J. Radioanal. Chem. 148, 201 (1991).
[7] H. Ch. Alt, Semicond. Sci. Technol. 3, 154 (1988).
[8] H. Ch. Alt and B. Dischler, Appl. Phys. Lett. 66, 61 (1995); H. Ch. Alt, Mater. Sci. Forum 196-201, 1577 (1995).
[9] D. A. Robbie, R. S. Leigh, and M. J. L. Sangster, Phys. Rev. B 56, 1381 (1997).

Semi-insulating properties of GaAs with artificially buried W discs

L.-E. Wernersson, A. Litwin, L. Samuelson, and W. Seifert
Solid State Physics/Nanometer Structure Consortium
Lund University, Box 118, S-221 00 Lund, Sweden

Abstract

In this paper we review our experiments in which a controlled formation of semi-insulating GaAs by introduction of buried metal discs is demonstrated. A tungsten disc matrix is embedded in an epitaxial layer by epitaxial overgrowth and the current transport perpendicular to the lattice is studied. The discs form Schottky barriers to the surrounding GaAs, which is depleted from free carriers. The conductance is demonstrated to vary by 7 orders of magnitude in structures as a function of varying disc separation. Using such a semi-insulating structure with a high disc density, the buried contacts have been characterised by photo-conductivity measurements and space-charge spectroscopy. The experimental results are in excellent agreement with an analytical model for depletion and charging of nano-scale Schottky barriers.

A. Introduction

The role of buried metallic clusters in semi-insulating semiconductors has attracted considerable attention during the last decade and these semi-insulating materials have been extensively investigated by different techniques [1,2,3]. Various schemes may be employed to create such materials, for instance precipitation of excess As in low-temperature grown GaAs using molecular beam epitaxy (MBE) [4], formation of Ga clusters by metalorganic vapour phase epitaxy (MOVPE) [5] and precipitation of diffused Cu in InP [6,7]. In addition to these fabrication techniques, it is attractive to achieve a controlled introduction of buried Schottky contacts. Thereby, the Schottky depletion model for semi-insulating materials can be directly tested and the physical properties of the buried contacts investigated.

In this paper, we describe how we have created a semi-insulating layer of GaAs by epitaxial overgrowth over arranged patterns of tungsten (W) nano-scaled discs. We demonstrate how the overlap of the depletion regions around the discs creates a barrier for the electron transport and how this barrier changes as a function of the separation between the discs. These measurements are made possible since the structures contain a single layer of arranged metallic inclusions. In addition, the internal semi-insulating region in these structures may be used for application of space-charge spectroscopy, performed on the actual buried nano-contacts. Thereby, the capture and emission process of buried metal nano-structures may be investigated. Even though these buried contacts are different from precipitates with respect to size and metal-semiconductor interface formation, we believe that the data presented here are model systems to describe the physical properties of nanometer-sized buried Schottky contacts, including precipitates.

B. Sample structure and current-voltage characteristics

The fabrication of the structures including the epitaxial overgrowth has been described in detail elsewhere [8] and only a brief discussion is included here. Matrices of W discs with a diameter of about 50 nm are fabricated by electron beam lithography and W evaporation using a lift-off process on an n-type (Si doped, $n=2\times10^{16}$ cm^{-3}) GaAs epitaxial layer. Structures with varying

     57

disc separation were included on the same substrate and they were simultaneously overgrown with an n-type GaAs layer by MOVPE. The same doping was used for the overgrown material. The fields of W discs were surrounded by a wide W frame, which masked the epitaxial growth and hence a mesa was directly formed by selective growth. Ohmic contacts were formed on the top of the mesas as well as to the back of the substrate. All measurements were done in a vertical configuration between the two ohmic contacts. A schematic illustration of the structure is shown in the inset of Figure 1. Measured room temperature conductance is shown as a function of the spacing between the discs in Figure 1. The reference level indicates measured values for structures without discs. From the presented data it is clear that only minor changes in conductivity are achieved for this doping level when discs are introduced with a separation of more than 400 nm. In contrast, a dramatic change is found for structures with 300 nm and 200 nm disc separation. In these cases, the current level drops by 3 and 7 orders of magnitude, respectively. This change in conductivity is attributed to the onset of overlapping depletion regions around the discs, as seen in the temperature dependence of the current-voltage characteristics shown below.

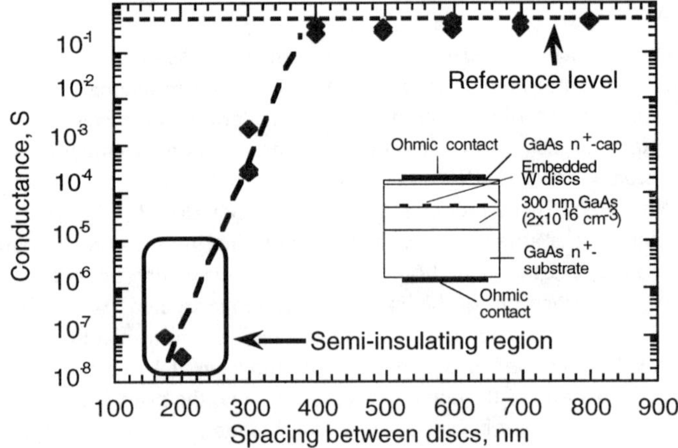

Figure 1. Measured room temperature conductance for structures with varying spacing between the discs. The sample structure is shown in the inset.

C. Temperature dependence on the current transport

The temperature dependence of the current transport was measured for the structures in Figure 1 and the data was plotted in Arrhenius plots according to the thermionic emission theory [8]. In this way, we obtained a measure of the metal induced barrier as shown in Figure 2a. These values are interpreted as the height of the internal barrier located in the centre of each unit cell with four different discs in the corners. For small biases, the measured barrier corresponds to the unperturbed overlap from the depletion regions and we see how this barrier increases as the discs are more closely spaced. The data are measured as various voltages applied to the structure and we see that the barrier decreases at the bias is applied. We relate this behaviour to an internal band bending in the depletion regions around the discs.

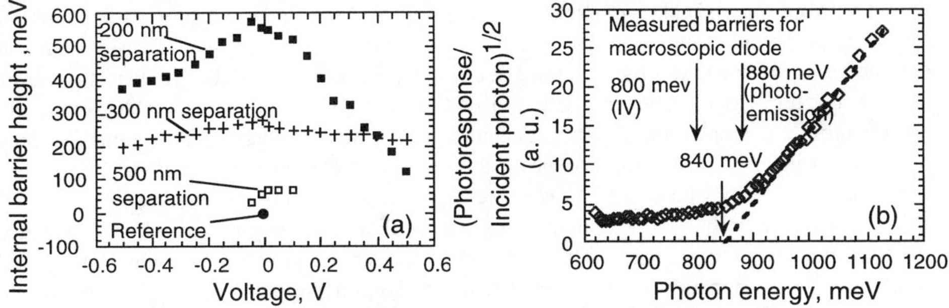

Figure 2. (a) Obtained activation energies from the temperature dependence of the current transport. The data represent the internal barrier height in the centre between four discs in the lattice. (b) Photo-response for varying photon energies in a semi-insulating structure.

D. Photo-emission

The Schottky barrier height for the buried contacts has been measured by internal photo-emission in an Fowler-like experiment in which a current, created by excitation of electrons from the metal discs, is measured. The photo-response at 77 K for a structure with 200 nm separation between the discs is shown in Figure 2b. In these measurements a voltage of 0.5 V was applied across the structure. From the indicated line, we obtain a threshold of about 840 meV, corresponding to the Schottky barrier height. As can be seen in the figure, good agreement between the barrier height for the buried contacts and macroscopic contacts (indicated by the upper arrows) is observed. In these measurements, we think that the major contribution to the measured photo-current is due to an indirect current *between* the discs. As the incident photons release electrons from the discs, the potential on the discs is changed and, as a consequence, the barrier for the electron transport between the discs is reduced. Thereby, the current level is increased as seen in the measurements. The mechanism resembles the control of an optically gated transistor.

E. Thermal capture and emission

Space-charge spectroscopy can be utilised to study the thermal capture and emission processes on the embedded contacts in the semi-insulating material [9,10]. We have applied deep-level transient spectroscopy (DLTS) to investigate the structure in Figure 1. In these measurements, the buried contacts capture electrons during an applied filling pulse and, thus, the width of the depletion region changes. The subsequent thermal emission of the captured electrons can be monitored by the associated changes in the capacitance at zero bias, when the system returns to thermal equilibrium. The number of captured electrons is related to the magnitude of the DLTS signal and the number of captured electrons at equilibrium, $\Delta C/C = \Delta Q/Q$ with Q=370 electrons. The equilibrium number of electrons per disc is determined by the magnitude of the capacitance at 0 V, 1.8 pF.

The Coulomb charging of the discs is important, both in the capture process and in the emission process. In Figure 3a, the signal magnitude, $\Delta C/C$, is plotted for various filling pulse lengths. The data is taken at 320 K with applied pulse magnitude of 0.5 V and with an emission rate window of 250 s^{-1}. In agreement with the studies of point defects in the vicinity of dislocations in Si [11], we interpret the observed logarithmic capture as a Coulomb repulsion of the subsequent electrons as they are captured onto the metal particles. In the emission process, we have observed that the emission rate depends on the number of electrons present on the discs, reflecting the added Coulomb energy, e^2/C [9]. Moreover, the combination of a temperature dependent capture process and the changes in the emission rates may lead to incorrect activation energies if the number of electrons on the metal discs is not accounted for. The difference for a Coulomb charging energy of only 4 meV may be as large as a few hundred meV, but with knowledge of the Coulomb charging energy, the one-electron emission rate may be calculated. This behaviour is illustrated in Figure 3b, which includes a correlation between measured and calculated one-electron emission rates. In addition, the measured activation energy from the DLTS can be correlated with and agrees well with optical data [10].

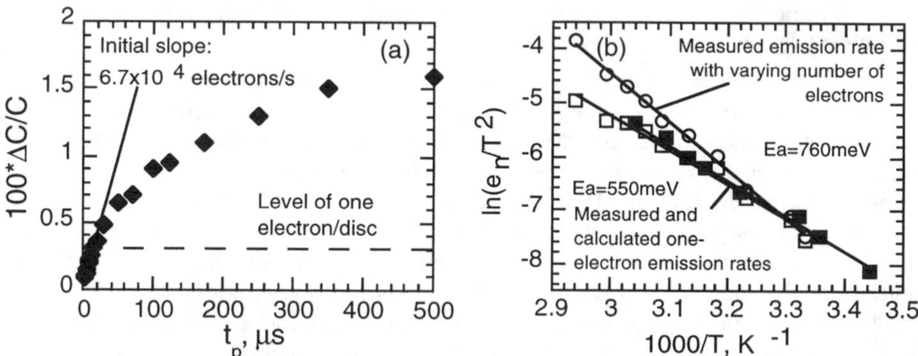

Figure 3. (a) Time dependence of the capture rate during 0.5 V filling pulses. (b) Arrhenius plot of the emission rates for measured one-electron emission together with measured emission rates with varying number of electrons on the discs. The data agree if the measured Coulomb charging energy of 4 meV is taken into account (calculated values) [10].

F. Modelling of the potential around buried Schottky contacts

We model the operation of the nm-sized Schottky contacts in a similar way to the method in Ref. 1. The model includes expressions for the potential around a buried metal, the Schottky barrier height, the number of captured electrons, and the Coulomb charging energy. Given the doping level and the geometrical size of the metal, all the quantities can be estimated and at the end of this section, calculated values will be compared with measured data.

The disc-shaped buried metals are modelled as spheres, with a radius of 30 nm. Since the depletion length is typically ten times longer than the radius, we assume that the exact shape of the discs is of minor importance. We further neglect the overlap between different depletion regions. First we calculate the charge, Q_m, which is collected from the space-charge region around the disc onto the metal. The radius of the metal is r_m and the radius of the space-charge region r_s. Then

$$Q_m = qN_d\left(\frac{4\pi}{3}r_s^3 - \frac{4\pi}{3}r_m^3\right) \tag{1}$$

for the doping level N_d in the semiconductor material. From Gauß theorem we obtain the electric field in the depleted semiconductor material and the potential around the sphere

$$V(r') = -\int_{r_s}^{r'} E(r)dr = \frac{qN_d}{6\varepsilon\varepsilon_0}\left(\frac{2r_s^3}{r'} + r'^2 - 3r_s^2\right) \qquad (2)$$

and, in particular, the barrier height Φ_b is given by

$$\Phi_b = \frac{qN_d}{6\varepsilon\varepsilon_0}\left(\frac{2r_s^3}{r_m} + r_m^2 - 3r_s^2\right). \qquad (3)$$

The capacitance is calculated as $C = \partial Q/\partial V$ and with the fixed metal radius r_m we find the well-known relation between the capacitance and the radius of the depletion region

$$C(r_s) = \frac{4\pi\varepsilon\varepsilon_0}{\dfrac{1}{r_m} - \dfrac{1}{r_s}}. \qquad (4)$$

Now we compare calculated values from these equations with measured data in Table 1. The calculated values are obtained for a doping level of 1×10^{16} cm^{-3} and a depletion length of 200 nm. These values are consistent with experimental data. The measured data are taken from this paper. The Schottky barrier height has been determined by optical emission, the barriers between the discs from the temperature dependence of the current transport, the number of electrons on the discs from the magnitude of the capacitance, and the Coulomb charging energy was measured by capacitance transient spectroscopy. The fact that an excellent agreement is found in the data is suggesting that this simple model can be used to describe the properties of a nm-sized Schottky contact. In addition, it demonstrates the advantage of our sample structure, which made a large set of different measurement techniques possible.

Table 1: Correlation between estimated values for a buried disc and data taken with a variety of measurement techniques.

	Estimated Values[a]	Measured Values
Schottky Barrier Height	0.95 eV	0.85-0.9 eV
Barrier in the Semiconductor between the Discs with 200 nm Disc Separation	0.66 eV[b]	0.55 eV
Barrier in the Semiconductor between the Discs with 300 nm Disc Separation	0.28 eV[c]	0.25 eV
Number of Excess Electrons on Each Disc	330	370
Coulomb Charging Energy	3-5 meV	3-4 meV

a) Based on a depletion region length of 200 nm and a doping level of 1×10^{16} cm^{-3}, which are consistent with measured values.
b) Calculated as the barrier height 140 nm from the metal.
c) Calculated as the barrier height 180 nm from the metal.

G. Concluding discussion

A semi-insulating GaAs layer was created by epitaxial overgrowth over a pattern of W discs. The layer has been characterised by various techniques, including the temperature dependence of the current transport, photo-conductivity measurements and DLTS. In structures with varying disc density, it is demonstrated that the conductance may be controlled in over 7 orders of magnitude. The Schottky barrier height of the buried contacts has been determined by optical emission and the electron capture and emission kinetics have been studied by DLTS. A good agreement between the experimental data and a model with overlapping depletion regions around the buried contacts is observed.

There are three important points with this fabrication technique. Firstly, we believe that the presented technology can be applied to any material in which a layer of buried Schottky contacts may be formed, for instance Si. Secondly, in separate investigations we have demonstrated the possibility to deliberately create vacant disc positions in the buried lattice, in which conducting channels are formed. These channels have been combined with the growth of a resonant tunneling structure, forming a vertical submicrometer resonant tunnelling diode [12]. Finally, one should be cautious in using the results to resolve the controversy over the origin of semi-insulating behaviour of low-temperature grown GaAs. What we have conclusively shown is that in samples with controlled introduction of buried metal nano-structures, ideal semi-insulating behaviour may be obtained and employed.

Acknowledgements

The authors are grateful for financial support from NUTEK, TFR, NFR and SSF and for stimulating discussions with H. Pettersson and L. Montelius.

References

[1] A. C. Warren, J. M. Woodall, J. L. Freeouf, D. Grischkowsky, D. T. McInturff, M.R. Melloch, and N. Otsuka, Appl. Phys. Lett. 57 (1990) 1331
[2] X. Liu, A. Prasad, W. M. Chen, A. Kurpiewski, A. Stoschek, Z. Liliental-Weber, and E. R. Weber, Appl. Phys. Lett. 65 (1994) 3002
[3] D. C. Look, Z.-Q. Fang, H. Yamamoto, J. R. Sizelove, M. G. Mier, and C. E. Stutz, J. Appl. Phys. 76 (1994) 1029
[4] F. W. Smith, A. R. Calawa, C.-L. Chen, M. J. Manfra, and L. J. Mahoney, IEEE Electr. Dev. Lett. 9 (1988) 77
[5] R. P. Leon, P. Werner, C. Eder, and E. R. Weber, Appl. Phys. Lett. 61 (1992) 2545
[6] R. P. Leon, M. Kaminska, K. M. Yu, and E. R. Weber, Phys. Rev. B 46 (1992) 12460
[7] W.-C. Chen, C.-S. Chang, and S. H. Chan, Appl. Phys. Lett. 69 (1996) 3239
[8] L.-E. Wernersson, A. Litwin, L. Samuelson, and W. Seifert, Jpn. J. Appl. Phys. 36 (1997) L 1628
[9] L.-E. Wernersson, A. Litwin, L. Montelius, H. Pettersson, and L. Samuelson, Appl. Phys. Lett. 72 (1998) 2610
[10] L.-E. Wernersson, A. Litwin, L. Montelius, H. Pettersson, and L. Samuelson, Phys. Rev. B 58 (1998) R4207
[11] P. Omling, E. R. Weber, L. Montelius, H. Alexander, and J. Michel, Phys. Rev. B 32 (1985) 6571
[12] L.-E. Wernersson, N. Carlsson, B. Gustafson, A. Litwin, and L. Samuelson, Appl. Phys. Lett. 71 (1997) 2803

Heteroepitaxial passivation of GaAs surfaces and its influence on the photosensitivity spectra and recombination parameters of GaAs epitaxial layers and semi-insulating materials

I.A. Karpovich[1], M.V. Stepikhova[1,2], and W. Jantsch[3]

[1] *Nizhny Novgorod State University, Gagarin aven. 23, 603600 Nizhny Novgorod, Russia*
[2] *Institute for Physics of Microstructures RAS, 603600 Nizhny Novgorod, GSP-105, Russia*
[3] *Institut für Halbleiterphysik, Johannes-Kepler-Universität, A-4040 Linz, Austria*

We investigate the influence of heteroepitaxial passivation of GaAs surfaces by a thin lattice matched InGaP layer on the spectra of photomagnetic effect, planar photoconductivity and capacitor photovoltage in conducting layers and semi-insulating substrates of GaAs. Reduction of the surface recombination rate by one or two orders of magnitude at passivation of layer surfaces simplifies significantly determination of their recombination parameters by photoelectric methods and enables one to get an insight into the nature of the phenomena causing the decrease of photosensitivity in a strong absorption region. We revealed a strong influence of the state of the surface on recombination parameters of semi-insulating GaAs: the ambipolar diffusion length and the magnitude and ratio of electron to hole components of photoconductivity. Surface passivation changes this ratio towards domination of the electron component.

High recombination activity of GaAs surfaces and its interfaces with insulators influences significantly the characteristics of electronic and photoelectronic devices. An effective method for reduction of surface recombination in GaAs is heteroepitaxial passivation. Formation of a fairly high-quality heterojunction on the surface by deposition of a thin layer of wide-gap material lattice-matched to GaAs, for example InGaP, limits considerably the recombination fluxes of carriers to the surface and results in a decrease of band bending [1,2]. In this contribution the effect of heretoepitaxial passivation of GaAs conducting layers and semi-insulating monocrystals on the spectra of photomagnetic effect (PME), photoconductivity (PC), capacitor photovoltage (CPV),and recombination parameters of materials is investigated.

Photoelectric spectroscopy is widely used for determination of recombination parameters of semiconductors, in particular, for Si. However, application of these methods to GaAs encounters some difficulties. For conducting epitaxial layers, they are concerned with a dominance of barrier mechanism in the photoconductivity [3] and enhanced anomalous drift component in PME [1]. For semi-insulating (s) GaAs, the main problems are associated with the necessity to take into account nonlinear recombination processes of excess carriers, with the dependence of electron (τ_n) and hole (τ_p) lifetimes, their relation $\gamma=\tau_p/\tau_n$, and the ambipolar diffusion length (L) on the state of the surface and on the photoexcitation level. Detailed studies of photoelectric effects in GaAs presented here have aimed at determining the recombination parameters: lifetimes of minority and majority carriers and surface recombination rates in these materials.

Experimental

Czochralski grown s-GaAs monocrystals and n-GaAs epitaxial layers, 3-7 μm thick, grown on (100) s-GaAs substrates by MOCVD technique were investigated. The semi-

"

insulating monocrystals used were commercial s-GaAs wafers (of "AG" type) differing by the method of doping: doped with Cr, Cr_2O_3, In and undoped. Heteroepitaxial passivation was performed by deposition (MOCVD) of a thin $In_{0.49}Ga_{0.51}P$ cover layer 15-25 nm thick.

The spectra of planar PC, PME and CPV were measured at 300 K in a low-signal, linear (relative to illumination intensity) regime and referred to the equal number of photons incident on the sample. For linearization of the photoeffects we usually used an additional white (background) illumination with maximal intensity of about $1 \cdot 10^{18}$ $cm^{-2}s^{-1}$ in a short-wavelength region. Depending on the design of a capacitor cell in CPV method [4], we were able to measure separately the photovoltage on surface and internal (layer-substrate) barriers, hereinafter referred to as surface and interface CPV, respectively.

Conducting layers of n-GaAs

A characteristic feature of *photomagnetic effect* in epitaxial GaAs layers with etched and real surfaces is a considerable reduction of photomagnetic current I_{PM} in the intrinsic absorption region at $h\nu > E_g$ (Fig.1, curves 3,5). This effect depends on the doping level of epitaxial layers and becomes more pronounced when n_0 is decreased, resulting in appearance of reverse-sign photomagnetic current at $h\nu > 1.7$ eV for layers with $n_0 < 5 \cdot 10^{15}$ cm^{-3} (curve 5). Detailed studies show the contribution of two mechanisms to the origin of this effect. First, an increase in the barrier region of the effective surface recombination rate, as the absorption coefficient $\alpha(h\nu)$ is increased (as predicted by the PME theory, taking into account variability of Fermi quasi-level [5]). Second, an increase in the same spectral region of anomalous drift component of photomagnetic current related to the presence of depletion layer and high recombination fluxes towards the GaAs surface. Calculation performed according to the theory of anomalous PME [5] shows that this component plays a significant role in layers with low carrier concentration, i.e. with a large barrier width W.

The observation of this reduction effect drastically complicates the application of PME methods for determination of recombination parameters in epitaxial GaAs layers. Here, we show that the situation can be improved in layers covered by a thin $In_{0.49}Ga_{0.51}P$ film. The spectra of photomagnetic current are presented in Fig. 1, curves 1,2. Heteroepitaxial passivation removes completely the effect of reduction of I_{PM}, resulting in an increase of current in a short-wavelength region by about one-to-two orders of magnitude. This effect is attributed to a strong reduction (by one-to-two orders of magnitude) of the surface recombination rate rather than to a decrease of

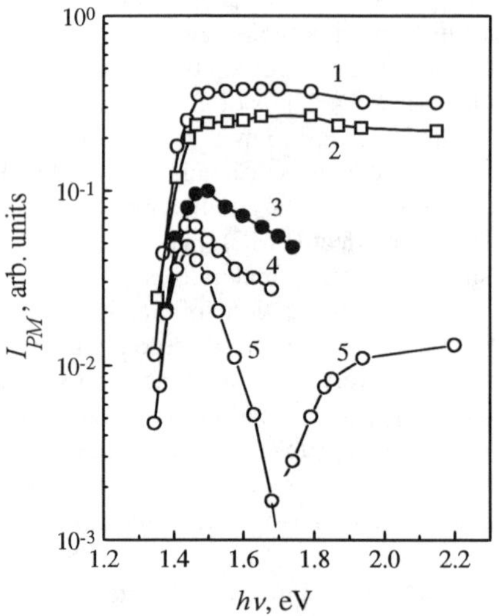

Fig. 1: Effect of surface passivation and layer doping on the photomagnetic spectra of epitaxial GaAs layers: 1,2 - passivated surface; 3,5 - real surface; 4 - real surface at background illumination; 1,4,5 - spectra of sample No 1; 2,3 - spectra of sample No 2.

Table 1: Some parameters of n-GaAs layers

No	n_0, 10^{16} cm^{-3}	μ_n, cm^2/ V·s	L_p, L^*, μm	τ, 10^{-8} s	S, 10^4 cm/s	Surface state
1	0.05	6700	3.2	1.4	<< 1	passivated
			1.1*	0.2*	5	real
2	12	3100	3.5	1.4	<<1	passivated
			2.1*	0.6*	1.6	real

NB: The asterisks mark the effective parameters calculated for maximal I_{PM}.

the surface barrier. The change of the surface barrier from ~0.6 eV down to ~0.2 eV by employing background illumination acts relatively weaker on I_{PM}. (Fig. 1, curve 4). On the other hand, as we have shown earlier [1], the decrease of the surface recombination activity is indicated also by enhancement of the photoluminescence. The photoluminescence at 300K increased in samples 1 and 2 by more than 100 and by more than 8 times, respectively.

Thus, for passivated layers we can ignore the contribution of barrier mechanisms and surface recombination, employing a conventional diffusion approximation of photomagnetic current to determine the hole diffusion length L_p (Table 1). For unpassivated layers, we determined, to the same approximation, the effective diffusion length L^* as a generalized characteristic of surface and volume recombination activity in the region $h\nu > E_g$ and calculated the surface recombination rate from L_p/L^*

The values of L_p and S are listed in Table 1. Our results indicate that the surface recombination rate depends slightly on carrier concentration, which contradicts the dependence $S \sim n_0$ found in [6].

In contrast to PME, that is affected by the barriers only indirectly and is mostly governed by the recombination processes on the surface, a presence of surface and interface barriers causes a dominance of barrier mechanisms in *photoconductivity*. PC in epitaxial GaAs layers can be described by the barrier model [3] and is determined by the variation of the width of quasi-neutral region in the layer due to the variation of depletion regions of barriers. The spectral behavior of PC correlated with the capacitor photovoltage measured on surface and interface barriers (Fig. 2). In the absence of background illumination, PC spectrum follows the spectrum of the surface CPV (Fig.2 curves 2,5) and also undergoes strong reduction by 2-3 orders of magnitude when the background illumination rises up to the maximal value (Fig.2 curves 7,9). Under this condition, i.e. when the surface barrier is suppressed, a decrease of photosensitivity can be seen in the short-wavelength region of the PC spectra. The

Fig. 2: Effect of surface passivation and background illumination on the photoconductivity σ_{ph} and capacitor photo-voltage V_{ph} spectra of epitaxial GaAs layers: 1,4,8 - interface CPV; 2,3,9 - surface CPV; 5,6,7 - PC. Surface treatment: 1,3,6 - passivated; 2,4,5 - real; 7,8,9 - spectra were measured under background illumination. The arrows show the change of the photosensitivity due to surface passivation.

spectrum coincides in this case with the CPV spectrum measured on interface barrier (Fig.2 curves 7,8), thus indicating the change of localization of barrierphotoconductivity.

It is interesting that the heteroepitaxial passivation leads to an increase of photomagnetic effect but slightly decreases PC and surface CPV (curves 3,6). The reason for that is evidently the decrease of surface barrier at passivation. The influence of a decrease of surface barrier on PC and CPV dominates over the opposite effect of reduction of surface recombination rate.

As we showed earlier [3], the lifetimes of majority carriers can be determined employing PC method only at considerably high photoexcitation conditions, when it is possible to distinguish a volume component of PC. The lifetime of electrons, determined at these conditions, in passivated samples are $\sim 3 \cdot 10^{-8}$s, comparable to the lifetimes of holes obtained from PME ($\sim 1.4 \cdot 10^{-8}$s).

Semi-insulating GaAs

A considerable spread in the recombination parameters obtained for s-GaAs by the methods developed on the basis of nonlinear theory of photoconductivity and photomagnetic effects can be found in the literature. The data presented by two scientific groups [6,7] differ in the values for τ_n by 2 and for τ_p by 4 orders of magnitude, and the values of parameter $\gamma = \tau_p/\tau_n$ were estimated to be ~ 0.2 [6] and ~ 10 [7]. Therefore, is not clear yet whether these discrepancies are due to the properties of the materials investigated or to the methods employed. For this purpose we studied here a series of s-GaAs monocrystals differing by the type of compensating doping. We used an alternative method [2] that allowed us to apply a linear theory of PC and PME. Note that both PME and PC in semi-insulating s-GaAs monocrystals can be described in terms of bulk effects.

The recombination parameters determined for the etched (100) surface of s-GaAs monocrystals are summarized in Table 2. One can see that we obtained, actually, the same values of parameters mentioned above, however they are strongly dependent on the doping of the material used.

Table 2: Recombination parameters of (100)s-GaAs

Material	Doping	γ	L, μm	τ_n, 10^{-9} s	τ_p, 10^{-9} s	S, 10^4 cm/s
AG-Cr	Cr	0.7	0.6	0.7	0.5	0.7
AG-4	Cr_2O_3	4.7	6.5	14	70	1.3
AG-6	undoped	6.7	2.4	1.6	12	5.0
AG-30	In	0.16	1.1	10	1.7	0.5

NB: Here we compare the parameters of s-GaAs monocrystals subjected to the same surface treatment, namely chemical etching.

Rather surprising is the influence of the state of the surface on the bulk parameters L, τ, γ revealed in s-GaAs (Table 3). The recombination parameters in the same sample differed markedly when determined for the real cleaved (110) and etched (100) surfaces. A cleaved (110) surface is characterized by a large value of the parameter $\gamma \sim 5$, i.e. the hole component in PC is dominant, and, as a consequence, by at least a 4 times larger value of ambipolar diffusion length L.

It would be interesting to understand in details the effect of heteroepitaxial passivation on the surface processes in s-GaAs. The effect of deposition of a thin $In_{049}Ga_{0.51}P$ layer on the surface of s-GaAs monocrystals is opposite to that in the case of epitaxial layers. We observe a remarkable increase of PC after passivation. At the same time, the photosensitivity of PME either decreases (for $\gamma > 1$ on the real surface) or remains almost unchanged (for $\gamma < 1$). Calculations of recombination parameters (Table 3) showed that heteroepitaxial passivation of s-GaAs resulted in a 2-10-fold increase of τ_n, depending on the type of material and surface, and in a decrease of the ratio γ that is always less than unity for a passivated surface. Obviously, this influence of the state of the surface on the electron-hole balance, described by parameter γ, is closely connected with the re-charging processes of the surface states under treatment. We have to take into account that the parameters determined in experiment characterize the near-surface region of semiconductor about L wide which is at diffusion-drift and recombination equilibrium with the surface.

Table 3: The influence of the state of the surface on recombination parameters determined for AG-Cr monocrystal

Surface treatment	γ	L, μm	τ_n, 10^{-9} s	τ_p, 10^{-9} s	S, 10^4 cm/s
(100), etched	0.46	0.7	1.3	0.6	~ 1
(110), cleaved	5.2	3.1	2.9	15	
(100), passivated	0.16	0.9	7.4	1.3	<<1
(110), passivated	0.8	1.8	5.1	4.2	<<1

Conclusions

The phenomenon of heteroepitaxial passivation is manifested in photomagnetic effect and photoconductivity of epitaxial layers as a decrease of surface recombination rate and surface barrier. Strong suppression of anomalous component of PME simplifies significantly determination of recombination parameters in epitaxial GaAs layers employing this method. We show that the bulk recombination parameters in semi-insulating GaAs, determined by PC and PME methods, depend strongly on the method of doping and on the surface state of materials. Heteroepitaxial passivation of s-GaAs monocrystals increases the contribution of the electron component to photoconductivity. The recombination parameters of n-GaAs epitaxial layers and semi-insulatingmonocrystals are presented.

[1] I.A. Karpovich, B.I. Bednyi, N.V. Baidus', L.M. Batukova, B.N. Zvonkov, M.V. Stepikhova, Sov.Phys.Semicond. 27 (1993) 958.
[2] I.A. Karpovich and M.V.Sepikhova, Sov.Phys.Semicond. 30 (1996) 934.
[3] I.A. Karpovich, B.I. Bednyi, N.V. Baidus', S.M. Plankina, M.V. Stepikhova, M.V. Shilova, Sov. Phys. Semicond. 23 (1989) 1340.
[4] I.A. Karpovich, V.Ya. Aleshkin, A.V. Anshon, N.V. Baidus', L.M. Batukova, B.N. Zvonkov, S.M. Plankina, Sov. Phys. Semicond. 26 (1992) 1034.
[5] V.A. Zuev, A.V. Sachenko, and N.B. Tolpygo, "Nonequilibrium Near-Surface Processes in Semiconductors and Semiconductor Devices" [in Russian] (Soviet Radio, Moscow, 1977).
[6] D.E. Aspens, Surf. Sci.. 132 (1983) 406.
[7] S.S. Li and C.I. Huang, J. Appl. Phys. 43 (1972) 1757.
[8] M.J. Papastamatiou and G.J. Papaioannou, J. Appl. Phys. 68 (1990) 1094.

The electrical properties of semi-insulating GaAs analysed as a relaxation semiconductor

B.K. Jones, J. Santana* and M. McPherson
Department of Physics, Lancaster University, Lancaster, LA1 4YB, UK.
** Departmento de Electronica Sistemas e Informatica, Instituto Technologico y de Estudios Superiores de Occidente (ITESO), Tlaquepaque, Mexico.*
b.jones@lancaster.ac.uk

Abstract

From experiments on semi-insulating (SI) GaAs diodes it is shown that the results can be analysed using relaxation semiconductor theory. This is most apparent for P-SI-N or Schottky-SI-Ohmic diodes and the agreement improves after irradiation when the ratio of generation-recombination (g-r) centres to deep levels increases.

A. Introduction

Nearly all commercial semiconductors are 'lifetime' materials. However by 1972 considerable work had been carried out on another class of semiconductor, 'relaxation' material typified by the work of van Roosbroeck and Casey [1] and the seminal review of Queisser [2]. There has been more recent work [3,4]. This approach has been neglected for semi-insulators, probably because most real samples have properties which are dominated by defect properties. We have been led to use relaxation semiconductor analysis for semi-insulating GaAs by the success of this approach on semi-insulating silicon [5] which has demonstrated the practical application of this analysis to experimental results. Much of this work has been carried out during the development of high energy particle detectors [6-12].

B. Relaxation Semiconductors

The conventional 'lifetime' material has $\tau_D \ll \tau_0$ while a 'relaxation' material has $\tau_D \gg \tau_0$. Here $\tau_D = \rho \varepsilon \varepsilon_0$ (ρ is the resistivity) is the dielectric relaxation time which is the time in which a space charge is neutralised by the flow of the free carriers. The minority carrier generation or recombination lifetime is τ_0 .

If a reverse bias is applied to a relaxation semiconductor P^+-N junction [4] the free carriers are not removed to cause a depletion region since new e-h pairs are created, on demand, to fill the space. This results in this region having a single quasi-Fermi level pulled to near mid-gap where the g-r process is most efficient. For current continuity of the created carriers in the steady state $n_0\mu_n = p_0\mu_p$ and with the equilibrium condition $n_0 p_0 = n_i^2$ we get $n_0 = p_0(\mu_p/\mu_n) = n_i(\mu_p/\mu_n)^{1/2}$. The resulting carrier densities, n_0 and p_0 are independent of the specific material, contacts or trap density or properties. The resistivity is then at its maximum value given by $\rho^{-1}_{max} = (2 \, e \, n_i \, (\mu_n\mu_p)^{1/2}) \sim (8 \times 10^8 \, \Omega$ cm$)^{-1}$ at room temperature for GaAs and $(3 \times 10^5 \, \Omega$ cm$)^{-1}$ for silicon. As well as this dynamic relaxation material there are other high resistivity materials in static equilibrium, Intrinsic material has high resistivity with $n = p = n_i$ and compensated material has a high resistivity which can reach intrinsic values

High resistivity semi-insulating GaAs with relatively few defects is found to behave like a (non-ideal) relaxation semiconductor. The features are obscured by the array of traps and inhomogeneities. Also the contacts are often not the optimum to see the effects. Irradiation enhances the relaxation properties and provides a useful experimental variable since apparently the main effect is to increase the number of g-r centres. The definitive experimental features are best seen in silicon and this is described elsewhere [5]. Clear experimental results has been seen on GaAs P-SI-N and Schottky-SI-Ohmic diodes which have also had their trap content measured by various methods.

C. Samples and Experiments

A large number of SI GaAs diode samples have been studied and many have been irradiated by high neutron fluences. To see clear relaxation effects it is necessary to use samples with asymmetric contacts so that one is not always reverse biased. The application of radiation damage enables a new experimental variable to be studied. The basic experiments performed were I-V and C-V(f,T) over a range of temperatures. For most samples a complete study of the traps was performed using PICTS, C-V and alpha particle excited DLTS [12].

D. Comparison with Relaxation Theory.

The I-V curves shown in Fig.1 illustrate the Ohmic resistivity at small forward and reverse bias. This Ohmic region extends beyond the kT/e voltage found in a lifetime diode [13]. The broken line indicates the value of the maximum resistivity, ρ_{max}, calculated from the normal mobility and intrinsic carrier density values and the full geometric dimensions. The agreement is therefore reasonable. In reverse bias, a non-Ohmic $V^{1/2}$ dependence is found due to the space charge of the traps, both donors/acceptors and deep levels. In forward bias the controlled breakdown is characteristic of relaxation effects. The C-V-f-T data of Fig. 2 show the expected complex effects due to deep levels. Between the end of the relaxation-like region and the Ohmic contact the bands bend so that different deep levels are ionised at different distances and are therefore active with respect to the AC drive signal. This is not an effect specific to relaxation material. However there is also a capacitance maximum at small reverse bias and a negative capacitance at forward bias which can only be explained by relaxation effects. This is caused essentially by the presence in relaxation material of a free carrier space charge near each contact so that changes in this appear as a very large capacitance [14].

E. Conclusions

We have shown that diodes made from semi-insulating GaAs, especially after they have been irradiated to generate more g-r centres than deep levels, show electrical properties that follow the theory of relaxation semiconductors. It is important therefore to analyse the results of all experiments on semi-insulating samples using the appropriate theory. There are many properties of commercial and experimental devices which can be readily explained by this interpretation. These include:(a) the detailed understanding of apparently Ohmic or blocking contacts, (b) the measured resistivities and Hall voltages, (c) the observed constant internal fields [15] (d) reduction in shot noise[16] and (e) the side-gating, back-gating and the kink-effect in GaAsFETs [11]. Since the current flow is still asymmetrical it is necessary to perform experiments on samples which have one Ohmic and one blocking contact in order to enable the analysis to be easy.

Acknowledgements

The authors would like to thank various collaborators for the supply of samples. We also appreciate the interactions within the members of RD8. Financial help has been obtained from PPARC, the National Council for Science and Technology (CONACYT) Mexico, the Government of Lesotho and CERN.

References

[1] W. van Roosbroeck and H.C. Casey Jr., *"Transport in relaxation semiconductors"*, Phys.Rev. **B5** (1972) 2154.

[2] H.J. Queisser. *"Semiconductors in the relaxation regime"*, Proc. ESSDERC 1972 145-88 IOP Conf Proc No 15 (1972)

[3] N.M. Haegel, *"Relaxation semiconductors: in theory and in practice"*, Appl.Phys. **A53** (1991) 1-7.

[4] J.C. Manifacier and R.Ardebili, *"Bulk and contact effects in p+-SI-n+ semi-insulating GaAs structures"*, J. Appl. Phys. **77** (1995) 3174-85.

[5] B.K. Jones, M. McPherson and J. Santana, *"The electrical properties of irradiated silicon: semi-insulating silicon"*, this conference.

[6] *"Irradiated silicon detectors"*, B.K. Jones, ROSE http://http1.brunel.ac.uk:8080/research/rose/rose.html

[7] B.K. Jones, J. Santana and M. McPherson, *"Semiconductor detectors for use in high radiation damage environments-Semi-insulating GaAs or Silicon?"*, Nucl. Instrum. and Methods. **A395** (1997) 81-7.

[8] B.K. Jones, K. Zdansky, J. Santana and T. Sloan. *"Relaxation Semiconductor Devices"*, Proc. GaAs and Related Compounds, San Miniato, March 1995, ed P.G. Pelfer, J. Ludwig, K. Runge and H.S. Rupprecht. 73-77.

[9] J. Santana, B.K. Jones, T. Sloan and K. Zdansky. *"GaAs Radiation Detectors in the Relaxation Regime"*, Proc. GaAs and Related Compounds, San Miniato, March 1995, ed P.G.Pelfer, J. Ludwig, K. Runge and H. S. Rupprecht. 61-6.

[10] J.M. Santana-Corte, *"GaAs diodes in the relaxation regime used for radiation detection"*, PhD Thesis, Lancaster University, 1997.

[11] M. McPherson, *"Irradiated silicon detectors as relaxation devices"*, PhD Thesis, Lancaster University, 1998.

[12] B.K. Jones, J.M. Santana amd T. Sloan, *"Defects in neutron irradiated LEC semi-insulating GaAs"*, Materials Science Forum, **258-63** (1997) 1039-44..

[13] B.K. Jones, J. Santana and M. McPherson, *"Ohmic I-V characteristics in semi-insulating semiconductor diodes"*, Solid State Comm. **105** (1998) 547-9.

[14] B.K. Jones, J. Santana and M. McPherson, *"Negative capacitance effects in semi-insulating semiconductor diodes"*, Solid State Comm. **107** (1998) 47-50.

[15] K. Berwick, M. R. Brozel, C. M. Buttar, M. Cowperthwaite and Y. Hou., *"Imaging of high field regions in SI GaAs under bias"* Defect Recognition and Image Processing in Semiconductors and Devices Conference, Santander, 1993. IOP Conference Series No. 135 304-10. IOP London, 1994.

[16] G.D. Marder, Th. Eich, N. Evans, R. Geppert, R. Goppert, R. Irsigler, J. Ludwig, M. Rogalla, K. Runge, Th. Schmid, *"Noise in GaAs detectors after irradiation"*, Nucl. Instrum. and Methods in Phys. Res. **A395** (1997) 141-4.

[17] J. Santana and B.K. Jones, *"Semi-insulating GaAs as a relaxation semiconductor"*, J. Appl. Phys. **83** (1998) 7699-705.

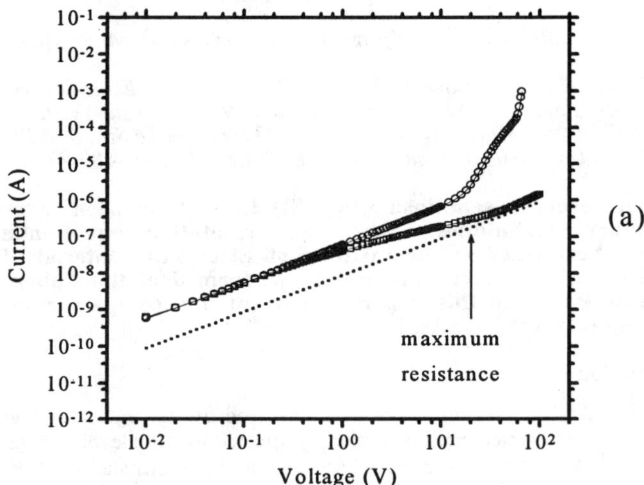

Fig. 1.   The I-V characteristics of a non-irradiated semi-insulating GaAs Schottky-Ohmic contact diode at 300K. The calculated value of the maximum resistance is shown. The Ohmic region due to the relaxation material and the $V^{1/2}$ region due to the space charge can be seen.

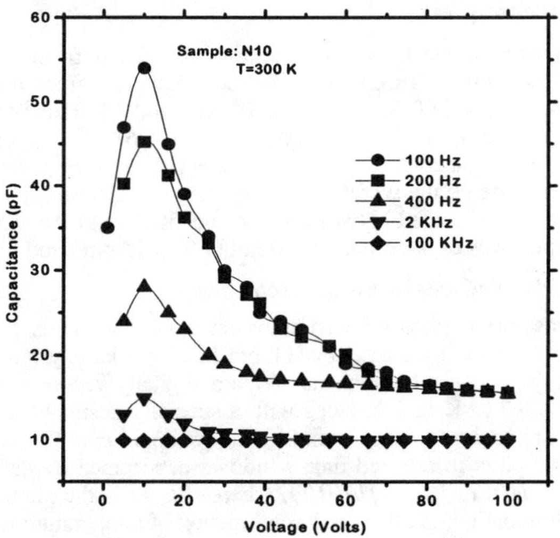

Fig. 2.   The C-V characteristics of a semi-insulating GaAs diode (N10) with frequency as parameter. The capacitance peak due to relaxation effects and the frequency dependence due to the particular deep levels can be seen.

New techniques for the characterization of defect levels in semi-insulating materials

C. Longeaud[1], J. P. Kleider[1], P. Kaminski[2], R. Kozlowski[2], M. Pawlowski[3], R. Cwirko[3]

[1]*Laboratoire de Génie Electrique de Paris, CNRS (URA0127), Ecole Supérieure d'Electricité, Universités Paris VI et XI, Plateau de Moulon, 91192 Gif sur Yvette cedex, France.*
[2]*Institute of Electronic Materials Technology, 133 Wolczynska Str., 01-919 Warsaw, Poland.*
[3]*Military University of Technology, 2 Kaliskiego Str., 01-489 Warsaw, Poland.*

Deep levels in semi-insulating (SI) GaAs have been investigated by two complementary techniques: the high resolution photoinduced transient spectroscopy performed at the Institute of Electronic Materials Technology, and the modulated photocurrent experiment performed at the Laboratoire de Génie Electrique de Paris. In this paper we present and compare results obtained by both techniques for SI GaAs.

A. Introduction

The quality of semi-insulating materials used as substrates for high speed electronic and optoelectronic integrated circuits is strongly affected by deep levels related to impurities and native defects. Although a number of studies aimed at the determination of deep-level defects in these materials have been performed, there is still a need to clarify and cross-check experimental data. So, the objective of the present paper is to show an approach towards fulfilling this need by application of two complementary techniques: the high resolution photoinduced transient spectroscopy (HRPITS) and the modulated photocurrent (MPC) measurements to study defect levels in the same samples of SI GaAs. The trap properties in the wafers of SI GaAs provided by two vendors, labeled as vendor A and vendor B, are compared.

B. Samples and techniques

The substrates received from both vendors originated from crystals grown by the liquid encapsulated Czochralski (LEC) method. For the wafers from the vendor A, the Hall concentration and mobility at 300 K were 1.1×10^7 cm^{-3} and 5450 cm^{2}V^{-1}s^{-1}, respectively. For the wafers from the vendor B, these were equal to 6.6×10^7 cm^{-3} and 5889 cm^{2}V^{-1}s^{-1}, respectively. Arrays of coplanar AuGe-Ni ohmic electrodes were formed on the mechano-chemically polished surface of the wafers. The gap between electrodes was 0.8 mm. In both techniques, the applied bias ranged between 10 V and 50 V and the wavelength of the light illuminating the samples was of the order of 650 nm (1.9 eV) from a red light emitting diode.

B1. *High resolution photoinduced transient spectroscopy*

The high resolution photoinduced transient spectroscopy is based on the classical photoinduced current transient spectroscopy [1], but has been largely improved. Following the light switch-off, the photocurrent transients $I(t)$ are digitally recorded for a wide range of temperatures from 30 to 320 K in 1-K steps with a specially designed high resolution (12 bit) digitizer. Then, they are analysed using a multi-window approach. The transient $I(t)$ is divided into single exponential parts in selected time windows determined by the times t_i and t_{i+1}. The PITS spectrum $I(T) = [I(t_i,T)-I(t_{i+1},T)]/I(0,T)$, where $I(0, T)$ is the photocurrent in the steady state under optical illumination, is calculated as a function of temperature for the emission rate e_i determined from equations shown in [2]. Usually we use 20 to 30 emission rate windows ranging between 5 and 50000 s^{-1} and we check the consistency of the plot of $\ln(T^2/e_i)$ versus $1/T$ with the Arrhenius formula $e_i = A_i T^2 \exp(-E_{ai}/kT)$, thus identifying a defect level by the values E_{ai} and A_i related to the energy position and the capture cross section, respectively. The flux of the excitation light was equal to 4.2×10^{15} cm^{-2} s^{-1}. A typical HRPITS spectrum obtained on an

     72

undoped SI GaAs sample of vendor A is presented in Fig. 1. The different detected peaks are labeled T1 (0.12 eV), T2 (0.15 eV), T3 (0.26 eV), T4 (0.36 eV), T5 (0.52 eV), T6 (0.64 eV) and T7 (0.83 eV).

B2. *Modulated photocurrent experiment*

The modulated photocurrent (MPC) technique, successfully applied to the study of the defect density of highly resistive amorphous materials [3], is also perfectly adapted to the study of SI crystalline materials. At a given temperature T, the sample is illuminated with a flux of light sinusoidally modulated at different frequencies f, $F=F_{dc}+F_{ac}\cos(2\pi ft)$. By means of a lock-in amplifier we measure the modulus of the ac photocurrent $|I_{ac}|$ and its phase shift Φ referred to the excitation. It was shown that the quantity NC/μ, where N is the density of states, μ the extended states mobility of the carriers interacting with the traps and C the capture coefficient of the traps, is proportionnal to $sin(\Phi)/|I_{ac}|$, the coefficient of proportionnality depending on known experimental parameters. The energy scaling is calculated by $\Delta E=E_{be}-E=kT\ln(\nu/2\pi f)$, where E_{be} is the energy position of the band of extended states, ν is the attempt-to-escape frequency of the trap $(\nu=AT^2)$ and k the Boltzmann constant. A plot of $sin(\Phi)/|I_{ac}|$ versus ΔE with a good choice of ν for different frequencies and temperatures reveals the peak position of the defect, its capture cross section being deduced from ν. Experimentally the temperature was varied from 330 K to 80 K in 3-K steps and for each temperature the frequency was varied in the range 12 Hz-40 kHz. F_{dc} was equal to 2×10^{12} cm^{-2}s^{-1} and F_{ac} was about 3 times lower.

We present in Fig. 2 the MPC spectra obtained for an undoped SI GaAs prepared by vendor A. In this figure each set of symbols corresponds to the variations of the frequency of the excitation at a given temperature. Clearly peaks are visible on these spectra that should correspond to peaks of trapping states. The energy scaling was achieved assuming a constant attempt-to-escape frequency $\nu=10^{12}$ s^{-1} at T=300 K for all the defect levels, a simplifying assumption which is used in a first step to reveal the presence of the peaks.

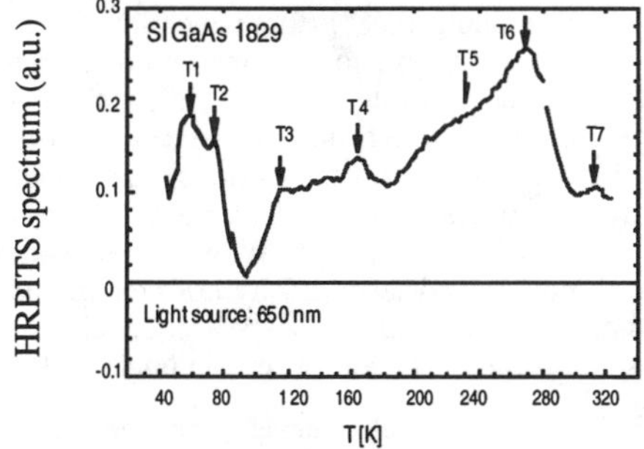

Fig. 1: HRPITS spectrum for an undoped SI GaAs sample of vendor A.

Fig. 2: MPC Spectra obtained for an undoped SI GaAs sample of vendor A.

In a second step, in order to determine the values of the attempt-to-escape frequency at T=300 K and the energy position of a given peak we adopted the following procedure. Since we varied the temperature in 3-K steps, a given peak is described by several MPC curves obtained at different temperatures, and all these curves must have their maximum at the same energy. Thus, the attempt-to-escape frequency we use to scale the energy must be adjusted in order to superimpose the maxima of these various MPC spectra. This procedure is illustrated in Fig. 3 where some spectra describing one particular level are plotted in Fig. 3a with $\nu(300$ K$)=2\times10^{11}$ s^{-1} and in Fig. 3b with $\nu(300$ K$)=10^{12}$ s^{-1}. The best agreement between the various spectra is obtained for the second value. We then have an order of magnitude of the attempt-to-escape frequency, and consequently of the capture cross section, as well as a good order of magnitude of the energy position of the level, 0.33 eV in the present case. The error in the determination of v can be then estimated as equal to a factor of five leading to an error in the energy position of the order of 30 meV. Finally, if the mobility values are known, an order of magnitude of the defect density can be further obtained from the values of NC/μ.

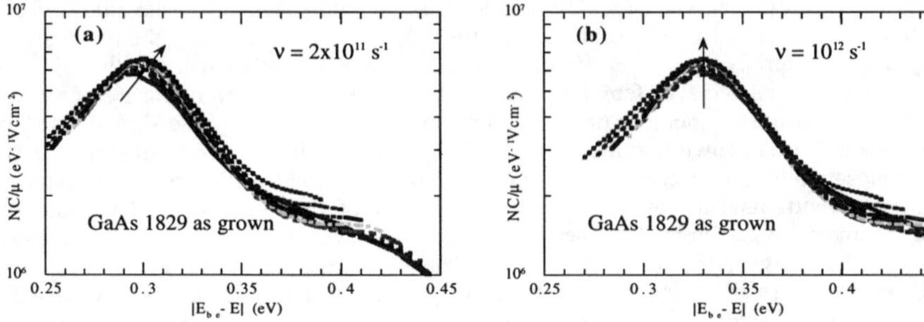

Fig. 3: Illustration of the procedure used to extract the attempt-to-escape frequency and the energy position of the defect. A good agreement between all the MPC spectra is obtained in (b).

C. Results and discussion

In Tables I and II we present the results obtained by both techniques applied to the same samples of undoped SI GaAs prepared by vendors A and B, respectively. We have tried to identify the detected peaks according to the results of the HRPITS and the traps properties found in the literature [4-11]. For the shallowest states, the MPC technique was unable to resolve the trap position and attempt-to-escape frequency due to the limitation in temperature (T>80 K) of our experimental set-up. However, these peaks can be seen on the spectrum in Fig. 2 at ~0.1 eV as indicated by arrows. It is worth noticing that, in both techniques, it is impossible to separate the contributions of holes and electrons to the measured current. Therefore, it is not possible to precise whether the peak energies detected by both techniques are referred to the valence or conduction band edge. For the same reason, we indicate the value of the attempt-to-escape frequency rather than that of capture cross sections. Of course, corresponding capture cross sections σ, for electrons or holes, can be calculated by means of the usual coefficients $\gamma_n = 2.0 \times 10^{20}$ or $\gamma_p = 1.8 \times 10^{21}$ cm^{-2}K^{-2}s^{-1} with A=$\gamma\sigma$. Within experimental errors, a rather good agreement is found in the peak positions determined by both techniques. For some peaks there is also a very good agreement for the attempt-to-escape frequencies, but some discrepancies that experimental errors cannot completely account for remain uncleared.

Table I: Parameters of traps detected by the HRPITS and MPC techniques for an undoped SI GaAs wafer prepared by vendor A.

HRPITS		MPC		Tentative
E_a (eV)	v(300 K) (s^{-1})	E_a (eV)	v(300 K) (s^{-1})	Identification
0.120 ± 0.005	4.0×10^{13}	Unresolved		V_{As} - related [4]
0.150 ± 0.005	3.6×10^{13}	Unresolved		Cu (0/-) [5]
0.26 ± 0.01	3.0×10^{14}	0.23 ± 0.03	5×10^{12}	HL12, Zn-related [6]
0.36 ± 0.015	1.6×10^{13}	0.33 ± 0.03	1×10^{12}	EL6,V_{As}-As$_{Ga}$ [7]
0.52 ± 0.02	2.6×10^{12}	0.50 ± 0.03	5×10^{12}	HL8 or EL2 (+/2+) [8]
0.64 ± 0.03	2.8×10^{14}	0.65 ± 0.03	1×10^{12}	Oxygen-related [9]
0.83 ± 0.04	4.0×10^{15}	not detected		EL12 [10]

Table II: Parameters of traps detected by the HRPITS and MPC techniques for an undoped SI GaAs wafer prepared by vendor B.

HRPITS		MPC		Tentative
E_a (meV)	$\nu(300\ K)\ (s^{-1})$	E_a (eV)	$\nu(300\ K)\ (s^{-1})$	Identification
0.100 ± 0.005	8.4×10^{11}	Unresolved		B_{As} (0/-) [11]
0.120 ± 0.005	3.6×10^{12}	Unresolved		V_{As} - related [4]
0.17 ± 0.01	1.8×10^{11}	0.17 ± 0.03	1×10^{11}	B_{As} (-/2-) [11]
0.28 ± 0.01	1.0×10^{13}	0.28 ± 0.03	1×10^{12}	HL12, Zn-related [6]
0.30 ± 0.02	1.2×10^{12}	0.32 ± 0.03	5×10^{11}	EB7 [10]
0.40 ± 0.02	4.8×10^{10}	not detected		EI1 [10]
0.45 ± 0.02	2.2×10^{14}	0.43 ± 0.03	1×10^{12}	HB4, Cu (-/2-) [5,6]
not detected		0.65 ± 0.03	1×10^{12}	Oxygen-related [9]

The results summarized in Table I and Table II show that defect centers common for the wafers from the both vendors are presumably related to As vacancies, Oxygen and Zn contamination. On the other hand, one can notice peculiar features which distinguish the wafers supplied by these vendors.

D. Conclusions

For the first time the MPC technique, developed previously for amorphous materials, has been employed to study deep traps in semi-insulating monocrystalline GaAs and the results are in good agreement with those obtained by HRPITS. The potentialities of the two methods are exemplified by comparing the properties of the grown-in defect centers in the wafers provided by two different vendors.

Acknowledgements

This research has been supported by the Polish Committee for Scientific Research and the French Ministry of Foreign Affairs under contract No. 7023

References

[1] C. Hurtes, M. Boulou, A. Mitonneau and D. Bois, Appl. Phys. Lett. **32** (1978) 821.
[2] P. Kaminski and H. Thomas, in "Defects in Crystals" (World Sci., Singapore, 1988) 438.
[3] J. P. Kleider and C. Longeaud, in Solid State Phenomena , Vols **44-46** (1995) 597.
[4] D. Pons and J. C. Bourgoin, J. Phys. C: Solid State Phys. **18** (1985) 3839.
[5] N. Kullendorff, L. Jansson and L. A. Ledebo, J. Appl. Phys. **54** (1983) 3203.
[6] A. Mittonneau, G. M. Martin and A. Mircea, Electron. Lett. **13** (1977) 666.
[7] P. Kaminski, G. Gawlik and R. Kozlowski, Mater. Sci. Eng. **B28** (1994) 439.
[8] J. C. Bourgoin, H.J. von Bardeleben and D. Stievenard, J. Appl. Phys. **64** (1988) R65.
[9] M. Müller, G. Gartner, G. Hirt, M. Jurisch, A. Kohler, G. Müller and B. Weinert, J. Cryst. Growth **166** (1996) 636.
[10] G. M. Martin, A. Mittonneau and A. Mircea, Electron. Lett. **13** (1977) 191.
[11] M. Bugajski, Electron Technology **28** (1995) 3.

Current-voltage characteristic of n-i-n structures made of semi-insulating GaAs

J. Wu

*Laboratory of Semiconductor Materials Science, Institute of Semiconductors,
Chinese Academy of Science, P.O. Box 912, Beijing 100083, China*

The current-voltage characteristic of semi-insulating GaAs is investigated by measuring the leakage current between ohmic contacts on SI GaAs. The measurement was performed on both a low-dislocation and a high-dislocation substrate grown by the liquid encapsulated Czochralski method and in addition, on a low-temperature (LT) GaAs layer grown by molecular-beam epitaxy (MBE) at very low temperature. It was found that the substrate current conduction in the low voltage range was related to the excess carrier lifetime rather than the resistivity of the substrate.

The current-voltage (I-V) characteristic of SI GaAs materials is usually investigated by measuring the current condition between the two ohmic contacts fabricated on the substrate (n-i-n). In general, the I-V curve of SI GaAs shows a linear or sublinear feature at low voltage, and the current suddenly increases at critical voltage. Usually, the steep current rise is explained by Lampert's carrier-injection model [1], and the linear behavior in the low voltage range is frequently interpreted as being simply and directly related to the bulk resistivity [2]. In this work, the I-V characteristic of SI GaAs material is investigated by measuring the current in the n-i-n structure fabricated on the three kinds of SI GaAs: a low-dislocation (LD) and a high dislocation (HD) material grown by the LEC method, and in addition, a low-temperature (LT) GaAs layer grown by molecular-beam epitaxy (MBE). It is found that the behavior of current conduction at low voltages is not directly determined by the bulk resistivity obtained in the Hall measurement and however, should be related to the excess carrier lifetime in materials.

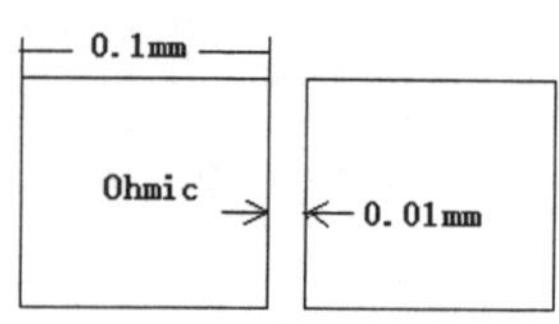

Fig. 1 n-i-n structure.

The LEC GaAs and MBE LT GaAs materials were grown in our laboratory. LD GaAs material is the conventional LEC SI GaAs with the dislocation density (EPD) of about $4 \times 10^4/cm^2$, while HD one with EPD of about $5 \times 10^5/cm^2$ is obtained by rapid cooling down at the end of the LEC growth. Both LD and HD GaAs are 2 inch in diameter. LT MBE GaAs of about 2 µm in thickness was grown at the substrate temperature of $250^\circ C$ and then annealed at $650^\circ C$ for ten minutes. Resistivities were determined by Hall measurement to be 5.0×10^7, 1.5×10^7 $\Omega \cdot cm$, and 2.0×10^6 $\Omega \cdot cm$ for LD, HD and LT materials, respectively. Two kinds of the ohmic contact were fabricated. One was formed by directly alloying on the SI material while another one was made on the n-layer formed with selective ion-implantation at energy of 90 keV and dose of 5×10^{12} cm^{-2}, respectively. The n-i-n structure is shown in Fig. 1. The area of the Au/Ge/Ni ohmic contact is 100x100 µm and displaced from each other by 10 µm in the structure. Leakage current IL was measured with an HP 4145B Transistor Analyzer.

     76

One of the main features of LT MBE GaAs is its very short minority carrier lifetime τ_{LT} (in the range of a few hundred fsec) [5] while in the conventional SI LEC GaAs, the minority carrier lifetime τ_{LD} is in the range of a few 10^{-8} sec [2] and the ratio of τ_{LD} and τ_{LT} is in the order of 10^5. It can be presumed that the minority carrier lifetime plays a significant role in determining the behavior of current conduction and the big difference between LT GaAs and LD GaAs in the I-V characteristic should be brought about by the difference in minority carrier lifetime. The shorter minority lifetime, the bigger the effective resistance R_{eff}. In addition, it is well known that dislocations induce a critical effect on the quality of minority carrier devices and has a significant influence on minority carrier lifetime in the material. Therefore, the minority carrier lifetime τ_{HD} in the HD GaAs should be much less than τ_{LD}, and R_{eff} of LD GaAs should be smaller than that of HD GaAs. The assumption is consistent with the result obtained in this work.

The characterization of the electrical properties of a semi-insulating material is much more complicated than that of a conductive one [6]. Carriers can be injected into the bulk of semi-insulating materials and produced nonequilibrium and changes in resistance [2]. According to van Roosbroeck criterion [7], the SI materials used in this work are dielectric relaxation semiconductors since their dielectric relaxation time τ_{DS} (ε/ρ is in the range of a few tens µs) are much bigger than the carrier lifetime τ (10^{-8} s), and the current conduction in these materials is space-charge-limited. However, the experimental results obtained in this work suggest that the I-V characteristic is correlated with the minority carrier lifetime.

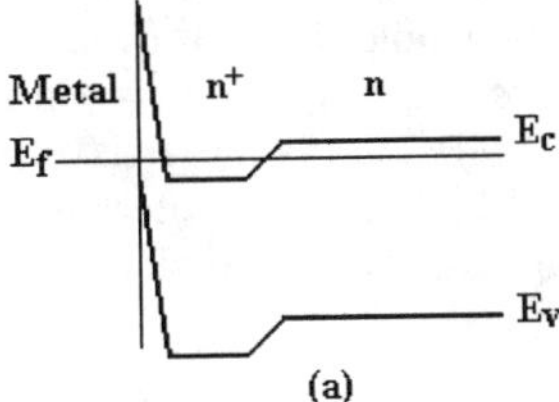

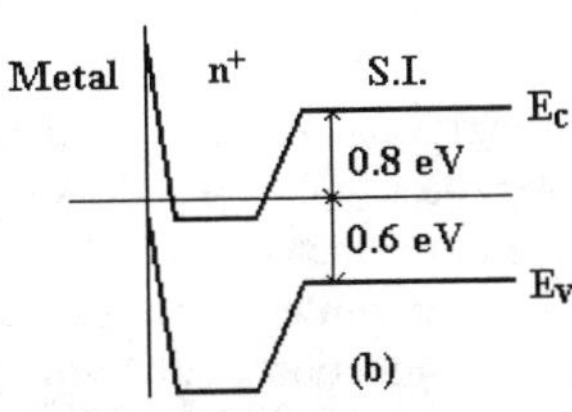

Fig. 3 Energy bands of the ohmic contact to (a) n-GaAs, (b) SI-GaAs

Fig. 3 shows the energy bands of the metal-semiconductor ohmic contacts for both an n-type and a semi-insulating material. Generally speaking, electrons are effectively injected at a metal-semiconductor ohmic contact for the n-type, as the energy barrier for an electron is much smaller than for a hole. However, the SI materials, the situation is different.

The Fermi-level pinning position for GaAs is approximately 0.8 eV below the conduction band minimum [8]. With such a case in mind, Wager and McCamant [9] pointed out that the hole injection was more likely at a metal-SI GaAs Schottky interface as the hole barrier (0.6 eV) is smaller than the electron (0.8 eV) at the interface. The situation seems to be applied to the metal-SI GaAs ohmic contact. When an electron is injected from the ohmic contact into the SI GaAs, it establishes a significant electric field in the dielectric relaxation material near the contact, as there is not enough holes available to neutralize it instantly as in a lifetime semiconductor. Therefore, two electric fields coexist and both have influences simultaneously on the electron near the contact in the opposite direction. One field is associated with the applied voltage, while another one is built by the electron itself which tends to induce a hole from the contact. The electron may drift away from the ohmic contact under the influence of the

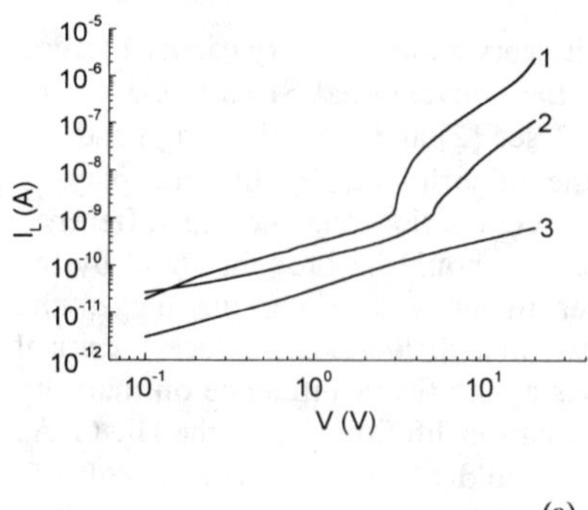
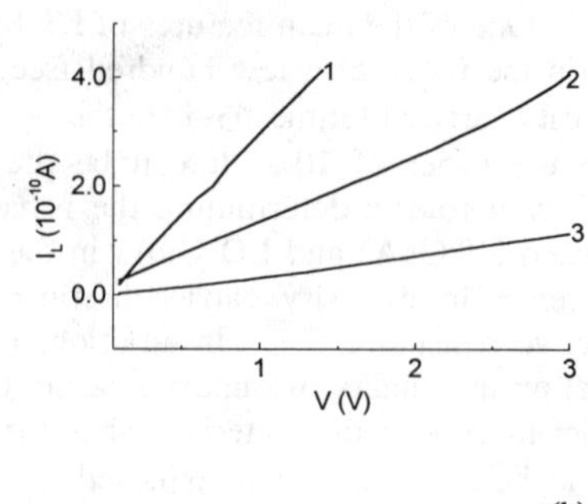

(a)   (b)

Fig. 2 I_L-V characteristics; (a) log-log, (b) linear-linear
1. LD GaAs, 2. HD GaAs, 3. LT GaAs

Fig. 2 (a) shows I_L versus voltage V applied at ohmic contact directly formed on the SI material (no apparent difference is observed between the two kinds of ohmic contacts in the I-V curves. The features of the I-V curve of the n-i-n structure in the LD and HD LEC GaAs are the same as usually reported: the current conduction is very small and linear or apparently ohmic at low voltages, but shows a sudden increase at a critical voltage, about 2.5 V and 4 V for the LD and HD materials, respectively. The critical voltage in the I-V curves corresponds to the onset of the sidegating effect [3]. I_L in the LT MBE GaAs remains small and no sudden increase in current occurs in the whole range of measurement, which is consistent with the well-known fact [4] that the sidegating effect is eliminated completely when MESFET devices are fabricated on an LT MBE GaAs buffer layer grown on SI GaAs. Fig. 2 (b) is the enlarged I-V curves below 3 V for the three materials, and all the curves within a low voltage range can be fitted well into the linear expression

$$I_L = A + V/B$$

where A and B are constants, and constant B corresponds to an apparent effective resistance R_{eff}. Table 1 gives B's of the three curves, together with the bulk resistivities ρ_b measured with the Hall method.

If the behavior of the current conduction in materials at low voltage is directly related to ρ_b, as usually assumed, B or R_{eff} should be proportional to it. However, the situation is reversed and the larger the value of B, the smaller the value of ρ_b. As shown in the table, although LT MBE GaAs exhibits the smallest value of ρ_b, the constant B is largest among the three materials. In addition, the value of ρ_b is smaller while B is larger for HD GaAs than for LD GaAs. Therefore, the behavior of current conduction at low voltages is not simply determined by the bulk resistivity. Obviously R_{eff} is quite different from ρ_b in behavior. The factors determine R_{eff} should be investigated to study the I-V characteristic of SI GaAs materials.

Table 1

Sample	B	ρ_b (Ω cm)
LD GaAs	3.8×10^9	5.0×10^7
HD GaAs	5.9×10^9	1.5×10^7
LT GaAs	2.6×10^{10}	1.7×10^6

applied field and contributes to the current. At the same time, the electron may induce the injection of a hole from the contact with ease and the hole recombines with the electron before it drifts away to contribute the current. The rate of recombination is determined by the minority carrier lifetime, and the longer the lifetime, the more electrons drift away and contribute to current. Therefore, the longer the minority carrier lifetime, the smaller the effective resistance R_{eff}, as observed in is work.

In conclusion, the experimental results demonstrate that the I-V characteristic of SI GaAs materials at low voltages is not directly and simply related to the bulk resistivity, but is related to the minority carrier lifetime in materials, and the effective hole injection at the metal-SI GaAs ohmic contact seems to be responsible for such a correlation.

Acknowledgment: This work is sponsored by the National Natural Science Foundation. The LT MBE and LEC GaAs were provided by Professors J.B. Liang and H.J. He at our institute.

[1] M. A. Lampert and P. Mark, "Current Injection in Solids" (Academic, New York, 1970).
[2] J. C. Manifacier and H. K. Henisch, J. Appl. Phys. **52** (1981) 5159.
[3] C. P. Lee, S. J. Lee and B. M. Welch, IEEE Electron Device Lett. **EDL-3** (1982) 97.
[4] F. W. Smith, A. R. Calawa, C. L. Chen, M. J. Manfra, J. and L. J. Mahoney, IEEE Electron Device Lett. **EDL-9** (1988) 77.
[5] D. C. Look, Thin Solid Films **231** (1993) 61.
[6] M. Eizenberg and H. J. Hovel, J. Appl. Phys. **69** (1991) 2256.
[7] W. Van Roosbroeck, Phys. Rev. **123** (1961) 474.
[8] S. M. Sze, "Physics of Semiconductor Devices" (Wiley, New York, 1981).
[9] J. F. Wager and A. J. McCamant, IEEE Trans. on Electron Dev. **ED-34** (1987) 1001.

Particle detector grade semi-insulating GaAs:
deep-level states studied by admittance transient spectroscopy

Juraj Darmo and František Dubecký

Institute of Electrical Engineering, Slovak Academy of Sciences
SK-84239 Bratislava, Slovakia

Abstract - Deep-level states in semi-insulating GaAs are analyzed from the viewpoint of their possible impact on the detection performance of particle detectors prepared from such material. Presence of deep-level states observed was correlated with the detection spectra of 122 keV photons.

A. Introduction

Semi-insulating gallium arsenide (hereafter SI GaAs) became a material for new generation of detectors of ionizing irradiation. It offers higher stopping power for photons and capability to operate at room temperature when compared to silicon or germanium, respectively. Furthermore, GaAs has better the costs-to-performance ratio than other semiconductor candidates like CdTe and CdZnTe. Therefore, GaAs-based detector systems are the most likely candidates for the digital X-rays imaging present under development.

Intensive researches of the SI GaAs based detectors have been launched several years ago, but many problems still continue (see [1]). Many of them relate to the presence of main deep-level state in SI GaAs responsible for compensation of shallow impurities. Concentration of these states and their charge control an electric field spread inside the detector, and lifetime of charge carriers. A question of the influence of other deep-level states present in material has been formulated as well [2].

In this contribution, we study deep-level states in the extensive range of semi-insulating GaAs materials and briefly evaluate their possible impact on the charge collection process in the particle detector made of SI GaAs. With regard to results obtained, basic tentative correlation are defined between deep-level state presence and detector performance.

B. Experimental

Our study includes eleven bulk semi-insulating GaAs materials from various producers. Materials are labeled A through L and details on their electrical parameters can be find in [3]. Thickness of all wafers was reduced by mechanical-chemical polishing down to about 200 μm. The detector structures with circular Schottky contact with diameter 2 mm with identical topology were prepared from materials. Detection performance of detectors i.e. the charge collection efficiency and the energy resolution was tested by 122 keV photons emitted from ^{57}Co. Range of these photons significantly exceeds detector thickness, so thus charge is deposited in whole detector volume. It allows also testing the width of the detector active region.

Study of deep-level states in high resistive materials like SI GaAs is complicated by difficulties with implementation of standard capacitance spectroscopy DLTS [4]. Namely, measurement of space charge capacitance is complicated by a presence of high resistance in series with it, and excitation of deep-level states by an bias pulse seems be ineffective. These problems were overcome when an optical excitation was used and the DC current instead of capacitance was measured. Spectroscopy techniques based on those approaches are called photo-induced current transient spectroscopy or optical DLTS.

Deep-level spectroscopy technique applied in our study is technique based on the analysis of admittance of measured structure done in time domain. This technique is called admittance transient spectroscopy (hereafter ATS) and was introduced by Dubecký et $al.$ [5]. Spectroscopy relies on an analysis of time dependence of conductance $G_m(t)$ and capacitance $C_m(t)$ of structure measured in the parallel equivalent circuit at high frequency (typically 1 MHz), hence expression

$$Y(t) = G_m(t) - j \cdot \frac{1}{\omega \cdot C_m(t)}$$

applies. If an effect of the series resistance is included, then time dependence of measured capacitance and conductance can be expressed as

$$\Delta C_m(t) = \Delta C_B(t) \cdot \frac{1 - D^2}{\left(1 + D^2\right)^2} \qquad \text{and} \qquad \Delta G_m(t) = \omega \cdot \Delta C_B(t) \cdot \frac{2D}{\left(1 + D^2\right)^2}$$

(D is $\omega C_B R_S$ with space charge capacitance C_B and series resistance R_S). Expression for measured capacitance when series resistance is considered was derived by Broniatowski [6]. He observed that change of the measured capacitance is less than a change of capacitance of space charge region due to deep-level states reoccupation following the derived formula. Moreover, if quality of circuit (quantity D) is higher than 1, then polarity of capacitance change, a tool to distinguish minority and majority traps, will change as well.

We extended the analysis of admittance of structure with series resistance and the relation between space charge capacitance ΔC_B and measured conductance ΔG_m was derived [7]. From expression for $\Delta G_m(t)$ follows that $i/$ a 1 pF change in a capacitance corresponds to about 1 µS change in conductance (at 1 MHz); $ii/$ polarities of measured changes in conductance and capacitance are always the same; and $iii/$ a $\Delta G_m(t)$ has maximum for circuit quality D values at which $\Delta C_m(t)$ is close to zero. Therefore, simultaneous measurement and analysis of both capacitance and conductance of structure allow to partially overcome problems with deep-level states study in the high resistive semiconductor, e.g. semi-insulating GaAs.

Problem of the effective reoccupation of deep-level states is at the ATS solved by an optical excitation used. Light with appropriate wavelength is applied in continuos as well as pulsed modes. Thus photogenerated charge carriers are injected into a structure and reoccupy deep-level states, or states are reoccupied by means of the photoionization process directly.

C. Results and discussion

We applied ATS on all eleven studied semi-insulating materials. Several deep-level states were observed when either an electrical excitation or optical excitations were used. One can see from the Arrhenius plot (Fig.1) that calculated apparent activation energies range from several tens of meV up to about 1 eV. Therefore, they lay within the whole GaAs energy band gap. Presence of particular deep-level states in each material studied are summarized in the table 1, where also probable origin of states is indicated. Generally, we can divide all states into three groups from viewpoint of their capture/emission rates and the collection time of particle detector [8,9]. First group includes main deep-level states present in material. They play crucial role in the compensation mechanism of semi-insulating GaAs and seem to control charge carrier lifetime. There are states DLE5 and DLE4 clearly related to Cr and antisite As_{Ga}, respectively. Other states DLO6 and DLO7 are believed to be hole emission from the same states.

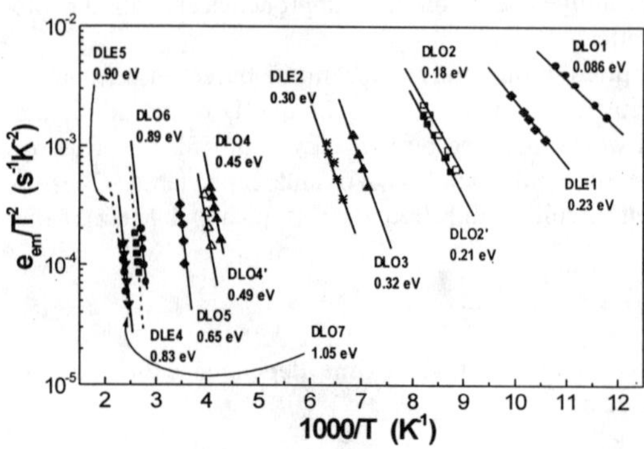

Figure 1: Arrhenius plot of deep-level states observed in semi-insulating GaAs

Second group of states consists of states DLO4, DLO4' and DLO5 with activation energy 0.65 – 0.45 eV, i.e. their emission rates at room temperatu-re are of the order of 10^{-5} s^{-1}, whiles capture rates are of order of 10^8 s^{-1}. Therefore, a capture of charge generated by ionizing particle can interfere with the collection process taking place in a particle detector. Third group of deep-level states is created with shallow states observed predomi-nately by the optical excitation are believed to have acceptor-like cha-rakteristics. Their position is about 0.2 ÷ 0.3 eV far from the top of valence band and they can contribute to the quantum charge noise of particle detector dark current, as well as collected charge.

Detectors made from all studied semi-insulating GaAs materials were tested on the detection of 122 keV photons. Shape of corresponding spectra, as well as resulting charge collection efficiency and energy resolution are presented elsewhere [3]. Here, we focus on main observed correlation between deep-level states content and detector performance.

Figure 2a, 2b shows ATS spectra measured at electrical excitation and photon spectra for detector structure made of materials labeled B1 and H. Material H is chromium doped semi-insulating GaAs and deep-level state found (DLE5) has signatures of Cr-related state. Both materials contain EL2 (here DLE4) and EL3 (here DLE3) states, while optical ATS spectra exhibited either presence of DLO6 or DLO7. That content of various main deep-level states has a consequence on detection capability of detectors made of these materials. While photon spectrum for material B1 contains peak at channel 200, spectrum for material H has no peak and signal reaches channel 140. Higher number of channel means larger collected charge, so thus detector made of material B1 exhibits better charge collection efficiency, or other words, lifetime of charge carriers is longer in the undoped semi-insulating GaAs (mat. B1) than in Cr-doped SI GaAs (mat. H).

Table 1: Deep-level states in the SI GaAs detected by the Admittance Transient Spectroscopy. DLEx and DLOx refer to states observed at electrical and optical excitations, respectively

	DLE4	DLE5	DLO6	DLO7	DLO5	DLO4	DLO4'	DLE2	DLO3	DLO2	DLO2'	DLO1
A1	X		X				X	X	X	X		
B1	X		X			X			X	X		
C	X					X		X	X		X	X
D1	X		X		X	X		X	X		X	
D2	X	X		X		X	X		X		X	X
E	X		X				X		X	X		
F	X		X				X		X	X		
G							X		X	X		X
H	X	X					X		X	X		
K	X	X			X		X		X	X		
L	X		X				X		X	X		
	EL2 electron emission	Cr-related	EL2 hole emission (?)	Cr-related	Cu-related (?)			EL6	V_{Ga}-related	V_{Ga}-related (?)	V_{Ga}-related (?)	

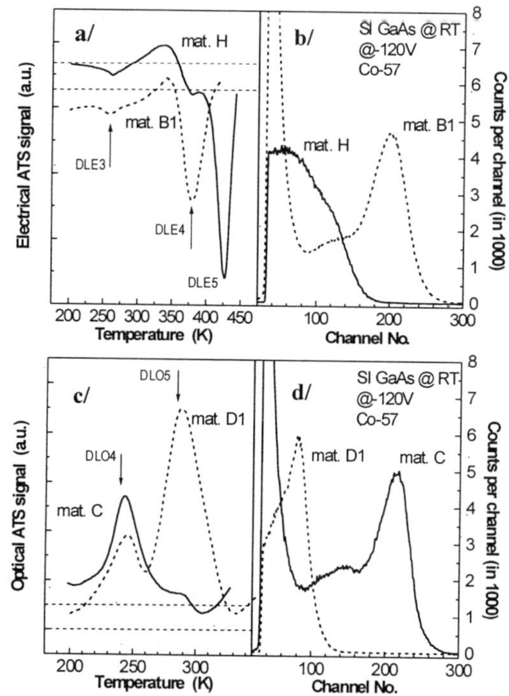

Figure 3: ATS (a) and photon (b) spectra for several semiinsulating GaAs materials (labels ref. to Tab.1).

Another correlation is presented in Figures 2c, 2d. Materials C and D1 exhibit otherwise very similar both optical and electrical ATS spectra except of part displayed here. Material C does not contain deep-level state labeled DLO5, present in D1 in the very high concentration. Position and capture cross section of DLO5 imply a possible significant participation of this level on the capturing of charge carriers generated in detector. Therefore, worse collection efficiency for detector made of material D1 in comparison to material C (peak position at channel No. 80 vs channel No. 220) is observed.

D. Conclusions

A correlation between deep-level states present in SI GaAs and perfor-mance of particle detectors was obser-ved. Namely, Cr-related states and acceptors states with position in the range 0.45÷0.65 eV far from the valence or conductance bands deteriorate collection process. These deep-level states directly contribute to charge carrier capture, or through additional ionization of EL2 level, they have direct impact on overall lifetime of the charge carriers. Clarifying of that requires further analysis.

Acknowledgements - Authors are grateful to M.Krempaský and M.Sekáčová for detector structures preparation. This work was supported in part by the Slovak Grant Agency though grants Nos. 2/5166/98 and 95/5305/509.

References

[1] Proc. of Int. Workshop on *GaAs and Related Compounds*, Nucl.Instr.Meth. **A395** (1997).

[2] F.Nava et al., Nucl. Instrum. and Methods **A349** (1994) 156.

[3] F.Dubecky, *et al*, this proceeding.

[4] D.V.Lang, J. Appl. Phys. **45** (1974) 3023.

[5] F.Dubecky *et al.*, SIMC-VII, eds. C.J.Miner, W.Ford, E.R.Weber (IOP, Bristol 1993) 269.

[6] A.Broniatowski *et al.*, J. Appl. Phys. **54** (1983) 2907.

[7] J.Darmo, PhD thesis (Institute of Electrical Engineering SAS, Bratislava 1996).

[8] M.Rogalla, *et al.*, Nucl. Instrum. and Methods **A395** (1997) 49.

[9] B.K.Jones, J.Santana, M.McPherson, Nucl. Instrum. and Methods **A395** (1997) 81.

III.

Non-Stoichiometric III-V Compounds

Change of electrical and structural properties of non-stoichiometric GaAs through Be doping

M. Luysberg, P. Specht*, K. Thul, Z. Liliental-Weber[†] and E. R. Weber*

Institut für Festkörperforschung, Forschungszentrum Jülich, 52425 Jülich, Germany
** Department of Materials Science, University of California, Berkeley, CA 94720*
[†] Materials Science Dev., Lawrence Berkeley National Laboratory, Berkeley, CA 94720

Abstract: The effect of Be doping on the electrical and structural properties of low temperature MBE grown GaAs (LT-GaAs) is investigated for high Be doping levels. As in undoped LT-GaAs As precipitates form upon annealing. However, from the Ostwald ripening of As precipitates a considerably larger activation energy for As diffusion is found in Be doped samples. Therefore, Be doping retards the growth of As precipitates and leads to a thermal stabilization. The electrical properties of as-grown and annealed samples can be explained by the residual point defect model. After short annealing times highly resistive material is obtained, i.e. the Fermi level is pinned at the deep As_{Ga} donor midgap states.

A- Introduction

Non-stoichiometric GaAs deposited at low temperatures by molecular beam epitaxy (MBE) has been successfully applied in electronic devices making use of its ultrashort carrier trapping times and its high resistivity. The key to understand and to control these unique properties is the incorporation of excess As accommodated by the formation of intrinsic point defects: As antisites As_{Ga}, gallium vacancies V_{Ga} and possibly also As interstitials As_i. In particular the As_{Ga} defect which was shown to be the dominant defect in low-temperature grown GaAs (LT-GaAs) [1] has a major effect on the electrical and optical properties owing to its double donor nature. In the case of undoped LT-GaAs the concentration of As_{Ga} was shown to be adjustable by properly choosing the growth temperature and the As/Ga flux ratio [2]. The As_{Ga} was found to dilate the lattice according to the longer bond length of the As-As bond when compared to the As-Ga bond. A linear correlation of the lattice mismatch and the As_{Ga} concentration was reported [3, 4]. Therefore, the lattice parameter can be used to estimate the As_{Ga} concentration within the non-stoichiometric layers, as long as a growth temperature range of 190°C to 300°C and a BEP ratio range of 10 to 25 is chosen [2]. By use of the magnetic circular dichroism of absorption, a few percent of the As_{Ga} were found to be in the positively charged state [1] requiring compensating acceptors to be present [5]. Indeed, positron annihilation studies revealed up to $10^{18} cm^{-3}$ V_{Ga} [6], known to be a triple acceptor. Recently, it has been demonstrated by comparing the V_{Ga} and the As_{Ga}^{+} concentrations, that gallium vacancies are the dominant acceptors in undoped LT-GaAs [2].

However, the As_{Ga}^{+}/As_{Ga}^{0} ionization ratio can not be controlled independently in undoped LT-GaAs. The charged As_{Ga}^{+} are important for electron trapping, and dominate the carrier trapping time in undoped LT-GaAs [3]. Therefore, the introduction of chemical acceptors should allow to achieve higher ionization ratios. This paper focuses on the effect of doping LT-GaAs with Be acceptors. Recently it has already been demonstrated that the carrier trapping time can be drastically reduced by Be doping [7]. Furthermore, Be was found to have the beneficial effect of thermally stabilizing the As rich material, i.e. despite a large amount of excess As present, no structural changes were observed upon annealing of samples grown at 250°C, in contrast to undoped LT-GaAs [7,8]. In this study, high Be doping levels are utilized to investigate the structural and electrical properties of as-grown and annealed non-stoichiometric GaAs. The effect of moderate Be doping levels on the resistivity and on the As antisite incorporation [9] as well as on the carrier trapping time [10] are also presented in this volume.

B - Experimental

Non-stoichiometric Be doped GaAs films were grown onto an AlAs buffer layer on n^+ GaAs wafers grown by AXT using the vertical gradient freeze technique. A Varian Gen II MBE system was used equipped with a diffuse reflectance spectroscopy setup to measure the substrate temperature. A substrate temperature of 190°C and BEP ratios of 10 and 20 were applied. The layers were doped with 10^{20} cm^{-3} Be. An undoped reference sample was grown for comparison. For isochronal and isothermal furnace annealing the samples were sealed into quartz ampoules containing an 1 bar of As partial pressure at the annealing temperature to prevent As loss during annealing. The annealing was performed in a temperature range from 600°C to 850°C for 30 min to 48 h. Structural characterization was performed by use of a JEOL 4000FX electron microscope operated at 400 kV. The temperature dependence of the electrical conductivity was measured using a standard van der Pauw configuration for thin films with indium contacts spot-annealed for 10 s at 300°C.

C - Results and Discussion

The effect of high Be doping levels (10^{20} cm^{-3}) on the structural changes upon annealing are shown in Figure 1. Undoped and Be doped LT-GaAs samples were annealed in the same quartz ampoule for 30 min at 700°C and 800°C, respectively. A homogeneous distribution of precipitates is observed in all cases. Precipitates were identified to consist of As by use of energy dispersive X-ray analyses and by measuring the Moiré fringes observed after annealing at 700°C. Higher annealing temperatures (e.g. 800°C) result in amorphous precipitates, showing a homogeneous dark contrast in Figs. 1 c and d. This behavior was found to be the same for undoped and Be doped samples. However, the size of the precipitates is significantly smaller in the Be doped case. A considerably smaller precipitate size in p-tye LT-GaAs was previously reported [11].

Quantitative analyses of the precipitate sizes revealed an Ostwald ripening mechanism underlying the growth of the precipitates with increasing annealing temperature. In Figure 2 the logarithm of the cube of the precipitate radius is plotted in an Arrhenius plot. The effective activation energy of As diffusion, deduced from the slope of the curves, is 1.3 eV for the undoped samples. This is in excellent agreement with previous studies [12]. A significantly higher activation energy is obtained for the Be doped samples. Therefore it can be concluded, that the As diffusion is retarded in Be doped material.

A similar conclusion was previously drawn from the annealing behavior of As_{Ga} in LT-GaAs:Be [13], where the IR absorption signal attributed to the As_{Ga} defect was recorded with increasing annealing time obtained from moderately doped samples. The reduced diffusion in the Be doped samples was attributed to a reduced V_{Ga} concentration governing the

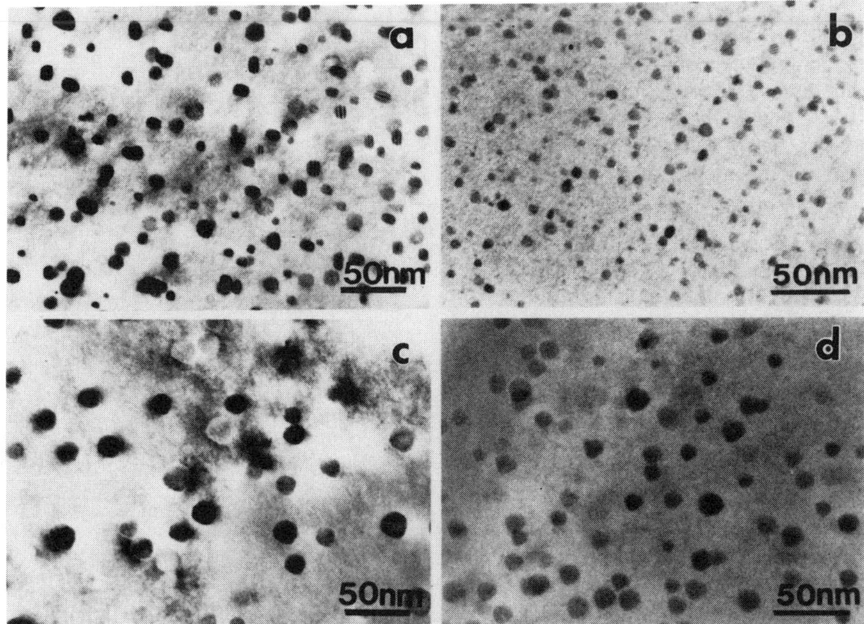

Fig. 1: TEM bright field images showing precipitates in undoped (left column, a and c) and Be doped (right column, b and d) LT-GaAs. Samples were annealed at 700°C (a, b) and 800°C (c, d) for 30 min. A considerably smaller precipitate size is observed in case of Be doping.

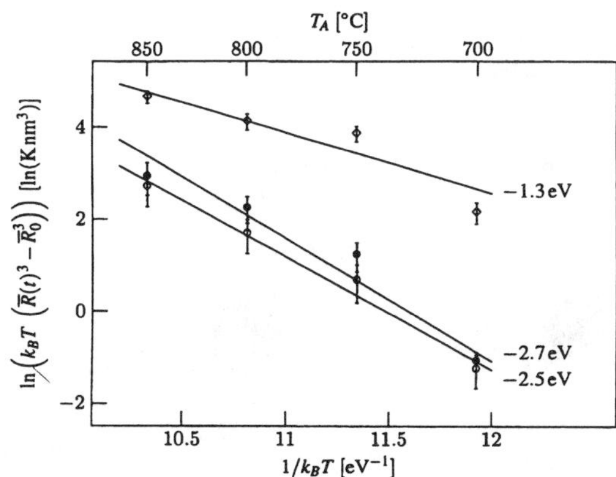

Fig.2: Ostwald ripening of As precipitates. The logarithm of the cube of R (precipitate radius) minus R_0 (initial prec. radius, after 10s rapid thermal annealing at 600°C) plotted as Arrhenius plot. Be doped samples have a considerably higher activation energy (2.6 eV) than the undoped counterpart (1.3 eV).

vacancy assisted diffusion of As_{Ga}. However, first results of positron annihilation studies on our Be doped samples do not indicate a significant lower V_{Ga} concentration compared to undoped material [14]. Therefore, other mechanisms may have to be considered to explain the suppressed As diffusion, like the formation of Be_{Ga}-As_{Ga} complexes. This would require of the complex to allow for the diffusion of the As_{Ga}. On the other hand, it has to be taken into account that a large amount of the As_{Ga} defects are in the positively charged state. Therefore, Coulomb repulsion between the As_{Ga}^+ and positively charged precipitates and/or other As_{Ga}^+ can also explain the low growth of the As precipitates.

The electrical properties obtained after different annealing conditions are shown in Figure 3. The as-grown sample shows a high conductivity. Temperature dependent measurements revealed the typical hopping conduction behavior [15]. Annealing at 600°C for short times (< 6h) leads to a decrease of the conductivity, whereas longer times and/or higher temperatures result in p-conductive samples. A similar annealing behavior of Be doped LT-GaAs was previously reported [11]. Following the Schottky barrier model [16], As precipitates were held responsible for this conduction behavior [11]. This model assumes the As precipitates being surrounded by depleted zones. The overlap of these depleted regions may ex-

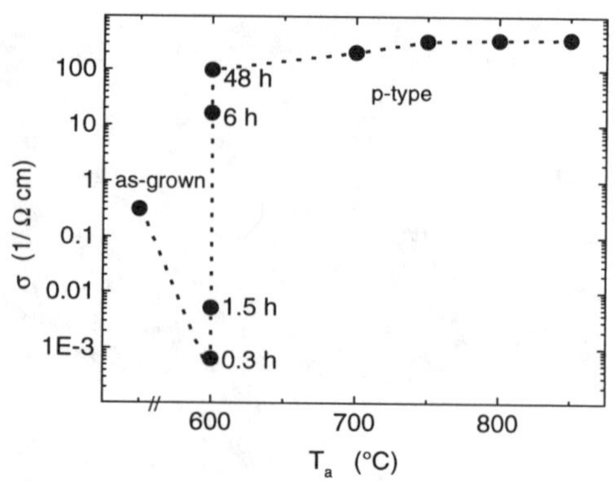

Fig.3: Change of room temperature conductivity of Be doped LT-GaAs upon annealing. The as-grown conductivity is high due to hopping conduction. Short annealing times turn the sample resistive before it finally becomes p-type at longer annealing times.

plain the resistive material obtained after short annealing times, where the precipitate density is highest [15]. However, by quantitatively measuring the sizes and densities of precipitates in our samples, we obtain the volume of the depleted zones to be at most 2.6% [15], considering

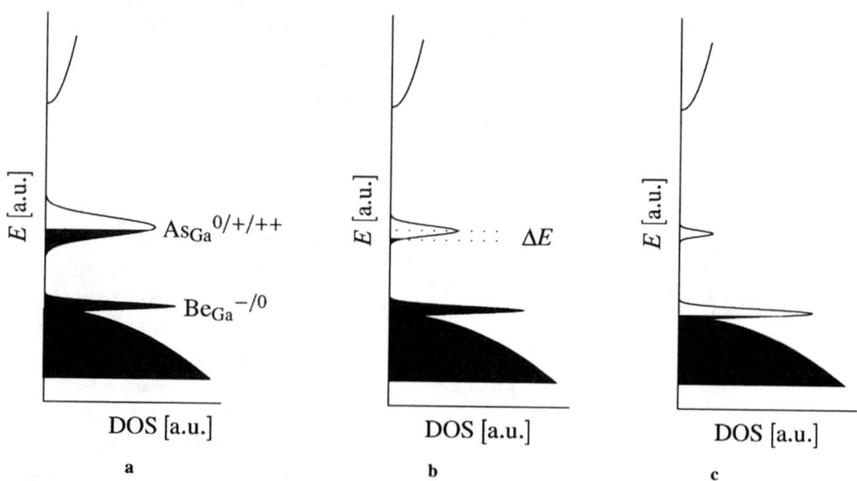

Fig.4: Density of states diagram for electrons for as-grown material (a), after short annealing times (600°C,<6h) (b), and after long annealing times and/or higher temperatures (c). The 0/+ and the +/++ band of the As_{Ga} donors are merged for simplicity, the V_{Ga} acceptor band is omitted, since $[V_{Ga}]$ is much smaller as [Be]. The Fermi level remains pinned by the deep midgap states until after long annealing times the sample become p-tye, i.e. the Fermi level is close to the valence band.

a maximum measured acceptor concentration of 5×10^{19} cm^{-3} and assuming a Schottky barrier height of 0.7 eV [16]. We conclude that the buried Schottky barrier model is not suitable to describe the conduction behavior in highly Be doped LT-GaAs.

Instead, residual defects have to be taken into account, as shown schematically in Figure 4. According to this model, the as-grown sample is characterized by a high As$_{Ga}$ concentration with a large fraction of As$_{Ga}^{+}$ (corresponding to the Be acceptors) pinning the Fermi level midgap (Fig.4a). Annealing the samples only shortly results in a decrease of the As$_{Ga}$ concentration (Fig. 4b). However, there are still enough midgap defects present to pin the Fermi level in the lower half of the band gap. According to the reduced defect concentration, hopping conductivity becomes less dominant and the resistivity rises. Only large annealing times and/or higher annealing temperatures make the material p-type, since finally the As$_{Ga}$ concentration becomes smaller than the Be acceptor concentration (Fig.4c).

The facts that the As diffusion is retarded by Be doping and that, despite very high Be doping levels, the material becomes resistive, has a major impact on the application of LT-GaAs:Be in devices. Using more moderate doping levels and/or higher growth temperatures it has been demonstrated, that semi-insulating material can be grown, which is thermally stable [8] (see also [9,10]). Furthermore, the high concentration of ionized antisites governed by the Be doping level leads to a reduction in carrier trapping time. Recently it was shown that a highly resistive ($>10^5$ Ωcm), thermally stable material with ultrashort carrier trapping times ($<$1ps) can be obtained [8].

D - Conclusions

Isothermal and isochronal annealing studies of highly Be doped non-stoichiometric GaAs revealed an higher activation energy of As precipitate formation following the Ostwald ripening mechanism. Therefore, Be retards the As diffusion in LT-GaAs. Whether a reduced V$_{Ga}$ concentration acting as diffusion vehicle, the formation of As$_{Ga}$-Be$_{Ga}$ complexes or Coulomb repulsion between As$_{Ga}^{+}$ and precipitates is responsible for this behavior can not be decided at present. The reduction in As$_{Ga}$ concentration upon thermal annealing has a drastic influence on the electrical conduction behavior. Whereas the as-grown material is characterized by hopping conduction, short annealing times turn the material highly resistive, i.e. the Fermi level is still pinned at the midgap defect states. Only longer annealing times or higher temperatures decrease the As$_{Ga}$ concentration below the Be acceptor level, resulting in p-type conduction. The thermal stabilization and the high resistivity make Be-doped non-stoichiometric GaAs offering new prospects in device applications, such as the elimination of AlAs diffusion barriers or improvement of the radiation hardness.

Acknowledgments

This work has been partly funded by the AFOSR grants no. F49620-97-1-0220 (electrical characterization) and AFOSR-ISSA-90-009 (TEM investigations). MBE growth was performed at the Integrated Materials Lab at UC Berkeley. One of the authors (M. L.) gratefully acknowledges the Feodor-Lynen fellowship granted by the Alexander von Humboldt Association.

References

[1] X. Liu, A. Prasad, W. M. Chen, A. Kurpiewski, Z. Liliental-Weber, and E. R. Weber, Appl. Phys. Lett. 65 (1994) 3002.

[2] M. Luysberg, H. Sohn, A. Prasad, P. Specht, Z. Liliental-Weber, and E. R. Weber, J. Appl. Phys. 83 (1998) 561.

[3] Z. Liliental-Weber, J. Ager, D. Look, X. W. Lin, X. Liu, J. Nishio, K. Nichols, W. Schaff, W. Swider, K. Wang, J. Washburn. E. R. Weber, and J. Whitaker, in: *Proc. of the 8th Conf. on Semiinsulating III-V Materials,* ed. M Godlewski (World Scientific, 1994) 305.

[4] M. Luysberg, H. Sohn, A. Prasad, H. Fujioka, R. Klockenbrink, and E. R. Weber in: *Semiconducting and Insulating Materials,* ed: C. Fontaine, (IEEE Piscataway NJ 1996) 21.

[5] X. Liu, A. Prasad, J. Nishio, E.R. Weber, Z. Liliental-Weber, and W. Walukiewicz, Appl. Phys. Lett., 67 (1995) 279.

[6] J. Gebauer, R. Krause-Rehberg, S. Eichler, M. Luysberg, H. Sohn, and E. R. Weber, Appl. Phys. Lett. 71 (1997) 638.

[7] P. Specht, S. Jeong, H. Sohn, M. Luysberg, A. Prasad, J. Gebauer, R. Krause-Rehberg, and E. R. Weber, Mater. Sci. Forum 258-263 (1997) 951.

[8] R. Lutz, P. Specht, R. Zhao, S. Jeong, J. Bokor, and E. R. Weber, accepted for publication in the Proceedings of the MRS Spring Meeting (1998).

[9] R. Lutz, P. Specht, R. Zhao, and E. R. Weber, this volume.

[10] R. Zhao, P. Specht , R. Lutz, E. R. Weber, S. Jong, J. Bokor, this volume.

[11] N. Atique, S. Harmon, J. C. P. Chang, J. M. Woodall, M. R. Melloch, and N. Otsuka, J. Appl. Phys. 77 (1995) 1471

[12] Z. Liliental-Weber,X. W. Lin, J. Washburn, and W. Schaff, Appl. Phys. Lett. 66 (1995) 2086.

[13] D.E. Bliss, W. Walukiewicz, J. W. Ager, E. E.Haller, K. T. Chan, and S. Tanigawa, J. Appl. Phys. 71 (1992) 1799.

[14] J. Gebauer, F. Börner, and R. Krause-Rehberg, priv. communication

[15] K. Thul, M. Luysberg, P. Specht, E. R. Weber, and Z. Liliental-Weber to be published.

[16] A. C. Warren, J. M. Woodall, J. L. Freeouf, D. Grischowsky, D. T. McInturff, M. R. Melloch, and N. Otsuka, Appl. Phys. Lett. 57 (1990) 1331.

LT GaAs delta-doped with In: enhanced As excess, In-Ga intermixing, As cluster array ordering

N.A. Bert, V.V.Chaldyshev, A.A. Suvorova, N.N.Faleev, A.E. Kunitsyn, Yu.G. Musikhin,
Ioffe Physical-Technical Institute, 194021 St.Petersburg, Russia,

V.V. Preobrazhenskii, M.A. Putyato, B.R. Semyagin,
Institute of Semiconductor Physics, 630090 Novosibirsk, Russia,

P. Werner,
Max-Planck-Institut fur Mikrostrukturphysik, D-06120 Halle, Germany

The effects of indium incorporation in GaAs grown by MBE at low substrate temperature (LT GaAs) have been studied. The 0.04% indium doping is shown to provide an increase in the As excess along with improved crystal quality. The In-Ga intermixing is found to enhance by almost two orders of magnitude due to a high point defect concentration in LT GaAs. The effective activation energy of the intermixing is determined from TEM measurements to be 1.1 eV. The As precipitation kinetics is found to be quicker in In-free regions than within the In-containing layers. Using these observations, we successfully formed a 30 nm period perfect superlattice consisting of thin (double cluster diameter) As cluster sheets separated by cluster-free LT GaAs spacers.

A. Introduction

Nonstoichiometric GaAs grown by MBE at low substrate temperature (LT GaAs) exhibits unique electrical and optical properties as well as device applications [1-4]. It contains up to 10^{20} cm^{-3} excess As atoms which precipitate on annealing in the form of nm-sized As clusters. Since both the excess-arsenic-related point defects and clusters are electrically active it is of importance to control their concentration and spatial distribution. We show that isovalent In impurity doping is a suitable tool to influence excess As concentration and to control the spatial distribution of As clusters.

B. Experimental

LT GaAs films undoped or doped with either Be or Si were grown in a dual-chamber "Katun" MBE system on 2-inch GaAs (001) substrates at 200 °C. Indium was additionally introduced in most of the films either as a randomly distributed isovalent impurity with a concentration of 0.04 or 0.2 at.% or as delta-layers. The indium content in the delta-layers was varied from 0.01 to 1 monolayer (ML). Each grown sample was cut into four parts, and three of them were annealed in the MBE chamber under As overpressure for 15 minutes at three different temperatures in the range 400 °C - 810 °C. The growth procedure is given in more detail elsewhere [5-6]. Electron-probe microanalysis (EPMA), X-ray diffraction (XRD), transmission electron microscopy (TEM), and near infrared optical absorption (NIRA) measurements were employed to characterize the samples.

C. Results and discussion

X-ray rocking curves recorded from the as-grown samples (fig. 1) showed an increased lattice expansion for the indium doped films as compared with those indium-free. Only a small amount of this increase can be attributed to In since its concentration is as low as 0.04 at.%. Near-infrared optical absorption (attributed to arsenic antisite defects) was observed in all the as-grown samples but it was markedly stronger in the In-doped films. Fig. 2 shows NIRA spectra for various samples. Comparing these spectra, we conclude that In doping leads to an increased As excess equivalent to a lowering by 20°C of the growth temperature. At the same

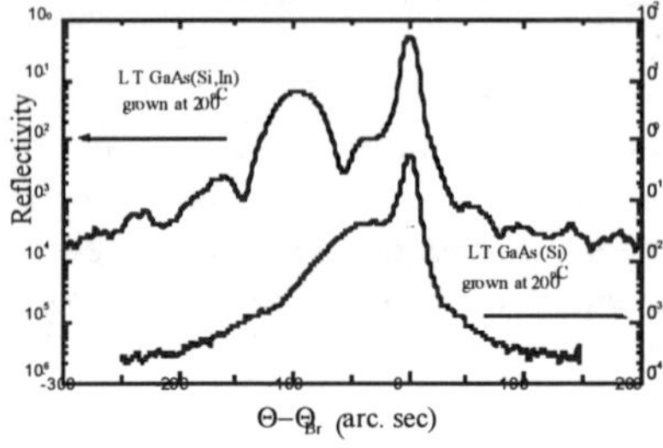

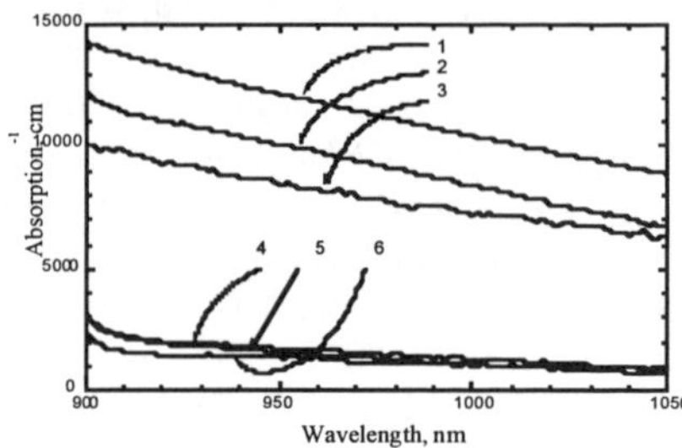

Fig.1. X-ray rocking curves near GaAs (004) CuK$_{\alpha 1}$ reflection for as-grown LT GaAs:Si films doped additionally with In and In-free.

Fig.2 Optical absorption spectra of LT GaAs:Si films grown at 150 (1) or 200°C (2, 3, 4, 5, 6) with (2, 4, 5, 6) and without (1, 3) indium. Anne-aling temperature: 500 (4), 600 (5), 700°C (6).

time, the interference pattern on the rocking curve (fig. 1) corresponding to In-doped sample is not seen on the In-free sample. This evidences an improved film quality with indium doping. Thus, the presence of isovalent indium impurity results in an enhanced As excess and a better crystalline quality of LT GaAs films.

In our previous experiments [5] we succeeded in obtaining preferable As precipitation on In delta-layers in LT GaAs films either undoped or doped with Si or Be. However, In-Ga intermixing has to happen in parallel with As diffusion and precipitation, and it may influence the efficiency of As cluster accumulation at In delta-layers. To evaluate the indium-gallium interdiffusion in LT GaAs delta-doped with In, the as-grown and annealed samples were studied by TEM. The micrograph taken from an as-grown sample in high-resolution mode is shown in fig.3. It reveals the In-containing layer thickness to be 4 ML (i.e. 1.12 nm) instead of 0.5 ML expected from the growth procedure. Since the In-Ga interdiffusion seems to be of low probability at a temperature as low as 200 °C, we attribute this apparent thickening to a fine-scale surface roughness of the growing film. The conclusion of negligible contribution of In-Ga interdiffusion is supported by the fact that the measured In-containing layer thickness is not affected by nominal indium content in the indium delta-layers (0.5 or 1 ML). After annealing, the In-containing layers became thicker due to In-Ga intermixing. The measured thicknesses for various annealing temperatures are given in the table. Under the assumption that the initial In distribution in the as-grown sample is described by Gauss error function the solution of the conventional diffusion equation is also Gaussian of which the dispersion σ is connected with the diffusivity $D_{In\text{-}Ga}$ through the initial dispersion σ_0 and annealing time t as

$$2D_{In-Ga}(T)t = \sigma^2 - \sigma_0^2 \; . \tag{1}$$

σ (and σ_0) can be calculated numerically from the measured delta-layer thickness w using the equation

$$\frac{w^{22}}{8\sigma} = \ln \frac{x_0}{x_{th}} - \ln \frac{\sqrt{2\pi}\sigma}{d_{002}} \tag{2}$$

(x_0 - nominal In mole fraction, d_{002} - monolayer thickness, i.e. 002 interplanar distance) if the In mole fraction x_{th} corresponding to the apparent layer boundary is known.

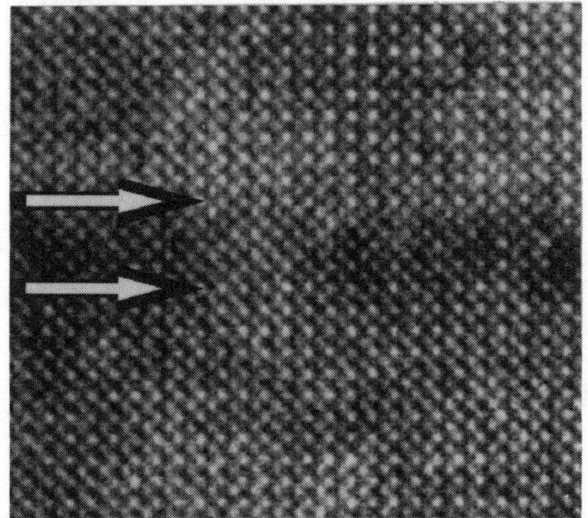

Fig. 3. HREM image of as-grown LT GaAs film with
nominal 0.5 ML indium delta-layer (arrowed).

Table. The measured In-containing layer thicknesses for various annealing temperatures

	Layer thickness, ML	
nominal	0.5	1.0
as-grown	4	4
annealed at 500°C	6	8
annealed at 600°C	12	15
annealed at 700°C	21-22	-

We evaluated the smallest indium mole fraction x_{th} observable in TEM to be 0.005. A sample with nominal indium deposit decreasing layer by layer from 0.5 ML down to 0.006 ML was used for this evaluation. Then, the In-Ga interdiffusion diffusivity was deduced from the indium-containing layer thicknesses measured for various annealing temperatures. Fig. 4 demonstrates an Arrenius plot of the In-Ga interdiffusion diffusivity versus annealing temperature. An effective activation energy of 1.1 eV is obtained. We propose the 0.7 eV difference between the Ga vacancy migration enthalpy [7-9] and experimental effective activation energy to be the Ga vacancy annihilation enthalpy following I. Lahiri et al. [10].

Despite the intermixing, In delta-layers continue to accumulate As clusters upon annealing, but precipitation kinetics for three-dimensional cluster arrays in between the delta-layers has been found to be faster than that for two-dimensional cluster arrays on delta-layers. This difference in the precipitation kinetics results in that the As cluster ensemble becomes random when annealing duration is too long or annealing temperature is too high. In our case, it happens when the annealing temperature is elevated to 700°C.

Using these observations, we successfully formed a 30 nm period perfect superlattice consisting of thin As cluster sheets located at In delta-layers which are separated by cluster-free LT GaAs spacers. Fig. 5 exhibits the LT As/GaAs superlattice formed by annealing at 500°C.

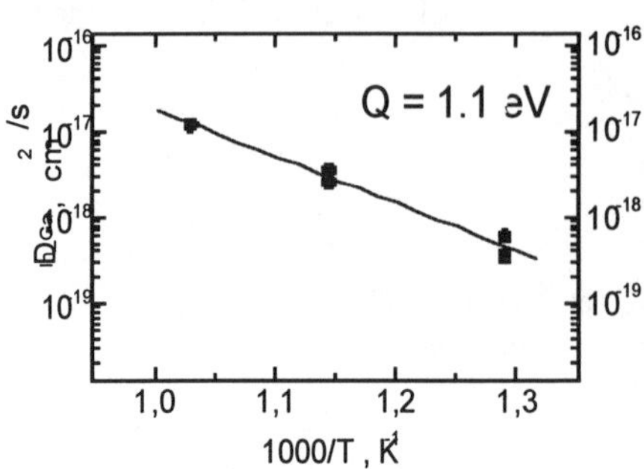

Fig. 4. Arrenius plot of effective In-Ga interdiffusion diffusivity versus temperature.

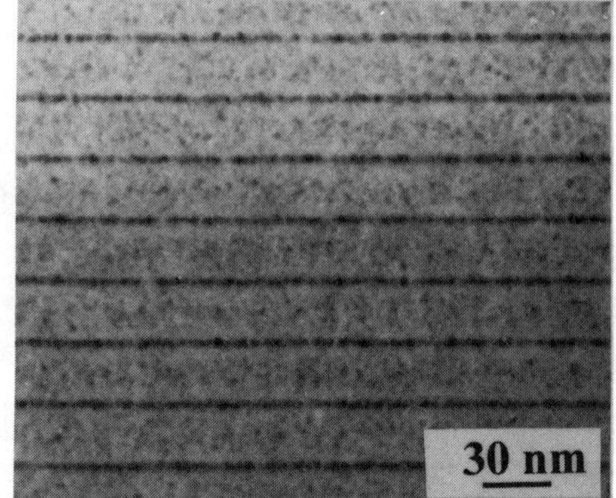

Fig. 5. TEM micrograph of LT As/GaAs superlattice grown at 200°C and annealed at 500°C for 15 min.

Acknowledgments

The work is carried out under the programs Physics of Solid State Nanostructures and Fullerens and Atomic clusters of the Ministry for Sciences and Technologies of Russia. The research is supported by RFBR. The DFG grant 436 RUS 113/322/0(R,S) is greatly acknowledged.

References

1. F.W. Smith, A. R. Calawa, Chang-Lee Chen, M. J. Mantra, L. J. Mahoney, IEEE Electron Devices Lett. 9, 77 (1988)
2. M. Kaminska, Z. Liliental-Weber, E. R. Weber, T. George, J. B. Kotright, F. W. Smith, B.-Y. Tsang, A. R. Calawa, Appl. Phys. Lett. 54, 1831 (1989)
3. M. R. Melloch, K. Mahalingam, N. Otsuka, J. M. Woodal, A. C. Warren, J. Cryst. Growth 111, 39 (1991)
4. M. R. Melloch, D. C. Miller, B. Das., Appl. Phys. Lett. 54, 943 (1989)
5. N. A. Bert, V. V. Chaldyshev, N. N. Faleev, A. E. Kunitsyn, D. I. Lubyshev, V. V. Preobrazhenskii, B. R. Semyagin, and V. V. Tret'yakov, Semicond. Sci. Technol. 12, 51 (1997).
6. N. A. Bert, V. V. Chaldyshev, A. E. Kunitsyn, Yu. G Musikhin, N. N. Fallev, V. V. Tretyakov, V. V. Preobrazhenskii, M. A. Putyato, B. R. Semyagin, Appl. Phys. Lett. 70, 3146 (1997)
7. J.-L. Rouviere, Y. Kim, J. Cunningham, J.A. Rentschler, A. Bourret, and A. Ourmazd, Phys. Rev. Lett. 68, 2798 (1992)
8. D. E.Bliss, W. Walukiewicz, J.W. Ager, E.E. Haller, K.T. Chan, and S. Tanigawa, J. Appl. Phys. 71, 1699 (1992)
9. S. Y. Chiang and G.L. Pearson, J. Appl. Phys. 46, 2986 (1975)
10. I. Lahiri, D.D. Nolte, M.R. Melloch, J.M. Woodal, W. Walukiewicz, Appl. Phys. Lett. 69, 239 (1996)

The influence of growth and annealing conditions on the structural quality of LT-GaAs layers grown on (111)B substrate by molecular beam epitaxy

C. Guerret-Piécourt[1], M. Toufella[2,3], E . Bedel[1], P. Puech[3], G. Benassayag[2], V. Bardinal[1], C. Fontaine[1], A. Claverie[2], R. Carles[3]

[1] *LAAS/CNRS, 7 Avenue du Colonel Roche 31077 Toulouse Cedex 4, France*
[2] *CEMES/CNRS, 29 Rue Jeanne Marvig 31055 Toulouse Cedex 4, France*
[3] *LPST, Université Paul Sabatier, 118 route de Narbonne, 31062 Toulouse Cedex 4, France*

Abstract : The structural quality of the layers grown at low temperature on (111)B GaAs is investigated by Raman spectroscopy and transmission electron microscopy. It is shown that both the amorphous layers grown at 150°C and the defect-rich layers (mostly based on multiple twinning) grown at 250°C can recover perfect crystalline quality upon annealing provided, i) they are not too thick (<350 nm) and, ii) the annealing temperature is high enough. Both conditions are necessary to allow the reordering of the layer initiated at the interface to propagate towards the surface without being stopped at dislocation nodes.

A) Introduction

The growth of LT (Low Temperature) GaAs on (100) surfaces is now quite well documented and various applications have been demonstrated. The large « excess As » concentrations found in this material either as antisite defects or within precipitates lead to unusual properties such as high resistivity and very short carrier lifetime. These properties are used for buffer layers, photoconductive switches and also for blocking layers in vertical cavity surface emitting lasers. On the opposite, much less is known on the growth at low temperature of GaAs layers on different surfaces such (111), (311)..., although specific physical properties are expected for such As-rich materials.

For this reason we have undertaken a systematic study of the structural properties of GaAs layers grown on semi-insulating GaAs(111)B substrates for a wide variety of growth temperatures (150-250°C), layer thicknesses (0.1-1µm) and annealing conditions (600-700°C). The results presented here contrast with the one published by Cheng et al. (1) and lead to a more optimistic view than the one previously reported by O'Hagan et al.(2) on the possiblity to obtain defect free As-rich GaAs grown on (111)B GaAs.

B) Experimental details

A series of 0.1-1 µm thick undoped LT-GaAs layers were grown on an optimized buffer layer (3) by Molecular Beam Epitaxy in a Riber 2300 system. The growth rate was fixed at 0.7µm/h with an As4/Ga Beam Equivalent Pressure (BEP) ratio of 24. Each sample was cleaved into several pieces subjected to different *in situ* annealings: 600°C for 30min., 700°C for 30 min. and 700°C for 2 hours. These samples and those in the « as grown » state were then analyzed by TEM, HREM, Raman and photoluminescence spectroscopies. Raman spectra were performed in the usual back-scattering configuration using a micro-Raman XY Dilor with 1800 groves/mm gratings. The samples were also prepared for « cross-sectional » TEM imaging using the standard mechanical thinning followed by ion milling procedure until electron transparency. The images were obtained either on a CM 30 Philips microscope with a resolution of 0.2 nm or on a CM 20.

C. Results

C1. as grown material

The structure of the layers was found to drastically depend on the growth temperature. The layers grown at 150°C are amorphous. This observation is quite similar to what has been reported for (100) LT-growth under similar conditions by Eaglesham et al.(4). The layers grown at 200°C are highly defective and show a columnar growth involving twinning on the different sets of {111} planes of the crystal which, for large thicknesses, turns the material into a polycrystalline state. More interesting are the layers grown at 250°C. Figures 1 (a-c) show the structure of these layers for increasing thicknesses. For thicknesses up to 100 nm, the as-grown layer consists of a random stacking of (111) planes parallel to the buffer surface. The electron diffraction pattern shows a series of lines characteristics of this type of « one-dimensional » disordering. When increasing the thickness up to 350 nm (Fig.1b), twinning of the other {111} planes occurs as evidenced by the electron diffraction pattern taken close to the surface. For 1 µm thick specimen, the region close to the surface consists of small disoriented grains as evidenced by the « rings » in the electron diffraction pattern.

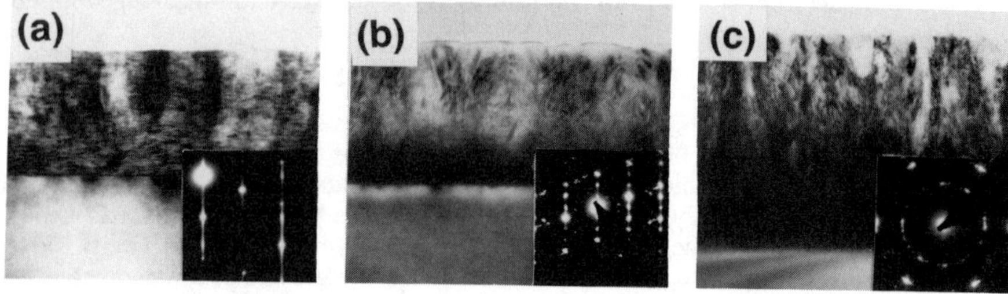

Fig. 1 : All as-grown layers at Tg=250°C. a) t=100 nm, b) t=350 nm and c) t=1 µm.
Inset are the associated diffraction patterns taken close to the surface

X-ray diffraction experiments show that the lattice expansion in the direction normal to the surface is 0.02 % for a growth temperature of 250°C. This agrees well with preceeding observations of a much lower incorporation rate of As on (111) than on (100) surfaces (1, 2).

C2. Annealed materials

Fig.2 is a set of Raman spectra obtained in crossed configuration on the 1 µm thick material grown at 250°C and various annnealing conditions. In this configuration, only the TO mode is allowed for (111) oriented defect free layers, the LO activation originates from a random misorientation of the scattering events (5). Calculations show that the polarization of the activated LO mode is in the plane of the layer and not in the growth direction. As the polarization of the TO and LO modes are in the same plane, the ratio of LO(forbidden)/ TO(allowed) intensities gives quantitative information about the amount of residual defects in the layers without any calibration. From Fig. 2, it is clearly seen that almost perfect recovering of the crystalline quality occurs for annealing at 700°C for 2h.

Fig. 3 is a set of XTEM images showing the same layers. Before annealing, the layer consists of (from the buffer to the surface) a region dominated by twinning of the (111)B planes parallel to the surface followed by an intermediate region where twinning also occurs on the

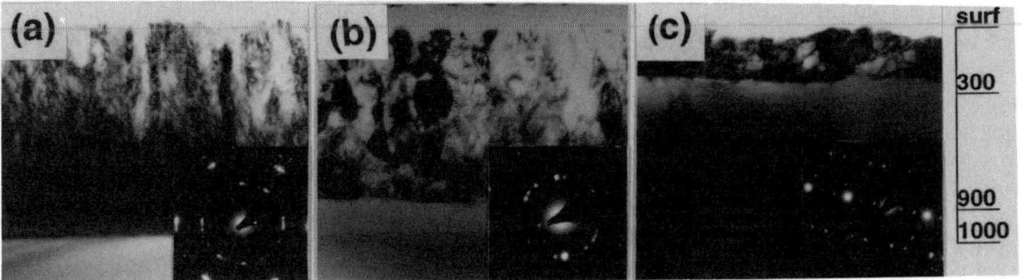

Fig. 3 : Tg=250°C, thickness = 1 μm. a) as grown, b)after 600°C, 30 min. and c) after 700°C, 2h.
Inset are the associated diffraction patterns taken close to the surface

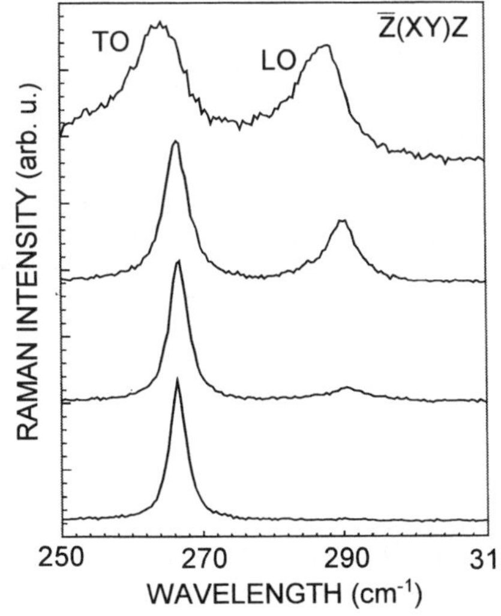

Fig. 2. : Set of Raman spectra obtained on the 1 μm-thick layers grown at 250°C. a) as grown, b) after 600°C, 30 min. c) after 700°C, 2 h and d) reference (buffer).

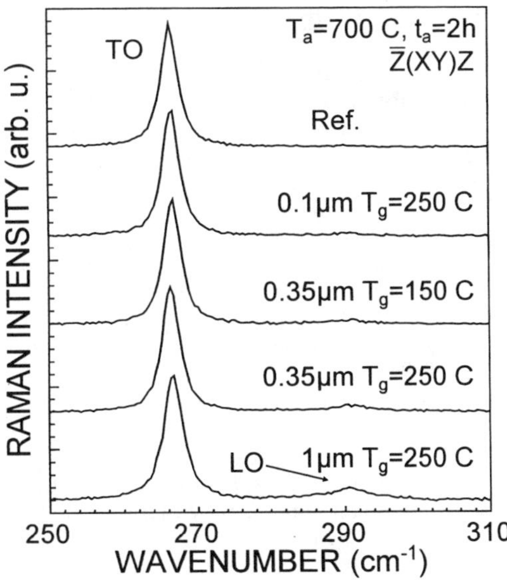

Fig. 4 : Set of Raman spectra showing the recovery of crystallinity obtained after specific annealing.

other {111} planes and finally, due to the non-homogeneous distribution of dislocations, to slightly misoriented grains and polycrystalline growth. Upon annealing, reordering of the layer occurs from the interface and towards the surface. After 30 minutes at 600°C, a 100 nm thick monocrystalline layer has regrown from the buffer. Increasing the annealing duration does not increase this thickness. Conversely, annealing at 700°C results in a 700 nm regrowth. However, the interfaces between the defective and the regrown layers always consist of multiple stacking faults systems bounded by complex dislocations. It appears that the recovery of the crystalline quality depends on the ability of the interface to anihilate or emit out the dislocations that prevent the re-stacking of the LT layer. This hypothesis is consistent with the observation that for thinner layers the full recovery of the LT layers can be obtained at some given temperature. After a 700°C 2h annealing, all layers except the 1 µm thick ones are monocrystalline : this is observed by TEM but also in the Raman spectra of Fig. 4. It is worth to note that annealing of thick layers at 600°C for increasing times results in the progressive de-twinning of the material within the remaining grains.

While the presence of As precipitates is an important characteristics of (100) LT-grown GaAs, they were rarely observed in (111) layers and always pinned on steps at some twin plane or at dislocation nodes consistently with the X-rays findings of small As excess concentrations.

D. Conclusions

In the as-grown state, we observe a clear transition from the amorphous state at 150°C to a polycrystalline structure at 200°C then to a twinned structure at 250°C. Stacking faults parallel to the growth front are the basic defects in these layers. X-ray measurements show that, as previously reported, the lattice expansion in the LT layers is much less than observed on (100) for the same growth conditions.

After annealing, the recovery of the perfect crystalline quality of the LT-layers has been found to be possible under specific conditions.

These results are discussed and specific features of the growth of (111)B LT-GaAs are evidenced. Arsenic incorporation is much efficient on (111) surfaces than on (100) and the breakdown of crystalline quality occurs for very small amounts of stress. While monocrystalline as-grown layers have not been produced, annealing of these layers can restore a perfect crystalline quality. It is shown that this recovery depends on the ability of the stacking faults to anneal out from the bottom to the surface of the material.

Finally, Raman spectroscopy can be used as a non-destructive technique for the characterization of (111) LT-GaAs layers since the ratio of the LO/TO intensities is a good indication of the recovery of these layers.

This work was partly supported by the Conseil Régional Midi-Pyrénées. Thanks to EPFL (Switzerland) and Prof. Ilegems for providing X-rays facilities.

References :
1) T. M. Cheng and C. Y. Chang and J. H. Huang, Appl. Phys. Lett. 66 (1), (1995), 55.
2) S. O'Hagan and M. Missous, J. Appl. Phys. 82(5), (1997), 2400.
3) C. Guerret and C. Fontaine, J. Vac. Sci. Technol. B, 16 (1998) 204
4) D. J. Eaglesham, L. N. Pfeiffer, K. W. West, Appl. Phys. Lett. 58, (1991), 65.
5) P. Puech, E. Daran, G. Landa, R. Carles, P. S. Pizani et C. Fontaine, J. Appl. Phys. 77, (1995), 1126.

Ultrafast response times and enhanced optical nonlinearity in annealed and Beryllium doped low-temperature grown GaAs

M. Haiml, A. Prasad, F. Morier-Genoud, U. Siegner, U. Keller, and E. R. Weber [a]

Swiss Federal Institute of Technology Zurich, Institute of Quantum Electronics,
ETH Hönggerberg HPT, CH-8093 Zurich, Switzerland
(a) Department of Material Science and Mineral Engineering,
University of California, Berkeley, CA 94720

We have quantitatively studied the optical nonlinearity in annealed and in Beryllium doped low-temperature grown (LT) GaAs. This study has been complemented by measurements of the temporal decay of the absorption modulation on the 20-fs time scale. Our results show that both Be doping and annealing of LT GaAs yield a material with superior properties for ultrafast all-optical switching applications.

A. Introduction

Ultrafast all-optical switching applications based on the absorptive nonlinearity of a semiconductor require a material with fast response time and high absorption modulation. When the semiconductor is used as a saturable absorber inside a laser cavity, another important aspect is the reduction of the absorption losses which remain even at the highest pulse fluences ("non-saturable losses") [1]. With regard to the optical nonlinearity, we will focus on the continuum states rather than on the excitonic nonlinearity [2], which does not provide enough bandwidth for switching with broadband 10-fs optical pulses.

The incorporation of arsenic antisites, As_{Ga}, in GaAs during low-temperature (LT) growth results in ultrafast recovery times of the absorption modulation due to enhanced carrier trapping [3]. However, as will be shown in this paper, the ultrafast time response is obtained at the price of small absorption modulation and large non-saturable losses in undoped as-grown LT GaAs. The electronic structure and the optical properties of LT GaAs can be changed by doping with Beryllium or by annealing. Doping with Be acceptors results in a reduction of the fraction of neutral As_{Ga} [4] while annealing reduces the total density of As_{Ga} [5] due to the formation of As precipitates [6]. We will demonstrate in this paper that Be doping or annealing of LT GaAs yields a material with ultrafast response, high absorption modulation, and small non-saturable losses. Therefore, LT GaAs:Be and annealed LT GaAs are superior to undoped as-grown LT GaAs in ultrafast all-optical switching applications. We will relate the improvement of the nonlinear optical properties to the changes of the electronic structure.

B. Experimental

We have measured the temporal decay of the absorption modulation for above-bandgap excitation in standard pump-probe experiments with 20-fs pulses centered at 810 nm or with 100-fs pulses centered at 830 nm. The modulation $\Delta T = T_s - T_0$ and the non-saturable losses $T_{ns} = 1 - T_s$ (T_0 linear transmission, T_s saturated transmission at very high pulse fluences) are obtained from single-pulse bleaching experiments where the incident and the transmitted fluence of a 100-fs laser pulse at 830 nm are measured over a wide range of incident fluences. The transmission T is given by the transmitted over the incident fluence. The measured range is extrapolated to the linear regime and to very high fluences using a fit based on the travelling wave rate equation for a two-level absorber [7, 8]. In Fig. 1, typical experimental and calculated curves are shown to illustrate the definition of the parameters. All experiments have been carried out at room temperature.

We have studied 1 μm thick GaAs samples grown by MBE at different growth temperatures and a BEP of 20. Annealing was performed in situ at 600 ^{0}C for 1 hour .

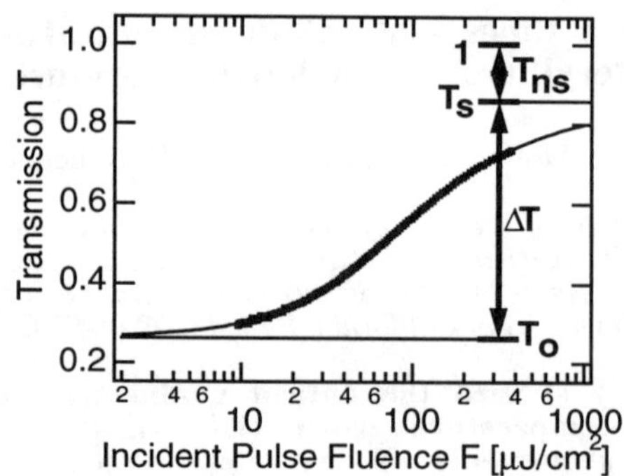

Fig. 1: Single-pulse bleaching experiment: transmission T vs incident pulse fluence. Dots: measured signal, thin line: calculated extrapolation. T_0 linear transmission, T_s saturated transmission, ΔT modulation, T_{ns} non-saturable losses.

C. Results and discussion

The results of pump-probe experiments with 20-fs excitation pulses are shown in Fig. 2a for different LT GaAs samples. We quantify the initial decay of the pump-probe signals by the 1/e decay time, which is plotted vs. growth temperature T_g in Fig. 2b. This plot clearly shows that Be doping of the 300 ^{0}C grown sample leads to a decay with an 1/e time of 110 fs which is much faster than the decay in the 300 ^{0}C grown undoped sample and almost as fast as the decay in the 250 ^{0}C grown undoped sample. Annealing makes the response slightly slower for T_g = 250 ^{0}C but yields a faster response for T_g = 300 ^{0}C. The decrease of the initial response time due to Be doping and its dependence on annealing show that the initial decay is strongly affected by carrier trapping besides contributions from intraband relaxation.

Pump-probe experiments with 100-fs pulses reconfirm that Be doping decreases the initial decay time and show the same dependence on annealing as the 20-fs experiments (data not shown). However, the decay times are in the picosecond range. We conclude that trapping is slowed down on longer time scales due to partial trap filling, which reduces the density of free traps giving rise to smaller trapping rates. Moreover, in this regime slow recombination processes start to affect the decay of the pump-probe signal in LT GaAs [9].

The results of the single-pulse bleaching experiments are summarized in Fig. 3. We find that, in undoped as-grown LT GaAs, the modulation drastically decreases while the non-saturable losses drastically increase if the growth temperature is lowered below 300 ^{0}C. This is the growth temperature range in which a fast 100-fs response is obtained, see Fig. 2b. Consequently, undoped as-grown LT GaAs with a fast response time will always suffer from large non-saturable losses and small modulation. Figs. 3a and 3b show that Be doping reduces the non-saturable losses and maintains a high modulation or even increases it. Importantly, the 300 ^{0}C grown Be doped sample with the ultrafast 100-fs response time has a high modulation and small non-saturable losses, much better than the 250 ^{0}C grown undoped sample with the 100-fs response time. This comparison shows that Be doped LT GaAs is superior to undoped LT GaAs in ultrafast all-optical switching applications.

Likewise, annealing of LT GaAs grown at T_g < 300 ^{0}C considerably improves the modulation and the non-saturable losses, as shown in Figs. 3c and 3d. This is accompanied by only a very small increase of the response time, see Fig. 2b. Therefore, annealing of LT GaAs grown at T_g < 300 ^{0}C yields an improved material for applications in ultrafast optics. For

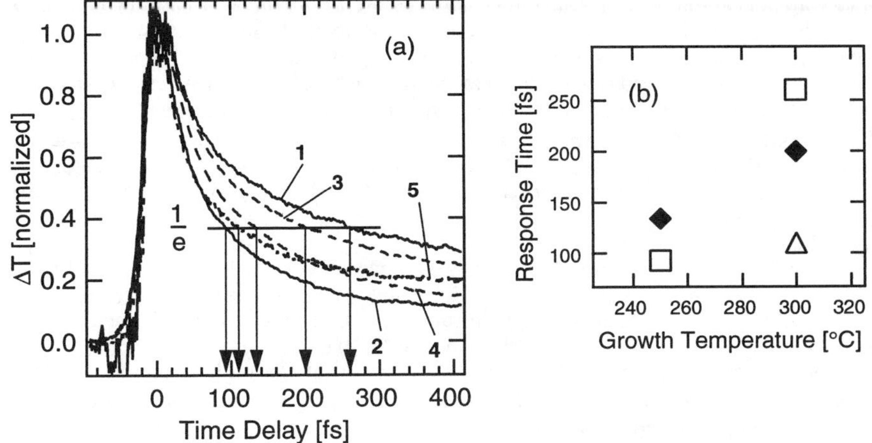

Fig. 2: (a) Pump-probe signal obtained with 20-fs pulses (carrier density 5×10^{17} cm^{-3}). (b) 1/e decay time vs. growth temperature T_g. (1, open square) $T_g = 300°C$, as-grown, undoped; (2, open square) $T_g = 250°C$, as-grown, undoped; (3, filled diamond) $T_g = 300°C$, annealed, undoped; (4, filled diamond) $T_g = 250°C$, annealed, undoped; (5, open triangle) $T_g = 300°C$, as-grown, [Be]= 3×10^{19} cm^{-3}.

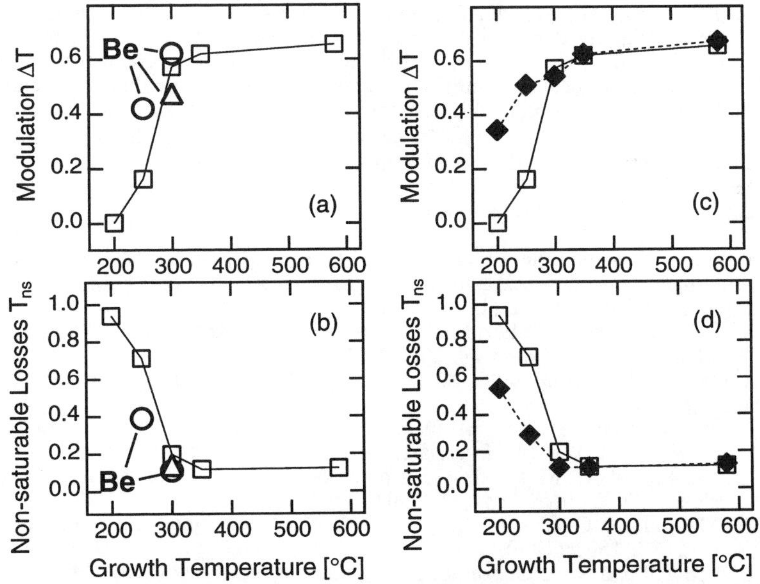

Fig. 3: Modulation and non-saturable losses vs growth temperature. (a), (b): as-grown LT GaAs, undoped (open squares), [Be]= 1×10^{19} cm^{-3} (open circles), and [Be]= 3×10^{19} cm^{-3} (open triangles); (c), (d): undoped LT GaAs, as-grown (open squares), annealed (filled diamonds).

higher growth temperatures, the modulation and the losses are not much affected by annealing.

Given the electronic structure of undoped as-grown LT GaAs [10], we attribute the non-saturable losses to the transition from the neutral As_{Ga}^0 to states about 700 meV above the bottom of the conduction band (CB). Due to the large density of the final states, we expect that this transition can hardly be saturated. This is substantiated by the fact that the density of As_{Ga}^0 [5], the strength of the As_{Ga}^0-CB transition, and the non-saturable losses increase with decreasing growth temperature. Importantly, ionized As_{Ga}^+ act as electron traps in undoped as-grown LT GaAs [9]. To obtain a faster response time, it is necessary to increase their density reducing the growth temperature [5]. This results inevitably in an increase of the neutral As_{Ga}^0 density and large non-saturable losses.

We have verified by near-infrared absorption measurements that the density of neutral As_{Ga}^0 decreases due to Be doping in our set of samples. This is the reason why Be doping reduces the non-saturable losses leading, in turn, to a high modulation. Magnetic circular dichroism of absorption measurements have shown that the density of singly ionized As_{Ga}^+ first increases with Be doping and then decreases again at the highest doping level. We conclude that doubly ionized As_{Ga}^{++} form at the highest doping level. The ultrafast trapping in the highly doped sample shows that the As_{Ga}^{++} are efficient electron traps, very much as the As_{Ga}^+. Summarizing this discussion, Be doping results in a reduction of the As_{Ga}^0 density reducing the nonsaturable losses and allowing for high modulation and, at the same time, increases the density of traps leading to a fast response time.

Likewise, As precipitate formation during annealing of LT GaAs reduces the neutral As_{Ga}^0 density resulting in reduced non-saturable losses due to point defects. The overall reduction of the losses demonstrates that As precipitates do not substantially contribute to the non-saturable losses. Consequently, a high modulation can be obtained. Our data show that the response time remains fast. Since the density of ionized As_{Ga}^+ electron traps is reduced by about one order of magnitude under our annealing conditions [5], we conclude that As precipitates have to be involved in carrier trapping in order to maintain a fast response time.

In conclusion, we have shown that Be doping or annealing improves the properties of LT GaAs for ultrafast all-optical switching applications. A fast response, high modulation, and low non-saturable losses are obtained at the same time. We attribute these findings to the reduction of the neutral As_{Ga}^0 density by Be doping or annealing, which, importantly, is not accompanied by a reduction of the density of useful carrier traps.

This work has been supported by the Swiss National Science Foundation. We gratefully acknowledge helpful discussions with P. Specht.

[1]   U. Keller, K. J. Weingarten, F. X. Kärtner, D. Kopf, B. Braun, I. D. Jung, R. Fluck, C. Hönninger, N. Matuschek, and J. Aus der Au, *IEEE J. Selected Topics in Quantum Electronics* **2**, 435 (1996)

[2]   D. S. Chemla, D. A. B. Miller, P. W. Smith, A. C. Gossard, and W. Wiegmann, *IEEE J. Quantum Electronics* **20**, 265 (1984)

[3]   S. Gupta, J. F. Whitaker, and G. A. Mourou, *IEEE J. Quantum Electronics* **28**, 2464 (1992)

[4]   P. Specht, S. Jeong, H. Sohn, M. Luysberg, A. Prasad, J. Gebauer, R. Krause-Rehberg, and E. R. Weber, *Mater. Sci. Forum* **258-263**, 951 (1997)

[5]   X. Liu, A. Prasad, W. M. Chen, A. Kurpiewski, A. Stoschek, Z. Liliental-Weber, and E. R. Weber, *Appl. Phys. Lett.* **65**, 3002 (1994)

[6]   M. R. Melloch, N. Otsuka, J. M. Woodall, A. C. Warren, and J. L. Freeouf, *Appl. Phys. Lett.* **57**, 1531 (1990)

[7]   G. P. Agrawal and N. A. Olsson, *IEEE J. Quantum Electronics* **25**, 2297 (1989)

[8]   L. R. Brovelli, U. Keller, and T. H. Chiu, *J. Opt. Soc. B* **12**, 311 (1995)

[9]   U. Siegner, R. Fluck, G. Zhang, and U. Keller, *Appl. Phys. Lett.* **69**, 2566 (1996)

[10]   G. D. Witt, *Materials Science and Engineering* **B22**, 9 (1993)

An ion-implanted anti-resonant Fabry Perot saturable absorber for passive mode-locking of solid state lasers

Max Lederer, Barry Luther-Davies

Laser Physics Centre, The Australian National University, Research School of Physical Sciences and Engineering, Canberra, ACT 0200, Australia

Hoe Tan, Chennupati Jagadish

Department of Electronic Materials Engineering, The Australian National University, Research School of Physical Sciences and Engineering, Canberra, ACT 0200, Australia

We have fabricated GaAs based anti-resonant Fabry-Perot Saturable Absorbers (A-FPSAs) for passively mode-locking near infrared solid state lasers, using Metal Organic Vapour Phase Epitaxy (MOVPE) growth followed by ion-implantation and optional thermal annealing. We present differential reflectivity measurements showing the effect of ion implantation and annealing. The devices were characterised for their large signal response including saturation fluence, modulation depth and non-bleachable losses – important parameters for passive mode-locking. Finally we demonstrate mode-locking using our samples within a Ti:Sapphire laser observing stable and reliable self-starting, pulses in the 100fs range, and 50nm tunability. Results of computer simulations are in good agreement with the experiments.

A. Introduction

Recently there has been a lot of interest in the creation of very short laser pulses in the near-infrared. Kerr lens mode-locking (KLM) was perhaps the most notable advance in this field. However, the disadvantage of the use of an ultra-fast Kerr lens is the need for starting aids except for highly optimised cavity designs where self-starting is possible. One starting mechanism that has been used successfully in KLM lasers is the saturable nonlinearity of a semiconductor Anti-resonant Fabry Perot Saturable Absorber (A-FPSA) [1]. More recently several different A-FPSA designs have been used to mode-lock broad band laser materials like Ti:Sapphire, Cr:LiSAF and Nd:glass by mechanisms other than KLM [2,3]. A successful scheme is the so-called soliton mode-locking where the A-FPSA recovery time can be more than an order of magnitude longer than the mode-locked pulse width without inducing instabilities. Recovery times from optical excitation in semiconductors (e.g. GaAs, InGaAs) grown at normal temperatures are typically greater than 100 ps, too long for the stable creation of ps and sub-ps mode-locked pulses. A common feature of all the A-FPSA designs has been, therefore, the use of low temperature (LT) MBE-growth of the absorber layer [4]. This creates ionised Arsenic anti-sites which act as traps both shortening the free carrier life time and hence the device recovery time and also increasing the CW saturation intensity. In our work we have used an alternative technique to make an A-FPSA fabricating the structure using MOVPE growth at normal temperature and tailoring the carrier life time using ion-implantation [5].

B. Design

In a laser cavity the A-FPSA is used as nonlinear saturable mirror substituting a dielectric mirror. Our structures (Fig. 1a) therefore consisted of an 830nm, 25 period, $Al_{0.15}Ga_{0.85}As/AlAs$ Bragg mirror on a wafer of semi-insulating GaAs followed by an AlAs spacer layer (800A or 150A thick), and a 450A GaAs absorber layer. The 150A spacer shifts the anti-node of the optical standing wave into the absorber, creating a resonant (R-)FPSA, whereas the 800A spacer shifts the node into the absorber creating the anti-resonant (A-) FPSA (see Fig. 1b). Whereas the A-FPSAs can be used for passive mode-locking the R-FPSAs would introduce too much loss and dispersion in the cavity and are therefore only used for measurement purposes here. The wafers were diced into 6x6mm^2 pieces and implanted with doses of $8 \cdot 10^{10}$ cm^{-2} $\rightarrow$ 10^{15} cm^{-2} of 40keV As-ions or $8 \cdot 10^{10}$ cm^{-2} $\rightarrow$ $5 \cdot 10^{11}$ cm^{-2} of 80keV As-ions. Monte Carlo simulations (TRIM) (Fig. 1c) showed that for 40keV ion energy the damage is restricted to the absorber layer, whereas at 80keV a small tail extends inside the spacer layer. The 40keV samples underwent thermal annealing at 600 °C for 20 min in an arsine atmosphere. These conditions have been found to remove much of the implantation damage caused by high ion doses and restore the transport properties in GaAs whilst still retaining short luminescence life times [6]. Additionally we found that the decrease in A-FPSA reflectivity caused by ion doses greater than $5 \cdot 10^{11}$ cm^{-2} could also be recovered completely by annealing the samples at 600°C for 20min. The 80keV samples remained unannealed.

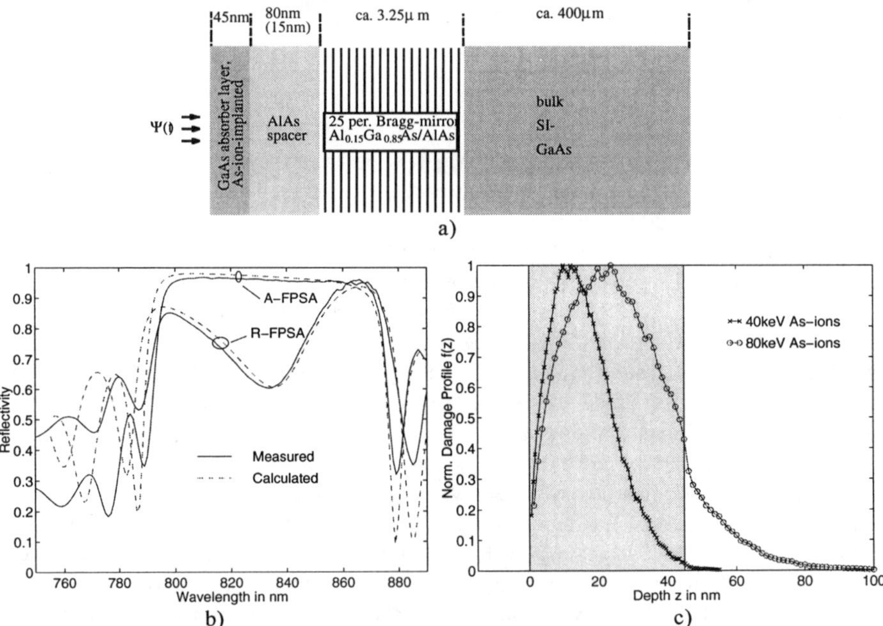

Fig. 1: Epitaxial design of A-(R-)FPSAs (a), calculated and measured specular reflectance of A-(R-)FPSAs (b), TRIM calculation of damage profile in absorber layer for 40keV and 80keV As-ion implantation (c).

C. Characterisation

We measured the temporal response of our ion-implanted samples (annealed as well as unannealed) using a wavelength degenerate pump-probe technique employing the 120fs, 76MHz pulse train of a KLM Ti:Sapphire laser. We deduced electron capturing times of < 500fs in the unannealed and < 1ps in the annealed case (Fig. 2a, b). A second process is associated with trap emptying where typical times are < 75ps in the unannealed and < 120ps in the annealed case. However, trap emptying appears incomplete since for long pump-probe delays residual bleaching decays rather slowly. For ion doses above $5 \cdot 10^{11}$cm^{-2} the annealed samples display induced absorption which is attributed to the long trap emptying times in combination with optical excitation out of the traps. In the unannealed samples this induced absorption only becomes visible for very high pump fluences indicating a more efficient trap emptying mechanism. From the differential reflectivity measurements we have further deduced the saturation fluence of the devices and their nonbleachable insertion losses which for the lowest dose annealed devices reached those of completely unimplanted samples.

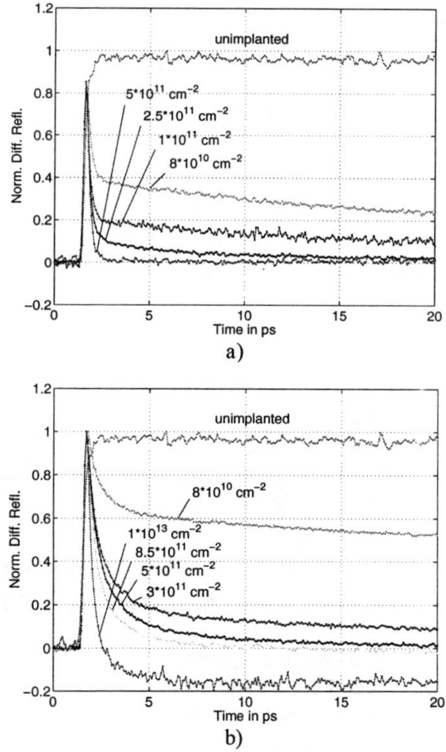

Fig. 2: Temporal response of unannealed (a) and annealed (b) R-FPSAs compared to the unimplanted. Pump Fluence $F_P = 1\mu Jcm^{-2}$, $\lambda_P = 840$nm, $n \cong 5 \cdot 10^{17}$ cm^{-3}.

D. Mode-locking

Finally, we have used the A-FPSAs as the sole mode-locking element in a Ti:Sapphire laser. Self-starting, stable CW-mode-locking was routinely achieved with build-up times between 30µs and 3ms over a wavelength range greater than 50nm (Fig. 3). We achieved pulses as short as 105fs. Inserting a plane dielectric mirror in place of the A-FPSA did not allow mode-locking to occur. For a certain parameter range we have also observed closely coupled multiple pulses. Both observations indicate that mode-locking was indeed due to the A-FPSA and not due to KLM by gain guiding. Further, we have performed numerical simulations of the Generalised Ginzburg-Landau Equation (GGLE) relevant for our laser which show that ultrafast (i.e. KLM) loss modulation is not necessary to stabilise the observed modelocked pulses.

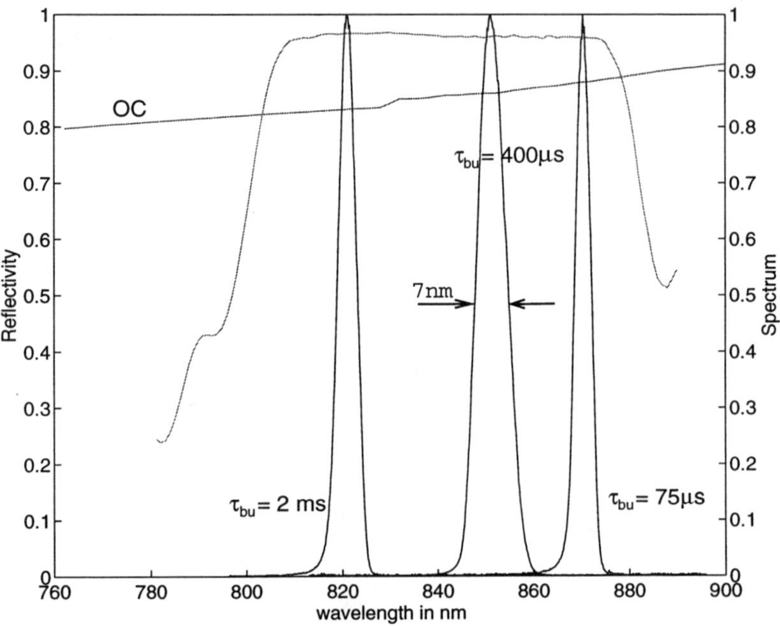

Fig. 3: Example spectra, with mode-locking build-up times annotated, achieved using the unannealed $2.5 \cdot 10^{11} cm^{-2}$ A-FPSA. Also shown: output coupler and A-FPSA reflectivities.

1.	I. D. Jung et al, Optics Letters, **20**, 1559, 1995
2.	S. Tsuda, et al, Optics Letters, **20**, 1406, 1995
3.	I. D. Jung, et al, Optics Letters, **20**, 1892, 1995
4.	L. Brovelli et al, J. Opt. Soc. Am B, **12**, 311, 1995
5.	M. J. Lederer et al, Appl. Phys. Lett., **70**, 3428, 1997
6.	C. Jagadish, et al, Appl. Phys. Lett., **67**, 1724, 1995

Influence of Growth Conditions on the As antisites, As_{Ga}^0 and As_{Ga}^+ concentrations in the Low Temperature GaAs MBE Growth: A First Theoretical Study

Natarajan Krishnan and Rama Venkat

Department of Electrical and Computer Engineering
University of Nevada Las Vegas, Las Vegas, NV 89154-4026
and
Donald L. Dorsey

Air Force Research Laboratory, Materials and Manufacturing Directorate,
Wright Patterson AFB, OH 45433-7707

Abstract

A novel theoretical investigation is employed to study the influence of surface dynamics in low temperature MBE growth of $GaAs$. A kinetic growth model along with the charge neutrality equation is used to find the distribution of As antisite between the neutral and charged species using the bulk Ga vacancy concentrations obtained from simulations for the first time. The experimentally observed temperature and As flux dependencies of the As antisite concentrations are successfully reproduced by the model. The As_{Ga}^0 and As_{Ga}^+ concentrations saturate with As flux for a given temperature due to saturation of an amorphous, physisorbed As surface layer which acts as the reservoir for As antisite incorporation. The As antisite concentration saturates at lower value for higher temperature due to larger evaporation of As antisite from the crystalline state. It is observed that both As_{Ga}^0 and As_{Ga}^+ concentrations decrease with increase in temperature. While the decrease of As_{Ga}^0 concentration with temperature is related to the direct evaporation of As antisite from the crystalline state, the decrease of As_{Ga}^+ is related to a decrease in the Ga vacancy concentration.

A. Introduction

The interesting optoelectronic properties of Low Temperature grown As rich non stoichiometric $GaAs$(LT-GaAs) are a result of excess As. The reliable and reproducible control of the optoelectronic properties requires a detailed understanding of the dynamics of formation of As antisites defects (As_{Ga}), and Ga vacancies (V_{Ga}). The As_{Ga} defect is a deep donor, a part of which is positively charged(As_{Ga}^+) and is compensated by the presence of Ga vacancies which are triple acceptors V_{Ga}^{3-}[1]. The ultrafast trapping characteristics of LT-GaAs are directly related to the presence and the concentration of As_{Ga}^+. The incorporation processes of neutral As antisites, As_{Ga}^0 and charged As antisites, As_{Ga}^+, in the LT GaAs MBE growth are not yet fully understood. In this article, a modified version of the stochastic model [2-4] is used to investigate the science behind the growth kinetics. The model addresses the influence of growth parameters such as temperature and flux ratio on the incorporation of As_{Ga}^+ and As_{Ga}^0. The theoretical formulation and the details of the model are presented in section B. Results and discussion are presented in section C and conclusions are in section D.

B. Rate equation Model of Growth for *LT GaAs*

The rate equation model of growth considers surface kinetic processes, namely, adsorption, evaporation and surface migration to describe the time evolution of the epilayer through the change of macrovariables normalized with respect to the number of possible atoms in the layer. A schematic picture illustrating the dynamic processes of the physisorbed state of *As* (PA) and *As* antisite is shown in Fig. 1. The macrovariables considered are: the layer coverage of atoms *As*, *Ga* and As_{Ga}, viz., $C_{As}(2n+1)$ in $(2n+1)^{th}$ layer, $C_{Ga}(2n)$ in the $2n^{th}$ layer and $C_{As_{Ga}}(2n)$ in the $2n^{th}$ layer, respectively, where n is the layer index. The in-register *As* belongs to the odd numbered layers while the in-register *Ga* and the *As* antisite occupy the even numbered layers. The rates of surface processes are assumed to be of Arrhenius type with frequency factors and activation energies. The time evolution equation for the layer coverage of *Ga* in the $2n^{th}$ layer, $C_{Ga}(2n)$, is given by:

$$
\begin{aligned}
\frac{dC_{Ga}(2n)}{dt} =~ & \left([C_{As}(2n-1) - C(2n)] \, J_{Ga} \right) (A) + [C_{As}(2n-1) - C(2n)] \\
\times~ & \left(R_0 e^{\frac{E_d(2n+2)}{kT}} \left(\frac{C_{Ga}(2n+2)}{C(2n+2)} \right) [C(2n+2) - C_{As}(2n+3)] \right. \\
+~ & \left. R_0 e^{\frac{E_d(2n-2)}{kT}} \left(\frac{C_{Ga}(2n-2)}{C(2n-2)} \right) [C(2n-2) - C_{As}(2n-1)] \right) (B) \\
-~ & R_0 e^{\frac{E_d(2n)}{kT}} \left(\frac{C_{Ga}(2n)}{C(2n)} \right) [C(2n) - C_{As}(2n+1)] \\
\times~ & \left([C_{As}(2n+1) - C_{Ga}(2n+2)] + [C_{As}(2n-3) - C(2n-2)] \right) (C) \\
-~ & R_0 e^{\frac{E_e(2n)}{kT}} \left(\frac{C_{Ga}(2n)}{C(2n)} \right) [C(2n) - C_{As}(2n+1)] \, (D)
\end{aligned}
\tag{1}
$$

where the terms A, B, C and D represent the change of *Ga* coverage in the $2n^{th}$ layer due to incorporation, migration into and out of the $2n^{th}$ layer and evaporation, respectively. J_{Ga} is the flux of atoms per site per second. Details of the fluxes and the procedure for obtaining fluxes from the experimental measurable such as BEP are given in section C. The R_0 is the frequency factor for the surface *Ga* atoms. E_d, and E_e, in Eq.1 are the activation energies of diffusion and evaporation of *Ga*, respectively, are given by:

$$
\begin{aligned}
E_d(2n) &= E_{d,iso} + 4E_{GaGa}C_{Ga}(2n) \\
E_e(2n) &= E_{e,iso} + 4E_{GaGa}C_{Ga}(2n)
\end{aligned}
$$

with e$E_{d,iso} = 0.4$V, $E_{e,iso} = 1.4$ $E_{GaGa} = 0.14$eV Similar time evolution equations are written for layer coverages of arsenic $C_{As}(2n)$, *As* antisite, $C_{As_{Ga}}(2n)$, and PA *As* layer, $C_{Phy,As}(2n)$ [Ref.2]. For the simulation of the growth 80 layers of *GaAs*, 3 x 80 differential equations given by Eq.1 and an additional equation describing the dynamics of the PA layer are solved simultaneously by a 4th order Runge Kutta method. The average concentrations of *Ga*, *As* and As_{Ga} in individual layers are obtained from the layer coverages. The separation of *As* antisites into neutral and charged, *i.e.*, As_{Ga}^0 and As_{Ga}^+ necessitates the consideration of Gallium vacancies, V_{Ga}. Total coverage of vacancies in the $2n^{th}$ layer, $C_{V_{Ga}}(2n)$, is $(1-C(2n))$, where $C(2n)$ is the sum of coverages of *Ga* and *As* antisite, which can be obtained from the solution to differential equations. The triply charged V_{Ga}^{3-} and the singly charged As_{Ga}^+ satisfy the charge neutrality equation, given by:

$$3C_{V_{Ga}^{3-}}(2n) \;=\; C_{As_{Ga}^{+}}(2n) \tag{2}$$

Since the total As antisite concentration, $C_{As_{Ga}}(2n)$ is the sum of $C_{As_{Ga}^{+}}(2n)$ and $C_{As_{Ga}^{0}}(2n)$, individual concentrations can be obtained by using the solution to differential equations in Eq. 2.

C. Results and Discussions

In this work, the growth direction is assumed to be [100]. The growth rate is $1\mu m$/hr. (0.983 monolayers/sec.) which is also the gallium (Ga) flux, J_{Ga}, used in the experiments[5]. The experimentally measured Beam Equivalent Pressure (BEP) is converted to flux in number of monomer atoms per site per second using Eq.1 of Ref.[7]. The conversion between J_{As}(monomer/cm^2.sec.) and J_{As}(monomer/site.sec.) is accomplished using the number of sites per cm^2, which in case of (100) $GaAs$, is 6.26 x 10^{14}[2]. Growth results for a growth time of 20 seconds were obtained. Antisite As concentration per cm^3 is calculated as per the procedure described in Ref.[2] and is converted into charged and neutral As antisite concentrations per cm^3 according to the procedure given in section B. For obtaining statistically reasonable averages, the concentrations of As_{Ga} and V_{Ga} from layers far from the substrate and surface are used. As charged and neutral antisite concentration versus BEP obtained from our simulations were fitted to four experimental data points of Luysberg *et al.*[5] to fix the model parameters accurately. Using the fixed model parameters, model predictions for the remaining growth conditions were obtained.

Plots of As_{Ga}^{0} and As_{Ga}^{+} versus BEP for 513°K and 473°K obtained using the model are shown along with the experimental data of *Luysberg et al.*[5] in Fig. 2. The agreement between the model predictions and the experimental results is good. Both As_{Ga}^{0} and As_{Ga}^{+} saturate beyond a BEP of 20 for $513°K$ and $473°K$, which can be explained as follows. For a given temperature, as BEP increases, the As flux in excess of Ga flux increases, resulting in increase in the PA layer concentration till the PA layer coverage reaches its maximum value of unity at a critical BEP. Beyond the critical BEP, any further increase in BEP does not change the PA layer coverage as it has attained its maximum. The As_{Ga} concentration incorporated in the crystal is dictated by two competing mechanisms, incorporation of As from the PA layer and evaporation of As_{Ga} from the crystal. For a given temperature, the saturation of As_{Ga} occurs because the incorporation and evaporation lifetimes and the PA layer coverage are all constant beyond the critical BEP. The saturation of As_{Ga} concentration appears to be lower for higher temperature because of higher evaporation rate of As_{Ga} from the crystal, even though data related to high BEP simulations is not reported here due to lack of experimental data to compare with. The decrease in As_{Ga}^{+} concentration with increase in temperature is attributed to its relation to the Ga vacancy concentration which decreases[5] with increase in temperature due to larger migration length of Ga. Both As_{Ga}^{+} and As_{Ga}^{0} exhibit the same dependencies on BEP and temperature, but the As_{Ga}^{0} is consistently one order of magnitude higher than the concentration of As_{Ga}^{+}.

D. Conclusions

This theoretical investigation of MBE growth of LT-GaAs using the stochastic model takes one step further in understanding the surface dynamics of incorporation of neutral and charged As antisites, As_{Ga}^{0} and As_{Ga}^{+}. The physics behind the dependence of concentration of As antisites on growth parameters is analyzed and explained. This improved knowledge of defect behavior will allow for better control of defect type and concentrations. The model developed in this work allows for the temperature- and BEP- dependent incorporation and evaporation of As antisites during LT-GaAs growth. The model predictions for As antisite concentrations agree well with the available experimental data[5].

Acknowledgment:

This work is supported by the Air Force Office of Scientific Research under DEP-SCOR, grant number F49620-96-0275. The authors gratefully acknowledge Drs. Eicke R. Weber and Petra Specht of University of California, Berkeley for the experimental data and many technical discussions.

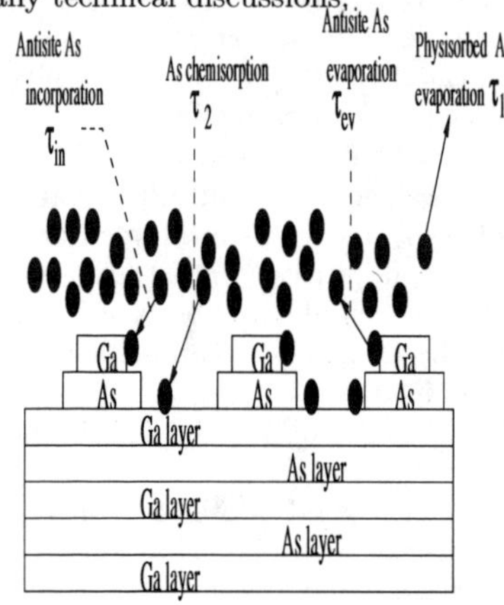

Fig. 1: A schematic picture of the surface processes.

Fig. 2: A plot of charged and neutral antisite concentrations versus BEP

References

[1] P. Specht, S. Jeong, H. Sohn, M. Luysberg, A. Prasad, J. Gebauer, R. Krause-Rehberg, E.R. Weber, *Proceedings of the Int. Conf. on Defects in Semiconductors ICDS 19, Aveiro, Portugal, Mater. Sci. Forum*, **258-263**, 951 (1997).

[2] S. Muthuvenkatraman, Suresh Gorantla, Rama Venkat, Donald L. Dorsey, *J. Appl. Phys.*, **83**, 5845, (1998).

[3] R. Venkatasubramanian, *J. Mater. Res.*, **7**, 1221 (1992).

[4] R. Venkatasubramanian, Vamsee K. Pamula and Donald D. Dorsey, *Applied Surface Science*, **104/105**, 448 (1996).

[5] M. Luysberg, H. Sohn, A. Prasad, P. Specht, Z. Liliental-Weber, E.R. Weber, J. Gebauer and R. Krause-Rehberg, *J. Appl. Phys.,*, **83**, 561 (1998).

[6] J. Gebauer, R. Krause-Rehberg, S. Eichler, M. Luysberg, H. Sohn and E.R. Weber, *Appl. Phys. Lett.*, **71**, 1 (1997).

[7] T.B. Joyce and T.J. Bullough, *J. Crys. Growth*, **127**, 265 (1993).

Electrical Properties and Thermal Stability of Be-doped Non-stoichiometric GaAs

R.C. Lutz, P. Specht, R. Zhao and E.R. Weber
Department of Materials Science and Mineral Engineering, University of California, Berkeley, CA 94720

Abstract

Beryllium doping of arsenic rich GaAs grown by MBE at low temperatures (LT-GaAs) allows one to grow GaAs with ultrashort carrier lifetimes at growth temperatures of 250° to 300°C. Hall effect and resistivity measurements showed that in as-grown samples hopping conduction can be suppressed in this growth temperature range. Semi-insulating behavior of as-grown LT-GaAs:Be can be achieved by careful adjustment of temperature and Be-doping. In addition, Be-doped LT-GaAs grown in this temperature range shows increased thermal stability of the arsenic antisite defects (As_{Ga}).

A. Introduction

Gallium arsenide grown by molecular beam epitaxy (MBE) at low temperatures (<400°C) is characterized by an excess of arsenic, most of which is incorporated as arsenic antisite defects (As_{Ga}) [1]. These As_{Ga} defects account for several unique properties of the material, including its short carrier lifetime and strong hopping conductivity. They are also thermally unstable in the high concentrations associated with low temperature growth, and precipitate into arsenic clusters upon annealing. The short carrier trapping time of this material cannot be fully exploited unless significant improvements in the resistivity and thermal stability are made. Beryllium doping offers a potential solution to both problems.

In undoped non-stoichiometric GaAs, the off-stoichiometry is incorporated mainly as arsenic antisite double donors and gallium vacancy (V_{Ga}) triple acceptors. No direct evidence for As interstitials has been found. The two As_{Ga} defect levels are centered at 0.75 and 1 eV below the conduction band [2], while the defect levels of the V_{Ga} are not yet agreed upon. Both can be present in high concentrations, increasing with decreasing growth temperature, but the $[As_{Ga}]$ is at least an order of magnitude greater than the $[V_{Ga}]$ under all growth conditions. Consequently, the V_{Ga} compensate only a fraction of the As_{Ga} and pin the Fermi level within the As_{Ga} defect band. The short carrier trapping time arises from the high concentration of ionized antisites acting as traps for free electrons. Hopping conduction results from charge transfer between neutral and charged As_{Ga} defects with overlapping electronic wavefunctions and increases exponentially with $[As_{Ga}]$ [3]. Beryllium doping has been introduced in order to increase the relative fraction of ionized antisites through controlled compensation, so that short trapping times can be obtained at higher growth temperatures where the overall antisite concentration (hence hopping conduction) is diminished. As an added bonus, beryllium doping appears to inhibit the As_{Ga} defects from precipitating upon annealing [4]. This paper will discuss the effects of Be on resistivity and thermal stability of LT-GaAs.

B. Experimental

For this study, GaAs films were grown at 250 to 320°C on semi-insulating GaAs substrates (AXT) in a Varian Gen II MBE machine after the deposition of a 0.5 μm undoped GaAs at 600°C. The deposition rate at a beam equivalent pressure (BEP) ratio of 20 was 1 μm/hr, and the overall non-stoichiometric film thickness ranged between 1.35 to 1.5 μm. We calibrated the growth temperature by measuring the c-lattice parameter mismatch between undoped films and the substrate [5], and the beryllium concentration by SIMS and Hall effect-derived carrier concentrations in high-temperature-grown films (600°C). Proximity annealing was performed at 600° and 700°C for 30 minutes.

We obtained resistivity and mobility values using the standard van der Pauw configuration for thin films with indium contacts spot-annealed for 10 sec at 300°C. We did not take into account the substrate conduction, a factor that can significantly influence measurements [6]. This is not a problem in as-grown films for two reasons: 1) the measured resistivities were low enough that the substrate would influence them little, and 2) carrier hopping in the film cannot extend into the semi-insulating substrate because the As_{Ga} defect concentration there is too low to support it. However, some of the annealed films showed resistivities greater than 1×10^5 Ω-cm, in which case the substrate would

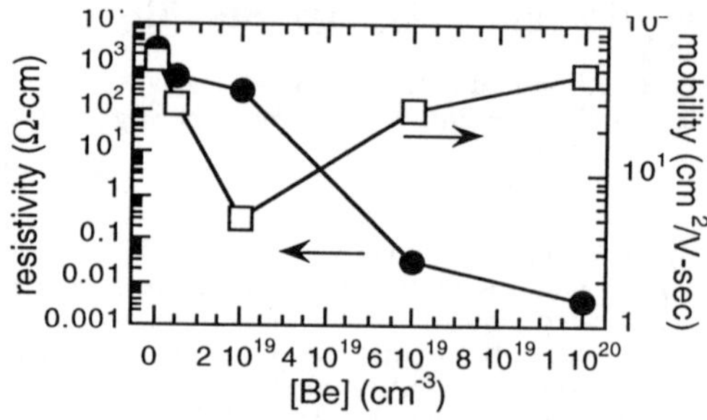

Figure 1: Resistivity and mobility of films grown at 250°C as a function of [Be].

have significant influence. In these cases, the reported resistivities represent lower limits. The substrate influences the measured mobilities more than the resistivities, hence these were not reported for annealed films.

Near-infrared absorption (NIRA) and magnetic circular dichroism of absorption (MCDA) were used to measure the neutral and singly ionized antisite concentrations, respectively. The neutral As_{Ga} concentration is determined from the absorption coefficient at 1.24 eV at 1.8 K, following the calibration given by Martin [7]. The ionized antisite concentration is determined from the MCDA signal at 0.94 eV at 1.8 K in accordance with the calibration proposed by Hofmann et al. [8]. A semi-insulating substrate was used as a reference, and its NIRA and MCDA spectra were subtracted from those of the films before determination of the $[As_{Ga}]$.

C. Results and Discussion
C.1. Electrical properties and their dependence on the As_{Ga} concentration

Figure 1 shows how the resistivity and mobility of films grown at 250°C vary with beryllium concentration. The resistivity decreases as [Be] is increased, while the mobility experiences a minimum at 2×10^{19} Be/cm^3. Hall effect measurements showed that the two most highly doped films were p-type, meaning that the beryllium has overcompensated the As_{Ga} defects and band conduction of holes is the dominant conduction mechanism. The films with lower [Be] appear n-type in Hall measurements, but the low mobilities indicate that charges hopping between localized defect states account for a significant portion of the current. This is understandable, considering that while some electrons can overcome the activation barrier of As_{Ga} defects at room temperature, the close proximity of the As_{Ga} defects and the Fermi level pinning in one of the As_{Ga} defect bands provide conditions that are ripe for hopping conduction. The mobility decreases with increasing [Be] in films that exhibit hopping, indicating that the hopping contribution gets stronger with increasing [Be], until overcompensation results in p-type behavior.

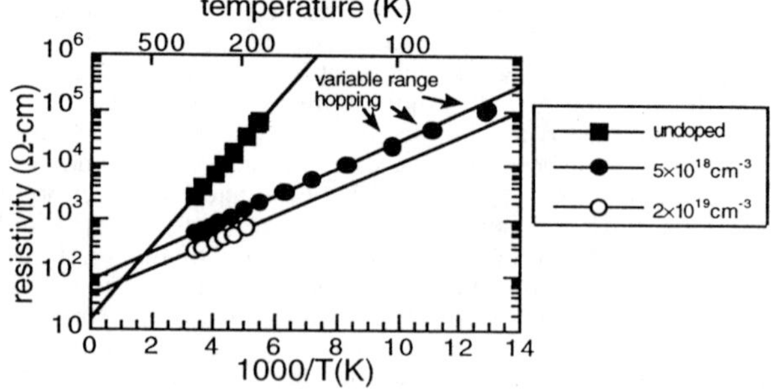

Figure 2: Temperature dependence of resistivity in films grown at 250°C.

Table 1: Activation energies and the pre-exponential factor for hopping of films grown at 250°C with various amounts of beryllium.

[Be] (cm^{-3})	activation energy (meV)	ρ_3 (Ω-cm)
0	130±1	16.1±0.9
5×10^{18}	51.1±0.06	75.8±0.3
2×10^{19}	49.1±0.1	39.4±0.3

We performed variable-temperature resistivity measurements on the n-type films to understand how beryllium effects the electrical properties of non-stoichiometric GaAs. The resistivity follows thermally activated behavior of the form

$$\rho = \rho_3 \exp\left(\frac{\varepsilon_3}{kT}\right) \tag{1}$$

where, ε_3 is the activation energy for hopping, k is Boltzmann's constant and T is the temperature. The pre-exponential, ρ_3, is given by

$$\rho_3 = \rho_{03} \exp\left(\frac{\alpha}{N^{\frac{1}{3}}a}\right) \tag{2}$$

where ρ_{03} is another pre-exponential factor, α is a constant, N is the defect concentration and a is the spatial extent of the defect wavefunction [9]. The subscript 3 is traditionally used to represent hopping conduction [9]. Figure 2 shows an Arrhenius plot of the resistivity data for the three n-type films from Figure 1, curve fitted with Equation 1. The curve fitting excluded the beginnings of variable range hopping seen at the lowest measurement temperatures. Table 1 shows the parameters determined by the curve fits. The most obvious effect of beryllium doping is the significant reduction in the activation energy for hopping. Additionally, ρ_3 increases initially upon beryllium doping, although it subsequently decreases between [Be] of 5×10^{18} and 2×10^{19}/cm^3. We determined the concentrations of neutral and ionized As$_{Ga}$ defects in order to analyze this behavior.

Figure 3 shows the neutral and ionized [As$_{Ga}$], as well as the total [As$_{Ga}$] as a function of [Be]. The addition of beryllium, as expected, increases the concentration of singly ionized As$_{Ga}$ defects. In the undoped film, the compensating defects are the triple-acceptor gallium vacancies [5], which must be present at a concentration of about 1×10^{18}/cm^3. Beryllium doping affects the overall concentration of As$_{Ga}$ defects as well, but these data yield no obvious relationship.

Figure 4 shows a plot of $\ln(\rho_3)$ vs. $N^{-1/3}$ using data from Table 1 and Figure 3. The curve fit with Equation 2 is extremely good, and is conclusive evidence that hopping between nearest-neighbor As$_{Ga}$ defects is the dominant conduction mechanism. It also supports the [As$_{Ga}$] measurements, which shows that beryllium has a complicated effect

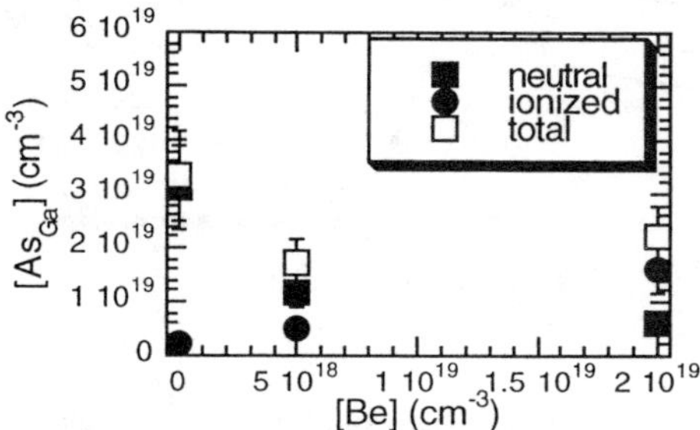

Figure 3: Effect of beryllium on [As$_{Ga}$] in films grown at 250°C.

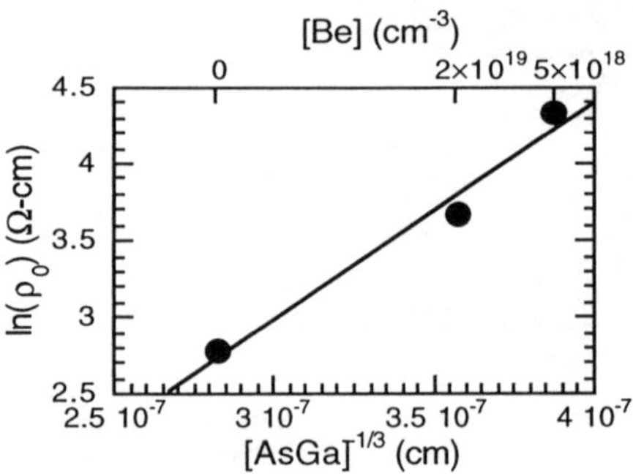

Figure 4: Inverse exponential dependence of the coefficient ρ_0 on [As$_{Ga}$].

on the overall [As$_{Ga}$]. The mechanism by which beryllium alters the total [As$_{Ga}$] requires further investigation. Shklovskii and Efros [9] have determined the constant α from Equation 2 to be 1.73. Using this value, the size of the As$_{Ga}$ defect, a, can be determined from the slope of the line in Figure 4 to be 8.6 Å. This radius is close to one determined earlier for As$_{Ga}$ by the same technique [3].

In conclusion, beryllium reduces the resistivity at room temperature is because it reduces the activation energy for hopping. It does this by compensating the As$_{Ga}$ defects, thereby lowering the Fermi level within the As$_{Ga}$ defect band. The fact that it does not necessarily increase the overall [As$_{Ga}$] is a promising result, as compensation should increase the resistivity if the overall [As$_{Ga}$] is too small to support hopping. Therefore higher resistivities may be achievable in Be-doped material than in undoped material grown at the same substrate temperature, likely above 300°C, where hopping is completely suppressed.

C.2. Thermal stability

Figure 5 shows how the resistivity of films grown under various conditions changes upon annealing for 30 minutes at 600 and 700°C. By properly matching the [Be] to the growth temperature, the amount of arsenic loss upon annealing can be inhibited. If the [Be] is too high, the material can become p-type upon annealing from arsenic loss. Otherwise, the resistivity increases upon annealing, but not as much as one would find in undoped material. This is evidence that there are still a lot of antisites remaining after annealing, as found in a previous study [4].

D. Conclusions

Beryllium doping has several interesting effects on the properties of non-stoichiometric GaAs. Doping actually lowers the resistivity through a lowering of the

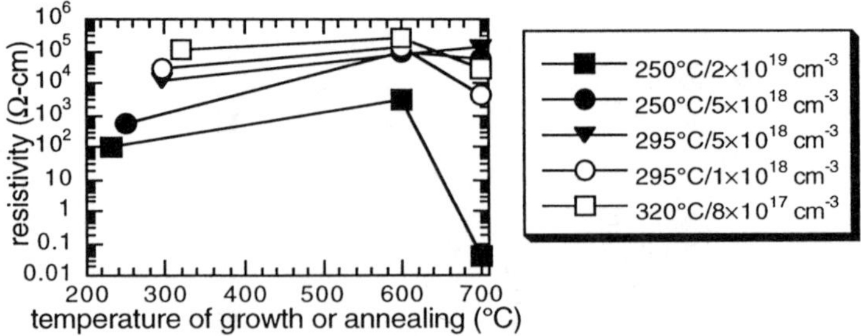

Figure 5: Change in resistivity upon annealing for selected films.

temperatures. Additionally, the beryllium has a stabilizing effect on the antisites, so that the material shows improved thermal stability.

Acknowledgements
This work has been supported by the Air Force Office of Scientific Research Grant #F49620-98-1-0135. Use of analytical equipment at the Lawrence Berkeley National Laboratory and the Integrated materials Laboratory at UC Berkeley is gratefully acknowledged.

References

[1] X. Liu, A. Prasad, J. Nishio, E.R. Weber, Z. Liliental-Weber and W. Walukiewicz, Appl. Phys. Lett. **67**, 279 (1995).

[2] E.R. Weber, H. Ennen, U. Kaufmann, J. Windscheif, J. Schneider and T. Wosinski, J. Appl. Phys. **53**, 6140 (1982).

[3] M. Kaminska and E.R. Weber, *20th Int. Conf. On the Phys. Of Semicond.* Thessaloniki, Greece, Ed. by E.M. Anastassakis and J.D. Joannopoulos, p.473 (1990).

[4] P. Specht, S. Jeong, H. Sohn, M Luysberg, J. Gebauer, R. Krause-Rehberg and E.R. Weber, Mater. Sci. Forum **258-263**, 951 (1997).

[5] M. Luysberg, H. Sohn, A. Prasad, P.Specht, Z. Liliental-Weber, E.R. Weber, J. Gebauer and R. Krause-Rehberg, J. Appl. Phys. **83**, 561 (1998).

[6] D.C. Look, D.C. Walters, M. Mier, C.E. Stutz and S.K. Brierley, Appl. Phys. Lett. **60**, 2900 (1992).

[7] G.M. Martin, Appl. Phys. Lett. **39**, 747 (1981).

[8] D.M. Hofmann, K. Krambrock, B.K. Meyer and J.M. Spaeth, Semicond. Sci. Technol. **6**, 170 (1991).

[9] B.I. Shklovskii and A.L. Efros, *Electronic Properties of Doped Semiconductors*, Springer-Verlag, New York (1979) pp. 74-143.

Vacancies in Low-Temperature-grown GaAs: Observations by positron annihilation

J. Gebauer[1], F. Börner[1], R. Krause-Rehberg[1], P. Specht[2], and E. R. Weber[2]

[1] *Fachbereich Physik, Martin-Luther-Universität Halle-Wittenberg, D-06099 Halle, Germany*
[2] *University of California and Lawrence Berkeley Laboratory, Berkeley, California 94720*

Abstract - Positron annihilation will be used to study vacancy defects in GaAs grown at low temperatures (LT-GaAs). The vacancies in as-grown LT-GaAs can be identified to be Ga monovacancies, V_{Ga}. The density of V_{Ga} is related to the nonstoichiometry in LT-GaAs, established by decreasing the growth temperature or increasing the As/Ga beam equivalent pressure (BEP) ratio. A maximum of about 2×10^{18} V_{Ga}/cm^3 was observed at the lowest growth temperature (200°C). Annealing at elevated temperatures ($\approx 600°C$) removes V_{Ga}, instead, larger vacancy agglomerates are found. They are most likely associated with the As-precipitates commonly found in annealed LT-GaAs.

A - Introduction

The unique properties of GaAs grown at low temperatures (LT-GaAs) are determined by intrinsic point defects [1]. As_{Ga} antisites are the dominant defect species in LT-GaAs [2]. They can account for the nonstoichiometric composition [3] as well as for the lattice expansion [2]. However, a part of the antisites exists as positively charged As_{Ga}^+ implying the presence of compensating acceptors. It is widely assumed that Ga vacancies, V_{Ga}, are these acceptors and account for the compensation of As_{Ga}^+ [2]. If this is the case, the density of V_{Ga} must track that of the antisites. Positron annihilation spectroscopy (PAS) indeed showed the existence of vacancy defects in LT-GaAs (e.g. [4]). Recently, these defects were shown to be related to V_{Ga} [5]. It was shown in the following that the V_{Ga} density can quantitatively account for the compensation of As_{Ga}^+ [6]. In the present work we will extend the earlier findings of PAS [5,6]. Samples grown with different As/Ga beam equivalent pressure (BEP) ratio will be investigated as well as annealed LT-GaAs.

B - Experimental

The LT-GaAs samples were grown by molecular beam epitaxy with a growth rate of typically 1 μm/h and a layer thickness of 1-2 μm. All samples were undoped. The growth temperature (T_G) varied from 200 to 350°C for samples grown with a BEP ratio of 20. An additional series with a BEP varying from 11-37 was grown at 200°C.

The investigations were carried out by means of slow monoenergetic positrons suitable for the detection of vacancies in thin layers by varying the incident energy between 0.1 to 40 KeV. Here, positron trapping in vacancy defects can be observed by the narrowing of the 511 keV annihilation peak. The annihilation peak was characterized by the lineshape parameters S and W. S (W) is defined as the number of counts in the central (wing) region of the peak divided by the total. Positron trapping in vacancies results in an increase (decrease) in S (W) compared to defect-free material. In principle, each defect type exhibits its own specific values S_V and W_V corresponding to complete annihilation in the defect. S_V usually increases with the defect size. In the following, S and W parameters will be normalized to the values S_b and

     118

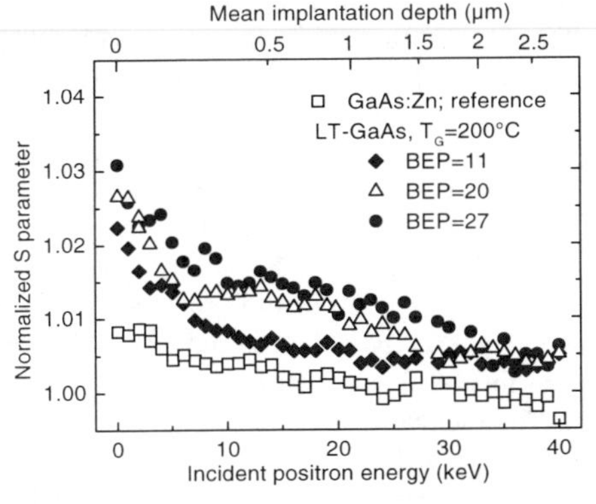

Fig. 1: S parameter as a function of incident positron energy for characteristic LT-GaAs samples grown with different BEP ratio compared to a GaAs:Zn reference free from positron trapping in defects.

Fig. 2: S- vs. W-parameter in LT-GaAs grown at T_G=200-400°C with BEP=20 (o) or with BEP=11-37 at T_G=200°C (▲) compared to Si-doped GaAs (■) and LT-GaAs annealed at 600°C (◇). Each point correspond to another sample.

W_b, measured in a GaAs:Zn reference sample free from positron trapping in defects. Further experimental details can be found in ref. [5].

C - Results

A distinct increase of the S parameter compared to a defect-free reference was observed in LT-GaAs (Fig. 1). The S parameter is nearly constant throughout the layer ($E_{Positron}$= 10 - 20 keV) indicating a homogeneous defect distribution. The results obtained on samples with varying BEP (Fig. 1) are very similar to those in samples grown at different temperatures [5]. The magnitude of the S parameter increases with increasing BEP or decreasing growth temperature. We conclude that PAS detects vacancy defects in LT-GaAs layers, apparently related to T_G and BEP. However, the S parameter alone does not allow to conclude on the defect type and, thus, on the defect density.

In order to obtain information on the defect type, the linearity between S and W parameter was analyzed. If only one type of vacancy defect trap positrons, the fraction η of positrons annihilating in defects can be written as: $\eta=(S-S_b)/(S_V-S_b)=(W_b-W)/(W_b-W_V)$ where S_V and W_V correspond to complete annihilation in the defect. S_b and W_b denote annihilation in the defect-free bulk. Than, S depends linearly on W if only the defect density changes (i. e. the fraction η of positrons annihilating in defects), but not the defect type characterized by S_V and W_V [7]. A linear variation was indeed observed in LT-GaAs (Fig. 2). The results agree for samples grown with varying T_G or BEP, respectively. Moreover, the straight line characteristic for as-grown LT-GaAs runs through (S_b,W_b). This allows to conclude that only one defect type is responsible for positron trapping in LT-GaAs.

To reveal the microscopic nature of the vacancies, the data are compared to results from highly Si doped GaAs (squares in Fig. 2). V_{Ga}-Si_{Ga} complexes were identified in these samples by correlated scanning tunneling microscopy and positron annihilation measurements [8]. The Doppler broadening parameters of V_{Ga}-Si_{Ga} complexes are experimentally known to agree with those of isolated Ga vacancies [9]. The data from LT-GaAs and GaAs:Si follows the same straight line (indicated by V_{Ga} in Fig 2.). This shows, that the vacancies in as-grown LT-GaAs are related to Ga monovacancies. Slow positron lifetime spectroscopy was additionally performed on some of our LT-GaAs samples by using a pulsed positron beam. These measurements revealed a monovacancylike character of the vacancies in LT-GaAs as well, strongly supporting the assignment to V_{Ga} [10]. Note, however, that the present data does not allow to conclude whether the vacancies are isolated or part of a defect complex.

Annealing of LT-GaAs at about 600°C is known to reduce the density of antisite defects drastically [2], however, the excess As forms precipitates [6]. Annealing results in highly resistive material, caused either by residual point defects (i.e. antisites) or by Schottky barriers at the As-precipitates. The PAS measurements shown in Fig. 2 revealed a strong increase in the S parameter in annealed LT-GaAs. The S-W data from annealed LT-GaAs define a straight line pointing towards the bulk parameters as well (indicated with V_x in Fig. 2). However, the linear behavior differs significantly from that for V_{Ga}. We have to conclude that the vacancy defects in annealed LT-GaAs are different from V_{Ga}. This may already be concluded from the high value of the S parameter (S=1.04 -1.05) typical for positron trapping at larger vacancy aggregates and much higher than for V_{Ga} (S_v=1.022-1.024, Fig 2.). A probable explanation is positron trapping at As precipitates or in vacancy clusters, sometimes associated with the precipitates [11]. These new defects dominate positron trapping in annealed LT-GaAs and Ga vacancies are no longer present in significant concentrations.

The vacancy concentration [V_{Ga}] in as-grown LT-GaAs can be calculated according to [V_{Ga}]=(λ_b/μ) (S_m−S_b)/(S_v−S_m) where λ_b=4.4×10^9 s^{-1} is the annihilation rate in defect-free GaAs. S_m denotes the S parameter measured in a particular sample at a depth of about 0.5 μm ($E_{Positron}$=10-15 keV), where almost all positrons annihilate within the bulk of the layer. A trapping coefficient μ= 1×10^{15} s^{-1} will be used for the calculation of vacancy concentrations which is the commonly accepted value for negative monovacancies in semiconductors at room temperature [12]. The resulting vacancy concentrations are shown in Fig. 3 (for samples grown at different temperatures) and Fig. 4 (for samples grown with different BEP ratio). The vacancy concentration increases with the growth temperature as well as with the BEP ratio (note the plateau at a BEP >20). A similar increase and saturation of the antisite concentration

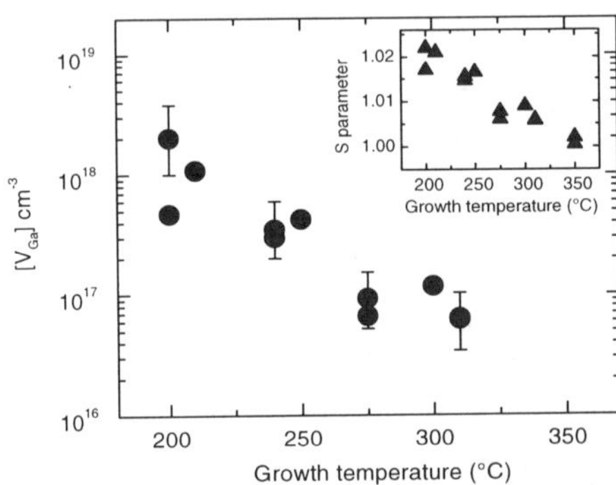

Fig. 3: Concentration of Ga vacancies [V_{Ga}] in LT-GaAs as a function of the growth temperature. [V_{Ga}] was calculated from the measured S parameter, shown in the inset. Typical errors for [V_{Ga}] are indicated.

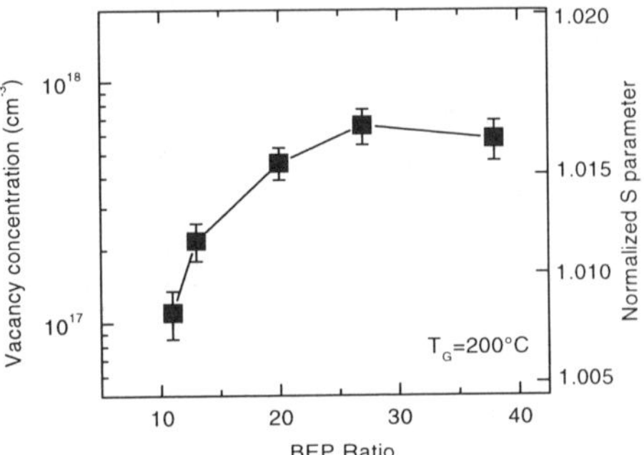

Fig. 4: Concentration of Ga vacancies in LT-GaAs grown at 200°C as a function of the BEP ratio during growth. Errors are due to the measured S parameter, S_m, only.

with increasing BEP was found by Luysberg et al. [6]. Moreover, it is well established that the antisite concentration increases with decreasing growth temperature [2]. [V_{Ga}] from Fig. 3 is in good agreement with 1/3 [As_{Ga}^+] measured on the same samples [6] (V_{Ga} is a triple acceptor according to theory [13]). The results provide further evidence that Ga vacancies are the dominating acceptors in as-grown LT-GaAs. The density of Ga vacancies indeed reflects that of the As antisites. More general, the appearance of Ga vacancies is dependent on the variation of the composition with changing growth conditions.

D - Conclusions

Positron annihilation gives strong evidence to identify the vacancy defects in as-grown LT-GaAs to be Ga monovacancies, either isolated or part of a defect complex. The vacancy concentration is sufficient to account for the compensation of As_{Ga}^+ antisites. This conclusion is true for a wide range of growth conditions, i.e. growth temperatures and BEP ratios. Ga monovacancies are no longer detectable after annealing at 600°C. Annealing produces larger vacancy aggregates, probably associated with As precipitates, which are then the dominant positron traps.

Acknowledgments

We acknowledge support from the Bundesland Sachsen-Anhalt. The work at Berkeley was supported by the Air Force Office of Scientific Research under Grant No. F49620-95-1-0091.

References

[1] M. Kaminska, Z. Liliental-Weber, E. R. Weber, T. George, J. B. Kortright, F. B. Smith, B.-Y. Tsaur, and A. R. Calawa, Appl. Phys. Lett. **54** (1989) 1881.

[2] X. Liu, A. Prasad, J. Nishio, E. R. Weber, Z. Liliental-Weber, and W. Walukiewicz, Appl. Phys. Lett. **67** (1995) 279 .

[3] K. M. Yu, M. Kaminska, and Z. Liliental-Weber, J. Appl. Phys. **72** (1992) 2850 .

[4] D. J. Keeble, M. T. Umlor, P. Asoka-Kumar, K. G. Lynn, and P. W. Cooke, Appl. Phys. Lett. **63** (1993) 87.

[5] J. Gebauer, R. Krause-Rehberg, S. Eichler, M. Luysberg, H. Sohn, and E. R. Weber, Appl. Phys. Lett. **71** (1997) 638.

[6] M. Luysberg, H. Sohn, A. Prasad, P. Specht, Z. Liliental-Weber, E. R. Weber, J. Gebauer, and R. Krause-Rehberg, J. Appl. Phys. **83** (1998) 561.

[7] L. Liszkay, C. Corbel, L. Baroux, P. Hautojärvi, M. Bayhan, A. W. Brinkmann, and S. Tatarenko, Appl. Phys. Lett. **64** (1994) 1380.

[8] J. Gebauer, R. Krause-Rehberg, C. Domke, P. Ebert, and K. Urban, Phys. Rev. Lett. **78** (1997) 3334.

[9] T. Laine, K. Saarinen, J. Mäkinen, P. Hautojärvi, C. Corbel, L. N. Pfeiffer, and P. H. Citrin, Phys. Rev. B **54** (1996) R11050.

[10] J. Gebauer, R. Krause-Rehberg, S. Eichler, W. Bauer-Kugelmann, G. Kögel, W. Triftshäuser, M. Luysberg, H. Sohn, and E. R. Weber, Mat. Sci. Forum **255-257** (1997) 204.

[11] S. Ruvimov, C. Dicker, B. J. Washburn, and Z. Liliental-Weber, Appl. Phys. Lett. **72** (1998) 226.

[12] R. Krause-Rehberg and H. S. Leipner, Appl. Phys. A **64** (1997) 457.

[13] S. B. Zhang and J. E. Northrup, Phys. Rev. Lett. **67** (1991) 2339.

Structural and Photoluminescence Analysis of Er Implanted LT–GaAs

R.L. Maltez[1], Z. Liliental-Weber[1], J. Washburn[1], M. Behar[2],
P.B. Klein[3], P. Specht[4], E.R. Weber[4]

[1] *Materials Science Division, Lawrence Berkeley National Laboratory, University of California, Berkeley, California 94720*
[2] *Instituto de Física, UFRGS, Porto Alegre, RS, Brazil 91501-970*
[3] *Naval Research Laboratory, Washington DC 20375-5347*
[4] *University of California, Department of Materials Science, Berkeley, California 94720*

Abstract

Characteristic 1.54 μm Er^{3+} emission has been observed from Er-implanted and annealed, low-temperature grown GaAs (Be doped and undoped samples). Cross-sectional transmission electron microscopy (TEM) studies reveal very little structural damage for elevated temperature implants up to an Er **total** fluence of 1.36×10^{14} Er/cm^2. No Er emission was observed from any of the as-implanted samples. Post-implantation annealing optimized the Er PL emission intensity near 650°C for Be doped samples, while for undoped samples it was near 750°C anneal. The beginning of Er precipitation was observed after 750°C annealing for the above Er fluence but, on increasing the Er concentration by higher implantation fluences, it was observed after annealing at 650°C. These precipitates are likely ErAs.

A. Introduction

Rare earth (RE) doped III-V semiconductors have potential applications in semiconductor light-emitting diodes (LEDs), lasers and as components for optical communication technology since they may emit a sharp, temperature stable and host independent luminescence. In particular, the doping with RE may result in a photon emission near 1.54μm. This is an important wavelength because it corresponds to the minimum absorption in standard silica-based optical fibers [1,2]. In our previous work [3], we have reported this Er emission for low temperature grown GaAs [4] doped with Be (LT-GaAs:Be). The doping with Be strongly suppresses As precipitation upon annealing [5]. Be doped layers grown at ~300°C show the same short photo-carrier lifetime that is obtained for undoped LT-GaAs grown at ~200°C [5]. In previous work [3] we studied samples Er implanted at room temperature (RT) and at 300°C temperature with an Er concentration around 10^{19} Er/cm^3. In the present work we investigate LT-GaAs:Be samples implanted to three different Er concentrations. We also studied undoped LT-GaAs and standard GaAs.

B. Experimental procedures

Undoped as well as Be doped ([Be] = 7×10^{19} cm^{-3}) LT-GaAs layers 1.5μm thick were grown by molecular beam epitaxy on (001) GaAs substrates with a fixed As/Ga beam equivalent pressure ratio of 20. The GaAs substrate temperature during growth was 200°C and 300°C for the undoped LT-GaAs and LT-GaAs:Be, respectively. Three Er plateau distributions in LT-GaAs:Be were performed at 300°C implantation temperature. In each case, three consecutive Er implants at energies of 480 keV, 155 keV and 40 keV were carried out. For the sample with the lowest **total** fluence, these implants were performed, respectively, to the partial fluences of 9.4×10^{12} Er/cm^2, 3.1×10^{12} Er/cm^2 and 1.1×10^{12} Er/cm^2, which were projected to result in an almost flat Er concentration of 1×10^{18} Er/cm^3 from 10 nm to 145 nm. These estimates were based on each individual profile as calculated by the TRIM [6] program.

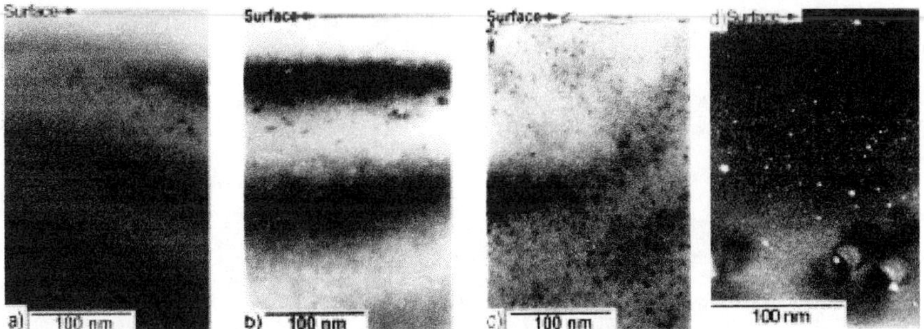

Figure 1. Cross-sectional TEM micrographs with the (001) surface at the top edge of each figure. Micrographs **a**, **b** and **c** correspond to the 10^{20} Er/cm^3 as-implanted (Be doped), 10^{19} Er/cm^3 annealed at 650°C (Be doped), 10^{19} Er/cm^3 annealed at 650°C (undoped), respectively, under BF $0\bar{2}2$ image condition. Picture **d** is a DF 200 image condition taken after a special annealing at 800°C for a half hour performed on a 10^{20} Er/cm^3 sample.

The intermediate Er concentration of 1×10^{19} Er/cm^3 was obtained by increasing each of the above fluences by a factor of 10. Similarly, the highest Er plateau concentration, estimated as 1×10^{20} Er/cm^3, was obtained by increasing the above fluences by a factor of 100. The implantation carried out for the undoped LT-GaAs layers, as well as for the standard GaAs, corresponds to the intermediate concentration, i.e., ~10^{19} Er/cm^3 along the plateau. Rapid thermal annealing (RTA) for 30 s at 650°C and at 750°C were performed. These samples were analyzed by cross-sectional TEM and photoluminescence (PL) measurements at 10 K.

C. Results

All of the as-implanted samples show almost no defects visible by TEM (extended defects) after implantation. This shows that the great majority of the ion damage is being recovered during implantation at 300°C. As an example, Fig. 1a is an image taken from a 10^{20} Er/cm^3 Be doped sample with no post-implantation annealing. Figure 1.b is a micrograph from a 10^{19} Er/cm^3 Be doped sample annealed at 650°C. There is no significant damage for samples up to this Er concentration, while a strong damaged area was observed for the sample 10^{20} Er/cm^3. This sample shows a high concentration of dislocation loops between the depth of 75 nm to ~240 nm below the sample surface (end of range area). A 750°C temperature anneal eliminated the small loops for the 10^{19} Er/cm^3 sample, but there was no marked change for the sample 10^{20} Er/cm^3, which just shows a sharper transition at the 90 nm depth (between the clean area and the end of range area).

Fig 1c is a micrograph from a 10^{19} Er/cm^3 undoped sample annealed at 650°C. Two strong differences can be seen when compared to the Be doped case (Fig. 1b): there is a very intense As precipitation and a complete absence of ion implantation damage. The average size of the As precipitates is about 5 nm (in diameter), after the 650°C annealing (Fig. 2c), and about 9 nm, after annealing at 750°C. The complete absence of extended defects is a surprising result that may be correlated to this strong As precipitation. Standard GaAs shows damage levels very similar to those observed for the Be doped case. Within the implanted area, after annealing, a different type of precipitate was observed from contrast analysis under dark field (DF) 200 image condition. Fig. 1d is an example of these special precipitates, which probably

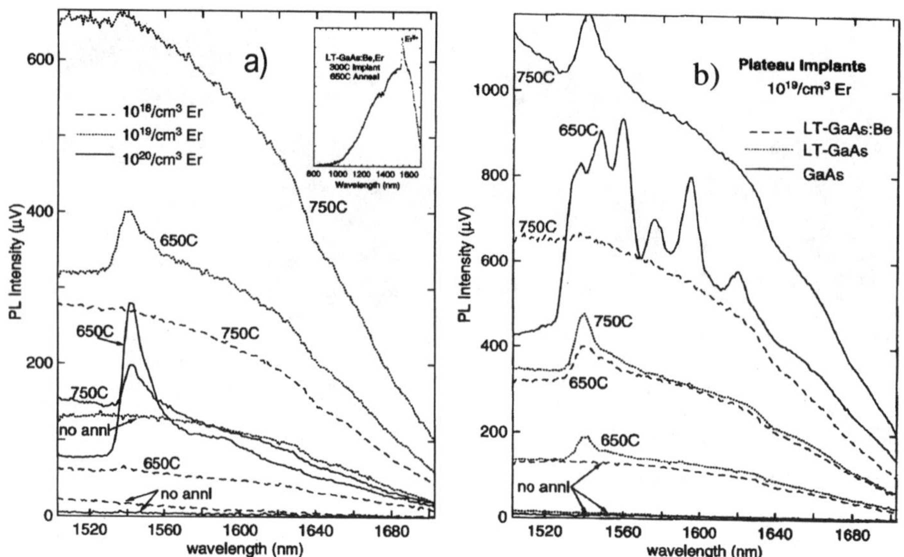

Figure 2. (a) PL spectra comparing 10^{18} Er/cm³ (dashed lines), 10^{19} Er/cm³ (doted lines) and 10^{20} Er/cm³ (full lines) Be doped samples. Insert: Er emission, from 10^{19} Er/cm³ sample annealed at 650°C, on top of the broad background PL over an extended wavelength range. (b) PL spectra from 10^{19} Er/cm³ samples comparing LT-GaAs:Be (dashed lines), undoped LT-GaAs (doted lines) and standard GaAs (full lines).

are ErAs precipitates. The figure is a DF 200 micrograph from the 10^{20} Er/cm³ sample after a special annealing at 800°C for 1/2 h. After RTA annealings, the precipitates with white contrast appear with a diameter not larger than 1.5 nm. In Fig. 1d, larger precipitates (~4 nm in diameter) can be seen at the end of range area. The 10^{19} Er/cm³ Be doped sample also showed this kind of precipitates, but only after annealing at 750°C and particularly at dislocation cores.

Figure 2a shows PL spectra comparing results from 10^{18} Er/cm³ (dashed lines), 10^{19} Er/cm³ (doted lines) and 10^{20} Er/cm³ (full lines) LT-GaAs:Be samples. The measurements were performed on as-implanted, annealed at 650°C and at 750°C samples. The emission in that spectral range is dominated by a broad background signal (except for the 10^{20} Er/cm³ sample annealed at 650°C) that increases monotonically with the annealing temperature. This background is apparently due to a very broad PL band (or bands) that appears to peak near 1500 nm, as shown in the insert in Fig. 2a for the 10^{19} Er/cm³ sample annealed at 650°C (the Ge detector fall-off for λ>1600nm distorts the spectrum). The exact origin of this background is not yet understood [3]. The optimum annealing temperature for the 1.54µm Er emission seems to be around 650°C for Be doped samples. Fig. 2a suggests, for Er concentrations up to 10^{19} Er/cm³, a linear relationship exists between the Er emission for the optimum cases and the **total** Er implantation fluence, in agreement to the previous work [3]. However, the Er peak from the 10^{20} Er/cm³ plateau implant exhibits a sublinear behavior, since it is just 2.4 times higher than that from the 10^{19} Er/cm³ plateau implant. As previously discussed [3], we have speculated that the formation of "Er-As complexes" is the reason for the drastic Er emission decrease observed for the 750°C anneal.

Figure 2b shows PL spectra comparing results from Be doped and undoped LT-GaAs as well as for standard GaAs used as reference. The background is present for all cases, even for the standard GaAs, indicating that its origin can not be explained just by the excess of point defects in LT-GaAs materials. Concerning the Er emission, after the 650°C anneal, the standard GaAs shows a highly structured spectrum that suggests a different type of Er site. After a 750°C anneal the line shape of the standard GaAs becomes similar to that of the LT-GaAs, suggesting similar site occupancy. As expected, the standard GaAs shows a much higher Er emission, since this material has a much lower concentration of point defects to compete, with the Er radiative centers, as trapping sources for the photo-carriers. Figure 2b exhibits a different anneal temperature dependence for Be-doped and undoped LT-GaAs. For undoped material, the Er emission from the 750°C annealed sample is almost twice than that from the 650°C annealed sample. This difference may well be explained just by taking into account that Be doped samples are more stable upon annealing [5], with regard to carrier lifetime changes, than undoped samples. From this point of view, longer carrier lifetimes for undoped samples annealed at 750°C might enhance the probability of photo-carriers being trapped at Er radiative centers, even if the concentration of these centers had decreased upon this anneal.

D. Conclusion

In this work we have observed the characteristic Er emission, the crystalline quality and the Er precipitation of Er implanted Be doped and undoped LT–GaAs. The 1.54 μm Er emission is optimized around 650°C annealing for LT-GaAs:Be but, undoped LT-GaAs shows higher emission intensity after a 750°C anneal. TEM measurements clearly show that samples implanted at 300°C have excellent crystalline quality up to a total fluence of 1.36×10^{14} Er/cm^2 (10^{19} Er/cm^3 plateau samples). The damage recovery of the undoped material is even better and does not show any evidence of extended defects, but does show intense As precipitation. Er precipitates, which probably are ErAs with the rock-salt structure, have been observed after annealing. The results show that Ion Implantation is able to incorporate Er in the matrix as a supersaturated solution.

Acknowledgements

This research was supported by AFOSR-ISSA-90-009. R.L. Maltez work was also supported by CAPES-Brasília/Brasil postdoctoral fellowship.

References:

[1] G.S. Pomrenke, H. Ennen and W. Haydl, *J. Appl. Phys.* **59** (1986) 601.
[2] A. Polman, *J. Appl. Phys.* **82** (1997) 1.
[3] R.L. Maltez , Z. Liliental-Weber, J. Washburn, M. Behar, P.B. Klein, P. Specht, E.R. Weber, submitted to publication.
[4] *Low Temperature (LT) GaAs and Related Materials*, edited by G.L. Witt, R. Calawa, U. Mishra and E. Weber, *MRS. Symp. Proc.* **241** (1991); F.W. Smith, p. 3 in previous reference.
[5] P. Specht, S. Jeong, H. Sohn, M. Luysberg, A. Prasad, J. Gebauer, R. Krause-Rehberg and E.R. Weber, *Int. Conf. on Defects in Semiconductors Proc.*, ICDS 19, Aveiro, Portugal, 1997, in press.
[6] J.F. Ziegler, J.P. Biersack and U. Littmark, *The Stopping and Range of Ions in Solids*, Vol. 1, Pergamon Press, Oxford, 1985.
[7] I. Poole, K.E. Singer, A.R. Peaker, *J. Crystal Growth* **121** (1992) 121.

Twinning of As precipitates in low-temperature GaAs during high temperature annealing

S. Ruvimov[1], Ch. Dieker[2], J. Washburn and Z. Liliental-Weber,
Lawrence Berkeley National Laboratory, Berkeley, CA 94720

Twinning of nano-scale As precipitates formed in low-temperature GaAs during high temperature annealing was studied by high resolution electron microscopy. Symmetrical $\{\bar{1}10\bar{4}\}$ twins were nucleated during crystallization of amorphous or liquid-like As precipitates. Nucleation of rhombohedral As appears to be most favorable on the short facet that is parallel to the $\{111\}$B plane of the GaAs. Twin formation then often initiates on the long facet bounded by the $\{111\}$A plane. The crystallization front terminates with a void because of the shrinkage of the As volume during solidification.

A. Introduction

Twinning in hexagonal metals is a well-known phenomenon [1]. Recently twin formation in small particles has attracted special attention [2] because twins effect the physical properties of these nano-scale objects. Twinning in As precipitates has been reported [3,4] in connection with thermal annealing of low temperature GaAs (LT-GaAs). LT-GaAs has unique properties (a high resistivity and short carrier lifetime (0.2-20 ps)) promising for device applications [5]. These properties are likely to be associated with the excess (up to 1.5%) of As and the high density (2×10^{20} cm^{-3}) of As_{Ga} antisite defects in LT-GaAs [6]. Thermal annealing is often required in device processing, but it effects the properties of LT-GaAs [7-11]. It is still not clear whether As precipitates [8] or the changes in the point defect density in LT-GaAs [12] are mainly responsible for the changes in physical properties associated with thermal annealing. Therefore, the detailed structure of twins in As precipitates and mechanisms of their formation are of interest . Here we report twinning of As precipitates in LT-GaAs layers during rapid thermal annealing at 850 and 950 °C as studied by high resolution electron microscopy and image simulation.

B. Experimental

0.5 μm-thick GaAs layers were deposited at 200 °C by molecular-beam-epitaxy (MBE) on (001) GaAs substrates with a 100 nm thick GaAs buffer layer. A 20 nm thick GaAs cap layer was then grown at 600 °C on top of the LT-GaAs. The growth rate of the LT-GaAs was 1μm/h and the beam equivalent pressure ratio was 10. The samples were annealed at 850 and 950 °C in a nitrogen atmosphere for 20, 40 and 60 min. HREM was carried out on a Topcon 002B electron microscope operated at 200 kV. Cross-sections, perpendicular and parallel to the major flat of the GaAs wafers, were prepared using mechanical polishing, dimpling followed by ion milling.

C. Results and discussion

Fig. 1 shows typical high resolution (a) and conventional (b) electron microscopy images of As precipitates formed in a LT-GaAs layer after annealing at 850 °C. All precipitates have a rhombohedral structure with a=0.376 nm and c=1.055 nm and a polyhedral shape with distinct facets. The facets are parallel to $\{111\}$, $\{001\}$, $\{110\}$ and $\{11\bar{2}\}$ crystallographic planes of GaAs. The long facets bounded by the $\{111\}$A planes of the GaAs are abrupt while

---

[1] On leave from A.F. Ioffe Physical Technical Institute, St. Petersburg 194021, Russia, e-mail: ruv@mh1.lbl.gov
[2] on leave from Institut fuer Schicht- und Ionentechnik (ISI) Foeschunszentrum Juelich GmbH, D-52425 Juelich, Germany

the others have a lot of steps. Therefore, each precipitate has the shape of a rounded tetrahedron that is close to the equilibrium shape of GaAs [13].

The fact that the shape of the As precipitate follows the crystallography of the GaAs matrix can be understood taking into account that the annealing temperature was higher than the melting point of As (810 °C at 24 atm.). Therefore, during annealing at 850 °C the As probably was liquid (or amorphous) and recrystallizes when the temperature falls. Because excess As in the LT-GaAs is mostly in the form of As antisite defects As_{Ga} the growth of precipitates appears to be due to diffusion of As_{Ga} toward the precipitate. When an As_{Ga} reaches the As/GaAs interface it disappears from the GaAs lattice and two As atoms are added to the precipitate. This also results in an elimination of one Ga-As pair at the precipitate facet. Solidification results in shrinkage of the As volume by 8-10 % and, hence, in a void formation [4,14]. Because the volume fraction of voids is typically less than 8-10 %, significant additional diffusion of As from the GaAs toward the precipitates during solidification appears to take place.

The precipitates in Figs.1 (a) and (b) contain at least one or two twins and a void, some have more. Voids were always observed in the twinned precipitates. This suggests that both twins and voids were formed during solidification of the As. Twins are visible on the TEM image of Fig.1 (b) when their planes are perpendicular to the image plane (this is the case for about 1/6 of all the precipitates). They are practically not visible if the twin plane is inclined. In contrast, voids are visible regardless of their location inside the precipitates. From one to four voids per precipitate were observed in our samples indicating multiple growth fronts and, possibly, multiple twinning.

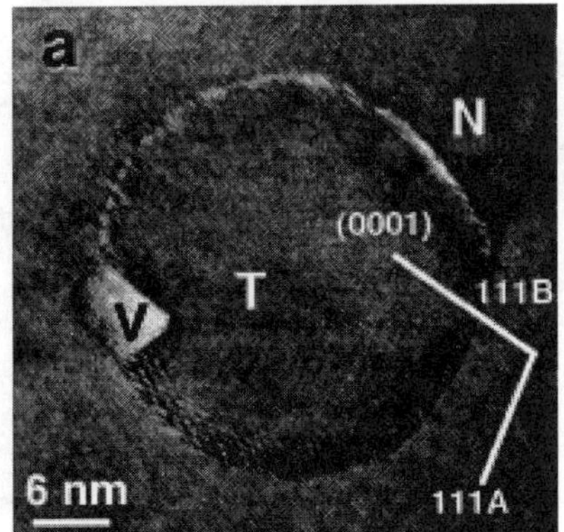

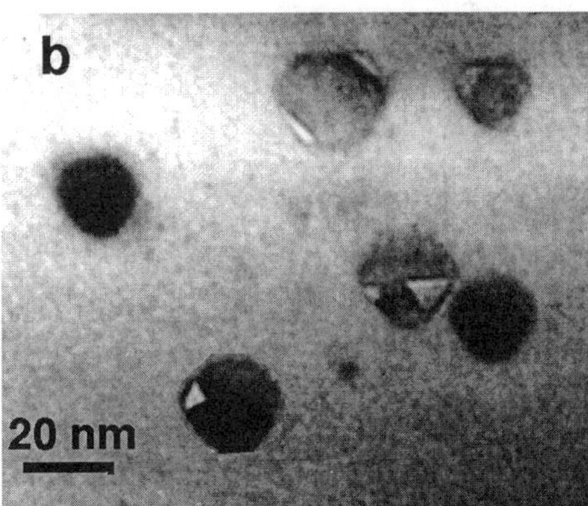

Fig. 1   HREM (a) and conventional (b) images of As precipitates in the GaAs matrix formed after annealing at 850 °C for 20 min. The nucleation of the As crystal at the (111)B facet is marked by N.

The twin in Fig. 1 is inclined by an angle $\alpha=16°$ with respect to the (001) surface of the LT-GaAs layer. Taking into account the 3m orientation relationship for the first As grain with the GaAs matrix this angle should be about 14°. This is close to the experimental value. The difference of 2° between experimental and theoretical values might be related to the variations in dislocation densities at different facets. Statistical analysis of twin orientation shows a large scatter in angle α between the twin plane and the layer surface [see Fig. 1(b)] indicating a variety of possible orientation relationships for the first As nuclei with the GaAs matrix.

The 3 possible orientation relationships reported in the literature [3,10,11] for As precipitates in a GaAs matrix give eight different angles, ±14, ±86, ±40 and ±50° [4], between the twin plane and the layer surface when the twin plane is parallel to

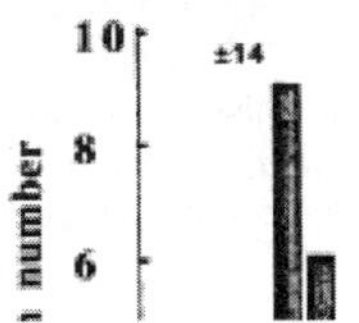

the transmitted beam. Among all the variants the angles ±14° is the most favorable (Fig. 2). This suggests that the 3m orientation relationship is the mostly energetically favorable in agreement with previous observations [10].

Fig. 2. The angular distribution of twin planes with respect to the layer surface.

The twin [Fig. 1 (a)] initiates at the {111}A facet and ends at the void which is bounded by the (0001) planes of As. A twin divides the precipitate into two grains which usually differ in size. The larger grain was probably nucleated first. It has an almost epitaxial orientation relationship with the {111}B facet of the GaAs: $(\bar{1}2\bar{1}0)_{As}$ // $(0\bar{1}1)_{GaAs}$ and $(0001)_{As}$// $(111)_{GaAs}$ [Fig. 1 (a)] indicating the nucleation of As on the short {111}B facet first. The fact that the nucleation of rhombohedral As is more favorable on the {111}B facet compared to on {111}A might be explained by the difference in their surface energies [14]. This orientation relationship which has 3m symmetry was typically observed for As precipitates in LT-GaAs after annealing [9]. It is also typical for epitaxial growth of rhombohedral or hexagonal layers on (111) oriented cubic substrates. Two other orientation relationships with 2mm and m symmetry [10] were observed, but relatively rare.

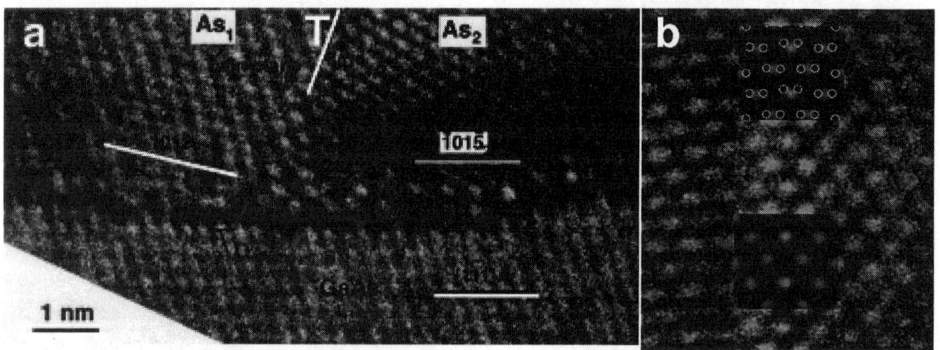

Fig. 3 Twin nucleation on the (111)A facet (a). The atomic steps at the interface are marked by arrows. D is a misfit dislocation, T is the twin. (b) is a HREM image of the twin. Inserted are the atomic model and the simulated image.

Due to the difference in structure between GaAs and As, the first As grain is misaligned with respect to the long {111}A facet: the $\{\bar{1}012\}_{As}$ planes are inclined to the $\{111\}_{GaAs}$ by 13° [Fig. 3 (a)]. However, the twinned grain is well oriented with the {111}A facet [$(\bar{1}105)_{As}$ //$\{111\}_{GaAs}$] and has a better match with the GaAs lattice at this facet (a higher density of coincidence sites for the As and GaAs lattices in the facet plane).

This observation suggests the following scenario of twinning. On cooling the As crystallization starts at the most rough, {111}B facet by epitaxial growth of the As nucleus resulting mostly in the 3m orientation relationship between the nucleus and the GaAs crystalline lattices. Its growth gradually increases strain energy due to the misfit in lattice parameters between As and GaAs which is periodically released by the formation of a misfit dislocation at the As/GaAs interface. When the solidification front reaches the {111}A facet, misalignment between $\{\bar{1}012\}_{As}$ and $\{111\}_{GaAs}$ crystallographic planes and the lattice mismatch cause a stress in the As crystal which is partially released by the formation of misfit dislocations. This stress may result in twinning which locally reduces the strain energy because the twinned As

128

grain has a better fit to the GaAs. Further crystallization then proceeds by growth of both the original and the twinned grains. Due to the difference in time of their formation, the first grain should be larger than the second twinned grain in agreement with the observations [see e.g. Fig.1 (a)].

The atomic structure of the twin that gives the best fit between experimental and simulated HREM images is shown in Fig. 3 (b). This structure has a mirror symmetry across the twin plane. The interatomic distances and the local atomic arrangement for first neighbors in the twin area are similar to those of the bulk. Therefore, taking into account the metallic character of bonds in As, one can expect the energy of the twin boundary to be rather low. The $(\bar{1}10\bar{4})$ crystallographic plane is one of the favorable twinning planes for many hexagonal metals [1].

In conclusion, the atomic structure and geometry of twins in As precipitates were investigated by HREM suggesting the twinning mechanism. Twin formation is associated with crystallization of amorphous or liquid As precipitates during cooling. The nucleation of rhombohedral As appears to be most favorable at the short $\{111\}$B facet while twin nucleation usually takes place at the long $\{111\}$A facet. Twinning results in a decrease of the strain energy that develops in the As crystal during its epitaxial growth. The crystallization usually terminates at a void because of the shrinkage of the As volume associated with solidification.

This research was supported by AFOSR-ISSA-90-0009. The use of the facility of the National Center for Electron Microscopy in Lawrence Berkeley National Laboratory supported by the U.S. Department of Energy under Contract No. DE-ACO3-76SF00098 is greatly appreciated.

1. H.S. Rosenbaum in *Deformation Twinning*, Ed. R.E. Reed-Hill, J.P. Hirth, H.C. Rogers, Gordon & Breach Sci. Publ., N.Y. 1964
2. J. Douin, U. Dahmen, K.H. Westmacott, Phil. Mag. B **63** 867 (1991)
3. Z. Liliental-Weber, A. Claverie, P. Werner, W. Schaff, and E. R. Weber, in Proc.of the 16th Int. Conf. on Defects in Semiconductors, Bethlehem, PA, 1991 (Materials Science Forum, 83-87), p.1045
4. S. Ruvimov, C. Dieker, J. Washburn, Z. Liliental-Weber, Appl.Phys.Lett. **72** 226 (1998).
5. F.W. Smith, H.Q. Lee, V. Diadiuk, M.A. Hollis, A.R. Calawa, S. Gupta et al., Appl. Phys.Let. **54** 890 (1989)
6. M. Kaminska, Z. Liliental-Weber, E.R. Weber, T. George, et al., Appl. Phys. Lett. **54** 1881 (1989)
7. M.R. Melloch, K. Mahalingham, N. Otsuka, J.M. Woodall, and A.C. Warren, J. Cryst. Growth **111**, 39 (1991); Melloch, et al Ann.Rev.Mat.Sci. **25** 547 (1995); E. Harmon, M.R.Melloch, J.M.Woodall, D.D. Nolte, N. Otsuka, C.L. Chang, Appl.Phys.Lett. **63** 2248 (1993); A.C. Warren, J.M. Woodal, F.L. Freeouf, D. Gruschkosky, D.T. McInturff, M.R. Melloch, and N. Otsuka, Appl. Phys. Lett. **57** 1331 (1990)
8. Z. Liliental-Weber, MRS Symp. Proc. **v. 241** 101 (1991)
9. A. Claverie & Z. Liliental-Weber, Phil. Mag. **65** 1003 (1992)
10. Z. Liliental-Weber, A. Claverie, J. Washburn, F. Smith, and R. Calawa, Appl. Phys. A **53** 141 (1991)
11. Z. Liliental-Weber, J. Ager, D. Look, X.W. Lin, X. Liu, J. Nishio, K. Nichols, W. Schaff, W. Swider, K. Wang, J. Washburn, E. R. Weber, and J. Whitaker, in Semi-Insulating III-V Material, ed. by M. Godlewski (World Scientific, N.Y., 1994), p.305
12. X. Liu, A. Prasad, W.M. Chen, A. Kurpiewski, Z. Liliental-Weber, and E.R. Weber, in Semi-Insulating III-V Materials, ed. by A. Claverie (World Scientific, Toulouse, 1996), p.347
13. N. Moll, A. Kley, E. Pehlke, and M. Scheffler, Phys. Rev. **54**, 8844 (1996)
14. A.G. Cullis, P.D. Augustus, and D.J. Stirland, J. Appl. Phys. **51**, 2556 (1980)

Time-Resolved Reflectivity Measurement of Thermally Stabilized Low Temperature Grown GaAs doped with Beryllium

R. Zhao, P. Specht, R. C. Lutz, N. W. Pu[*], S. Jeong[†], J. Bokor[*] and E. R. Weber

Department of Material Science and Mineral Engineering, University of California, Berkeley, CA 94720
[*]Department of Electrical Engineering and Computer Sciences, University of California, Berkeley, CA 94720
[†] Material Science Dev., Lawrence Berkeley National Laboratory, Berkeley, CA 94720

Abstract

The influence of Be-doping on carrier lifetime in low temperature grown (LT) GaAs has been studied using time-resolved reflectivity measurement. The ultrafast photogenerated carrier lifetime decreases with increasing Be doping concentration, because the total amount of excess As, i.e., arsenic antisites (As_{Ga}), and the ionized arsenic antisites are increased due to mechanical and electrical compensation between As_{Ga} and Be acceptor. It was observed that the carrier lifetime in Be-doped LT-GaAs can be even shorter after high temperature annealing, which is contrary to the behaviors of undoped samples.

A. Introduction

In the last ten years, GaAs grown by molecular beam epitaxy (MBE) at low substrate temperature has attracted extensive attention [1-4]. Low temperature (LT) GaAs, characterized by the excess As incorporation, shows high resistivity, ultrafast carrier lifetime and relatively high carrier mobility. It has been reported that LT-GaAs can improve significantly the behaviors of electronic and optoelectronic devices.

Upon annealing at high temperature, most of the excess As forms As precipitates and/or diffuses out of LT layer [4], which can be detrimental to the neighbor layers and lead to a degradation of devices. Undoped LT-GaAs has to be annealed to achieve high resistivity, which can be explained by the midgap pinning of Fermi Level due to high concentrations of remaining As_{Ga} [5]. We have explored Be-doped LT-GaAs to enhance the thermal stability of As_{Ga} [6, 7]. Be acceptors, with smaller atomic radius than Ga, compensate the larger As_{Ga} both mechanically and electrically. Thus, the As diffusion can be suppressed and also large fractions of ionized As_{Ga} (As_{Ga}^{+} or As_{Ga}^{++}) can exist, which determine the carrier lifetime in this material. This work focuses on the influence of Be-doping on the carrier lifetime.

B. Experiments

As-rich and Be-doped GaAs films were grown on semi-insulating GaAs substrate using a Varian Gen II MBE system in the temperature range of 240-320°C. The growth temperature was calibrated by the diffused reflectance spectroscopy (DRS, Thermionics

NW). The 1.5µm thick epilayers were deposited at the growth rate of 1µm/h with a fixed beam equivalent pressure (BEP) ratio at 20. Be concentrations were calibrated with SIMS. Proximity annealing at 600°C and 700°C for 30 minutes was performed in Ar atmosphere.

The carrier lifetimes were measured using time-resolved transient reflectivity with a pump-probe setup. A mode-locked Ti-sapphire laser is operating at λ=800nm with a pulse width 100fs at 100MHz repetition rate. The 100mW pump and 5mW probe were cross-polarized and focused down to spot sizes of 50µm and 30µm, respectively. The carrier excitation level is estimated to be 10^{18}/cm^3. The standard lock-in detection was used to measure the transient reflectivity.

C. Results and Discussions

Fig. 1 shows the lattice mismatch and resistivity versus Be doping concentration for samples grown at 250°C. The dual compensation is seen: as the [Be] increases, the samples show decreasing resistivity and turn from n-type to p-type, accompanied by the transition from positive lattice mismatch tc negative mismatch [7]. After annealing above 600°C, the lattices are totally relaxed in the undoped and lightly Be-doped samples, while no change for samples with high Be-doping was observed. Be doping increases As incorporation and suppresses the formation of As precipitates [7]. It is possible to grow lattice matched samples by tuning the growth temperature and Be-doping level, for example, if an LT-GaAs layer is grown at Tg=250°C, 2×10^{19}/cm^3 Be are needed to obtain lattice matching.

The transient reflectivity results are shown in Fig. 2 from the above discussed as-grown samples. The photogenerated carrier lifetime is defined as the time it takes for the

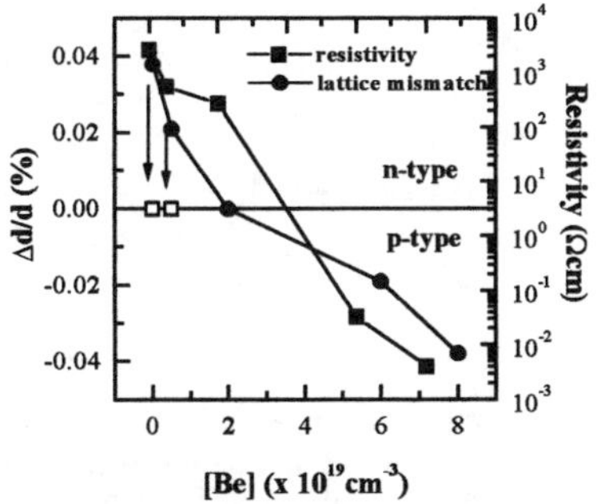

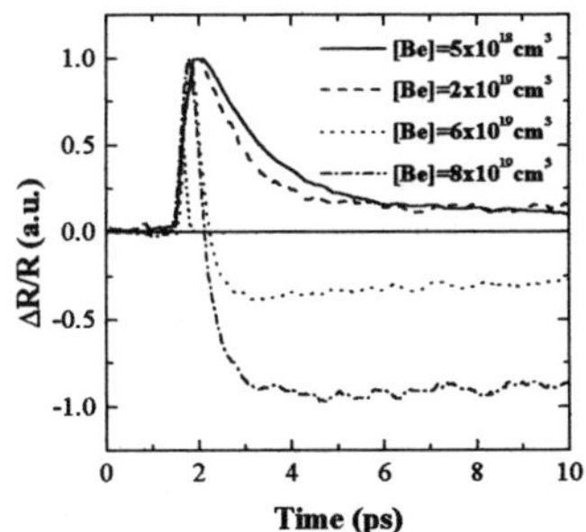

<table>
<tr><td>

Fig.1 lattice mismatch and resistivity change with Be-doping level. The arrows indicate lattice relaxation after annealing.

</td><td>

Fig. 2. Time-resolved transient spectra from as-grown samples, T$_g$=250°C. The pulse autocorrelation is also shown.

</td></tr>
</table>

reflectivity change to drop to 1/e times its maximum value. From Fig. 2, the transient signals decay faster as the [Be] increases, and large negative signals are observed for high Be concentrations. Another feature is the delay of the maximum signal at low Be concentrations. It is known that the ionized As$_{Ga}$ is the effective carrier trapping center in as-grown undoped LT-GaAs [8]. With increasing [Be], more As$_{Ga}$ become ionized due to electrical compensation, which further shortens carrier lifetime. For very high Be-doping level, the sample which is p-type due to Be overcompensation may also contain significant concentration of doubly charged antisites, As_{Ga}^{++}, which have even larger electron capture cross section because of the strong Coulombic attraction. Therefore, it is expected that the fastest trapping occurs in highly Be-doped samples, as the very sharp peaks in Fig.2 indicate. The trapped carriers built up in the midgap levels can again be excited by the probe beam, which accounts for the negative signal. The slow decay of the induced absorption shows that the filled traps have very slow final recovery processes with the recombination either between free carriers and trapping carriers or between the trapped carriers [9].

The carrier lifetime is summarized in Fig. 3. Notice that though high Be-doping level favors the short lifetime, the sample is p-conducting. However it is possible to simultaneously achieve high resistivity and sub-picosecond carrier lifetime in as-grown LT-GaAs:Be by adjusting growth temperature and Be doping [7].

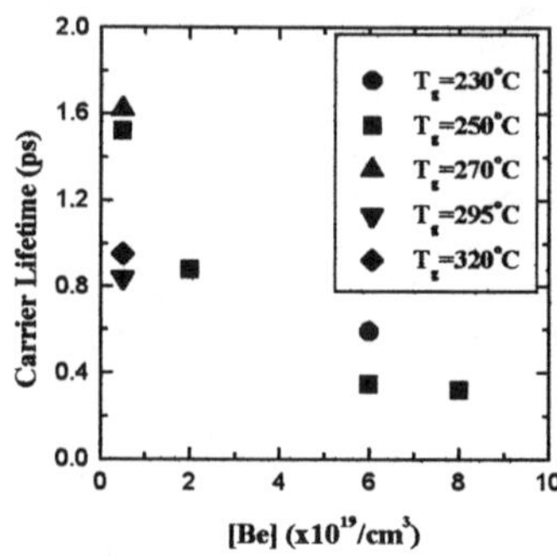

Fig.3. Carrier lifetime as a function of [Be] for samples grown at different temperatures. Notice that the experimental error, ~0.09ps, is determined by the pulse width.

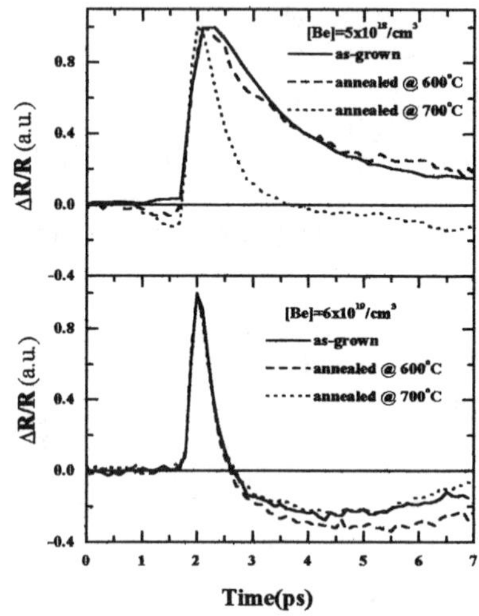

Fig.4. Influence of annealing on transient reflectivity. The samples were grown at 250°C with different [Be]: 5×10^{18}/cm^3 (*upper*) and 6×10^{19}/cm^3 (*lower*).

The influence of thermal annealing on the carrier lifetime is shown in Fig. 4. Undoped LT-GaAs is known to exhibit longer carrier lifetimes after annealing because of the As_{Ga} loss due to As precipitate formation [3, 5]. The carrier lifetime of the Be-doped LT-GaAs with 5×10^{18}/cm^3 Be concentration, however, shows little change after 600°C annealing, while the carrier lifetime gets shorter after 700°C annealing. For the sample with high Be concentration, above 6×10^{19}/cm^3, no annealing effect is seen from transient reflectivity measurement, even after annealing at 700°C. In the sample with high Be-doping level, a large fraction of As_{Ga} might be already doubly positively charged due to Be compensation, and the As precipitates are prevented because of the Coulombic repulsive force between them. But in the sample with less Be doping, the As_{Ga} decreases by As precipitate formation upon thermal annealing and leads to an increase of doubly charged arsenic antisites, As_{Ga}^{++}, which can shorten the carrier lifetimes.

D. Conclusion

Time-resolved reflectivity measurements show that the carrier lifetime is shortened in Be-doped LT-GaAs. The annealing behavior confirms the thermal stabilization in LT-GaAs:Be with adjusted Be concentration dependent on the growth temperature. The compensation mechanism between acceptors are important characteristics for the properties of LT-GaAs:Be. Further studies on the compensation mechanism between As_{Ga} and Be are necessary to understand the unique properties of this material. We have already succeeded in growing optimized LT-GaAs with high resistivity, short carrier lifetime and enhanced thermal stability.

Acknowledgement

The authors would like to thank Dr. M. Luysberg for fruitful discussion. This work has been funded by the Air Force Office of Scientific Research under grant No.s F49620-98-1-0135, and F49620-94-1-0464 (JSEP). We appreciated MBE and XRD support from Integrated Material Laboratory (IML), UC-Berkeley.

Reference

[1] F. W. Smith, A. R. Calawa, C. L. Chen, M. J. Manfra and L. J. Mahoney, *IEEE Electron Device Lett.*, **9** (1988) 77.
[2] D. C. Look, *Thin Solid Films*, **231** (1993) 61.
[3] Z. Liliental-Weber, H. J. Cheng, S. Gupta, J. F. Whitaker, K. Nichos and F. W. Smith, *J. Electron Mater.*, **22** (1993) 1465.
[4] A. J. Lochtefeld, M. R. Melloch, J. C. P. Chang and E. S. Harmon, *Appl. Phys. Lett.* **69** (1996) 1465.
[5] X. Liu, A. Prasad, J. Nishio, E. R. Weber, Z. Liliental-Weber and W. Walukiewicz, *Appl. Phys. Lett.*, **67** (1995) 279
[6] P. Specht, S. Joeng, H, Sohn, M. Luysberg, A. Prasad, J. Gebauer, R. Klause-Rehberg and E. R. Weber, Mater. Sci. Forum 258-263, 951(1997).
[7] R. C. Lutz, P. Specht, R. Zhao, S. Joeng, J. Bokor and E. R. Weber, to be published.
[8] Z. Liliental-Weber, et al, in Proc. Of 8th Conf. On Semi-insulating III-V materials, ed. M. Godlewski (World Scientific, 1994), p305.
[9] U. Siegner, R. Fluck, G. Zhang and U. Keller, *Appl. Phys. Lett.*, **69** (1996) 2556.

Stoichiometry-dependent vacancy formation in n-doped GaAs

J. Gebauer, M. Lausmann, and R. Krause-Rehberg

Fachbereich Physik, Martin-Luther-Universität Halle-Wittenberg, D-06099 Halle, Germany

Abstract - The formation of vacancies in n-doped GaAs with variable stoichiometry was studied by means of positron annihilation. Highly Te-doped (n=1.5-5×10^{18} cm^{-3}) GaAs samples were annealed at high temperatures under variable arsenic vapor pressure (p$_{As}$) and rapidly quenched to room temperature. We found that monovacancies exist in Te-doped GaAs regardless of the annealing conditions. In contrast, the vacancy *concentration* was found to be dependent on p$_{As}$, i.e. on the stoichiometry. The vacancy concentration increases like $p_{As}^{1/4}$ and saturates at high As pressures (3-6 bar). This is a proof that the dominant vacancies in Te-doped As-rich GaAs are related to Ga vacancies, in agreement to theoretical results. Moreover, the vacancies were independently identified by using the Doppler broadening coincidence technique.

A - Introduction

A compound semiconductor like GaAs will always exhibit a deviation from ideal stoichiometry. This deviation is, by necessity, governed by native defects. First principle calculations [1] and equilibrium thermodynamics [2] showed that in highly n-doped, As-rich GaAs Ga vacancies should be the dominant native defects. There is, however, conflicting evidence on As vacancies [3] as well as on Ga-vacancy related defects [4-6] in n-doped GaAs by means of positron annihilation. Moreover, a direct relationship between stoichiometry (or, equivalent, As-pressure in equilibrium with the crystal during high temperature annealing) and vacancy concentration has never been confirmed experimentally, although such annealing is known to influence electrical and structural properties of GaAs [7]. In the present study we will use annealing of GaAs under defined As vapor pressure to change the stoichiometry and study the changes of the vacancy density by means of positron annihilation. It is our aim to directly confirm a relation between As pressure and vacancy density. On the other hand, information on the microscopic nature of vacancies may be obtained.

B - Experimental details

The samples used in this study were Te-doped GaAs, grown by the LEC method. Three crystals with carrier densities of 1.5,1.5-2, and 5×10^{18} cm^{-3} were investigated. Annealing was performed in sealed quartz-glass ampoules arranged in a two-zone furnace. The sample temperature (T_S) was independently varied from the temperature of the metallic As source ($T_{As} < T_S$). Then, the As vapor pressure (p_{As}) depends only on T_{As} and can be obtained from Gokcen´s review [8]. The ampoules were fast quenched into water after annealing.

Positrons may be trapped by vacancies during diffusion in a crystal. This results in an increasing positron lifetime and a narrowing of the annihilation momentum distribution. An increase in the average positron lifetime (τ_{av}) compared to the lifetime in defect-free material (τ_{bulk}) is a clear sign for positron trapping in vacancies. A decomposition of the lifetime spectra provides the vacancy-related lifetime (τ_{vac}) specific for a given defect. The defect density (c_{vac}) can be obtained according to $c_{vac}=1/(\tau_{bulk}\,\mu)\,(\tau_{av}-\tau_{bulk})/(\tau_{vac}-\tau_{av})$ where μ is the trapping coefficient. μ is about 1×10^{15} s^{-1} for negatively charged monovacancies in semiconductors at 300 K [9] and will vary like $T^{-1/2}$. However, μ is independent of temperature for

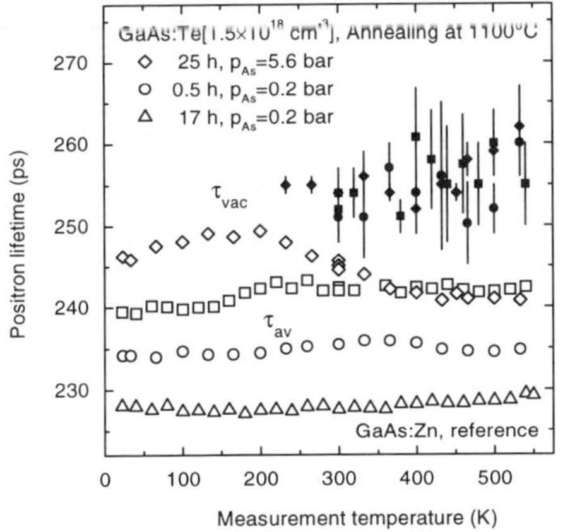

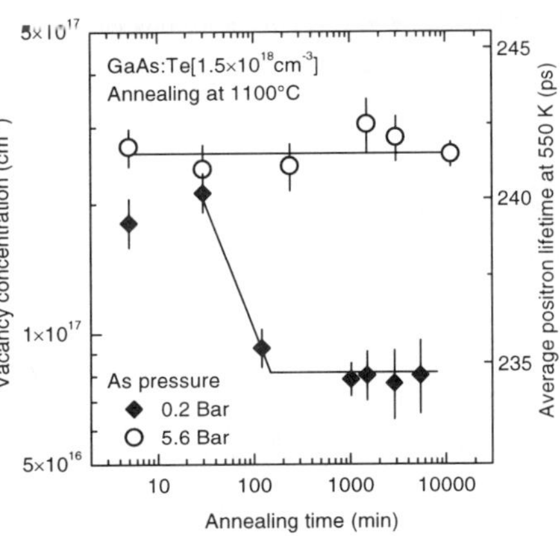

Fig.1: Average positron lifetime (open symbols) and defect related lifetime (solid symbols) vs. the measurement temperature in GaAs:Te annealed at 1100°C, compared to a GaAs:Zn reference.

Fig. 2: Concentration of monovacancies in GaAs:Te vs. the annealing time for As pressures of 0.2 and 5.6 bar. Error bars correspond to the reproducibility of the average positron lifetime.

neutral vacancies [10]. Positron lifetime spectroscopy was performed with a conventional fast-fast system (resolution 250 ps) in a temperature range of 20-550 K. Coincident Doppler-broadening-spectroscopy was done with two Ge-γ-detectors. This technique allows observation of annihilation with high-momentum electrons due to a reduction of the background. This can be used to identify the chemical surrounding of a vacancy [11].

C - Results and discussion

Fig. 1 shows typical positron-lifetime results in Te-doped GaAs annealed at 1100°C for different times and under different As pressures. τ_{av} is well above τ_{bulk} measured in a GaAs:Zn reference, indicating the presence of vacancies. A defect-related lifetime of about 254 ps was obtained, compatible with monovacancies and in agreement to results in as-grown GaAs:Te [4]. The increase in τ_{av} with decreasing T indicates a negative vacancies. The decrease of τ_{av} at $T < 100K$ shows the additional presence of negative ions, which trap positrons in their shallow potential only at low temperatures [12]. The density is, however, lower than that of the vacancies, since the decrease in τ_{av} is weak. A distinct overall decrease in the average positron lifetime was observed for a longer annealing time and reduced p_{As} (Fig. 1). The vacancy density is, thus, indeed related to the annealing conditions, i.e. duration and As pressure. Both were varied to obtain the desired relation between stoichiometry and vacancies.

In Fig. 2, the concentration of monovacancies (n_{vac})in GaAs:Te[1.5×10^{18} cm^{-3}] is shown versus the annealing time for As pressures of 0.2 and 5.6 bar. n_{vac} was calculated from τ_{av} at 550 K using a trapping coefficient of 0.74×10^{15} s^{-1} as extrapolated by the $T^{1/2}$ law. The use of the data at 550 K allows more safe interpretation, since the influence of shallow positron traps on the calculation of vacancy concentrations is negligible at high temperatures. The vacancy concentration reaches constant levels for long annealing times. Almost no changes were observed for p_{As}=5.6 bar. In contrast, n_{vac} was distinctly decreased at p_{As}=0.2 bar. The carrier density, however, increases to be about $1.8-1.9\times10^{18}$ cm^{-3} after annealing with p_{As}=0.2 bar but is nearly constant for the samples annealed with p_{As}=0.2 bar. The increase in the car-

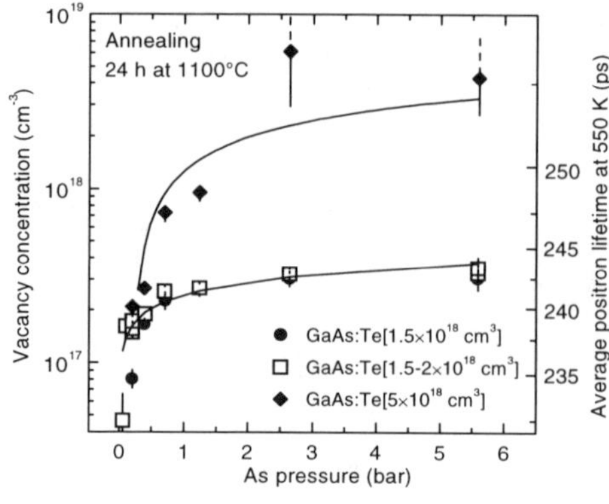

Fig. 3: Concentration of monovacancies in Te doped GaAs as a function of the As vapor pressure during annealing at 1100°C for 24 h.

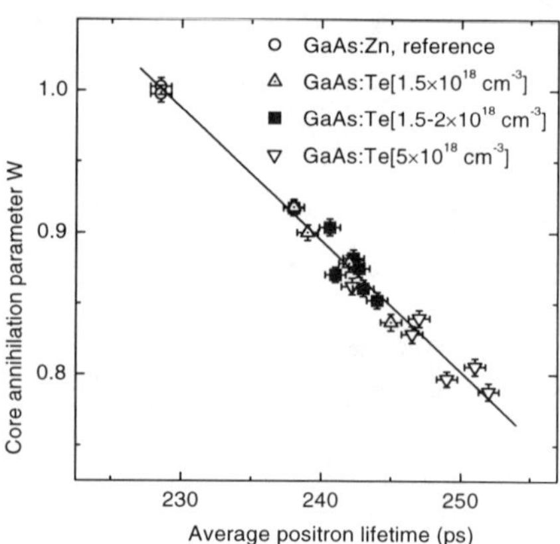

Fig. 4: Core annihilation parameter W (relative to defect free GaAs) vs. τ_{av} in annealed GaAs: Te. Each point correspond to another sample.

rier density is compatible with the decrease in the vacancy concentration as observed by positrons, since vacancies are known to reduce the carrier density in Te-doped GaAs [4]. We conclude, that As vapor pressure and vacancy concentration reaches equilibrium during annealing at 1100°C within a time of some 100 minutes. An annealing time of about 1440 min (1 day) was applied in the following.

The next step was the investigation of samples annealed under different As vapor pressures. The concentration of monovacancies for differently Te-doped GaAs· is shown versus the As vapor pressure during annealing in Fig. 3. We observed in all three crystals an increase in the vacancy concentration with increasing p_{As}, i.e. if the stoichiometry becomes more As-rich. This is a strong indication for Ga vacancies. If As vacancies would be the dominant defect, a decrease of the vacancy concentration with increasing As pressure is expected. The drop of n_{vac} at p_{As}<1-2 bar occurs in the same range of p_{As} where an increase of the resistivity is observed in annealed semi-insulating GaAs [7].

A defect-related lifetime, τ_{vac}, of 254±5 ps was found in all samples under investigation, indicating the same type of monovacancies to be present. However, V_{As} as well as V_{Ga} related defects in GaAs should exhibit the same defect-lifetime of about 260 ps. In order to check independently for the defect type, we applied the Doppler broadening coincidence technique to our samples. A linear relation is found in Fig. 4 between τ_{av} and the core annihilation parameter W, characterizing the intensity of annihilation with high momentum core electrons (p=15-20 m$_0$c). Such a linear variation is expected if only one type of vacancies trap positrons. Thus, the same type of monovacancies is present in all samples. A W parameter W_v=0.77 for the vacancies in Te-doped GaAs can be deduced from the data in Fig. 4, much lower than the theoretical value for V_{As} (W=0.96, [13]). However, the value is in good agreement with experimental (W=0.74, [5]) and theoretical (W=0.69, [13]) values for V_{Ga}. A detailed investigation of the momentum distribution allows a direct identification of the vacancies to be Te$_{As}$-V_{Ga} complexes [14].

For Ga vacancies, a kinetic equilibrium reaction like ¼ As$_4$ (gas) $\leftrightarrow$ V_{Ga}+GaAs should apply (As$_4$ is the dominating As vapor species, [8]). Thus, the concentration of Ga vacancies is expected to be proportional to $p_{As}^{¼}$ [2]. Indeed, we found the vacancy density after

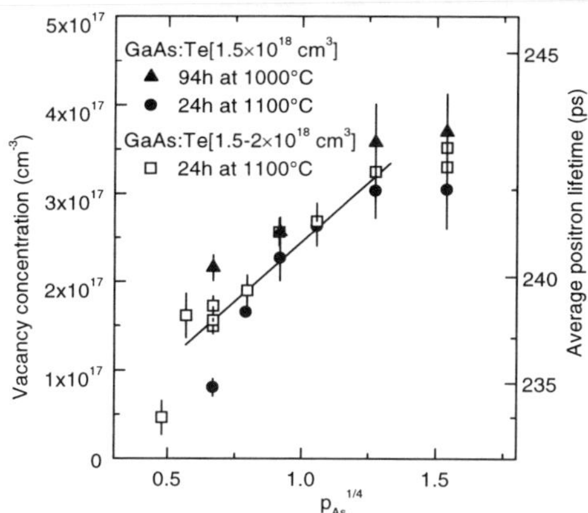

Fig. 5: Vacancy concentration in annealed GaAs:Te ($n=1\text{-}2\times10^{18}$ cm^{-3}) as a function of $p_{As}^{1/4}$.

annealing to be proportional to $p_{As}^{1/4}$ (Fig. 5). We found a saturation of the vacancy density at $3\text{-}4\times10^{17}$cm^{-3} for high As pressures ($p_{As} \geq 2.6$ bar), most probably related to the As-rich phase boundary in highly Te doped GaAs. The vacancy concentration is in accordance to theoretical calculations [1,2].

In summary, we observed the formation of monovacancies during annealing of highly Te-doped GaAs under defined As vapor pressure. The vacancy density is directly related to p_{As} during annealing. The vacancies were identified to be Ga vacancies and should be the dominant defect in as-rich, n-doped GaAs. The observation of V_{Ga} related defects in highly n-doped, as-grown GaAs [4-6], can thus be understood to be related to As-rich material.

Acknowledgements

We are indebted to Dr. M. Jurisch (Freiberger Compound Materials) for the support with the 10^{18} cm^{-3} doped sample material. We thank Dr. G. Lippold (University Leipzig) for Hall-effect measurements. The work was supported by the Deutsche Forschungsgemeinschaft.

References

[1]   S. B. Zhang and J. E. Northrup, Phys. Rev. Lett. **67** (1991) 2339.

[2]   T. Y. Tan, H. M. You, and U. M. Gösele, Appl. Phys. A **56** (1993) 249.

[3]   K. Saarinen, P. Hautojärvi, P. Lanki, and C. Corbel, Phys. Rev. B **44** (1991) 10585.

[4]   R. Krause-Rehberg, H. S. Leipner, A. Kupsch, A. Polity, and T. Drost, Phys. Rev. B **49** (1994) 2385.

[5]   T. Laine, K. Saarinen, J. Mäkinen, P. Hautojärvi, C. Corbel, L. N. Pfeiffer, and P. H. Citrin, Phys. Rev. B **54** (1996) R11050.

[6]   J. Gebauer, R. Krause-Rehberg, C. Domke, P. Ebert, and K. Urban, Phys. Rev. Lett. **78** (1997) 3334.

[7]   O. Oda, H. Yamamoto, M. Seiwa, G. Kano, T. Inoue, M. Mori, H. Shikamura, and M. Oyake, Semicond. Sci. Technol. **7** (1992) A215.

[8]   N. A. Gokcen, Bulletion of Alloy Phase Diagrams **10** (1989) 11.

[9]   R. Krause-Rehberg and H. S. Leipner, Appl. Phys. A **64** (1997) 457.

[10] M. J. Puska and R. M. Nieminen, Rev. Mod. Phys. **66** (1994) 841.

[11] M. Alatalo, B. Barbiellini, M. Hakala, H. Kaupinen, T. Korhonen, M. J. Puska, K. Saarinen, P. Hautojärvi, and R. M. Nieminen, Phys. Rev. B **54** (1996) 2397.

[12] K. Saarinen, P. Hautojärvi, A. Vehanen, R. Krause, and G. Dlubek, Phys. Rev. B **39** (1989) 5287.

[13] S. Pöykkö, M. J. Puska, M. Alatalo, and R. M. Nieminen, Phys. Rev. B **54** (1996) 7909.

[14] J. Gebauer, M. Lausmann, T.E.M. Staab, R. Krause-Rehberg, and M. J. Puska, to be published (1998).

IV.

Defects in III-V Compounds

Comparison of Single and Double Barrier Pseudomorphic Doped-Channel GaInP/GaInAs/GaAs HFET's: Performance and Isolation Properties

Sheldon McLaughlin, Xiangang Xu, S. P. Watkins, and C. R. Bolognesi

Compound Semiconductor Device Laboratory (CSDL)
Department of Physics & School of Engineering Science
Simon Fraser University, Burnaby BC, V5A 1S6, CANADA

Doped-channel GaInP/GaInAs/GaAs pseudomorphic Heterostructure Field-Effect Transistors (HFETs) were fabricated both with and without a GaInP barrier layer under the GaInAs channel. Two sets of alloy compositions were investigated: $Ga_{0.85}In_{0.15}As$ channels with $Ga_{0.51}In_{0.49}P$ barriers, and $Ga_{0.70}In_{0.30}As$ channels with $Ga_{0.62}In_{0.38}P$ barriers. For each composition, 1.1 µm gate length devices with single and double barrier structures showed similar maximum drain currents (I_{dmax}), and peak transconductances (g_m), while the double barrier devices showed reduced output conductances (g_{ds}) compared to the single barrier devices, resulting in a 30-50% larger voltage gain. RF measurements showed a higher maximum frequency of oscillation (f_{max}) for the double barrier devices of both compositions. For mesa etch isolation of the double barrier devices, it was found to be necessary to remove the bottom GaInP barrier layer to achieve satisfactory device isolation. We speculate that a parasitic 2-dimensional electron gas may be formed at the interface between the bottom GaInP barrier and the GaAs buffer layer.

A. Introduction

Pseudomorphic Heterostructure Field Effect Transistors (HFETs) fabricated from the AlGaAs/GaInAs/GaAs materials system are well established as microwave circuit elements. Barrier layers of GaInP have several advantages over AlGaAs, including higher breakdown voltages, a lack of DX centers, lower noise performance, and simplified processing due to the greater resistance of GaInP to oxidation in air. Several groups have reported promising results for GaInP/GaInAs/GaAs HFET's. [1-5] To date, most investigations of this material system have used heterostructures with the GaInAs channel grown directly on a GaAs buffer layer. From previous studies on other materials systems, it is expected that a second GaInP barrier layer inserted under the channel will provide additional electron confinement and further improve the device peformance. For modulation doped structures, the bottom GaInP barrier also allows a second donor layer to be inserted under the channel, thus increasing the sheet carrier concentration in the channel. Wang et al. [5] have taken this approach, and demonstrated a transistor with improved performance as a result of the additional doping. However, the different doping levels make it more difficult to determine whether differences in performance between the single and double barrier devices are due to the better confinement provided by the bottom barrier, or simply due to the increased doping level. In this work, we use doped channel rather than modulation doped structures, and compare the performance of transistors fabricated from heterostructures which are identical except for the absence or presence of a bottom GaInP barrier layer.

B. Material Growth and Device Fabrication

Two pairs of single/double barrier structures were grown by metal-organic chemical vapour deposition (MOCVD). Details of the growth will be published elsewhere. The first pair of structures consisted of a GaAs buffer layer, a 200 Å undoped $Ga_{0.51}In_{0.49}P$ (lattice matched) bottom barrier layer on one of the structures, a 150 Å Si-doped $Ga_{0.85}In_{0.15}As$ channel, a 200 Å undoped $Ga_{0.51}In_{0.49}P$ top barrier, and a 400 Å Si-doped (3×10^{18} cm^{-3}) GaAs ohmic contact layer. For the second pair, the channel composition was increased to $Ga_{0.70}In_{0.30}As$, and the barriers were Ga-rich strained $Ga_{0.62}In_{0.38}P$. The channel thickness was reduced to 80 Å, and undoped GaAs smoothing layers with thicknesses of 30-40 Å were inserted between the GaInAs channel and the GaInP barriers. For all structures, the channel doping was adjusted to give a Hall sheet concentration (with the doped GaAs cap removed) of 3×10^{12} cm^{-2}. For each pair, the single and double barrier structures were grown in consecutive runs.

HFET's were fabricated using a conventional liftoff and mesa isolation process. The fabrication processes for the single and double barrier devices in each pair were similar, but there were slight process modifications between the $Ga_{0.85}In_{0.15}As$ channel devices and the $Ga_{0.70}In_{0.30}As$ channel devices. Ohmic contacts were formed by electron-beam evaporation of AuGe/Ni/Au and subsequent annealing. The annealing steps consisted of either 60 seconds at 395°C in a heated reaction chamber, or 20 seconds at 390°C in a Rapid Thermal Processing (RTP) unit. Device isolation was achieved by a series of selective mesa etches. Arsenide layers (GaAs and GaInAs) were etched with solutions of either 20:1 Citric Acid:H_2O_2 (where the citric acid solution was formed by mixing 1 g of anhydrous citric acid with 1 mL of water) or 3:1:50 H_3PO_4:H_2O_2:H_2O. The GaInP barriers were etched with 1:1:1 H_3PO_4:HCl:H_2O. Prior to gate metal deposition, a gate recess was etched in the GaAs cap using a selective (GaAs over GaInP) etch as noted above. Ti/Au gate metal was deposited by electron-beam evaporation. It is expected that optimization of the ohmic contacts and the gate recess etch will yield further improvements in device performance.

C. Device Results and Discussion

On-wafer dc measurements on 1.1 μm gate length devices were made with an HP-4145 Semiconductor Parameter Analyzer. Figures of merit were averaged for five to ten devices per wafer. The average values of maximum drain current (I_{dmax}), peak extrinsic transconductance (g_m), output conductance (g_{ds}), and maximum voltage gain (g_m/g_{ds}) are shown in Table 1. The peak drain current typically occurred at a gate bias of Vgs=+2 V, where the gate leakage current (I_g) was greater than the conventionally defined breakdown value of 1 mA/mm. We have chosen to define the drain current at a gate bias which gives I_g=1 mA/mm as the "maximum drain current".

It is clear that the single and double barrier devices have nearly identical maximum drain currents, and similar values of transconductance for both the $Ga_{0.85}In_{0.15}As$ and $Ga_{0.70}In_{0.30}As$ channel structures. The major difference between the single and double barrier devices is in the ouptut conductances. In both cases, the double barrier devices have lower output conductances than the single barrier devices, resulting in a higher voltage gain for the double barrier devices. Output conductance effects in HFET's have been attibuted to current flowing from drain to source through the sub-channel layers. [6] Since the double barrier structure presents a larger potential barrier to electron injection into the sub channel layers, it should result in smaller currents in the sub-channel layers and thus a lower output conductance.

Table 1: DC performance of 1.1 μm gate length devices at V_{ds}=5 V.

Wafer	I_{dmax} (mA/mm)	g_m (mS/mm)	g_{ds} (mS/mm)	g_m/g_{ds}
Single Barrier, $Ga_{0.85}In_{0.15}As$ channel	360	230	4.4	52
Double Barrier, $Ga_{0.85}In_{0.15}As$ channel	360	180	2.7	69
Single Barrier, $Ga_{0.70}In_{0.30}As$ channel	620	200	13.6	15
Double Barrier, $Ga_{0.70}In_{0.30}As$ channel	630	190	8.7	23

An HP-8510B Network Analyzer with GGB Picoprobes was used to characterize the on-wafer microwave performance of 0.6 μm gate-length HFET's. The extrinsic unity current-gain frequency, f_T, and the unilateral maximum frequency of oscillation, f_{max}, were obtained from measured scattering parameters, then corrected for gate contact pad capacitance effects. At a drain-source bias of 5 V, we obtain f_T=16 GHz and f_{max}=45 GHz for single barrier $Ga_{0.85}In_{0.15}As$ devices, f_T=19 GHz and f_{max}=63 GHz for double barrier $Ga_{0.85}In_{0.15}As$ devices, f_T=15 GHz and f_{max}=49-55 GHz for single barier $Ga_{0.70}In_{0.30}As$ devices, and f_T=18 GHz and f_{max}=74-78 GHz for double barrier $Ga_{0.70}In_{0.30}As$ devices. The ratio f_{max}/f_T is consistently higher for the double barrier devices of both compositions.

The presence of the bottom GaInP barrier layer also affects device isolation. With a selective mesa isolation etch to the bottom of the GaInAs channel on the double barrier structure, there was a significant leakage between ohmic contact pads on adjacent devices. This leakage was not present in the single barrier structures, and was eliminated in the double barrier structures when the mesa etch was extended to the bottom of the GaInP layer. Since the breakdown properties of the devices indicate a high resistivity GaInP layer, we speculate that the leakage may be due to the formation of

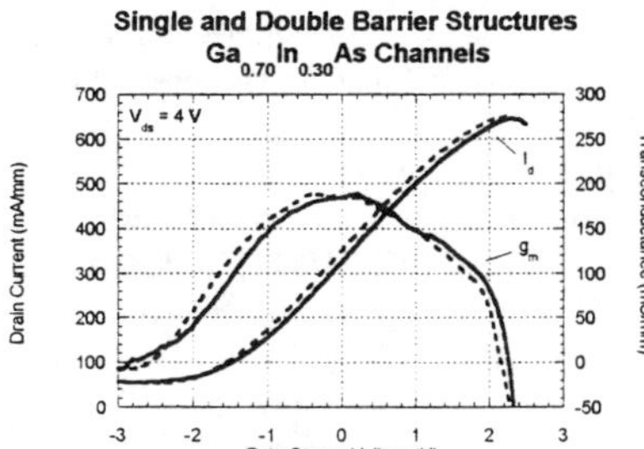

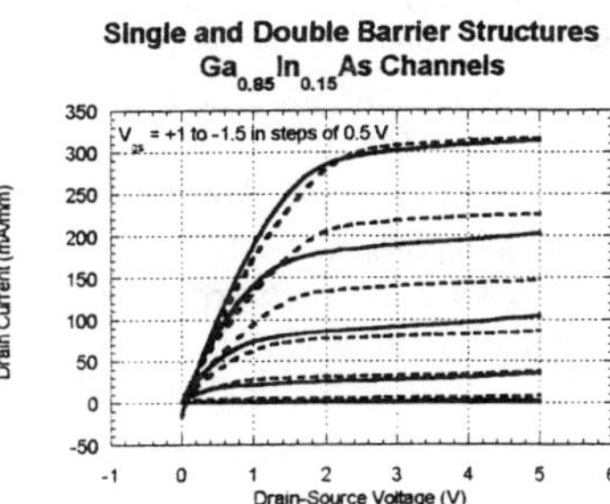

Figure 1 DC performance of 1.1 μm gate length HFET's. Solid lines indicate single barrier structures, and dashed lines indicate double barrier structures.

an additional 2-dimensional electron gas at the interface between the bottom GaInP layer and the GaAs buffer layer.

D. Conclusions

We have demonstrated doped channel GaInP/InGaAs/GaAs HFET's with single and double GaInP barriers, for two combinations of channel and barrier compositions: $Ga_{0.85}In_{0.15}As$ channels with $Ga_{0.51}In_{0.49}P$ barriers, and $Ga_{0.70}In_{0.30}As$ channels with $Ga_{0.62}In_{0.38}P$ barriers. For both compositions, the single and double barrier variations exhibit similar maximum drain currents and peak transconductances, but the double barrier devices have a lower output conductance at the transconductance peak, resulting in a higher maximum voltage gain g_m/g_{ds}. This result can be explained by assuming that the output conductance is due to drain to source currents in the sub-channel layers. The GaInP/GaInAs heterointerface below the channel in the double barrier devices presents a larger barrier to electron injection into the sub-channel layers than the GaAs/GaInAs interface in the single barrier devices. Microwave measurements show that the double barrier devices have a significantly higher f_{max}, and a larger f_{max}/f_T ratio than the single barrier devices. To achieve sufficient device isolation by mesa etching, both the GaInAs channel and the bottom GaInP barrier in double barrier structures must be removed.

References

1. Geiger, D., *et al.*, *InGaP/InGaAs HFET with high current density and high cut-off frequencies.* IEEE Electron Device Letters, 1995. **16**(6): p. 259-261.
2. Lin, Y.-S., T.-P. Sun, and S.-S. Lu, *$Ga_{0.51}In_{0.49}P/In_{0.15}Ga_{0.85}As/GaAs$ pseudomorphic doped-channel FET with high current density and high breakdown voltage.* IEEE Electron Device Letters, 1997. **18**(4): p. 150-153.
3. Pereiaslavets, B., *et al.*, *GaInP/InGaAs/GaAs graded barrier MODFET grown by OMVPE: design, fabrication, and device results.* IEEE Transactions on Electron Devices, 1996. **43**(10): p. 1659-1663.
4. Pereiaslavets, B., *et al.*, *Narrow-channel GaInP/InGaAs/GaAs MODFET's for high-frequency and power applications.* IEEE Transactions on Electron Devices, 1997. **44**(9): p. 1341-1348.
5. Wang, Y.C., *et al.*, *Single- and double-heterojunction pseudomorphic $In_{0.5}(Al_{0.3}Ga_{0.7})_{0.5}P/In_{0.2}Ga_{0.8}As$ high electron mobility transistors grown by gas source molecular beam epitaxy.* IEEE Electron Device Letters, 1997. **18**(11): p. 550-552.
6. Han, C.J., *et al.* *Short channel effects in submicron self-aligned gate heterostructure field effect transistors.* in *IEDM.* 1988. San Francisco, CA: IEEE.

Formation of a quasi-neutral region in Schottky diodes based on semi-insulating GaAs

Markus Rogalla[*], Kay Runge

*Faculty of Physics, Albert-Ludwigs-University, Hermann-Herder-Str. 3,
79104 Freiburg, Germany*

A model is developed for the electric field distribution beneath the Schottky contact in semi-insulating (SI) GaAs particle detectors. The model is based on a field-enhanced electron capture of the EL2 defect. The influence of the compensation mechanism in SI GaAs on the field distribution and leakage current density of the detectors will be discussed. The detailed understanding allows a device optimisation.

A. Introduction

There has been wide interest in semi-insulating (SI) GaAs as detectors material for X-ray and particle detectors operated at room temperature. It has been shown that Schottky diodes made on SI-GaAs work well as radiation detectors [1-3]. According to Ramo´s theorem the charge collection is mainly determined by the mobility, lifetime of the charge carriers and the electric field distribution in the detector. Experimental evidence was found by various techniques, among others the Franz-Keyldysh [4] method and OBICs (Optical Beam Induced Currents) [5] for a formation of a quasi-neutral region beneath the Schottky contact resulting in a linear increase of the active region with bias voltage. In addition the leakage current density and the full active voltage (where the high electric field reaches the back contact of the device) strongly depends on the compensation ratio between the EL2 and the shallow acceptors [6,7].

A detailed analysis of the charge collection process of the electrons in the detectors as a function of the ionisation ratio of the EL2-defect indicates a field-enhanced electron capture by the EL2. This limits the electron lifetime and therefore the charge collection efficiency of the device [8]. In this work a detector model based on the drift-diffusion model, Shockley-Read-Hall occupation factor and field-enhanced electron capture of the EL2 will be presented which explain the formation of the quasi-neutral region. The model is applied to the substrate resistivity dependence of the leakage current density and the full active voltage.

B. The model

Previously published calculations of the electric field distribution in particle detectors made of SI-GaAs are based on a simultaneous solution of the transport and Poisson equation coupled through the quasi Fermi-levels of the charge carriers [9,2]. These models can not quantitatively explain the experimentally observed formation of the quasi-neutral region with electric fields of about one $V/\mu m$ [4]. McGregor [10] was the first to propose that a field enhanced electron capture of the EL2 present in SI-GaAs may allow the formation of the quasi-neutral region similar to that described for n^+-contacts on Au-doped Silicon [11]. But with his ad hoc extension of the drift-diffusion model it was not possible to quantitatively verify the field distribution with the defect densities present in the SI-GaAs.

---

[*] Tel. +49 761 203 5897, Fax. +49 761 203 5931, e-mail rogalla@ruhpb.physik.uni-freiburg.de

The charge state of a defect can be generally described by the Shockley-Read-Hall (SRH) occupation factor:

$$f_{\mathrm{SRH}} = \frac{e_p + c_n\, n}{e_n + c_n\, n + e_p + c_p\, p} \tag{1}$$

where the occupation of the defect is determined by the electron/hole emission rates (e_n/e_p) and capture coefficient (c_n/c_p). In equilibrium, and for low electric fields the SRH occupation factor is equivalent to the occupation determined via the quasi-Fermi levels [7]. In SI-GaAs where the Fermi-level distance to the conduction band is between 0.5 and 0.7 eV, the SRH occupation factor of the most important defect the EL2 can be approximated by

$$f_{\mathrm{SRH}}(n, c_n, e_n) \approx \frac{1}{1 + c_n\, n/e_n} \tag{2}$$

because of the small hole capture coefficient ($c_p = 3.4\,10^{-11}$ s^{-1}cm^3) and hole emission rate ($e_p = 5\,10^{-5}$ s^{-1}) [7] of the EL2 and negligible hole concentration p in the valence band. Therefore the charge state of the EL2 is mainly determined by the electron concentration n in the conduction band, the electron capture coefficient and emission rate. In equilibrium values are given as $c_n = 3.8\,10^{-8}$ s^{-1}cm^3 and $e_n = 0.045$ s^{-1} [12]. The electron emission rate was reported to be approximately independent of the field strength up to several 10 V/µm. In contradiction to that, the electron capture coefficient of the EL2 increases with the electric field strength [8,13]. This means less electrons are needed in the conduction band to preserve the same ionisation of the EL2 under high and low electric fields.

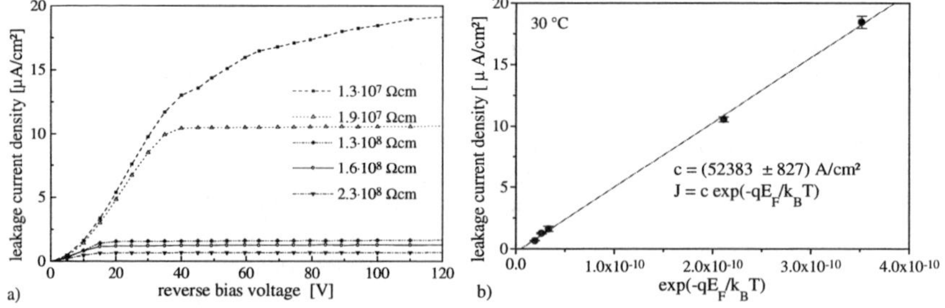

Figure 1: a) The IV characteristic of Schottky diodes made on SI-GaAs with different resistivities between 1.3 - 23·10^7 Ωcm. b) The leakage current density in the saturation region at 100 V bias as a function of exp($-qE_F/k_B T$) where E_F denotes the equilibrium Fermi-level position in the corresponding semi-insulating material.

Experimental evidence was found that the formation of the quasi-neutral region beneath the Schottky contact takes place independently of the substrate resistivity (between $1\,10^7$ and $2.3\,10^8$ Ωcm) [6], which means that it is independent of the equilibrium electron concentration. In addition it was found that the leakage current density is independent of the contact material and varies only with the substrate resistivity. Figure 1 a) show the IV-characteristics of reverse biased Schottky diodes (for a detailed description of the device geometry and technology see ref. [8]). All diodes show a saturation of the leakage current density with increasing reverse bias voltage. The saturation leakage current density increases with decreasing bulk resistivity from 0.6 to 18.4 µA/cm². Assuming that the increase of the EL2 capture cross-section in the high electric field region of the diodes is independent of the material, the electron concentration n_{hf} beneath the contact is according to the eq. 2 proportional to the equilibrium electron concentration n given by the Fermi-level in the bulk material.

Neglecting diffusion of electrons through the high field region and generation and recombination currents, the leakage current density J only depends on the electron drift velocity v_d and concentration n_{hf} and is given by

$$J \approx q\, v_d\, n_{hf}. \tag{3}$$

The saturation of the drift velocity for electric fields higher than 1 V/μm explains the experimentally observed saturation of the leakage current. The saturation value of the leakage current density is then proportional to the Boltzmann factor of the equilibrium electron concentration (see Fig. 1b)). A comparison of the equilibrium and high field electron concentration, assuming a saturation drift velocity of $1\cdot10^7$ cm/s results in a factor 15.2 higher electron capture coefficient in the high field region than under equilibrium. The corresponding capture cross section is $5.8\cdot10^{-14}$ cm^2 and only 30% smaller than the values given in refs. [8] and [13].

Unfortunately, no data are readily available describing the EL2 electron capture cross-section dependence on the electric field between 0 and 1 V/μm. Hence no empirical formulation of such a field dependence is derivable. It is only known that the capture cross-section shows a rapid increase between electric fields E of $2\cdot10^3$ and $1\cdot10^4$ V/cm. In order to calculate the field distribution in the detector the increase of the capture coefficient of the EL2 is approximated by a Boltzmann growth according to

$$c_n(E) = c_n(E>10^4 \text{ V/cm}) - \frac{c_n(E=0) - c_n(E>10^4\text{V/cm})}{\left(1 + \exp\left(\dfrac{E-5\cdot10^3\text{V/cm}}{500 \text{ V/cm}}\right)\right)} \tag{4}$$

where $c_n(E=0) = 3.8\cdot10^{-8}$ s^{-1}cm^3 and $c_n(E>10^4$ V/cm$) = 5.8\cdot10^{-7}$ s^{-1}cm^3. For the numerical solution of the continuity equation of electrons and the Poisson equation, including the SRH occupancy factor with the field-enhanced capture in addition the coupling of the leakage current density to the bulk resistivity (see Fig. 1b)) is taken into account. The following results are based on an EL2-conentration of $1.2\cdot10^{16}$ cm^{-3} and 300 V bias voltage. Fig. 2a) shows the calculated electric field distribution beneath the Schottky contact as a function of the substrate resistivity between $2.5\cdot10^6$ and $1.2\cdot10^8$ Ωcm. For all materials the electric field increases rapidly at the edge of the active region and is approximately constant for electric fields higher than $1\cdot10^4$ V/cm. The maximum field strength in the detector increases with increasing substrate resistivity from $1.1\cdot10^4$ to $1.6\cdot10^4$ V/cm. The corresponding active layer width as a function of the bias voltage is shown in Fig. 2b). Independently of the substrate resistivity the active layer width increases linearly with bias voltage. The corresponding slope of the curve decreases with increasing substrate resistivity. In particular, the fully active voltage of a 200 μm thick detector was determined as 187 V for the lowest resistivity and 277 V for the highest resistivity. These values of the fully active voltage are slightly higher than the measured values but confirm the experimentally observed increase of the active layer width with decreasing substrate resistivity at the same bias voltage.

The highest field gradient is at the position, where the electron drift velocity reaches its maximum. Under these conditions fewer electrons will be needed to maintain the continuous current flow, resulting in the highest degree of ionisation of the EL2 or largest space charge density in the detector. The resistivity dependence of this maximum of space charge density can be explained by the fact that in high resistivity materials the equilibrium of the Fermi-level is closer to the ionisation energy of the EL2, where the slope of the occupation factor is steeper, whereas in the low resistivity material the Fermi-level is at about 0.5 eV. Therefore the same change in the electron concentration causes a larger change of the EL2 ionisation in the high resistivity than in the low resistivity SI-GaAs.

In comparison to the experimental results of Berwick [4], the electric field increases faster at the edge of the active region. This is due to the narrow space charge maximum obtained from the simulation which depends strongly on the shape of the drift velocity and capture coefficient of the EL2 as a function of the electric field. Also it is only possible to simulate the detectors for high bias voltages corresponding to the saturation region of the leakage current, because no model exists for the leakage current which describes the continuous change in the Schottky barrier height under low field conditions. Nevertheless the extension of the drift diffusion model by the field-enhanced electron capture of the EL2 is the first model which explains the formation of the space charge region together with the leakage current density in the saturation region. In particular, the influence of the compensation mechanism in SI-GaAs on the detector performance is shown.

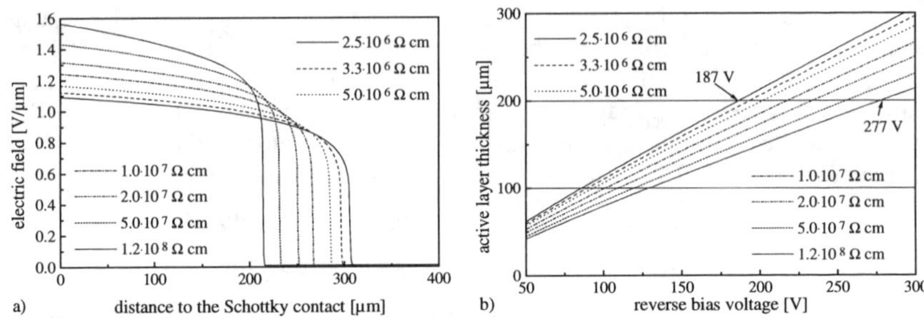

Figure 2: a) The caluculated electric field distribution benath the Schottky contact for different substrate resistivities with an EL2-concentration of $1.2\cdot10^{16}$ cm^{-3} and 300 V reverse bias voltage. b) The width of the active layer determined for different substrate resistivities as a function of bias voltage with an EL2-concentration of $1.2\cdot10^{16}$ cm^{-3}.

C. Summary and Conclusion

A model was proposed which explain the formation of a quasi-neutral region beneath Schottky contacts made on SI-GaAs with the field enhanced electron capture of the EL2-defect. In addition the validity of the model is supported by the dependence of the leakage current and full active voltage on the substrate resistivity.

References

[1] D.S. McGregor et al., Nucl. Instr. and Meth. A 343 (1993) 527.
[2] M. Rogalla et al., Nucl. Instr. and Meth. A 365 (1995) 273.
[3] C.M. Buttar, Nucl. Instr. and Meth. A 395 (1997) 1.
[4] K. Berwick et al., Mat. Res. Soc. Symp. Vol. 302 (1993) 363.
[5] M. Alietti, et al., Nucl. Instr. and Meth. A 355 (1993) 420.
[6] R. Irsigler et al., Nucl. Instr. and Meth. A 395 (1997) 71.
[7] Systematic Investigation of Gallium Arsenide Radiation Detectors for High Energy Physics Experiments, Phd. Thesis, M. Rogalla, Shaker Verlag, Aachen 1998.
[8] M. Rogalla et al., accepted for publication in Nucl. Instr. and Meth. A (1998).
[9] Th. Kubicki et al., Nucl. Instr. and Meth. A 345 (1994) 468.
[10] D.S. McGregor et al., J. Appl. Phys. 75 (12) (1994) 7910.
[11] P.T. Panousis et al., Appl. Phys. Lett. Vol. 15 No. 3 (1969) 79.
[12] G.M. Martin, The mid-gap donor level EL2 in GaAs, in ST. Panteldes (Ed.), Deep Centres in Semiconductors, 2nd ed., Gordon & Breach, New York (1992) 399.
[13] V. Ya. Prinz and S.N. Rechkunov, Phys. Stat. Sol. (b) 118 (1983) 159.

Study of key physical parameters of bulk semi-insulating GaAs for radiation detector fabrication

F. Dubecký[1], J. Darmo[1], M. Krempaský[1], V.Nečas[2], P.G. Pelfer[3],
P. Boháček[1], and M. Sekáčová[1]

[1]*Institute of Electrical Engineering, Slovak Academy of Sciences*
Dúbravská cesta 9, Bratislava, SK-842 39 Slovakia; (elekfdub@savba.sk)
[2]*Faculty of Electrical Engineering and Informatics*
Ilkovičova 3, Bratislava, SK-813 15 Slovakia
[3]*University of Florence, Dept. of Physics and INFN*
Largo E. Fermi 2, Firenze, I-50125 Italy

Abstract - Study of selected physical properties of bulk semi-insulating (SI) GaAs grown by LEC and VGF techniques, undoped and Cr-doped is presented. Conductivity, Hall, GDMS, I-V and C-V techniques were used for material and radiation detector characterisation. Detection performances of detectors have been tested using ^{57}Co source of 122 keV gamma rays. Correlation between the physical properties of base materials and performances of detectors is presented. Physical parameters suitable for estimation of the detector grade bulk SI GaAs are discussed.

A. Introduction

Bulk semi-insulating (SI) GaAs has become important semiconductor for fabrication of low cost, fast and radiation-resistant detectors of charged particles and photons operated at room temperature [1]. Such detectors have many important applications, e.g. in monitoring of on-line industrial processes, defectoscopy, security control and in medicine for progressive digital X-ray radiography.

Published results show that optimised bulk SI GaAs particle detectors have improved parameters at room temperature (RT): the charge collection efficiency for charged particles and X-rays >90%, and the energy resolution <3% for α-particles [1, 2] and <10% for X-rays [1, 5, 6]. A strong dependence of these parameters on physical properties of base materials was generally observed [1, 4-6]. However, until now study neither on the role of chemical purity of the base material, nor space charge inhomogeneities was published. Quality of the base material become especially important in the case of fabrication of sufficiently thick detector with high enough detection efficiency (for stopping of 90% of 60 keV photons it is necessary to have 2.1 mm wide active region of GaAs detector).

In this contribution a study of selected *"key"* physical properties of bulk SI GaAs from 11 various producers grown by LEC and VGF techniques, undoped and Cr-doped is presented. Conductivity, the Hall, GDMS (glow discharge mass spectrometry), I-V and C-V methods were used for material and detector characterisation. Detection performances of detectors have been tested using ^{57}Co source of 122 keV γ-rays. Correlation between the physical properties of base materials and performances of detectors is presented and discussed. For conclusion, the most important physical parameters suitable for estimation of the radiation detector grade bulk SI GaAs are discussed.

B. Experimental

Detectors were fabricated from bulk SI GaAs wafers with crystallographic orientation (100) using the same technology. Wafer fragments were polished down to 200 (±20) μm. Silicone nitride passivation layer with thickness of 100 nm was sputtered onto both sides of the samples. A back-to-back Schottky barrier sandwich structure with symmetrical circular contacts (ϕ=2 mm) formed by evaporation of Ti/Au metallization was fabricated using both-sided photolithography. Before evaporation circular windows were opened in the passivation layer by dry etching in O_2 plasma. Detector chips (2.6x2.6 mm^2) were mounted onto the TO5 transistor holder using a silver paste.

Tabe 1. Information about bulk SI GaAs wafers and detection performances to 122 keV γ-radiation.

Sample label	Growth Method	Doping, contamination	EPD cm⁻²	Resistivity Ωcm (RT)	Hall mobility cm²/Vs (RT)	Detection performances (RT)		
						CCE, %	HWHM, %	P/V
A1	LEC	Non	$<6x10^4$	$3.9x10^6$	7464	79	18.5	4.2
B	LEC	Non, Ti	$<4x10^4$	$1.15x10^7$	7227	59	24	2.5
C	LEC	Non	$<4x10^4$	$2.44x10^7$	6040	65	14	2.6
D1	VGF	Non, Cu, Fe, Ti	$<5x10^3$	$8.8x10^7$	5400	28	35	2
D2	VGF	Non, Cu, Fe	$<4x10^3$	$4.63x10^7$	6203	43	21	2.5
E	HP LEC	Non	$<6x10^5$	$2.95x10^6$	6940	73	22.5	2.9
F	LP LEC	Non	$<2x10^5$	$1.06x10^7$	5816	72	21.6	2.6
G	LEC	Non	$<8x10^4$	$2.8x10^8$	5122	42	no photopeak detected	
H	LEC	Cr	$<1x10^5$	$1.2x10^8$	5770	51	25	1.4
K	LEC	Non	$<2x10^5$	$9.65x10^6$	7517	57	no photopeak detected	
L	LEC	Non	$<6x10^4$	$2.6x10^7$	6915	72	12.5	3.8
M	LEC	Non	$<8x10^5$	$1.4x10^8$	4830	32	no photopeak detected	

Galvanomagnetic technique using the Van der Pauw method was applied for determination of the resistivity and the Hall mobility at 300 K (RT). Values obtained are summarised in Table 1 together with other information about wafers and performance of detectors tested by 122 keV photons using ⁵⁷Co source. Investigated samples of various material producers are labelled hereafter as A1-M. The RT resistivity and the Hall mobility ranges between $(0.3÷28)x10^7$ Ωcm and (4830÷7517 cm²/Vs), respectively. Performances of the radiation detectors are characterised by the charge collection efficiency (CCE, %, considering an energy of 4.2 eV for the electron-hole pair creation in GaAs [1]), the relative energy resolution in half width at half maximum (HWHM, %) and the peak to valley (P/V) ratio. The values (Table 1) were calculated from the detected pulse-height spectra (Fig. 1) at a bias voltage V_b (120÷250 V), at about 10 V lower than a breakdown voltage for each sample. No distinct photopeak was observed with detectors fabricated from materials labelled G, K, and M.

Content of typical chemical impurities generally observed in GaAs (30 elements) in all tested

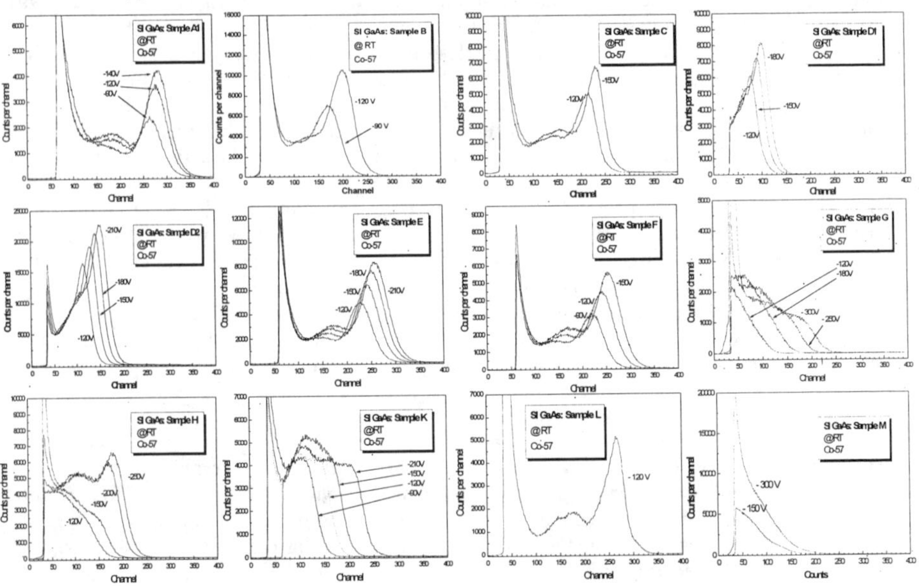

Fig. 1. Pulse-height spectra obtained using 122 keV gamma rays from Co-57 with detectors fabricated from the bulk SI GaAs materials under study (labelled A1 – M).

Table 2. Results of study of chemical impurities by the GDMS method (units in ppb atomic).

Element	Sample label											
	A1	B	C	D1	D2	E	F	G	H	K	L	M*
B	190	303	51	1041	966	271	301	71	470	n/a	20	5600
Na	<1.5	<2	<1.7	4	<2	8	3	<2	4		<1.5	3
Mg	<2	<2	<2	<2	<2	4	<1.9	<1.9	5		<1.5	<2
Al	<1	3	<1	14	<1.4	13	3	2	3		<1	9
Si	<3	11	<3	142	4	212	5	20	11		<3	445
P	<3	<3.5	4	11	<2.8	12	<3	12	20		<3	110
S	<3	21	9	7	12	25	10	10	102		<3.5	77
Cl	14	13	9	13	16	<25	12	4	13		7	11
Ti	<0.4	2	<0.4	3	<0.5	<0.5	1	<0.4	1		<0.4	<0.4
Cr	<1.2	<1.2	<1.2	<1	<1	<1	<1.2	<1.2	70		<0.9	<1.5
Fe	<0.4	1	<0.5	1.8	0.7	1.4	1.5	0.8	0.8		<0.4	8
Cu	<2.5	<3	<2.5	26	8	<3	<3	<2.5	<3		<2.5	<2.7
Total:	<447	<594	<312	<1515	<1258	<806	<576	<360	<953		<257	>6655

Following impurities were obtained in all samples under given detection limit: F<25, Li<6, Be<5, K<25, Ca<20, Mn<0.5, Ni<1.1, Zn<4, Ge<40, Se<13, Mo<1.8, Cd<0.5, In<100, Sn<4, Te<2, Sb<2, Pb<0.5, Bi<0.5. NOTICES: In the analysis there are not included C, N_2, O_2 as the background contaminants in GDMS and host atoms, Ga and As. *Content of other important impurities (ppb at.) in the sample M is following: F 35, Mn 5, and Te 32.

samples (exception of sample K) was determined by the GDMS. This technique is one of the most sensitive basic tool for determination of the content of chemical impurities used also for GaAs [7]. The results of the analysis are summarised in Table 2. Total content of the detected impurities (in ppb at.) ranges between <257 ppb at. for sample L and >6655 ppb at. for sample M.

Finally, samples were tested by the C-V analysis [8] where particular interest was concentrated to the capacitance (C_0) measured at temperature of 400 K , ac signal frequency <1 kHz, and at zero bias voltage. Obtained results for all samples are listed in Table 3. The measured values ranges between 10 pF (G) and 1050 pF (B). For sample M no C-V dependence was observed, with a value of C_0<10 pF, out of sensitivity limit of the range of apparatus used in the analysis (HP 4192A LF Impedance Analyser, V_{ac}=50 mV, average mode).

C. Discussion

All tested samples could be divided into 3 groups: i/ the best quality materials (A1, C, L), ii/ medium quality materials (B, E, F), and iii/ unacceptable materials (G, K, M with no detected photopeak and D1, D2, H with worse detection parameters). The best CCE and P/V gives A1 and the best HWHM samples L and C. For all the best samples is typical high chemical purity with a value of total detected impurities content <450 ppb at., RT resistivity $<2.5 \times 10^7$ Ωcm, the RT Hall mobility >6000 cm^2/Vs and C_0>260 pF. On the other hand, unacceptable materials show usually a higher content of chemical impurities (>950 ppb at.), higher resistivity $>9 \times 10^7$ Ωcm, a low value of the Hall mobility <5500 cm^2/Vs, and C_0 (<50 pF). As follows from a more detail analysis of the GDMS contamination of SI GaAs by impurities creating deep acceptor states in the band gap (such as Cr (H), Cu (D1, D2), Fe (B, D1, D2), Ti (B, D1, F, H), etc.), plays an important role in degradation of detector performance. Materials D1 and D2 were grown by VGF with typical low dislocation density (Table 1). These materials show a nice shape of the photopeak, however related to a low value of the CCE due to a high trapping by deep acceptors. So, bulk SI materials overcompensated by deep acceptors, e.g. Cr-doped (H), are not suitable for fabrication of radiation detectors. Typical correlation observed between resistivity and detector performances: the lower the resistivity the better detection confirms previous observations [4, 8]. Values of the RT Hall mobility >6000 cm^2/Vs observed for most of the investigated samples confirm improvement in material technology during the past decade. However, acceptable values of the resistivity and the Hall mobility alone do not give guaranty of good enough detection performances (e.g. in sample K).

Observed values of detector capacitance C_0 measured at suitable conditions [8] show an excellent

Table 3. Results of capacitance measurements performed at T = 400 K, V_b = 0 and f < 1 kHz.

	Sample label											
	A1	**B**	**C**	**D1**	**D2**	**E**	**F**	**G**	**H**	**K**	**L**	**M**
Measured capacitance, pF	780	1050	580	16	22	200	800	10	850	220	260	<10

correlation with observed detection performances. Its value in the ideal case is determined by the series connection of two barriers with space charge controlled by the net ionized impurities close to the EL2 concentration [8]. Taking into account a value of 1.5×10^{16} cm^{-3}, the same geometry of both barriers (this study), one obtains a value of $C_0 \cong 1100$ pF (1/2 of the single barrier capacitance) in our case. A large drop in C_0 observed (the base length of about 200 µm) indicates presence of the space-charge inhomogeneities inside the volume of investigated materials. The unacceptable materials D1, D2, G and M show such a typically low value of C_0 that confirms an important role of the space-charge inhomogeneities in degradation of detection performance of the radiation detectors. Their creation in the bulk SI GaAs is related to the material nonstochiometry, structural inhomogeneities, and metallic impurities (or their complexes, precipitates, clusters) surrounded by a space charge.

The material labelled K with acceptable RT resistivity ($<10^7$ Ωcm) and the highest value of the Hall mobility (>7500 cm^2/Vs) does not correspond to the presented correlation. Possible explanation of worse detection performance could be related to the presence of space-charge inhomogeneities indicated by a lower value of the C_0 (220 pF) or some drawbacks in detector technology.

A more detail correlation of observed results with the deep-level and structural study of selected investigated materials is presented in another papers [11, 12] published in this proceedings.

D. Conclusions

It was shown that bulk SI GaAs materials grade for radiation detectors could be well estimated by the measurement of: chemical purity, resistivity, Hall mobility and detector capacitance (at zero bias, 400 K, f<1 kHz). It is shown that the measured capacitance is sensitive tool for indication of space-charge inhomogeneities that play an important role in degradation of detection performances of detectors. Values obtained for the best materials give content of a typical impurities <500 ppb at., $<2 \times 10^7$ Ωcm and >6000 cm^2/Vs for the RT resistivity and the Hall mobility, respectively, and >250 pF for the detector capacitance.

Acknowledgements - Authors acknowledge D. C. Look (Wright State University) for unique SI GaAs materials, R. Senderák, and J. Ivančo (Institute of Physics, SAS) for assistance in technology. The work was supported by the Slovak Grant Agency (grants No. 2/5166/98 and 95/5305/509).

References:

[1] T.E. Schlesinger, R.B. James: *Semiconductors for Room Temperature Nuclear Detector Applications,* Semiconductors and Semimetals, **43**, treatise eds. R.K. Willardson, A.C. Beer, and E.R. Weber, (Academic Press, San Diego 1995).
[2] M. Alietti et al., *Nucl. Instr. Meth.* **A 362** (1995) 344.
[3] F. Dubecký et al., *Nucl. Instr. Meth.* **A 377** (1996) 475.
[4] F. Dubecký et al., *Nucl. Phys.* **B 54 B** (1997) 368.
[5] D.S. McGregor et al., *Nucl. Instr. Meth.* **A 380** (1996) 165.
[6] Properties of GaAs, 3rd Edition, eds. M.R. Brozel and G.E. Stillman, EMIS Datareview, Ser. No. 16, INSPEC IEE, London 1996, p. 942.
[7] *Ibid.,* p. 373.
[8] F. Dubecký, et al., *Semicon. Sci. Technol.* **9** (1994) 1654.
[9] K. Berwick, et al., Nucl. Instr. Meth. **A 380** (1996) 46.
[10] M. Rogalla, et al., Nucl. Instr. Meth. **A 380** (1996) 14.
[11] J. Darmo, and F. Dubecký, *this Proceedings*.
[12] D. Korytár, et al., *this Proceedings*.

Neutron and proton irradiation: effects on the active layer width and electric field distribution in semi-insulating GaAs Schottky diodes

A. Castaldini*, A. Cavallini*, L. Polenta*, C. Canali[#], and F. Nava[##]

*INFM and Dipartimento di Fisica, Università di Bologna, Italy,
[#]INFM and Dipartimento di Scienze dell'Ingegneria, Università di Modena, Italy,
[##]INFN and Dipartimento di Fisica, Università di Modena.

Irradiation (with neutrons or protons) markedly changes the values of the active layer width as well as the electric field distribution of Schottky diodes made on semi-insulating GaAs. However, the behaviour of both them as a function of the reverse biasing keep almost the same as before irradiation, either neutrons or protons are used. Spectroscopic measurements revealed that not only the dominant defect EL2 increases in density but also other shallower level traps, which must be accounted for to explain the changes of both active layer width and electric field observed.

A. Introduction

In semi-insulating (SI) gallium arsenide (GaAs) Schottky diodes the distribution of the electric field $\mathscr{E}$ as well as the width W of the active layer as a function of the reverse bias applied V_a are of major importance since they determine the diode performance.

In recent years a few theories have been developed to model the behaviour of both $\mathscr{E}$ and W [1,2,3], while a large body of work has been done to determine them by different experimental methods [4,5].

Since SI GaAs, liquid encapsulated Czochralski (LEC) grown, recently has set off great interest because of its radiation hardness for future uses in host environment (such as in energy physics experiments and medical physics applications), it is of particular interest to investigate how high dose irradiation influences both $\mathscr{E}$ and W. In addition, irradiation at room temperature seems to be a powerful tool for studying complex defects [6].

This paper deals with the effects induced on deep-level defects and, consequently, on the reverse bias dependence of active layer width W and distribution of the electric field $\mathscr{E}$, by irradiation with protons and neutrons as a function of the irradiation dose.

B. Experimental

Semi-insulating GaAs wafers grown by Liquid Encapsulated Czochralski (LEC) method, supplied by Nippon and Freiberger, were analyzed. All the wafers were undoped, (100) oriented, with the n-type resistivity $\rho = 1 \div 3 \times 10^7$ Ωcm. Nippon wafers were 100 μm thick, while the Freiberger ones were 250 μm thick. Samples of 5×5 mm^2 were cut. The carbon content was [C] = 4×10^{15} cm^{-3} and 4×10^{14} cm^{-3} for the Nippon and the Freiberger materials, respectively. The Schottky contacts (diameter $\phi = 3$ mm) were obtained by Au metallization (150 Å thick). The ohmic contacts were realized in two different ways [7,8]. Irradiations were carried out at room temperature either with 24 GeV protons with ion fluences equal to 5.6×10^{13} and 1.2×10^{14} cm^{-2} or with fast neutrons of 1 MeV with fluences ranging from 4.2×10^{13} to 4.2×10^{14} cm^{-2}.

All measurements reported in the following were performed as a function of the neutron equivalent fluence F in order to assess the irradiation effects.

The total density of the EL2 defects (the As_{Ga} related defects at E_c-0.75eV, dominant in SI GaAs) was determined by capacitance-voltage (C-V) characteristics measured at a frequency f = 80 Hz and at temperature high enough (T = 420 K) to determine the net effective charge in the Schottky barrier space charge region [9]. Current-voltage (I-V) characteristics were also measured to deduce [10] from the reverse current density $j = n\sigma_n v_{th}$ (with n free carrier concentration, σ_n capture cross section and v_{th} thermal velocity) the ratio of ionized centers EL2$^+$ to total density of EL2 by means of the expression $N_{DD}^+ \cong N_{DD}/(1 + n\sigma_n v_{th}/e_n)$ and, in turn, the density of EL2$^+$.

To detect defects other than EL2 and to assign the trap types, spectroscopic methods were applied. Photo-Induced Current Transient Spectroscopy (PICTS) [10] and Photo-Deep Level Transient Spectroscopy (P-DLTS) [11] were utilized [12]. The density of the most effective deep level defects was obtained by NTSC (Normalized Thermally Stimulated Conductivity) [13] and reported elsewhere [7]. The depletion layer width W and the behaviour of the electric field J as a function of the biasing voltage V_a were investigated by surface potential (SP) [4,5] and optical beam induced current (OBIC) methods [5] at room temperature.

C. Results and Discussion

The density of the EL2 defects increased under irradiation. Figure 1 shows the dependence of the total and ionized parts of the equivalent neutron fluence, as determined by high temperature C-V and I-V characteristics, for proton and neutron irradiated materials. As also reported in literature [14], the EL2 defect density increases after neutron irradiation even if the introduction rates for its neutral and ionized components are significantly different, 80 cm^{-1} and 3.5 cm^{-1}, respectively. A similar behaviour occurs also after proton irradiation but with much higher introduction rates, namely 170 cm^{-1} and 7 cm^{-1}, respectively. This strongly suggests that by either neutron or proton irradiation arsenic antisite defects are generated [14,15], as indicated also by the marked increase of the level (Ec-0.15eV) [7], attributed, too, to the defect As_{Ga} [16,17].

Figure 2 reports the histogram of the levels mainly affected by proton irradiation at the different fluences used in this work, as resulting by PICTS, P-DLTS and NTSC analyses [7]. For the sake of clarity in the comparison among defect density increase vs irradiation, also the results relevant to the EL2 defect are reported in Fig. 2. It emerges that the electron traps E2 [16] at $(E_c - 0.15eV)$ (absent in the as-grown material) and EL6 at $(E_c - 0.37eV)$, both known from literature [18] to be donor levels, as well as the acceptor center at $(E_v + 0.52eV)$ [19] and the EL2 double donor defect $(E_c - 0.79eV)$ increase with irradiation. In particular, the shallower donor

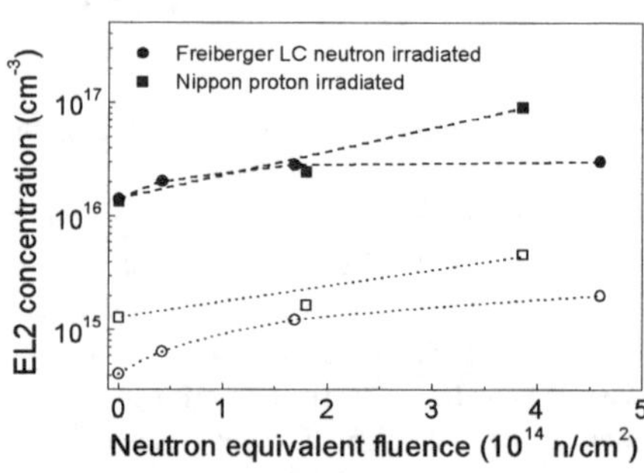

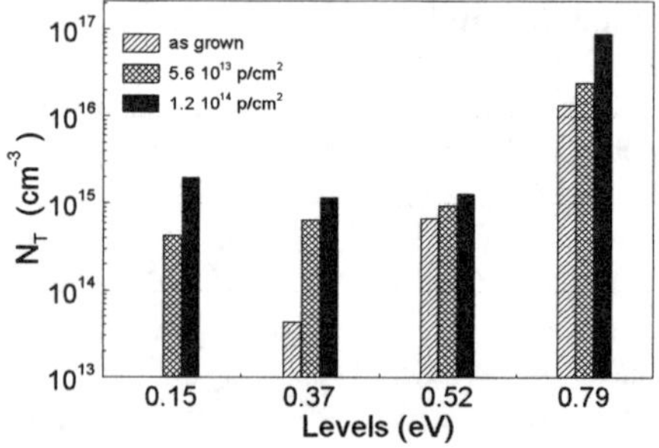

Fig. 1 Density of ionized (dotted lines) and total (dashed lines) EL2 defects vs equivalent neutron fluence for both proton (squares) and neutron (circles) irradiated materials.

Fig. 2 Hystogram of the deep levels mainly affected by irradiation. The data refer to proton irradiation.

concentrations at the highest fluence are on order of a few 10^{15} cm^{-3}. A similar behaviour was observed in neutron irradiated samples where, however, the two donor centres (E$_c$ - 0.25eV) and (E$_c$ - 0.37eV) are the mostly affected ones by the irradiation [8]. This is very significant when considering, on the basis of Look's compensation model, that "defects shallower than EL2 can swing the E$_F$ position by large amounts" [17] and, in turn, the compensation ratio k, defined as $k = (N_A - N_D)/N_{EL2}$, where N$_A$ and N$_D$ are the sum of the all acceptor and donor centers, respectively.

SP and OBIC investigations allowed us to analyze the behavior of W, $\mathscr{E}$ and total positive charge per unit area Q.

In recent years the behaviors of both W and $\mathscr{E}$ vs biasing voltage V_a were differently modeled, but some substantial discrepancies were anyway present between experimental data and theoretical predictions. Only recently Hu and coworkers [20] very well fitted the experimental findings on the electric field and modeled the buried space charge distribution, slightly modifying McGregor's model by accounting for an enhancement of the EL2 capture cross section occurring also at low electric fields and of the consequent rapid increase of EL2$^+$ neutralization.

In irradiated material, due to the prevalent generation of donors shallower than EL2 and the consequent change in the compensation ratio k, the behavior of active region depletion width W, electric field $\mathscr{E}$ and net effective charge density N$_{eff}$, due to all (shallow and deep) ionized acceptor and donor levels, are greatly affected by irradiation. Even if some details are observed to depend on the impinging particle kind, neutrons or protons, the most relevant findings are common: the electric field plateau extends vs V_a and moves from the Schottky contact towards the ohmic one much more rapidly than in unirradiated material. Correspondingly, the width W of the depletion region faster widens (Fig. 3a) and, in turn, the buried net positive ionized charge N$_{eff}$ density distribution lowers (Fig. 3b). As a consequence, the total positive charge per unit area Q decreases.

Concerning the physical meaning of these findings, they are strongly related to the generation of donors other than EL2 and the level dynamic properties. We observed the increase of the density of the defect EL2 both neutral and ionized but at the same time a significantly larger increase of the donor density. Hence, when the electric field is high enough to enhance the capture cross section of EL2, the electrons thermally released from irradiation induced donors shallower than EL2 are captured by the ionized EL2$^+$ defects [21]. This process enhances the neutralization mechanism,

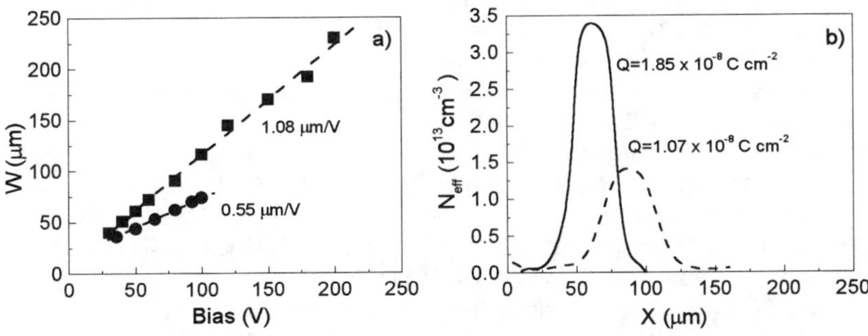

Fig. 3. a) Depletion layer width W vs reverse bias. Circles refer to as-grown material, squares to irradiated material. b) Net effective positive ionized charge N$_{eff}$ density of as-grown samples (continous line) and irradiated ones (F = 4.6 H 10^{14} n/cm^2) (dashed line). Bias V$_a$=80V. The total positive charge per unit area Q is also reported in the figure. Data refer to Freiberger samples.

thus inducing a closer compensation [17]. This explains the expansion of the "quasineutral region" which corresponds to the plateau in the electric field distribution, since in these conditions electrons injected into the high field region over the Schottky barrier can move further on towards the ohmic contact. The increase of the diode depletion rate, expressed by W vs V_a, is a straightforward consequence of this process as well as the decrease of the net ionized positive charge density N_{eff}, resulting from the transition of electrons between EL2 and shallower levels.

D. Summary and Conclusions

The behavior of the active region width and electric field in SI GaAs were studied as a function of either proton or neutron irradiation. It was found that both kinds of irradiation increase the density of EL2 (ionized and neutral) and also the density of shallow levels. As a consequence, a faster depletion of the diode vs reverse biasing occurs, due to the closer compensation. The electric field plateau vs V_a extends deeper into the substrate than in as-grown material and its maximum value lowers, while the buried total ionized positive charge density decreases.

Acknowledgements

Drs. F. Lemeilleur and B.Dezillie from CERN are kindly acknowledged for their support in carrying out the proton irradiation as well as Prof. K. Luebelsmeyer and Dr. M. Toporowsky for providing the neutron irradiated samples.

References

[1]   D.S.McGregor, R.Rojeski, G.F.Knoll, F.L.Terry Jr., J.East, Y.Eisen, J.Appl.Phys. **75** (1994) 910

[2]   T.Kubicki, K.Lübelsmeyer, I.Ortmanns, D.Pandoulas, O.Syben, M.Toporowsky, W.J. Xiao, Nucl. Instr. & Meth. Phys. Res. A **345**, (1994) 468

[3]   A.Cola, L.Raggiani, L.Vasanelli, J.Appl.Phys. **81** (1997) 997

[4]   K.Berwick, M.Brozel, C.Buttar, M.Copperthwaite,P.Sellin,Y.Hou, Mater.Sci.Eng. **28** (1994) 485

[5]   A.Castaldini, A.Cavallini, L.Polenta, C.Canali, C.del Papa, F.Nava, Phys.Rev. B**56** (1997) 9201

[6]   D. Pons and J.C.Bourgoin , J.Phys. C: Solid State Phys. **18** (1985) 3839-3871

[7]   A.Castaldini,A.Cavallini,L.Polenta,C.Canali,F.Nava,R.Ferrini, M.Galli, MRS Proc. 1997 in press

[8]   L.Polenta, PhD Thesis (1998) University of Bologna

[9]   L.S.Berman, Sov. Phys. Semicond. **9** (1975) 406

[10]   D.C.Look, "Electrical Characterization of GaAs Material and Devices",(Wiley & S., UK 1989)

[11]   P.J. Mooney, Appl. Phys. **54** (1983) 208-213

[12]   A.Castaldini,A.Cavallini,P.Fernandez,B.Fraboni, J.Piqueras, L.Polenta, MRS Proc.41 (1996) 177

[13]   D.C. Look, Z-Q Fang, J.W. Hemsky and P.Kengkan, Phys Rev. **55** (1997) 2214

[14]   J.Kruger,Y.Shin, L.Xiao, C.Wang,J.Morse, M.Rogalla, K.Runge and E.Weber, "Semiconducting and Semi-Insulating Materials Conf." SIMC-9 Ed. Claverie (1996 Toulouse, France) 347

[15]   M.R.Brozel, Nucl.Instr. & Meth. Phys.Res. A 395 (1997) 88

[16]   D.C.Look, Z-Q Fang, J.R.Sizelove Phys. Rev. B 49 (1994) 16757

[17]   D.C.Look, Chapter 3 "Semiconductors and Semimetals" **38** (1993) Academic Press Inc. U.S.A.

[18]   G.M.Martin, A.Mitonneau and A.Mircea Electronics Letters **13** (1977) 191

[19]   J.Lagowski, D.G Lin, T.P.Chen, M.Skowronski, H.C.Gatos Appl.Phys.Lett **47** (1985) 929

[20]   Y.F.Hu, C.C.Ling, C.D.Beling and S.Fung , J.Appl.Phys. **82** (1997) 3891

[21]   M.Kaminska, J.M.Parsey, J.Lagowski and H.C.Gatos, Appl.Phys.Lett. **41** (1982) 989

Small Angle Scattering Experiments on Annealed Gallium Arsenide Single Crystal Wafers

M. Herms[a,e], G. Goerigk[b], V. Klemm[c], G. Meier[d], and G. Zychowitz[e]

[a] *Kyoto Institute of Technol., Dpt. of Electr. and Inform. Science, Kyoto 606-8585, JAPAN*
[b] *Forschungszentr. Jülich, Inst. für Festkörperforsch., PF 1913, D-52425 Jülich, GERMANY*
[c] *TU Bergakademie Freiberg, Institut für Metallkunde, D-09596 Freiberg, GERMANY*
[d] *Max Planck Institut für Polymerforschung, D-55021 Mainz, GERMANY*
[e] *TU Bergakademie Freiberg, Institut für Experiment. Physik, D-09596 Freiberg, GERMANY*

The minimum concentration of excess atoms precipitated in GaAs and InP being necessary to cause observable Small Angle Scattering of X-rays and Neutrons (SAXS and SANS, respectively) was estimated. Consequently, undoped melt-grown GaAs wafers were annealed at high arsenic pressure. Areas of a strongly increased density of precipitate were localized by Transmission Electron Microscopy. SAXS experiments have revealed a very weak intensity being dependent on energy at values of the scattering vector $Q > 0.02$ Å^{-1}. Different annealing procedures have resulted in distinguishable curves of SANS ($Q < 0.03$ Å^{-1}) in dependence on temperature and arsenic pressure.

A. Introduction

Precipitates influence a lot of essential properties of semiconductor crystals. As revealed by several techniques, the shape, size distribution and number density of native precipitates are determined by time and temperature of the thermal treatment and by the excess concentration of the elemental constituents being available for precipitation. The spatial distribution of precipitates inside the crystal is strongly affected by other defects such as the cellular dislocation network in melt-grown GaAs. The better knowledge about native precipitates helps to understand the phase diagram of semiconductor compounds and to improve the reliability of microelectronic and photonic devices [1]. In the case of melt-grown substrates, spatial distribution, size and density of precipitates must be more homogenized and reduced, respectively [1,2]. On the other hand, thin layers of extremly high densities of precipitate produced either by Molecular Beam Epitaxy or by ion implantation are under discussion because of their promising electrical properties [3].

The Small Angle Scattering of X-rays and Neutrons (SAXS and SANS, respectively) is a technique well-established in material science to investigate inhomogeneities in a size range about between 1 and 10^3 nm. SANS has been proved to study the oxygen precipitation in silicon (see [4] and ref.). Only a few SAXS studies on GaAs are known [5,6]. In the only one paper about SANS on GaAs crystals it was finally concluded „that the neutron scattering remaining" after annealing „arises from some compositionally modified regions ..."[7].

B. Estimation of small angle scattering of precipitates in semiconductors

For a small concentration of identical particles in a matrix, the cross section Σ of SAS is related to the total number N_p and the squared volume V_p of the particles multiplied by the scattering function S of a single particle and the squared contrast $\Delta\rho$. $\Delta\rho$ is the difference of the average densities of the atomic scattering factors in the particle and in the matrix, respectively. In view of polydisperse systems we substitute N_p by the concentration of atoms N being available for precipitation and have to introduce the distribution of a characteristical size $c(R)$:

$$d\Sigma/d\Omega\ (Q,E) \sim (\Delta\rho)^2\ N\ \xi_p^{-1}\ \xi_m^{-1} \int V_p^2(R)\ c(R)\ S(Q,R)\ dR\ /\ \int V_p^2(R)\ c(R)\ dR \qquad (1)$$

where ξ_p and ξ_m are the average atomic densities of precipitate and matrix, respectively.

As the critical detection limit of SAXS experiments Porai-Koshits et al. [8] estimated the minimum value of the mean squared fluctuation of electron density $\langle(\Delta\rho)^2\rangle_{crit}$ to 1×10^{41} cm^{-6}. Thus, we can introduce a minimum concentration of excess or doping N_{crit} (eq. 2) given in Tabl. 1

$$N_{crit\ (SAXS)} = \xi_p\ \langle(\Delta\rho)^2\rangle_{crit}\ (\Delta\rho)^{-2} \qquad (2)$$

for several systems of semiconductors in case of SAXS far from absorption edges. But if the energy of X-ray E is chosen close to a suitable absorption edge of the elemental constituents of the material the contrast increases about up to one order of magnitude (Anomalous SAXS) [5]. ξ_p is partly insufficiently known and differs probably in dependence on size.

For SANS, to our knowledge, there is no generalized limit. Therefore, we compare the compounds in Tabl. 1 with the system Si-SiO$_2$ and assume the lowest concentration of oxygen N_O measured in antimony doped material to 2.0×10^{16} cm^{-3} [4] is the inevitable minimum value. Thus,

$$N_{crit\ (SANS)} \approx N_O\ \xi_{SiO_2}\ \xi_p^{-1}\ (\Delta\rho_{Si\text{-}SiO_2})^2\ (\Delta\rho)^{-2}. \qquad (3)$$

Matrix	Precipitate	Reference	$\xi_p\ /\ 10^{22}$ cm^{-3}	$N_{crit\ (SAXS)}\ /$ cm^{-3}	$N_{crit\ (SANS)}\ /$ cm^{-3}
GaAs	As	[16]	4.60	5×10^{17}	5×10^{19}
	C	[9]	11.33	2×10^{16}	2×10^{15}
	FeAs	[10]	7.21	2×10^{16}	9×10^{15}
InP	P	[11]	3.54	2×10^{18}	8×10^{21}
	C	[9]	11.33	3×10^{16}	1×10^{15}
	FeP$_2$	[12]	7.84	2×10^{17}	6×10^{15}
Si	SiO$_2$	[4]	7.22	3×10^{18}	2×10^{16}

Tab. 1: Minimum concentration of atoms N_{crit} necessary for observation as precipitates in SAS experiments in different semiconductor materials

C. Experimental

In order to increase the arsenic excess and to vary the size distribution of precipitate in dependence on temperature, pressure and cooling rate, undoped Liquid Encapsulated Czochralski (LEC) grown GaAs wafers (Freiberger Compound Materials Ltd) were treated in different annealing experiments. For this, abbreviations are used as follows: H1 (1100 °C, 10 h), H2 (1000 °C, 10 h), T1 (850 °C, 2h), T2 (950 °C, 2h), T3 (750 °C, 2h), Q1 (rapid cooling > 60 K s^{-1}), SC (slow cooling < 5 K s^{-1}), P1 (0.75 bar), P2 (0.8 bar), P3 (2.5 bar), P4 (3.0 bar), P5 (3.5 bar).

The SAXS experiments were performed at the JUSIFA equipment (Hasylab, B1), which covers a range of Q about between 0.01 and 1 Å^{-1} [13]. Because of the high mass attenuation the samples were thinned and chemically polished down to about 60 μm. The SANS measurements were carried out on 1 cm wafer stacks at the ELLA-KWS2 beamline [14] what allows to observe a Q range between 0.002 and 0.2 Å^{-1}. Before, the annealed wafers were etched in H_2SO_4-H_2O_2-H_2O (3:1:1) in order to minimize surface-related effects. Additionally, the samples were examined by Transmission Electron Microscopy (TEM) (see also [15]).

D. Results and discussion

The TEM image shown in Fig. 1 was recorded in a sample annealed at high arsenic pressure. We have found a (diagonally) long area around a cluster of dislocation where the density of precipitate is strongly increased. Assuming the avarage diameter of precipitate about to 10 nm, the local arsenic excess was estimated to 3×10^{20} cm^{-3}, what is much higher than the total data usually discussed between 1.3×10^{15} cm^{-3} [16] to 2×10^{19} cm^{-3} [1,19].

Fig. 1: TEM image of a cluster of dislocation surrounded by arsenic precipitates localized in a GaAs wafer annealed at high arsenic pressure (H1T3SCP5); the black bar is 100 nm.

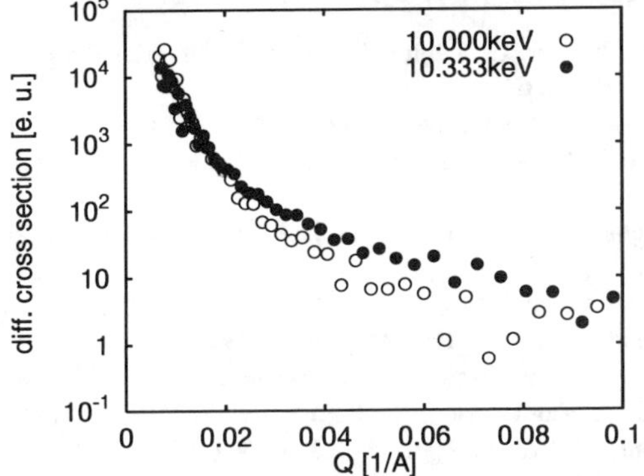

Fig. 2: Radial averaged small angle X-ray scattering of an annealed GaAs wafer (H1T1SCP3) measured at different energies of synchrotron radiation close to the absorption edge of gallium (10.367 keV).

The SAXS curves shown in Fig. 2 are representative of both as-grown and annealed samples. We found a weak dependence of the intensity on energy, for annealed samples at Q > 0.02 Å^{-1} only, what points to the existence of arsenic-rich inhomogeneities.

The SANS experiment has distinguished between different annealing procedures. As shown in Fig. 3, the double-logarithmic plot reveals that the slope of curves increases with increasing temperature before cooling what could be caused by a change of size distribution. Furthermore, a variation of arsenic pressure changing N results in a parallel shift.

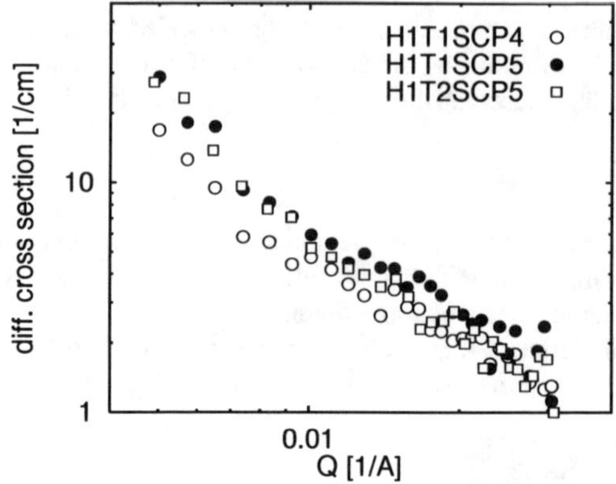

Fig. 3: Radial averaged SANS of GaAs wafers etched after different two-step annealing procedures.

Conclusion

The study of arsenic precipitates in melt-grown GaAs by SAS experiments is a challenge close to the lower detection limit of the method. In the case of SAXS, the variation of contrast by energy shifting close to the absorption edge of Ga is mandatory and has shown an energy dependent scattering probably caused by arsenic-rich inhomogeneities. On the other hand, at the moment, SANS seems to be more promising to reveal effects due to different annealing procedures. However, effects of surface degradation should not completely exclude despite etching.

References

[1] O. Oda, H. Yamamoto, M. Seiwa, G. Kano, M. Mori, H. Shimakura and M. Oyake; Semicond. Sci. Technol. 7 (1992), A215

[2] B. Hoffmann, M. Jurisch, G. Kissinger, A. Köhler, G. Kühnel, T. Reinhold, W. Siegel and B. Weinert; Proc. of IEEE SIMC-9, Toulouse (1996), 63

[3] A. Claverie, F. Namavar and Z. Lilienthal-Weber; Appl. Phys. Lett. 62 (1993), 1271

[4] S. Gupta, S. Messoloras, J. R. Schneider, R. J. Stewart, and W. Zulehner; Semicond. Sci. Technol. 7 (1992), 443

[5] V. Vezin, H. Okuda, K. Osamura, Y. Amemiya, K. Kitahara and N. Nakajima; Journ. de Physique IV, 3, C8 (1993), 381

[6] R. Gebhardt; Berichte d. Forschungszentrums Jülich 2851 (1993)

[7] S. Gupta, E. W. J. Mitchell, R. J. Stewart and G. Kostorz; Phil. Mag. A 37 (1978), 227

[8] E.A. Porai-Koshits, V. V. Golubkov and A. P. Titov; Borate Glasses, Plenum, N. Y. (1978), 183

[9] A. J. Moll, E. E. Haller, J. W. Ager III and W. Walukiewicz; Appl. Phys. Lett. 65 (1994), 1145

[10] M. W. Bench, C. B. Carter, Feng Wang and P. I. Cohen; Appl. Phys. Lett. 66 (1995), 2400

[11] A. Claverie, J. Crestou and J. C. Garcia; Appl. Phys. Lett. 62 (1993), 1638

[12] N. A. Smith, I. R. Harris, B. Cockayne and W. R. MacEwan; J. Cryst. Growth 68 (1984), 517

[13] H.-G. Haubold, K. Gruenhagen, M. Wagener, H. Jungbluth, H. Heer, A. Pfeil, H. Rongen, G. Brandenburg, R. Moeller, J. Matzerath, P. Hiller, H. Halling; Rev. Sci. Instrum. 60 (1989), 1943

[14] D. Schwahn, G. Meier and T. Springer; J. Appl. Cryst. 24 (1991, 568

[15] V. Klemm, U. Martin, U. Muehle and H. Oettel; Pract. Metallogr. 33 (1996), 386

[16] P. Schlossmacher, K. Urban and H. Rüfer; J. Appl. Phys. 71 (1992), 620

[17] N. Chen, Y. Wang, H. He, Z. Wang, L. Lin and O. Oda; Appl. Phys. Lett. 69 (1996), 3890

Effects of Cu diffusion on electrical properties of GaAs

D. Seghier and H.P. Gislason

Science Institute, University of Iceland, Dunhagi 3, IS-107 Reykjavik, Iceland

We investigated semi-insulating GaAs samples produced by Cu diffusion into n-type starting material by means of thermally stimulated current (TSC) and photoconductivity (PC) measurements. We show that the Cu-diffusion changes the EL2 centre into a deep donor (T3) with a lower activation energy, 0.7 eV. Similar effects have been observed in MBE GaAs grown at low temperature. The PC quenching and recovery of the T3 trap are similar to those usually observed for EL2. We conclude that the deep donor is a complex centre involving As_{Ga}. In addition, we observe the usual Cu acceptor levels Cu_A and Cu_B at $E_v + 0.15$ and $E_v + 0.4$ eV. The dynamics of optical quenching and thermal recovery of the TSC signal from the Cu levels suggest that they are neither related to EL2 nor two levels of the same Cu-related double acceptor.

A. Introduction

In previous work [1] we reported the observation of a deep electron trap T3 at $E_c - 0.7$ eV, in addition to the well known copper acceptor levels at $E_v + 0.15$ eV and $E_v + 0.4$ eV usually denoted Cu_A and Cu_B, respectively (Fig. 1). The centres were found in conducting n-type GaAs:Cu and semi-insulating GaAs samples which were produced by Cu-diffusion into n-type GaAs. No signal from undistorted EL2 was detected in these samples. The nature of the deep electron trap T3 is not known. In this work we observe photo-quenching and thermal recovery of the photoconductivity (PC) similar to the fingerprint of EL2. We deduce that T3 is a complex centre including As_{Ga}. Several authors have attributed the Cu_A and Cu_B levels to the two ionisation levels of a double Cu-related acceptor [2]. Others claim that the two levels are independent [3,4]. By checking the quenching and thermal recovery of the thermally stimulated current (TSC) signal from the Cu_A and Cu_B levels, we find that their behaviour is not controlled by EL2, and conclude that they are not likely to belong to a double Cu acceptor defect.

B. Experimental

The samples were cut from a horizontal Bridgman GaAs material with $n = 1.5 \times 10^{16}$ cm^{-3} and diffused with copper [1]. For TSC and PC measurements a semi-transparent Ohmic contact was made on the front side of the samples and another one on the back side. Both the PC and TSC signals were measured under a bias of 10 V where metallic contacts show good Ohmic behaviour.

C. Results

The Cu-doped samples were illuminated with light of sub-bandgap photon energy (1.2 eV) at 90 K. Fig. 2 shows the time evolution of the PC signal during the illumination, using different light intensity. For strong photon flux densities a quenching of the PC is observed. Fig. 2 also shows the PC signal from an n-type GaAs sample without Cu-diffusion, measured with a high light intensity. The quenching of the photoconductivity in the two samples is similar. Next we heated the sample to different temperatures, each time measuring the resulting PC and TSC signals. These experiments involved several steps: (a) Quenching of the PC at low temperature (90 K) by prolonged illumination at 1.2 eV, using a photon density of about 5×10^{16} cm^{-2} s^{-1}. (b) Heating the sample to a given temperature, annealing for 30 s, thereafter measuring the PC. (c) Cooling down to 80 K and illuminating the sample again for a short time with light of above-bandgap photon energy in order to fill the traps. (d) Finally, recording a TSC spectrum while heating the sample up to 320 K. The cycle was then repeated from step (a) after cooling

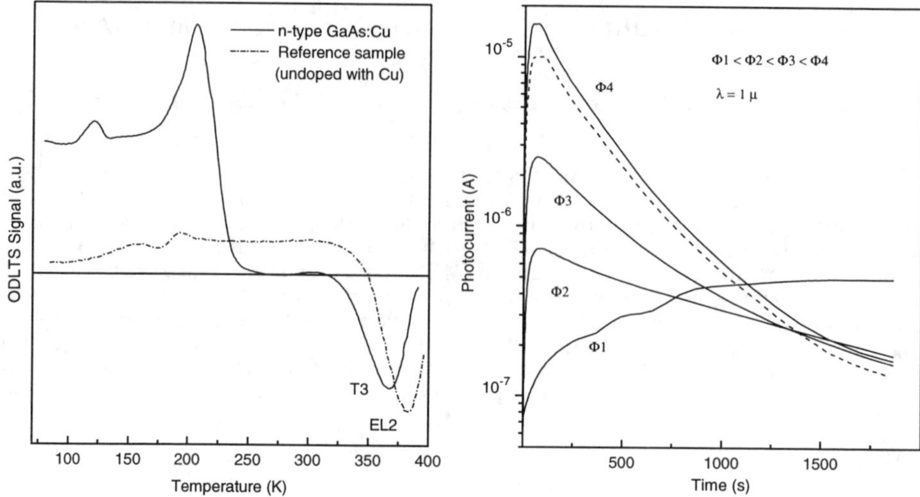

Fig. 1: Optical DLTS spectra from n-type GaAs. The rate window was 30 s^{-1} [1].

Fig. 2: Time evolution of the PC at 80 K illuminating with photons at 1.2 eV, for different photon flux densities Φ. For comparison, quenching of the PC from an n-type GaAs reference sample is presented as a dashed line.

down. Each time a new annealing temperature was used in step (b). The results are summarised in Fig. 3. We observe that the PC signal recovers completely after heating to around 125 K. Such a recovery is also a characteristic of the EL2 centres. We conclude from these observations that the Cu-diffused semi-insulating samples exhibit similar photo-quenching and recovery of the photoconductivity as the GaAs reference samples. Fig. 3 also shows the $I_{TSC}/I_{TSC}(0)$ ratio for Cu_A and Cu_B as a function of annealing temperature. Here I_{TSC} is the thermally stimulated current intensity with the PC having been quenched at low temperature and $I_{TSC}(0)$ the normal intensity. The TSC signal from both levels seems to be almost unaffected by the quenching of the photoconductivity. Indeed, even when the PC is fully or partially quenched at the beginning of a TSC scan, the $I_{TSC}/I_{TSC}(0)$ ratio from Cu_A and Cu_B is close to 1.

We also performed thermally stimulated current scans by illuminating the samples at 1.2 eV for different time duration. In contrast to the procedure of Fig. 3 the samples were not illuminated with above-bandgap light. Fig. 4 shows several typical spectra. The curve labelled *5 minutes* was recorded after short illumination before quenching of the PC occurs and shows the largest TSC signal. Illumination of longer duration below the bandgap quenches the PC and thereby limits the generation and subsequent capture of free carriers into the Cu_A and Cu_B centres. This decreases the corresponding TSC peaks and distinguishes between them, since the Cu_B signal decreases more rapidly than that of Cu_A. While the signal from Cu_B in the TSC spectra is always stronger than that of Cu_A, the situation is inverted after quenching of the PC for longer time periods.

D.   Discussion

The commonly accepted explanation for the photoconductivity quenching in GaAs is the optically induced transformation of EL2 centres into their metastable states, EL2* [5]. Since EL2* interacts only weakly with light, this causes an interruption of the main path of electron transfer from the valence to the conduction band, which decreases the generation rates of free carriers

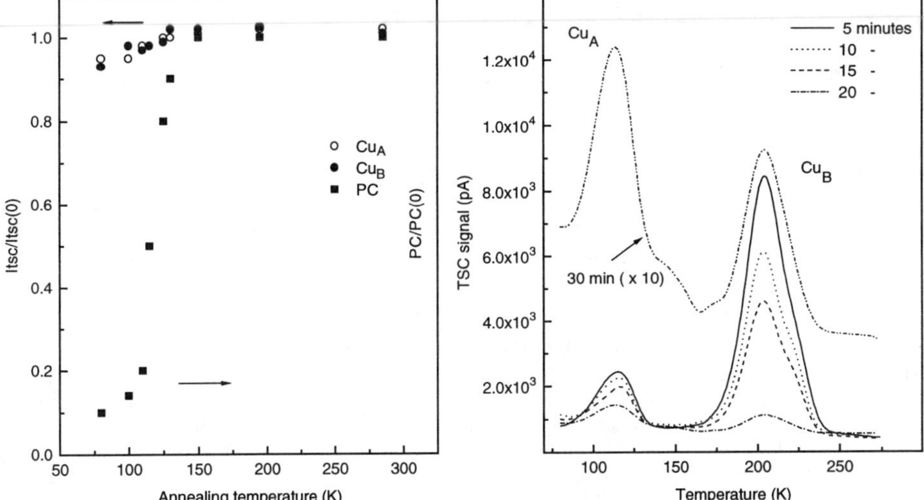

Fig. 3: Recovery of thermally stimulated current peaks of Cu_A and Cu_B as a function of annealing temperature. $I_{TSC}/I_{TSC}(0)$ is the ratio between the quenched and unquenched peaks. The thermal recovery of the photoconductivity is also presented.

Fig. 4: Thermally stimulated current for various degrees of photoconductivity quenching at 90 K corresponding to increasing illumination times.

and consequently quenches the photoconductivity. The recovery of the photoconductivity after annealing at 125 K is caused by the recovery EL2* $\rightarrow$ EL2 [6]. In our samples, a deep electron trap T3 is observed instead of EL2. Measurements suggest that T3 has similar optical characteristics as EL2. Although T3 is apparently different from EL2 (it has different apparent activation energy and capture cross section), we nevertheless deduce that the structure of this defect presumably includes EL2. A possible structure is As_{Ga} in a complex with another defect induced by the copper diffusion. Hence, we conclude that Cu-diffusion produces changes in EL2 similar to those observed in low-temperature MBE GaAs [7]. The results presented in Fig. 3 clearly show that the quenching of the photoconductivity is not associated with a quenching of the TSC signal from Cu_A and Cu_B. Indeed, the TSC signal is almost the same irrespective of the PC being quenched or not. This indicates that the transformation of EL2 into EL2* does not affect Cu_A and Cu_B. Here we assume that EL2 is responsible for the metastable properties of T3. Hence, we conclude that there is no correlation between these two levels and EL2. If a level is associated with EL2 a strong correlation between the quenching and recovery of the PC and the TSC signal from the defect is expected [8]. On the basis of TSC experiments similar to ours it has been shown that the well-known native defects EL3, EL5, EL6 and EL14 have EL2 as a common constituent [9].

The representative TSC curves in Fig. 4 show a constant decrease in the peak heights when the sample is illuminated for longer periods of time. The longer the sample is illuminated the higher fraction of the EL2-related centres transform to their metastable state. Then there is a smaller concentration of photo-created free carriers in the bands, which lowers their availability for trapping. Thus, illumination may involve electrons from deep traps, thereby affecting their occupancy. Under illumination at low temperature the steady state occupancy of a hole trap is (neglecting the thermal emission rate e_p^{th}):

$$p_t^o(\infty) = N_t\left(c_p + e_n^o\right)\big/\left(c_p + e_p^o + e_n^o\right) \tag{1}$$

where symbols have their usual meanings. The optical emission rate e_p^o is defined as

$$e_p^o = \sigma_p^o(\lambda)\phi(\lambda) \tag{2}$$

where $\sigma_p^o(\lambda)$ is the optical cross section and $\phi(\lambda)$ the photon flux at wavelength λ. Hence the steady state occupancy of the trap is mainly a competition between (*i*) hole capture and (*ii*) hole optical emission. Cu_A and Cu_B may have different optical cross sections and then the balance between the two processes is different for each level. When the PC is quenched there are less free carriers available for capture and c_p decreases. Therefore the decrease rate in the final occupancy of each trap will be defined by the corresponding value of e_p^o. A high value of $\sigma_p^o(\lambda)$ may cause a drastic decrease in $p_t^o(\infty)$ and consequently in the TSC signal. For the first four illumination times in Fig. 4 the TSC peak height from Cu_B is higher than that of Cu_A. When the sample has been illuminated for 30 minutes or longer, the Cu_B peak becomes lower than the Cu_A peak. If the two energy levels belong to the same double Cu-related acceptor, the shallower level Cu_A (the 0/- level) cannot trap a hole unless Cu_B (the -/-- level) is already filled. Since the TSC signal is proportional to the final occupancy of the traps after illumination, one can deduce that in case of a double acceptor the TSC signal from Cu_B should always be stronger than that from Cu_A. Obviously this hypothesis does not fit with the TSC spectrum for illumination longer than 30 minutes. Hence, we suggest that Cu_A and Cu_B are two disconnected levels as already deduced from the passivation behaviour of the two levels [4]. This conclusion agrees with previous observations that although the Cu_A and the Cu_B acceptor levels can be passivated simultaneously by hydrogen and lithium, the Cu_B level can be reactivated independent of the Cu_A level after passivation [4]. If these levels correspond to the two ionisation levels of a double acceptor this situation cannot occur.

E.   Conclusion

A new electron trap at $E_c - 0.7$ eV is observed instead of EL2 in copper-diffused GaAs samples. Quenching of the PC at low temperature has been observed from these materials after illumination. We deduce that the new centre may be a complex defect including EL2 with probably a Cu-induced defect. By checking the dynamics and recovery of the TSC signal from the Cu acceptors Cu_A and Cu_B, we deduced that these two levels are not correlated to EL2 and do not belong to the same Cu-related double acceptor .

Acknowledgements

This work was supported by the Icelandic Research Council and the University Research Fund.

References

[1]   B.H. Yang, D. Seghier and H.P. Gislason, in *Semi-insulating III-V Materials*, ed. Ch. Fontaine (IEEE, Toulouse, France, 1996) p.163.
[2]   G. Hofmann, Appl. Phys. Lett. **61**, 2914 (1992).
[3]   R. A. Roushe, D.C. Stoudt, and M.S. Mazzola, Appl. Phys. Lett. **62**, 2670 (1993).
[4]   K. Leosson and H.P. Gislason, Physica Scripta T**69**, 196 (1997)
[5]   U.V. Desnica, Dunja I. Desnica and B. Santic, Appl. Phys. A**51**, 379(1990).
[6]   Y.N. Mohaparta and V. Kumar, J. Appl. Phys. **64**, 956(1988).
[7]   D.C. Look, Z-Q Fang, J.R. Sizelove and C.E. Stutz, Phys. Rev. B**70**, 465 (1993).
[8]   B. Santic and U.V. Desnica, Appl. Phys. Lett. **56**, 2636(1990).
[9]   S. Chichibu, N. Ohkudo, and S. Matsumo, J. Appl. Phys. **64**, 3987 (1988).

Be - related traps in Al$_{0.5}$ Ga$_{0.5}$ As MBE layers

J.Szatkowski, E.Płaczek-Popko, K.Sierański ,O.P.Hansen[1]
Institute of Physics, Technical University of Wrocław,
Wybrzeże Wyspiańskiego 27, 50-370 Wrocław, Poland.
[1]Oersted Laboratory , University of Copenhagen, Universitetparken 5,
DK-2100 Copenhagen,Denmark.

Abstract

In this paper we report on studies of deep levels in Be doped Al$_{0.5}$Ga$_{0.5}$As by deep level transient spectroscopy measurements. The aim of the studies was to investigate which of observed hole traps may be Be related. For this purpose samples of four different Be concentration ranging from 10^{15} cm^{-3} to 10^{18} cm^{-3} were prepared. For the samples of the highest Be concentration the DLTS measurements reveal the presence of four hole traps whereas in the others six traps are observed. Four of observed traps can be attributed to Be incorporation.

A.Introduction

Beryllium is a preferable effective mass acceptor dopant in Al$_x$Ga$_{1-x}$As grown be molecular beam epitaxy (MBE) method. There are some suspicions however that incorporation of Be accompanies appearance of deep levels, especially for higher Al contents [1,2]. We have already presented some results of our studies on deep hole traps in Be-doped MBE AlGaAs by deep level transient spectroscopy (DLTS) method [3]. In this paper we report on results of investigations of Be concentration influence on the concentration and distribution of deep levels in the Be doped Al$_{0.5}$Ga$_{0.5}$As by DLTS.

B.Sample preparation.

Al$_{0.5}$ Ga$_{0.5}$ As samples were grown on a semi-insulating [001] GaAs substrate in a Varian Gen II MBE-machine. Four types of samples with different Be concentration of 10^{18} cm^{-3} (A samples), $2*10^{17}$cm^{-3} (B samples), 10^{16}cm^{-3} (C samples) and 10^{15} cm^{-3} (D samples) were prepared. Metal-semiconductor structures were realised in order to perform DLTS studies. The layer structure applied to the samples was described in details elsewhere [3,4]. Thermal evaporation of Au/Zn/Au followed by alloying was used for the ohmic contact, and Al/Ti/Au for the Schottky contact.

C.DLTS results

DLTS measurements were carried out with the help of DLS-82E System based on a 1 MHz capacitance bridge and lock-in integration averaging. Samples were mounted in a liquid nitrogen bath type sample holder provided in a heater enabling measurements within 77K-400K-temperature range. Exemplary DLTS signals versus temperature, the so-called DLTS temperature scans are shown for samples A,B and D in Fig.1 and for samples C and D in Fig.2. DLTS signals resulted in Arrhenius plots shown in Fig.3 (triangles – A samples, squares - B samples, circles- C samples and open squares- D samples).

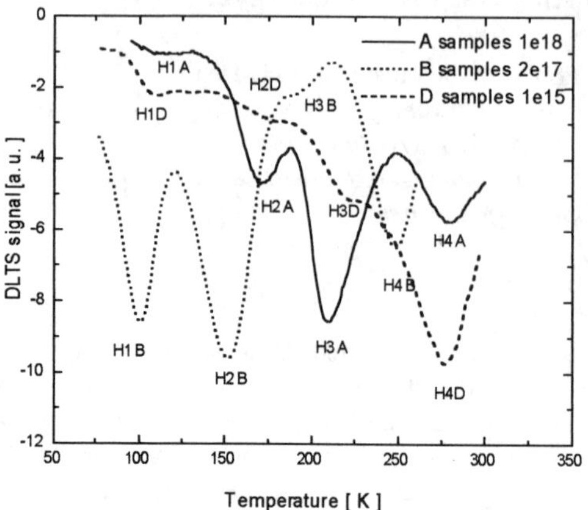

Fig.1. DLTS temperature scans for the samples A(solid curve), B(dotted curve) and D (dashed curve). Lock-in frequency was equal 55Hz.

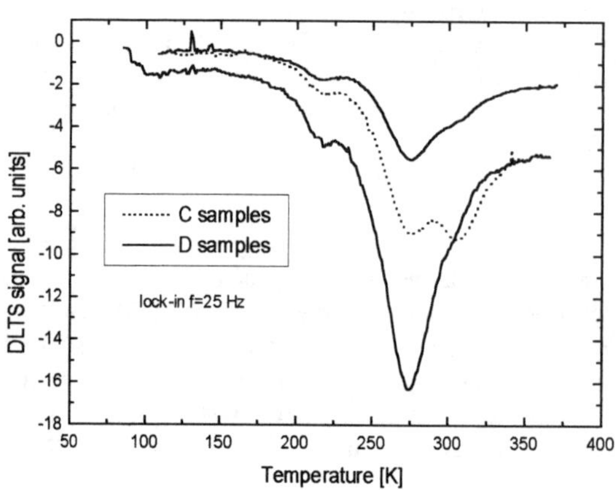

Fig.2. DLTS signal temperature scans for the samples C (dotted curves) and D (solid curve). Lock-in frequency was equal 55Hz.

For the samples A the DLTS measurements reveal the presence of four hole traps (labelled as H1A-H4A) whereas in the B, C and D samples six traps are observed (labelled respectively as H1B-H6B, H1C-H6C and H1D-H6D). Activation energies and apparent capture cross sections obtained from Arrhenius plots given in Fig.3 are collected in Table I. In case of electric field dependent emission rates activation energies extrapolated to zero electric field are given in it.

From profile measurements for the samples A,C and D trap concentration for the traps H2-H4 and H6 was of the order of 0.01 of the net acceptor concentration N. For the B samples the trap concentration for all of the traps beyond the trap H5 was equal 0.001*N. In the case of the H5 trap its concentration was of the order of $10^{13} cm^{-3}$ in all samples.

In the case of A samples only four traps were observed because the higher temperature range of measurements was not available due to increasing leakage currents. DLTS data were determined exactly for H2A-H4A as the data for H1A were scattered for different samples from 0.2eV to 0.4eV. Electric field did not affect emission rate from the traps. The defect linked to the level H4A had thermally activated capture cross section yielding the value of energetic barrier for the process equal about $E_B=0.2$ eV and high temperature cross section $\sigma_{p\infty}=1.6*10^{-16} cm^2$.

For the B samples the fifth, H5B ,DLTS signal peak was observable only for very short filling pulses. The electric field dependence of emission rate from the trap H2B has been observed and explained in the terms of the Frenkel-Poole mechanism [5]. Thermally activated capture cross sections were observed in the case of the levels H1B (E_B= 0.04eV and $\sigma_{p\infty}$ = $3*10^{-18}$cm^2) and H4B (with E_B =0.18eV and $\sigma_{p\infty}$ =$2*10^{-16}$cm^2).

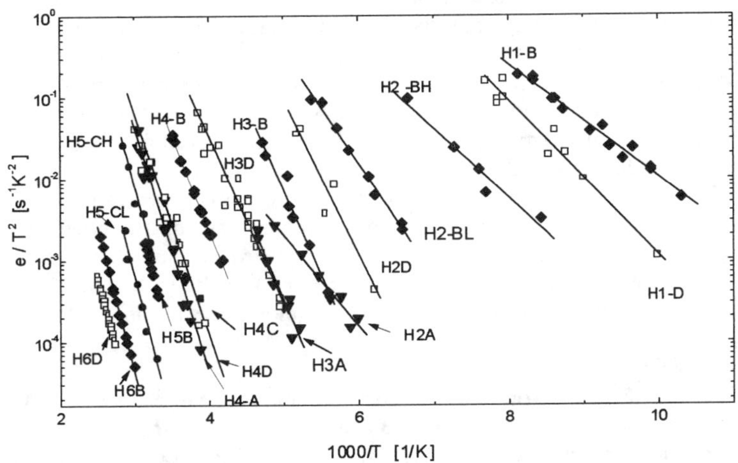

Fig.3. Arrhenius plots for the traps observed by us in the Al$_{0.5}$Ga$_{0.5}$As samples: (triangles - A samples, squares- B samples, circles- C samples and open squares-D samples). Solid lines are the best least square to the experimental data.

Table I. Activation energies E_T and apparent capture cross sections for the Al$_{0.5}$Ga$_{0.5}$As samples.

	Samples A		Samples B		Sample C		Sample D	
trap label	E_T [eV]	σ_p [cm^2]	E_T [eV]	σ_p [cm^2]	E_T [eV]	σ_p [cm^2]	E_T [eV]	σ_p [cm^2]
H1	0.2-0.4	-	0.15	$5*10^{-17}$	-	-	0.19	2.210^{-15}
H2	0.27	$4.1*10^{-19}$	0.37	-	-	-	0.38	$1.4*10^{-13}$
H3	0.41	$2.9*10^{-15}$	0.40	$3*10^{-14}$	-	-	0.41	$2.9*10^{-15}$
H4	0.59	$7*10^{-15}$	0.46	$6*10^{-14}$	0.53	$3*10^{-15}$	0.53	$3*10^{-15}$
H5	-	-	0.77	$1.4*10^{-12}$	0.80	-	-	-
H6	-	-	0.71	$1.5*10^{-15}$	-	-	0.68	$1.3*10^{-13}$

In the C samples the fifth peak was much better resolved (cf. Fig.2,) so we focused our studies on this trap in these samples. It was found that electric field affects emission rate from corresponding trap level which was also explained in the terms of Frenkel-Poole mechanism.

For the D samples the peak H5B was hardly seen due to high background from the peak H4C so its properties were not determined. DLTS signals resemble those for the samples C.

D. Discussion

The results obtained for the samples A,B and D are the same only for the trap H3: activation energies are the same and apparent capture cross sections are the same for the traps in the A and D samples but in the B samples it is a little higher (cf. Fig.3 : the plots for H3A and H3D are shifted parallel towards higher temperatures as compared to the H3B which means the same activation energy but slightly different capture cross sections).

The samples A,C and D resemble each other with respect to the levels H4: activation energies differ but Arrhenius plots are located close indicating close values of apparent capture cross sections. Additionally the results for H1 are scattered for both the A and D samples.

On the other hand there are some similarities in results obtained for the samples B and D as regards the trap H2 and the high temperature peaks H5 and H6:
- as for the trap H2, electric field enhanced emission rate was observed in the B samples. For the D samples, their shallow acceptor concentrations are on the order of 10^{15}cm^{-3} and electric field is too low to affect emission rate. However if the two peaks come from the same origin the activation energy of H2B measured in the bulk should be equal to the activation energy of the trap H2D and Arrhenius plot corresponding to the H2D should be shifted towards higher temperatures with respect to the H2B plot, which actually takes place (cf.Table I);
- for the trap H5 its electric field influence on emission rate explains again difference in activation energies observed in samples B and C,D: in Fig.3 Arrhenius plot for H5B (higher acceptor concentration means higher electric field) fits within experimental error to the one for H5CH measured at high electric field.
-   Arrhenius plots are close for the trap H6 in both B and D samples.

The lattice relaxation processes observed in both samples A and B yield close values of energetic barrier heights equal about 0.2eV.

For the traps H2-H4 and H6 trap concentration is of the order of 10^{15}cm^{-3}, 10^{14}cm^{-3} and 10^{13}cm^{-3} in the samples B, C and D, respectively. As a result the trap H5 of concentration 10^{13}cm^{-3} is observed better in the C and D than in the B samples. In the case of samples A it is not observed as it may be covered by the H4 peak, about 100 times more intensive in these samples. Surprising is that properties of some of deep levels existing in the samples A and C,D differ from those observed in B samples. This particularly concerns the traps responsible for the low temperature DLTS peaks. It seems that the defects labelled by us as H2-H4 and H6, following Be doping concentration are somehow connected with Be incorporation. The acceptor-like trap H5 may be oxygen-related [6].

Acknowledgements. MBE-growth and sample processing was made by C.B.Soerensen at the III-V Nano-Lab in Copenhagen.

[1] P.Krispin and R.Hey, Materials Science Forum **143-147**, 359-364 (1994).
[2] A.Mitonneau, G.M.Martin, and A.Mircea, Electronics Letters **13**, 666-667 (1977).
[3] J.Szatkowski, E.Płaczek-Popko, K.Sierański, O.P.Hansen, Materials Science Forum **258-263**, 1653 (1997).
[4] O.P.Hansen, J.Szatkowski, E. Płaczek-Popko, K.Sierański, J. Cryst. Res. Technol. **31**, 313-316 (1996).
[5] J.Frenkel, Phys.Rev. **54**, 647 (1938).
[6] K.Yamanaka, S.Naritsuka, K.Kanamoto, M.Ishi; Appl.Phys.Lett. 61,5062 (1982).

Photoluminescence at 1.5 µm of heavily Er-doped insulating films on Si

S. Lanzerstorfer[a], J. D. Pedarnig[b], R. A. Gunasekaran[b], D. Bäuerle[b], and W. Jantsch[a]

[a]*Johannes Kepler Universität Linz, Institut für Halbleiterphysik, A-4040 Linz*
[b]*Johannes Kepler Universität Linz, Institut für Angewandte Physik, A-4040 Linz*

Abstract.
A comparison of the photoluminescence (PL) properties of Er-doped SiO_2, soda-lime glass, and ZBLAN glass films fabricated by pulsed-laser deposition (PLD) is given. The PL yield depends strongly on the host material. Under identical growth conditions and the same erbium concentrations in the targets, films deposited from the soda lime and ZBLAN glass show more than an order of magnitude higher luminescence yield. This enhancement is attributed to a higher concentration of optically active erbium in the multicomponent glass environment. Temperature quenching of the PL yield is observed for SiO_2: Er. The PL yield of soda-lime glass: Er and ZBLAN: Er increases with increasing temperature. A room temperature external quantum efficiency of 2×10^{-4} was obtained for Er-doped soda-lime glass.

PACS: 78.55.Hx, 78.60.Fi, 81.15.Fg
Keywords: Thin film, Pulsed-laser deposition, Photoluminescence, Erbium, SiO_2: Er

A. Introduction

Er ions in the trivalent charge state emit light at 1.54 µm independent of the host material. Since this emission is due to an intra-atomic transition within the well shielded 4f shell ($^4I_{13/2} \rightarrow {}^4I_{15/2}$), the emission wavelength is temperature independent. This specific wavelength coincides with the minimum loss window of optical fibers used in tele-communication systems. Silica as a host for Er is commonly used in optically pumped fiber amplifiers for a 1.54 µm signal within long distance communication systems [1]. Such fiber amplifiers contain Er at ppm concentrations. Therefore, long interaction paths of the order of tens of meters are required for high gain. High gain waveguide amplifiers with short paths allowing integration into microelectronic technology would require much higher Er content in the optically active medium.

In this communication we report on the fabrication and characterization of highly Er-doped films by pulsed-laser deposition (PLD) [2]. Photoluminescence (PL) spectra, their temperature dependence, PL yields, and the external quantum efficiency were measured for different host materials.

B. Experimental

Laser ablation targets were fabricated from powders of SiO_2, soda-lime glass (70% SiO_2, 12% CaO, 10% Na_2O, and a few percent of K_2O, Al_2O_3, Li_2O), and ZBLAN glass (53% ZrF_4, 20% BaF_2, 4% LaF_3, 3% AlF_3, 20% NaF) doped with 1.5 mol% Er_2O_3 with a purity of 99.9%. In order to obtain dense targets, the SiO_2: Er pellets were sintered at 1400 °C for 3 h, the glass: Er pellets at 600 °C for 3 h. As a substrate material chemically cleaned Si (100) wafers mounted on a heated substrate holder were used. Laser ablation was performed by KrFexcimer laser radiation ($\tau\sim20$ ns, repetition rate 5 Hz, and spot size ~2 mm^2). In all cases we used 6000 laser pulses with a fluence $\Phi=2.5$ J/cm^2 which is well above the threshold fluence for ablation. In order to obtain strongly sticking films a substrate temperature of 700 °C was employed. For homogenous ablation, the targets were rotated and the laser focus was moved across the target. The substrate to target distance was 3.5 cm. The reaction chamber was pumped to a base pressure of 1×10^{-2} mbar before filling back with an O_2 pressure of 0.6 mbar for SiO_2: Er and soda-lime glass: Er ablation. ZBLAN: Er films were deposited in an N_2-atmosphere at a pressure of 0.05 mbar. PL emission was excited by an Ar^+-laser and recorded by a Bomem DA8.22 Fourier transform spectrometer equipped with a liquid nitrogen cooled NorthCoast Ge-detector. PL decay times were measured by a Peltier-cooled InGaAs detector with a system response time of 20 µs. The quantum efficiency was measured with Ge-photodiode calibrated at 1.54 µm.

C. Results and discussion

We fabricated luminescent Er-doped silica, soda-lime glass, and ZBLAN glass films by PLD. The PL spectra of laser-deposited Er-doped SiO_2 films are similar to those obtained from Er ion-implanted silica [3] and Er chemical vapor doped silica fibers [4]. The emission spectra of laser-deposited films are different for the three host materials, silica, soda-lime glass, and ZBLAN glass [5], as shown in Fig. 1. PL intensities and spectra did not depend on the substrate temperature during deposition. PL decay times of all films were ~10 ms which is comparable with results obtained on ion-implanted silica and soda-lime glass [3]. Fig. 1a shows the normalized emission spectra of Er-doped silica, soda-lime glass and ZBLAN glass films at 5 K. Common to all spectra are additional lines at higher wavelengths than the maximum emission line. These lines arise from different Er sites in the host material and hence the different interaction of the crystal field of the nearest neighbors with the ground state of the Er^{3+}. The main peak of the SiO_2: Er spectrum occurs at a wavelength of 1.531 µm with a shoulder at 1.537 µm, which coincides with the maximum emission of the glass films at 5 K. The ZBLAN: Er spectrum show a rich fine structure in the wavelength region between 1.525 µm and 1.625 µm, indicating a stronger crystal field interaction in a fluoride environment. The full width at half maximum (FWHM) decreases from 16 nm for ZBLAN glass to 10 nm for

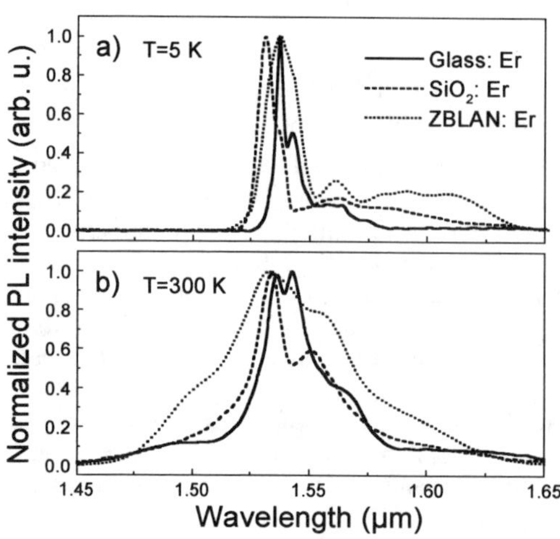

Fig. 1a) Comparison of the normalized emission spectra of PLD grown Er-doped films of soda-lime glass, SiO_2, and ZBLAN glass at 5 K and b) at 300 K.

SiO_2 to 5 nm for soda-lime glass as the Er host. The normalized spectra at 300 K of SiO_2 and the glass hosts for Er are shown in Fig. 1b. Compared to the low temperature spectra, all emission lines are broadened at 300 K due to phonon interactions. Additional lines occur at shorter wavelengths, mainly for the glass-type hosts. These so-called hot lines arise from the emission of thermally populated higher lying states of the $^4I_{13/2}$ multiplet due to crystal field interaction. The main line of SiO_2: Er at 1.531 µm at 5 K is quenched and the initially shoulder at 1.537 µm remains up to room temperature. The emission maxima of the Er-doped glasses change slightly with temperature. The broadest room temperature emission with a FWHM of 57 nm is obtained for ZBLAN: Er, which make ZBLAN a possible candidate for broad band amplifiers. The sharpest emisson, which is needed for 1.54 µm emitters, is observed for soda-lime glass: Er with a FWHM of 24 nm at 300 K. The temperature dependence of the PL yield of the Er hosts investigated is shown in Fig. 2. All intensities are normalized to the low temperature values. The temperature dependence is drawn for the maximum emission line at room temperature for each host. The luminescence yield of SiO_2: Er show a monotonic decrease by a factor of 5 from 5 K to room temperature with an activation energy of 15 meV. At low temperatures the usual quenching is observed for the glass-type hosts. The PL yield of ZBLAN: Er, in contrast to SiO_2: Er, starts to increase with increasing temperature above 150 K, which can be explained in terms of a thermally activated energy transfer of the exciting light to the Er ions. This behavior is much more pronounced for the soda-lime glass host. The luminescence yield increases by more than one order of magnitude between 60 K and 300 K. The applicability of a host material for Er based emitters or amplifiers depends on its quantum efficiency and the output power. Fig. 3 gives a comparison of the output power of 100 µm thick Er-doped films of soda-lime glass, ZBLAN glass, and SiO_2 grown under identical conditions. The Er solubility in pure SiO_2 is rather low. The addition of Al and P has been found to reduce the formation of non-

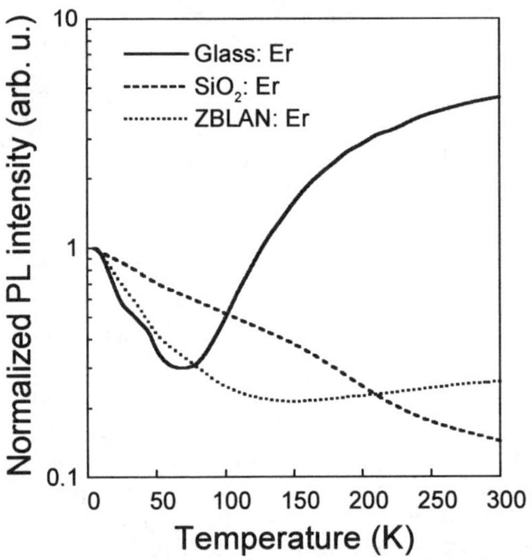

Fig. 2. Temperature dependence of Er-doped films of soda-lime glass, SiO_2, and ZBLAN glass.

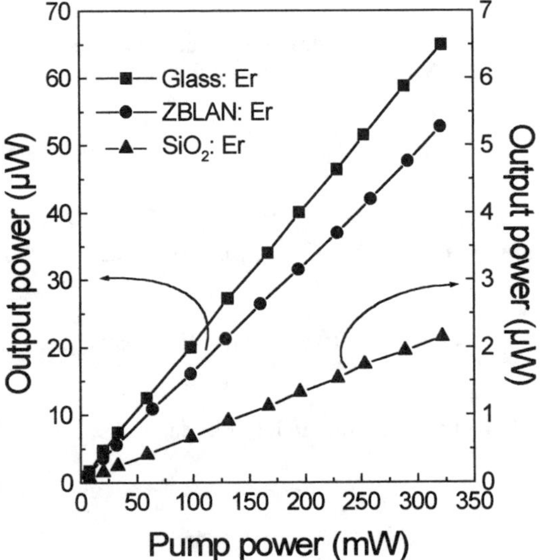

Fig. 3. Output power of Er-doped 100 µm thick films of soda-lime glass, ZBLAN glass, and SiO_2 as a function of the pump power.

luminescent Er clusters [6]. A concentration ratio of Al or P to Er of about 50 is used in low Er-doped fiber amplifiers [5]. Such a high co-doping concentration could not be achieved here as 1.5 mol% of Er_2O_3 was used in the target material. This low Er solubility in SiO_2 is reflected by the low output power and an external quantum efficiency of 7×10^{-6}. We therefore used multicomponent glass-type hosts which are known to have a strongly enhanced Er solubility. The output power is by a factor of 30 higher for the multicomponent glass films. The external quantum efficiency of soda-lime glass: Er is 2×10^{-4}, the quantum efficiency of ZBLAN: Er is by 20% lower, mainly due to the broad emission spectrum. Taking the refractive index of soda-lime glass of 1.45 at 1.54 µm into account, the internal quantum efficiency is calculated to be 6×10^{-4}.

D. Conclusions

Photoluminescence at 1.54 µm of pulsed-laser deposited Er-doped films was found to depend strongly on the Er host material. PL spectra demonstrate the incorporation of isolated Er^{3+} ions in the film and rule out the formation of Er_2O_3 precipitates. The Er emission around 1.54 µm is substantially broadened in all host materials at room temperature compared to the FWHM at 5 K. Usual PL temperature quenching occurs in SiO_2: Er with an activation energy of 15 meV. A thermally activated energy transfer process is proposed in order to explain the increase of the PL yield with increasing temperature of Er-doped soda-lime glass and ZBLAN glass. For these films, the luminescence was almost two orders of magnitude higher compared to SiO_2: Er. This increase is due to a much higher amount of optically active Er^{3+} ions in the glass host. The external quantum efficiency of soda-lime glass: Er is 2×10^{-4}.

Acknowledgements

Part of the work was supported by the Gesellschaft für Mikroelektronik, the Fonds zur Förderung der Wissenschaftlichen Forschung, Vienna, and the Jubiläumsfonds der Österreichischen Nationalbank, Vienna.

References

[1]   R. J. Mears, L. Reekie, I. M. Jauncey, and D. N. Payne, Electron. Lett. **23** (1987) 1026

[2]   D. Bäuerle in *"Laser Processing and Chemistry"*, 2nd edition, Springer Verlag Berlin (1996)

[3]   A. Polman, J. Appl. Phys **82** (1997) 1

[4]   A. M. Jurdyc, B. Jacquier, J. C. Gacon, J. F. Bayon, and E. Delevaque, J. Luminescence **61&62** (1994) 89

[5]   W. J. Miniscalco, J. Lightwave Technol. **9** (1991) 234

[6]   K. Arai, H. Namikawa, K. Kumata, T. Honda, Y. Ishii, and T. Handa, J. Appl. Phys. **59** (1986) 3430

Electron-irradiation-induced disordering of CuPt-ordered GaInP studied by TEM-mode optical spectroscopy

Y. Ohno, Y. Kawai, and S. Takeda

*Department of Physics, Graduate School of Science, Osaka University,
1-16, Machikane-yama, Toyonaka, Osaka 560*

Electron-irradiation-induced disordering in CuPt-ordered GaInP has been examined by *in-situ* photoluminescence and cathodoluminescence spectroscopy in a transmission electron microscope. A decrease of luminescence intensity by an electron-irradiation in the energy range above 120 keV has been observed, and we have shown that the decrease is due to the Frenkel-type defects on the Ga and In sublattices generated by electron-irradiation. We have proposed that an electron-irradiation-induced migration of the Ga- and In-vacancies dominates the disordering in the dose range below 2×10^{20} cm^2.

A. INTRODUCTION

The ternary semiconductor GaInP, which is of interest for the fabrication of semiconductor lasers, is known to form the CuPt-type ordered structure, and such atomic ordering strongly affects the band gap energy. From a practical point of view, much efforts have been made to fabricate artificial order/disorder structures; as an example, a wire-structure with a width down to 35 nm is fabricated by this ion-implantation method [1]. For applications that require a narrower width, disordering by electron-irradiation [2] may be more suitable than the method, since an electron-beam can be focused on an area of nanometer size.

Electron-irradiation-induced disordering in CuPt-ordered GaInP is caused by the migration of the group-III (Ga and In) interstitials and vacancies generated by electron-irradiation. We have investigated the kinetics of point-defect-reactions under electron-irradiation by *in-situ* cathodoluminescence (CL) and photoluminescence (PL) spectroscopy in a transmission electron microscope (TEM) method [3], and proposed a new disordering model based on an electron-irradiation-induced migration of group-III vacancies.

B. EXPERIMENTS

Specimens were CuPt-ordered GaInP grown on a GaAs substrate by metal-organic vapor-phase epitaxy (MOVPE) at 700 °C; the substrate was 2° off from (001) towards [1$\bar{1}$0]. They were irradiated with an electron-beam in a TEM (equal to the irradiation temperature of 110 K); the direction of the incident-beam was kept parallel to the [110] zone axis. Incident-electron energy E ranged from 100 to 170 keV, and electron-dose D, i.e. electron-flux ϕ multiplied by irradiation-time t_{ir}, was up to 1.0×10^{23} cm^{-2}. The irradiated specimens were characterized by transmission electron diffraction (TED) [2] and CL and PL spectroscopy [3]. Figure 1 shows examples of CL spectra.

The degree of atomic ordering in GaInP is characterized by an order parameter S, and S was estimated by [4]

$$S = \left\{ \left(2.017 - E_{CL(110K)} \right) / 0.471 \right\}^{0.5}, \qquad (1)$$

where $E_{CL(110K)}$ represents CL peak energy in electron volt obtained at 110 K. S was also estimated by TED [2]; S is approximately given by $S = S_0 \{ I_{TED}(D) / I_{TED}(0) \}^{0.5}$ where S_0 and $I_{TED}(D)$ denote the order parameter for an as-grown specimen and the dose dependent TED intensity of an ordered spot, respectively.

C. RESULTS AND DISCUSSION
C1. Generation and annihilation of Frenkel-type defects

The CL peak intensity I_{CL} of an electron-irradiated specimen decreased to a lower value compared to as-grown one [as an example, see Fig. 1(a) and 1(b)]. We found that I_{CL} was well expressed by a function

$$I_{CL} = 100/\{1 + \sigma_1 D\},\qquad(2)$$

where σ_1 is a fit parameter. Figure 2 shows results for $E =$ (a) 120 and (b) 140 keV, respectively. $\sigma_1 = 0$ for $E \leq 120$ keV, and σ_1 increased with increasing E when $E > 120$ keV; σ_1 was independent of ϕ. In order to understand quantitatively the decrease, we applied a recombination-center model; i.e. localized energy levels of irradiation-induced defects act as non-radiative recombination centers, and I_{CL} is given by a function [3]

$$I_{CL} = 100/\left\{1 + \Sigma_i \gamma_{(i)} C_{(i)}\right\},\qquad(3)$$

where $C_{(i)}$ represents the concentration of i-th non-radiative center. $\gamma_{(i)}$ is related to the carrier-lifetime for the non-radiative recombination due to electron-phonon interactions, and it may remain a constant value during the present experiments. Comparing the experimental results given by Eq. (2) with the theoretical one by Eq. (3), we conclude that $C_{(i)}$ increases linearly with increasing D; i.e. the non-radiative centers are related to the primary defects generated by electron-irradiation.

The above-mentioned centers are not related to the defects on the P sublattice, since the expected threshold electron energy for the displacement of P atoms (100 keV [2]) is smaller than 120 keV. We consider that Frenkel-type defects on the Ga and In sublattices are related to the non-radiative centers. The change of the concentration of each defect with respect to t_{ir} is theoretically expected by an equation

$$\frac{dC_{IV(\xi)}}{dt_{ir}} = \sigma_{(\xi)}(1 - f)\phi - K_{IV(\xi)} C_{IV(\xi)}^{2},\qquad(4)$$

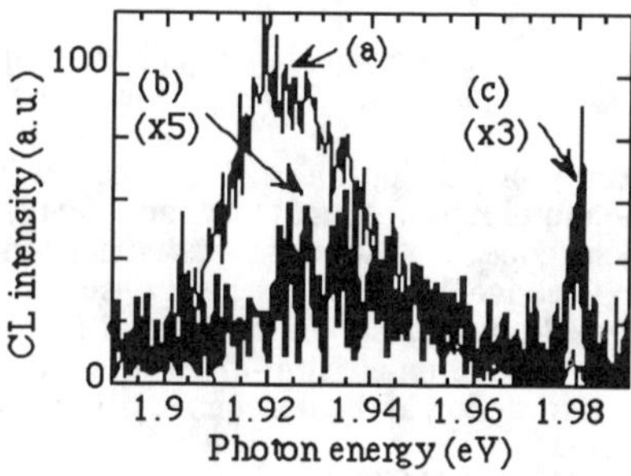

Fig. 1 CL spectra of (a) as-grown and (b) e⁻-irradiated ($D = 9.3 \times 10^{20}$ cm⁻²) specimens. $E = 160$ keV. (c) a CL spectrum of a specimen post-annealed (at 300K for 64 h) after e⁻-irradiation (D $= 6.0 \times 10^{22}$ cm⁻²). The CL peak intensity of as-grown specimen is normalized to 100.

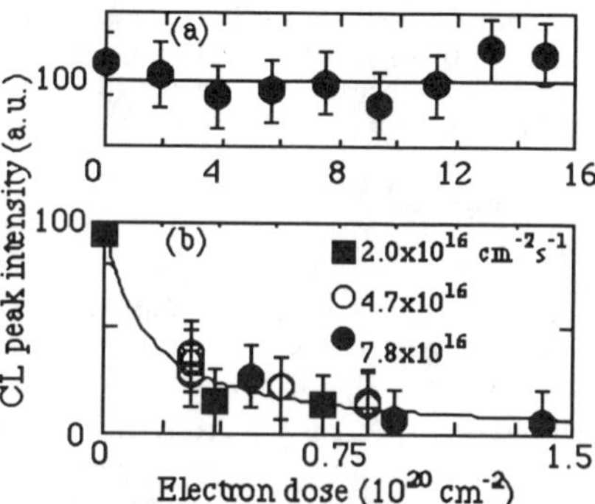

Fig. 2 CL peak intensity vs. electron dose for electron energies of (a) 120 and (b) 140 keV, respectively. Electron fluxes are denoted in the figure. Solid lines represent fits of Eq. (2).

where $C_{IV(\xi)}$ represents the concentration of the Frenkel-type defect on the ξ sublattice ($\xi =$ Ga and In). f is the correlated recombination factor, $\sigma_{(\xi)}$ the cross sections for displacement damage of ξ-atoms; $K_{IV(\xi)}$ is proportional to the sum of the mobility of ξ-interstitials and that of ξ-vacancies. Numerical calculations of Eq. (4) provide the result that $C_{IV(\xi)}$ approximates to $\sigma_{(\xi)}(1 - f)D$ for $D < 10^{20}$ cm⁻². Suppose all defects act as non-radiative centers, σ_1 is

$$\sigma_1 = (1-f)\sum_\xi \gamma_{(\xi)}\sigma_{(\xi)}. \qquad (5)$$

The electron-energy-dependence of σ_1 was well expressed by the equation under the assumption that the threshold electron energies for the displacement of Ga and In atoms, $E_{d(Ga)}$ and $E_{d(In)}$ are 145±2 and 120±2 keV, respectively (Fig. 3). We thus conclude that I_{CL} decreased owing to the Frenkel-type defects on the Ga and In sublattices generated by electron-irradiation.

Figure 1(c) shows a CL spectrum of a specimen post-annealed after a high-dose electron-irradiation (the peak energy of 1.98 eV). Before the post-annealing, the specimen showed no CL emission. Annealing of some irradiation-induced non-radiative centers presumably results in the recovery of the CL intensity. The annealing process is, however, still uncertain.

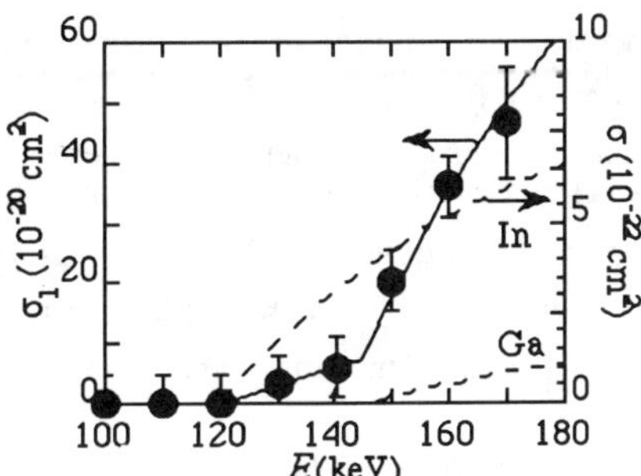

Fig. 3 Fitting parameter vs. incident-electron energy. Broken lines are the theoretical cross sections for displacement damage of Ga and In atoms. The solid line is a fit of Eq. (5) to the data.

C2. *Disordering process under electron-irradiation [4]*

Figure 4 shows the relationships of S with D for several incident electron energy. We found that S decreases gradually with increasing D when $E > 140$ keV; two disordering stages in the dose ranges $D < 2\times10^{20}$ cm^{-2} and $D > 5\times10^{21}$ cm^{-2} were clearly observable. Order-disorder reactions under electron-irradiation are generally explained by three kinds of effects; i.e., spontaneous recombinations of an interstitial and a vacancy, sequential collisions, and the migration of vacancies. The disordering in the latter stage is attributed to the spontaneous recombinations of a group-III interstitial and a vacancy generated by electron-irradiation [2]; the change of S is expressed by

$$\left[\frac{dS}{dt_{ir}}\right]_{spontaneous} = -\frac{\sigma_{(Ga)}\sigma_{(In)}}{\sigma_{(Ga)} + \sigma_{(In)}}(1-f)\phi S. \qquad (6)$$

Fig. 4 Order parameter vs. electron dose. Incident-electron energies are indicated in the figure. Curves denote the theoretical calculations by Eq. (9) [$K_{(Ga)} = K_{(In)} = 7.0$, and $K_{IV(Ga)} = K_{IV(In)} = 0.2$]

The spontaneous recombination model cannot yields the dose-dependence of S in the range $D < 2\times10^{20}$ cm^{-2}, since the order parameter estimated by Eq. (6) is much larger than the experiments. The effect of sequential collisions may be disregarded in the present study, since the incident-electron energies used in the experiments were very close to the threshold electron energies for atomic displacements. We have hence considered that the disordering in the range $D < 2\times10^{20}$ cm^{-2} is attributed to the migration of vacancies introduced by electron-irradiation. The vacancy-migration-mediated disordering can be expressed by

$$\left[\frac{\partial S}{\partial t_{ir}}\right]_{vacancy} = -\sum_\xi\left[\alpha_{(\xi)}D_{V(\xi)}\nabla^2 C_{V(\xi)} + \beta_{(\xi)}D_{V(\xi)}C_{V(\xi)}\right]S, \qquad (7)$$

in which $C_{V(\xi)}$ represents the concentration of the ξ-vacancy, $\alpha_{(\xi)}$ and $\beta_{(\xi)}$ are disordering-efficiency-factors. $D_{V(\xi)}$ is the diffusion constant of the ξ-vacancy under electron-irradiation. Since the onset temperature for the thermal migration of group-III vacancies in GaInP ($= 923$ K [1]) is much higher than 110 K, the value of $D_{V(\xi)}$ must be small in the present experiments: the second term in Eq. (7) may be negligible. Assuming $D_{V(\xi)}\nabla^2 C_{V(\xi)}$ is proportional to the differential of $C_{V(\xi)}$ with respect to irradiation-time, $[dS/dt_{ir}]_{vacancy}$ may be proportional to $dC_{IV(\xi)}/dt_{ir}$. Hence the change of S can be expressed by

$$\frac{dS}{dt_{ir}} = -\frac{\sigma_{(Ga)}\sigma_{(In)}}{\sigma_{(Ga)}+\sigma_{(In)}}(1-f)\phi S - \sum_{\xi}K_{(\xi)}\frac{dC_{IV(\xi)}}{dt_{ir}}S, \tag{8}$$

where the first and second terms are related to the disordering by the spontaneous recombinations and the migration of group-III vacancies, respectively; the phenomenological factor K should be proportional to the number of atomic sites around a vacancy for the migration of the vacancy. Analytical calculation of Eqs. (4) and (8) provides the following equation,

$$\ln\left[\frac{S}{S_0}\right] = -\frac{\sigma_{(Ga)}\sigma_{(In)}}{\sigma_{(Ga)}+\sigma_{(In)}}(1-f)D$$
$$-\sum_{\xi}K_{(\xi)}\left\{\frac{\sigma_{(\xi)}(1-f)\phi}{K_{IV(\xi)}}\right\}^{0.5}\tanh\left[\left\{\frac{\sigma_{(\xi)}(1-f)K_{IV(\xi)}}{\phi}\right\}^{0.5}D\right]. \tag{9}$$

The dose-dependence of S (Figs. 3 and 4) were well expressed by Eq. (9) when the values of K and K_{IV} were on the order of 10^0 and 10^{-1}, respectively; solid lines in the figures show simulated curves of S vs. D. The value of the first term is negligible in the range $D < 2\times10^{20}$ cm^{-2}, and hence the migration of group-III vacancies under electron-irradiation dominates the disordering in the dose range. For $D > 5\times10^{21}$ cm^{-2}, the value of the second term comes to a small steady-state value and thus Eq. (9) almost corresponds to Eq. (6).

As shown in Fig. 4, all experimental data have consistently been expressed by Eq. (9), and thus we have concluded that group-III vacancies can migrate at the temperature of 110 K under electron-irradiation. So far, the thermal migration of the vacancies has been observed only in the temperature range above 923 K [1]. We showed that CL peak intensity decreases owing to a non-radiative electron-hole recombination (Sec. C1), and the energy of the recombination may enhance the motion of the group III vacancies (recombination-enhanced-migration). Such recombination-enhanced effect has aroused broad interest, since electronic and optical properties of final products may vary drastically owing to the effect; as an example, a recombination-enhanced rapid-migration of impurities causes the degradation of some GaAs-based laser diodes [5]. We have first observed a recombination-enhanced migration of vacancies in GaInP, and the present experimental data may be useful for a general understanding of point-defect-reactions in III-V compound semiconductors.

References

[1] M. Burkard, A. Englert, C. Geng, A. Mühe, F. Scholz, H. Schweizer, and F. Phillipp, J. Appl. Phys. **82**, 1042 (1997).
[2] N. Noda and S. Takeda, Phys. Rev. **B53**, 7197 (1996).
[3] Y. Ohno and S. Takeda, J. Electron Microsc. **45**, 73 (1996).
[4] Y. Ohno, Y. Kawai, and S. Takeda, unpublished.
[5] M. Uematsu and K. Wada, Appl. Phys. Lett. **60**, 1612 (1992).

Multi-Layer Conduction in Epitaxial InSb Grown on GaAs Substrates

Mohamed Ahoujja(1), William C. Mitchel(1), Eric Michel(2), and Manijeh Razeghi(2)
(1) Materials Directorate, Air Force Research Laboratory, AFRL/MLPO, W-PAFB, OH 45433-7707
(2) Center for Quantum Devices, Electrical and Computer Engineering Department, Northwestern University, Evanston, IL 60208

Abstract

Electron transport properties of molecular beam epitaxy grown InSb on GaAs substrates are studied using the conventional Hall effect and the mobility spectrum technique. The latter technique reveals the existence of at least three different conducting channels in InSb. Our analysis of the Hall data and the mobility spectrum shows that the conduction in InSb is determined by a highly dislocated interfacial layer between the InSb layer and the GaAs substrate, a high mobility bulk-like InSb layer, and, possibly, a surface layer.

A. Introduction

Epitaxial InSb layers grown on GaAs substrates have been studied with great interest due to their potential application for infrared sources and detectors[1] as well as high speed circuit elements[2]. Several research groups have reported apparently "anomalous" behavior in the transport properties of epitaxial InSb on GaAs as due to a multi-layer conduction. Some groups have argued that conduction is due to bulk-like InSb and the surface layer[3]. Others have suggested that it is due to bulk-like InSb and the interface region between InSb and GaAs in light of their 14% lattice mismatch[4,5]. Recently Besikci *et al.* [6] proposed a three-layer model, by taking into account both the surface electron accumulation layer and the defects at the interface region, to qualitatively describe the electrical properties of the InSb films. Therefore the question of whether the InSb/GaAs interface layer or the surface layer or both layers contribute to the conduction in addition to bulk-like InSb is still unresolved. In this report we use the conventional Hall effect and the "mobility spectrum" (MS) [7] technique to show a presence of three different conducting channels in InSb.

B. Experimental Details

2.0, 5.0, and 9.2 µm thick InSb epilayers were grown on semi-insulating (001) GaAs substrates using molecular beam epitaxy (MBE). The optimum growth conditions were determined and described elsewhere[8]. Electrical measurements were done on cleaved samples of Van der Pauw patterns in a variable temperature liquid helium dewar and in magnetic fields generated by an electromagnet whose poles were such that the magnetic field direction was perpendicular to the surface of the sample. The conventional Hall measurements were done in a constant perpendicular magnetic field (usually 5 KG) over a temperature range of 10-300 K. For the variable magnetic field data, Hall measurements were done at 20 field values which were equally spaced in inverse field between 0.025 and 2 T.

     177

C. Results

The Hall concentration and mobility as a function of temperature for the 2, 5 and 9.2 μm thick samples are displayed in Figure 1. The carrier concentration is temperature independent at temperatures below 50 K but it increases sharply at temperatures above ~ 200 K. This indicates the onset of intrinsic conduction in the bulk InSb layers. The fact that both the Hall concentration and mobility are constant with temperature at low temperatures is a signature of two dimensional electron gas system most likely at the interface between InSb and GaAs and/or at the surface. The sharp drop-off of the Hall mobility at temperatures below the peak in mobility suggests that the mobility measurement is dominated by a low mobility parallel conduction channel, most likely the highly dislocated region near the InSb/GaAs interface. We found that both the resistivity and Hall coefficient depend on magnetic field, indicating the presence of multiple species.

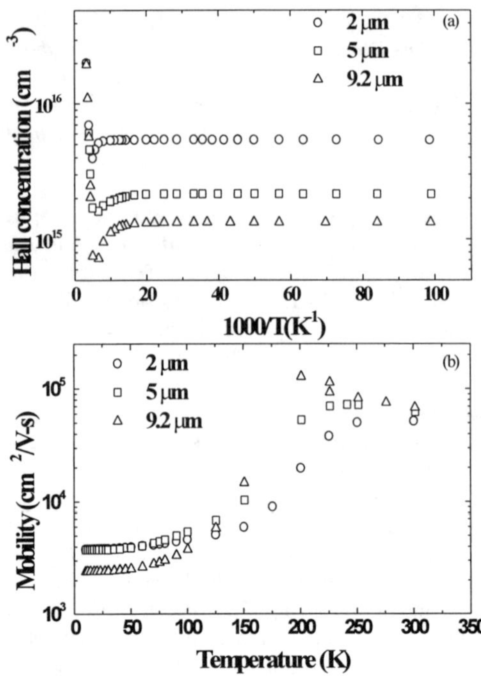

Fig.1: A plot of (a) Hall concentration and (b) mobility as a function of temperature for a 2 μm (circle), 5μm (square), and 9.2 μm (triangle) thick InSb layers.

The (MS) is a novel technique which generates a "mobility spectrum" where the maximum carrier conductivity is determined as a continuous function of mobility. The peaks in this mobility spectrum correspond to the different conducting layers. In figure 2 we show a plot of the MS for the 2 and 5 μm samples at 10 K. A discrete carrier is seen as a distinct peak whose abscissa is the carrier mobility and whose amplitude is the conductivity. The presence of two distinct peaks at low temperature is the evidence of at least two different carrier species. In figure 3 the MS for the 9 μm sample at 300 K is displayed and three distinct peaks are observed. The higher mobility peak which is not observed at low temperatures is attributed to the intrinsic conduction.

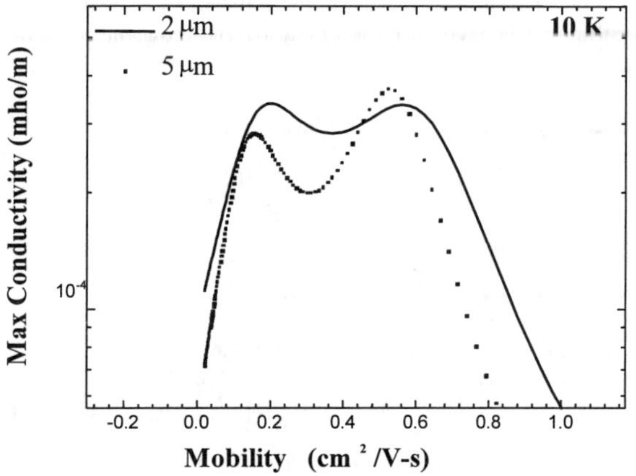

Fig.2: The maximum conductivity as a function of mobiltiy for the 2 μm (solid) and 5 μm (dash) samples at 10 K. Two distinct peaks corresponding to two different conducting layers are observed.

D. Discussion

The values of the carrier concentration and mobility obtained from the conventional Hall measurements at 10 K are shown in Table 1. The sheet concentration, n_s, in the last column is obtained by multiplying the Hall concentration with the respective InSb layer thickness. Table 2 shows the concentration as well as the mobility corresponding to the two peaks for all the samples at 10 K. Using a two-layer model we calculated the overall carrier concentration and mobility and the results are shown in the last two columns of Table 2. Clearly the calculated n_H and μ_H, as shown in the last two columns of table 2, agree well with the values of the measured Hall density and mobility displayed in Table 1. We interpret the lower mobility peak as being associated with the highly dislocated interfacial layer and the other one with the surface layer. Both of these peaks cannot correspond to the bulk-InSb layer because at low temperature the carrier density in this layer is only of the order of 10^9 cm^{-2}. In conclusion our analysis of the Hall data and the MS shows that the conduction in the InSb layer is determined by a highly dislocated interfacial layer, a high mobility bulk-like InSb layer and a surface layer. To the best of our knowledge, this is the first quantitative analysis of a three layer conduction in InSb layer grown on GaAs substrates.

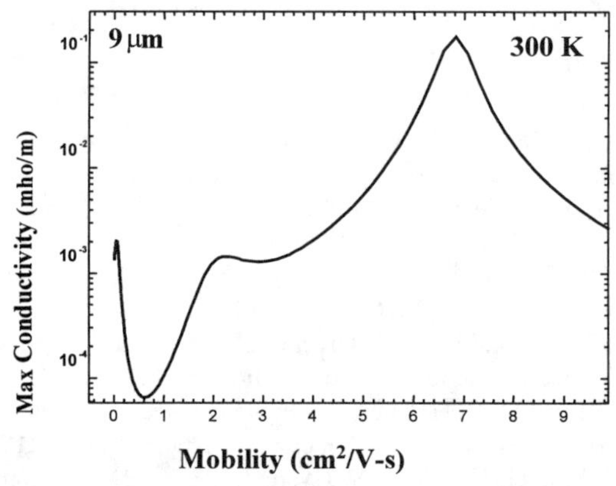

Fig.3: The MS for the 9 μm sample at 300K. One can clearly see the high mobility peak corresponding to the bulk-like InSb layer. This peak is absent at low temperatures.

179

Table 1. The Hall mobility and Hall concentration for the three different thickness values of InSb on GaAs at 10 K. n_s is the sheet concentration obtained by multiplying the Hall concentration with the InSb layer thickness.

Sample thickness (μm)	μ_H (cm^2/V-s)	n_H (10^{15} cm^{-3})	n_s (10^{11} cm^{-2})
2	3803	5.40	10.8
5	3749	2.14	11
9.2	2400	1.33	12.19

Table 2. n_1 (n_2) and μ_1 (μ_2) are the sheet carrier density and mobility corresponding to the two different peaks in the mobility spectrum for the three different thickness values of InSb on GaAs at 10 K. The Hall mobility, μ_H, and Hall concentration, n_H, are calculated using the values of n_1, n_2, μ_1 and μ_2 in a two band model.

Sample thickness (μm)	n_1 (10^{11} cm^{-2})	μ_1 (cm^2/V-s)	n_2 (10^{11} cm^{-2})	μ_2 (cm^2/V-s)	n_H (10^{11} cm^{-2})	μ_H (cm^2/V-s)
2	3.51	5799	11.87	1721	10.87	3756
5	4.43	5225	11.69	1500	11.23	3620
9.2	2.43	5225	13.04	1700	11.68	2983

Acknowledgments: One of the authors (M. A.) is supported by the National Research Council (NRC) and the Air Force of Scientific Research (AFOSR).

References:

[1]   G. C. Osbourn, J. Vac. Sci. Technol. **B2** (1984) 176 .

[2]   R. A. Stradling, Semicond. Sci. Technol. 6 (1991) C52 .

[3]   J. R. Soderstrom, M. M. Cumming, J- Y Yao, and T. G. Andersson, Semicond. Sci. Technol. **7** (1991) 337.

[4]   J. I. Chyi, D. Biswas, S. V. Iyer, N. S. Kumar, H. Murkoc, R. Bean, K. Zanio, H. Y. Lee and H Chen, Appl. Phys. Lett. **54** (1989) 1016.

[5]   T. D. Golding, S. K. Greene, M. Pepper, J. H. Dinan, A. G. Cullis, G. M. Williams, and C. R. Whitehouse, Semicond. Sci. Technol. **5** (1990) S311.

[6] C. Besikci, Y. H. Choi, R. Sudharsanan, and M. Razeghi, J. Appl. Phys. **73** (1993) 5009.

[7]   W. A. Beck and J. R. Anderson, J. Appl. Phys. **62 (1987)** 541.

[8] E. Michel, J. D. Kim, S. Javadpour, J. Xu, I. Furgeson, and M. Razeghi, Appl. Phys. Lett. **69** (1996) 215.

X-Ray Studies of $Al_xGa_{1-x}As$ Implanted With 1.5 MeV Se Ions

W. Wierzchowski[1], K. Wieteska[2], A. Turos[1], W. Graeff[3], R. Groetzschel[4]

[1] *Institute of Electronic Materials Technology, Wólczyńska 133, 01-919 Warsaw, POLAND*
[2] *Institute of Atomic Energy, 05-400 Otwock-Świerk POLAND*
[3] *HASYLAB at DESY, Notkestraße 85, D-22603 Hamburg, GERMANY*
[4] *Forschungszentrum ROSSENDORF, D-01314 Dresden, GERMANY*

Abstract

$Al_xGa_{1-x}As$ epitaxial layers with x=0.45 and low dislocation density, implanted with 1.5 MeV Se^+ ions to doses of 6×10^{13}-1×10^{15} ions/cm^2 were studied with X-ray diffraction methods using conventional and synchrotron X-ray sources. The implantation caused considerable increase of lattice parameter reaching the maximum at the dose of 2×10^{14} ions/cm^2. It was also found that the implantation produced a relatively thick layer with almost constant lattice parameter. The applied doses did not produce lattice amorphization at room temperature.

A. Introduction

Despite wide use of ion implantation the structural changes caused by this process are yet not well known. That concerns in particular multielement semiconductors, where the new applications of implantation concern introduction of insulating layers and compositional disordering of multilayer structures. A certain number of papers describes the effect of implantation in $Al_xGa_{1-x}As$, particularly its higher resistance against amorphization for large Al concentration [1-3].

In the present work the $Al_xGa_{1-x}As$ epitaxial layers with x=0.45 implanted with 1.5 MeV Se^+ ions to a wide range of doses are studied with a wide set of complementary X-ray diffraction methods. The important point of the investigation was the use of epitaxial layers with low dislocation density to improve the detection of implantation induced effects. The present work was supported from the State Committee for Scientific Research (grant 8 T11B 049 13) and from Polish-German Agreement on scientific cooperation POL-GER-N-100-95/BO.

B. Experimental

The $Al_xGa_{1-x}As$ epitaxial layers were deposited with MOCVD method on (100) oriented substrates cut out from Bridgman grown GaAs crystal doped with 0.3% indium and contained less than 10^3 cm^{-2} dislocation. The layers were 3 μm thick and 0.2 μm of undoped GaAs buffer layer was initially deposited. The substrates allowed growth of the layers with similar low dislocation density partly because of lower misfit factor due to the lattice parameter significantly increased by high indium concentration. The value of lattice parameter, determined with method described by Godwod *et al.* [4] was 6.5344 Å. As it was revealed by X-ray topography only single misfit dislocations occurred locally in some areas of the samples. The implantation with 1.5 MeV Se^+ ions was performed at Rossendorf Research Center at room temperature. The ion doses covered a wide range from 2×10^{13} to 1×10^{15} ions/cm^2. The 6° inclination of the sample was applied to eliminate channeling effects.

The samples were studied with various complementary X-ray diffraction methods realized both with synchrotron and conventional X-ray sources. The synchrotron investigations were performed using both white and monochromatic beam. The important point of white beam experiments was taking Bragg-case section topographs [5] with very narrow beam limited by 5 μm slit and at a glancing angle close to 4°. In view of the small layer thickness the method provided mainly the information concerning lattice deformation, visible at large film to crystal distances. Minor depth location effects were easily separated by comparison with topographs taken at smaller distance. The information about ion dose variation along the beam was also obtained. The method is not directly sensitive to lattice parameter changes. In addition to the section experiments the pin-hole micro-Laue patterns were taken in Bragg-case geometry using a point beam of a diameter 10-30 μm. The method enabled the determination of diffraction vector orientation changes and monitoring the formation of extended dislocation loops by observation of asymptotic diffuse scattering.

The conventional double-crystal spectrometer with Cu Kα radiation was used for the realization of an extended method of measurement providing the information about lattice parameter differences and evaluation of strain relaxation. The method consisted of recording three rocking curves respectively in symmetrical 400 reflection and the two asymmetrical 511 reflections differing by rotation about diffraction vector through 180°. All curves were recorded using and 511 reflection at (111) oriented Ge monochromator. The angular difference between the maxima corresponding to the thin layers and the substrate with misfit $\Delta d/d$ may be written generally as:

$$\Delta\Theta = tan\Theta \left\{ \frac{\Delta d}{d} - \left[sin^2\varphi \left(1 + \frac{2v}{1+v} \right) - \frac{2v}{1+v} \right] \xi_{xx} \right\} - \frac{1}{2} sin\,2\varphi \left(1 + \frac{2v}{1+v} \right) \xi_{xx} \quad (1)$$

where $\xi_{xx} = \xi_{yy}$ is the strain in the plane of the sample and v is the Poisson ratio, φ is the inclination angle. The upper sign correspond to the glancing angle Θ-φ and lower to Θ+φ. Using $\Delta\Theta$ from the three curves the formula (1) can be used for the indication of strain relaxation or, when the relaxation may be excluded, for the evaluation of the Poisson ratio. The curves with better visibility of interference fringes, suitable for strain profile analysis, were obtained by means of a multicrystal equipment using a synchrotron beam with size reduced to 50 μm. These curves were recorded in symmetrical 400 reflection using X-ray beam with 1.1 Å wavelength.

The synchrotron investigation included also projection topography in back-reflection and transmission geometry and plane wave topography. In the last case due to crystal curvature we obtained distinct fringes corresponding to the interference maxima in the rocking curves.

C. Results and discussion

The present investigation confirmed a significant change of lattice spacing and tetragonal deformation caused by Se$^+$ implantation, which was 3-4 times smaller than lattice spacing difference between the epitaxial deposit and the substrate. Most of the Bragg-case section topographs have three separate stripes representatively shown in Fig. 1, corresponding to the implanted layer, non penetrated part of the epitaxial layer and the substrate respectively. The separation of the stripes is caused by the tetragonal deformation induced by the lattice parameter changes in subsequent layers. A set of three conventional rocking curves for a representative sample implanted with a dose of 1.8×10^{14} ions/cm^2 is shown in Fig. 2. It is noticed that the distance between the maxima is much larger for the reflection with glancing angle $\Theta - \varphi$ than for the reflection with the glancing angle $\Theta + \varphi$ due to respective adding and subtracting of the component coming from the parameter difference and the reflecting planes disorientation.

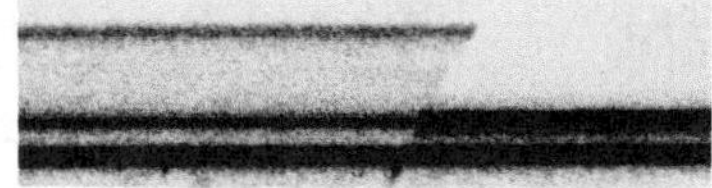

Fig. 1: Representative narrow slit pattern for the $Al_xGa_{1-x}As$ epitaxial layer implanted with Se^+ ions to the dose 1.8×10^{14} ions/cm^2.

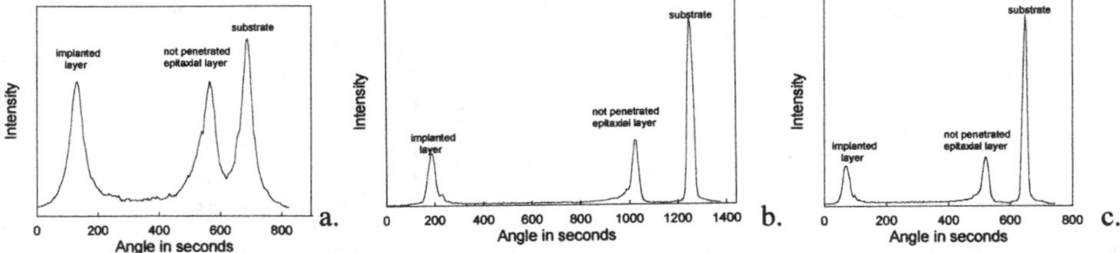

Fig. 2: A set of three double crystal rocking curves in Cu Kα with 511 Ge monochromator for symmetrical 400 reflection (nonparallel setting) – a, and two asymmetrical reflections with glancing angles $\Theta - \varphi$ and $\Theta + \varphi$ for the sample implanted to 1.8×10^{14} ions/cm^2.

The measured values of the lattice parameter changes are plotted in Fig. 3. The lattice parameter change increased with the ion dose reaching its maximum at 2×10^{14} cm^{-2} and then decreased with further increase of ion dose.

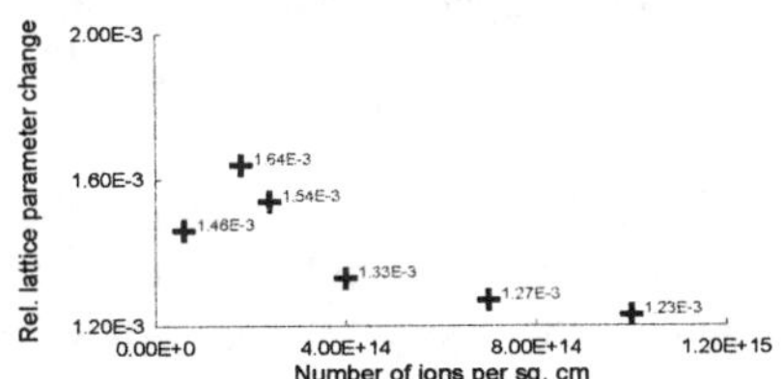

Fig. 3: The plot of relative lattice changes versus ion dose for the $Al_xGa_{1-x}As$ implanted with 1.5 MeV Se ions.

Fig. 4: The double crystal topographs of sample implanted to 4.0×10^{13} ions/cm with "snow –like" defects which may be attributed to the amorphization centers.

In the presently examined samples we did not found any lattice amorphization and strain relaxation. The sole exception where some effects may be attributed to amorphization was the snow-like pattern visible in the synchrotron plane wave topograph of the sample implanted with 5×10^{14} ions/cm^2, shown in Fig. 4. The evaluations using curves shown in Fig. 2 confirmed that the Poisson coefficient is very close to that of GaAs, i.e. 0.32. The implantation significantly decreased the height of the epitaxial layer maximum dominating in the curve taken from the non implanted region. The height of the maximum coming from the non penetrated part of the epitaxial layer systematically decreased with the ion dose suggesting increasing thickness of layer with lattice parameter change.

The rocking curve simulation, illustrated by Fig. 5, pointed that the thickness of the layer with distinct lattice parameter change is more then 1.5 times thicker than it comes from the TRIM 95 program using Monte Carlo method. The evident point is that the lattice parameter seems be almost constant in the near surface region. The profile providing the proper shape of the rocking curve was obtained by transforming the initial vacancy distribution by two operations. The first one was realized by numerical integration of the diffusion equation, providing suitable extension of the profile. The second was flattening of the maximum assuming local saturation of the lattice parameter change similar to that coming from Fig. 3, approximated by a sine function. That result points that the final strain distribution is formed with participation of point defect diffusion. The character of the strain profiles was confirmed by the RBS/channeling measurement which will be published separately.

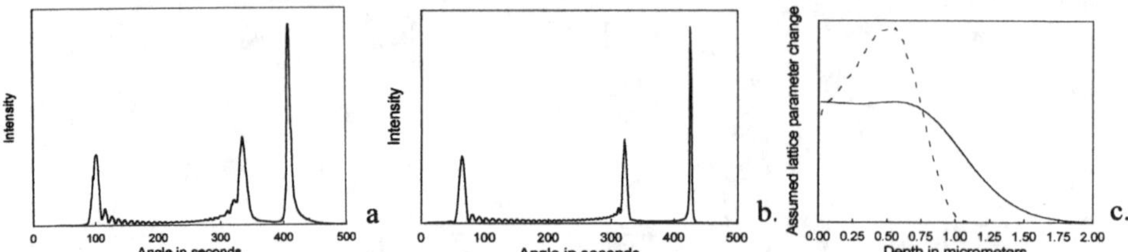

Fig.5: The synchrotron 400 rocking curve in 1.1Å for the sample implanted with 2.4×10^{14} ions/cm^2–a with simulated rocking curve –b and assumed strain distribution profile plotted together with TRIM-95 calculated vacancy distribution profile (dashed line) –c.

The micro-Laue investigation confirmed that the diffraction vector change connected with tetragonal deformation corresponded to the rotation of the vector about the intersection of the reflecting plane with the plane of the sample. No distinct increase of diffuse scattering was found. The experimental tests seem to prove that the sensitivity of the method is sufficient to exclude the formation of distinct dislocation loops in the investigated samples.

D. Conclusions

The room temperature implantation with 1.5 MeV Se$^+$ into Al$_x$Ga$_{1-x}$As with $x = 0.45$ did not produced the amorphization even at the doses exceeding 8×10^{14} cm^{-2}. The implantation causes lattice parameter change achieving the maximum at the dose 2×10^{14} cm^{-2} while the larger dose causes smaller lattice parameter changes. The recorded rocking curves point that the depth range of lattice parameter changes is larger than coming from the Monte Carlo calculation and systematically increasing with the dose. Also the near surface region of the implanted layer seems to have almost constant lattice parameter.

References

[1] P. Partyka, R.S. Averback, J.J. Coleman, P. Erhart, W. Jäger: Mat. Res. Soc. Symp. Proc. **354** (1995) 219.

[2] H.H. Tan, C. Jagadish, J.S. Wiliams, J. Zou, D.H.J. Cockayne, A. Sikorski: J. Appl. Phys. **77** (1995) 87.

[3] B.A. Turkot, B.W. Lagow, I.M. Robertson, D.V. Forbes, J.J. Coleman, L.E. Rehn, P.M. Baldo: J. Appl. Phys. **80** (1996) 4366.

[4] K. Godwod, R. Kowalczyk, Z. Szmid: phys. stat. sol. (a) **21** (1974) 227.

[5] K. Wieteska, W. Wierzchowski, W. Graeff:: J. Appl. Cryst.. **31** (1996), 238.

Electrical properties of SiGe layers grown by LPE and CVD

Olaf Krüger[*], Winfried Seifert[*], Martin Kittler[*], Astrid Gutjahr[§], Inge Silier[§],
Mitsuharu Konuma[§], Khalid Said[+], Matty Caymax[+], and Jef Poortmans[+]

[*] *Institute for Semiconductor Physics, Walter-Korsing-Str. 2, 15230 Frankfurt (Oder), Germany*
[§] *Max-Planck-Institut für Festkörperforschung, Heisenbergstr. 1, 70569 Stuttgart, Germany*
[+] *Interuniversity Micro-Electronic Centre (IMEC) vzw, Kapeldreef 75, 3001 Leuven, Belgium*

A. Introduction

High-quality, thin SiGe layers on Si substrates can be obtained by using advanced epitaxy methods [1]. These improvements have stimulated the research on silicon-based heterostructure devices for electrical and optoelectronic applications. Despite the recent potential applications of SiGe in heterojunction bipolar transistors and solar cells, there is still a lack of systematic knowledge about electrical parameters. In this paper we present electrical properties of a few microns thick, relaxed SiGe layers for solar cell application. The layers were characterized by means of electron beam induced current (EBIC), X-ray microanalysis (EDX), backscattered electrons, and Hall effect analysis.

Due to the enhanced absorption of red and near-infrared light by SiGe as compared to pure Si, the application of SiGe in solar cell structures is of increasing interest. Recently, Möller et al. have demonstrated that solar cells processed from multicrystalline SiGe material show enhanced short-circuit currents when compared to Si reference cells [2]. Another promising configuration for photovoltaic application of SiGe alloys would be a crystalline thin-film cell. Active layer thicknesses in the order of 10 µm are required for efficient light collection. To study if such structures are technically feasible for solar cells, epitaxial SiGe layers between 8 and 17 µm were grown by LPE and CVD techniques on monocrystalline p^+-Si substrate. These layers contain 5-20 at-% Ge. We obtained layers of high quality and demonstrate that passivation by hydrogen has only a small impact on the recombination activity of misfit dislocations at room temperature. Deposition of a p^+-SiGe buffer layer improves the minority carrier diffusion length in the active epilayer.

B. Experimental

The <u>LPE layers</u> were deposited from In solution on 100 mm diameter, (111) oriented Si wafers. The (111) orientation of the substrate allows the deposition of layers of homogeneous thicknesses and relatively flat surfaces without inhomogeneous surface morphology as observed for (100) substrates. During LPE growth the Ge concentration increases in the SiGe layers due to the increasing difference in solubility of Si and Ge in In solution with decreasing temperature. The initial Ge concentration dissolved in In was typically 1-6 at-%. <u>CVD layers</u>, which contain 10 at-% Ge, were deposited on 6" Si (100) in H_2 atmosphere at 750 °C and 800 °C using $Si_2H_2Cl_2$ and GeH_4 as precursors and B_2H_6 as p-type doping gas. CVD and LPE layers were also grown on p^+-SiGe buffer layers to reduce threading dislocations in the active SiGe films caused by lattice mismatch between the substrate and the epitaxial top layer. The epitaxial growth conditions used to obtain the layers for this study have been described in detail recently [3, 4].

Remote plasma hydrogenation was carried out at 400 °C for 60 min. [5]. The minority carrier diffusion length was extracted from measurements of the EBIC collection efficiency as a function of the electron beam energy between 2 and 40 keV [6, 7].

C. Results
C.1. Properties of the LPE (111) SiGe layers

EDX analysis revealed that, depending on the initial Si/Ge composition of the In solution, the Ge content in typical LPE layers varied from 1-7 % at the epilayer-substrate interface to 7-12 % on top of the 8-17 μm thick layer. The gradient of the Ge concentration in the LPE layers, which is a consequence of the growth process, may act as a Ge graded buffer layer.

The dislocation density of the LPE layers (etch pit density, EPD) caused by the lattice mismatch between SiGe and Si is about 10^6-10^7 cm^{-2}. Only low recombination activity of dislocations at room temperature has been observed. This indicates that most of the defects remain electrically inactive.

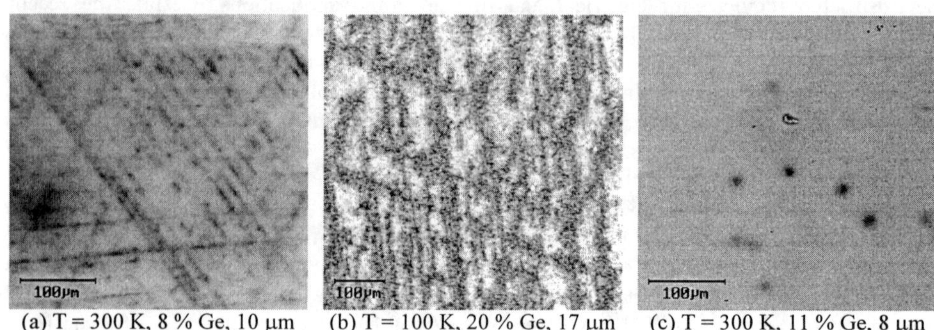

(a) T = 300 K, 8 % Ge, 10 μm   (b) T = 100 K, 20 % Ge, 17 μm   (c) T = 300 K, 11 % Ge, 8 μm

Fig. 1: EBIC images of LPE (111) SiGe layers, which illustrate the different dislocation patterns observed. (beam energy: 30 keV, beam current: 60 pA)

As shown in Fig. 1, the dislocations can appear differently in the EBIC images. The well-known dislocation features for (111) oriented substrates, i.e., intersecting lines at angles of 60°, were observed for as-grown layers as well as after hydrogenation (displayed in Fig. 1a for a 10 μm, 8 % Ge active layer grown on a 19 μm, 5 % Ge buffer layer). At low temperatures broad bands of dislocation contrasts could be observed as shown in Fig. 1b for a 17 μm layer with a Ge content of 20 at-%. Most recently grown layers exhibit a low recombination activity at room temperature. Only a few randomly distributed defects/dislocation segments are observed by EBIC, as demonstrated in Fig 1c which was taken from a solar cell structure with an 8 μm active layer containing 11 % Ge (on 40 μm, 4-9 % Ge buffer layer). In addition, EBIC images of some as-grown LPE SiGe layers and bulk SiGe solar cells (not shown) revealed dark regions. The backscattered-electron technique and X-ray microanalysis have proved the dark regions to be caused by lateral changes in the alloy composition by SiGe inclusions with about 1 at-% higher Ge content.

The dependence of the collection efficiency on the electron beam energy indicates large minority carrier diffusion lengths of 90 μm to ≥120 μm reaching the upper detection limit of our method and exceeding by far the thickness of the layers. An impact of the hydrogenation on the diffusion length was not detectable because the diffusion length is already large before the treatment. The diffusion length of holes was measured at different temperatures. It decreases upon cooling and, simultaneously, the recombination activity of the misfit dislocations increases with decreasing temperature. This proves that the diffusion length in the layers is determined by dislocations. The temperature dependence of the dislocation activity is characteristic of shallow trap levels and indicates a low degree of metal contamination of the dislocations [8, 9].

C.2. Properties of the CVD (100) SiGe layers

High-quality SiGe layers with EPDs in the order of 10^5 cm^{-2} may also be obtained by CVD. Fig. 2 presents EBIC micrographs that demonstrate the influence of the remote plasma hydrogenation on the temperature behavior of the dislocation contrast.

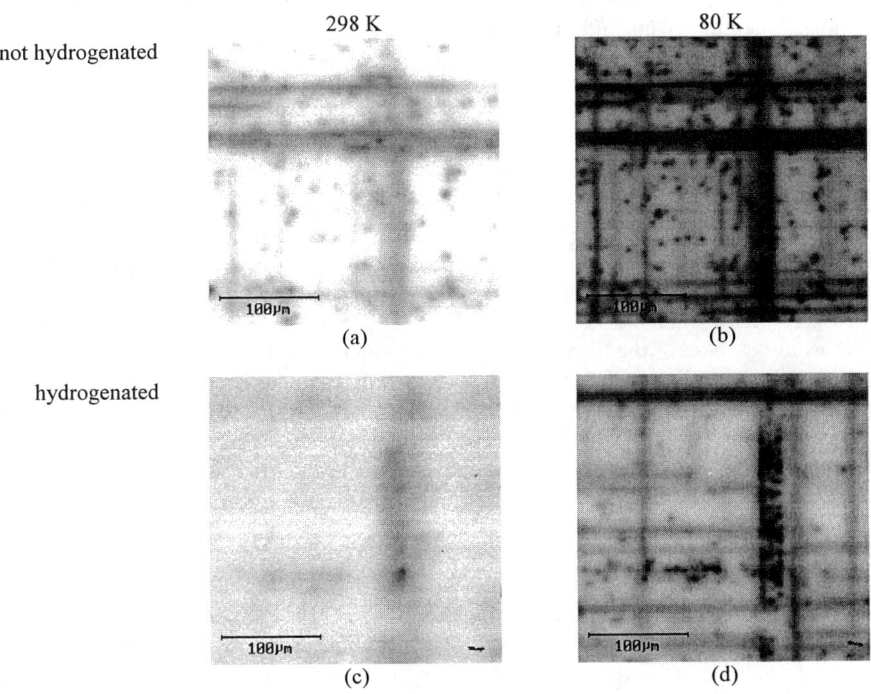

Fig. 2: EBIC images of CVD (100) SiGe layers without (a, b) and after remote plasma hydrogenation (c, d). Micrographs are taken at 298 K (a, c) and 80 K (b, d) at a beam energy of 30 keV and a beam current of 50 pA.

The typical dislocation features for CVD films grown on p$^+$-Si (100), i.e., lines intersecting at 90°, are barely visible at room temperature irrespective of the hydrogen treatment. At T = 80 K the dislocation contrast is clearly enhanced. This indicates a low level of contamination of the dislocation and can be explained by shallow recombination states [8, 9]. The diffusion length of holes extracted from efficiency measurements as a function of beam energy is about 65 μm and much larger than the layer thickness of 15 μm. After hydrogenation the diffusion length increased to about 75 μm. Although the recombination activity (contrast) of the dislocations at 80 K can be significantly reduced by hydrogenation, the intrinsic behavior (i.e., only very weak contrast at 80 K [10, 11]) cannot be restored.

The diffusion length in the active SiGe layer could be further improved to about 85 μm when an underlying p^+-SiGe layer was used to buffer the lattice mismatch. It reduces the density of threading dislocations in the active films. For these buffered structures no improvement after hydrogenation was detectable.

C.3. Carrier mobility in SiGe

Carrier mobility is one of the most essential parameters for solar cell simulations but experimental data on mobility in lightly doped unstrained SiGe are rare and scatter over a wide range [12]. The accuracy of modeling solar cells, however, largely depends on accurate electrical data. Of course, the mobilities obtained for the SiGe crystals depend on the growth conditions. Therefore, Hall effect measurements were carried out on identically processed LPE and bulk SiGe solar cells as well as on Si reference cells to keep at least the cell processing comparable.

The electron mobilities and electron densities in the n-type emitter layers of LPE cells containing 10 % Ge were about 100 $cm^2(Vs)^{-1}$ and $(7\text{-}8)\times10^{18}$ cm^{-3}, respectively. For bulk SiGe samples with the same electron density and a Ge content of 4-5 % the electron mobility ranged from 80-110 $cm^2(Vs)^{-1}$. For comparison, the mobilities in the n-type emitter $(n_0 = (8\text{-}9)\times10^{18}$ $cm^{-3})$ of the Si reference samples were 115-125 $cm^2(Vs)^{-1}$. Due to the missing electronic barrier in the p-SiGe/p^+-Si cells, hole mobilities in the p-type base could be determined for the bulk SiGe cells only. The p-doping level of these samples ranged from $(1.8\text{-}3.6)\times10^{16}$ cm^{-3} and the hole mobility was 180-225 $cm^2(Vs)^{-1}$. For comparison, hole mobilities in the base of the Si reference cells (doping of about 1.5×10^{16} cm^{-3}) were 250-260 $cm^2(Vs)^{-1}$. Due to alloy scattering, the drift mobilities obtained for the SiGe solar cell structures are about 20 % lower than the values obtained for Si reference cells. By reduction of the forward current, the lower mobility in the alloy helps to compensate for the loss of the open-circuit voltage in solar cells that results from the smaller band gap (higher intrinsic carrier concentration) of SiGe [13].

D. Concluding remarks

High-quality, homogeneous SiGe layers on p^+-Si were obtained by CVD and LPE. SiGe epitaxial layers can be applied as the base of solar cells. EBIC micrographs of the as-grown layers and thin-film solar cell structures show that the recombination activity of the dislocations at room temperature is rather low. Thus, it is not surprising that the remote plasma hydrogenation has only a minor impact on the contrast of the dislocations and the minority carrier diffusion length at room temperature. The diffusion length is much larger than the active layer thickness and can be further improved using SiGe buffer layers. We observed that both the electron and hole mobilities in our SiGe samples are about 20 % smaller compared to the bulk-Si reference samples.

The decrease of the open-circuit voltage of SiGe solar cells compared to Si cells can partially be compensated by the lower carrier mobility in the alloy. On the other hand, there is a gain in the short-ciruit current due to enhanced absorption in the red. Efficiencies of about 15 % have already been reported for 200 μm bulk $Si_{0.9}Ge_{0.1}$ cells [13]. In the present study, using thin SiGe layers grown by LPE and CVD on highly doped p-Si substrate, cell efficiencies of up to 10.4 % have been reached without any surface texturization or optical confinement.

Acknowledgments

We acknowledge Dr. Peter Gaworzewski (Institute for Semiconductor Physics) for the Hall effect analysis. The authors thank the European Commission for support of the BANSIS project, JOR3-CT96-0109.

References

[1] „Properties of Strained and Relaxed Silicon Germanium", E. Kasper (ed.), EMIS Datareviews Series No. 12, INSPEC, London (UK), 1995.

[2] P. Raue, D. Yang, L. Long, H.J. Möller, and E. Buhrig, „Multicrystalline Si_xGe_{1-x} Alloys for Solar Cell Applications", Proc. 14th European Photovoltaic Solar Energy Conference and Exhibition, Barcelona, 1997.

[3] M. Caymax, R. Loo, K. Said, J. Poortmans, A. Daami, G. Brémond, O. Krüger, W. Seifert, and M. Kittler, Appl. Phys. Lett., submitted for publication.

[4] G. Brémond, A. Daami, A. Laugier, W. Seifert, M. Kittler, J. Poortmans, M. Caymax, K. Said, M. Konuma, A. Gutjahr, and I. Silier, „SiGe Thin-Film Structures for Solar Cells", Paper presented at MRS Fall Meeting, Symposium G, Boston (MA), December 1-5, 1997.

[5] G. Beaucarne, K. Said, T. Vermeulen, F. Duerinckx, J. Poortmans, and J. Nijs, Proc. 14th European Photovoltaic Solar Energy Conference and Exhibition, Barcelona, 1997.

[6] C.J. Wu and D.B. Wittry, J. Appl. Phys. 49 (1978) 2827.

[7] M. Kittler, J. Lärz, G. Morgenstern, and W. Seifert, J. de Phys. IV 1 (1991) C6-173.

[8] M. Kittler and W. Seifert, Mater. Sci. Engin. B24 (1994) 78.

[9] M. Kittler, C. Ulhaq-Bouillet, and V. Higgs, J. Appl. Phys. 78 (1995) 4573.

[10] V. Higgs and M. Kittler, Appl. Phys. Lett. 65 (1994) 2804.

[11] M. Kittler, W. Seifert, and V. Higgs, Mat. Res. Soc. Symp. Proc. 378 (1995) 989.

[12] P. Gaworzewski, K. Tittelbach-Helmrich, U. Penner, and N.V. Abrosimov, J. Appl. Phys., (1998) accepted for publication.

[13] K. Said, J. Poortmans, M. Libezny, M. Caymax, J. Nijs, R. Mertens, D. Vyncke, W. Seifert, M. Kittler, I. Silier, and M. Konuma, „Low-Temperature Passivation for SiGe-Alloy Solar Cells", Proc. 14th European Photovoltaic Solar Energy Conference and Exhibition, Barcelona, 1997.

V.

Low-Dimensional Structures

Tuning the luminescence emission in III-V self-forming quantum dots: influence of structure, material system and dot/barrier interface.

R. Leon

Jet Propulsion Laboratory ,California Institute of Technology, 4800 Oak Grove Drive, Pasadena, CA 91109-8099

Tuning the emission in semiconductor quantum dots [QDs] has be attained using different strategies: by changing the material composition in the dot/barriers; interdiffusion or intermixing; using a graded growth rate; and changing the quantum box average dimensions and concentrations for fixed ternary compositions. The effects of these structural and compositional variations on the luminescence properties of these structures is presented and discussed.

A. INTRODUCTION

Issues of interest in the growing field of self-assembled nanostructures include the ability to tune island dimensions and their surface densities. This in turn means control of their luminescence wavelength for certain technologically desirable target emissions. Tunability in zero-dimensional semiconductor technology thus offers obvious advantages in extending the range of possibilities for devices.

In III-V semiconductor quantum dots (QDs), besides the most obvious dimension dependent variations in the ground state and excited state emissions, there are subtler effects of size variations. These include the temperature stability of their luminescence emission [1] and the effects of interdiffusion on the magnitude of blue shifts in InGaAs/GaAs QD's [2]. Other expected effects include variations in non-linear optical properties and lifetimes. All of these dimension dependent changes will have repercussions on device implementation and design.

The capability of tuning the emission wavelength by varying material systems has been demonstrated. Changing the semiconductor composition from different bulk bandgaps in the dot and barrier materials results in different wavelength emissions for islands of similar sizes, as expected [3].

Tuning dot dimensions within a fixed ternary composition is examined by three different approaches. First, changes with growth temperature (T_g) will be discussed. Variations in island size with temperature are presented for MOCVD growth of InGaAs/GaAs and InAlAs/AlGaAs QDs. Changes in AsH_3 partial pressure and miscut angle can be used (via step edge surface density) to tune islands dimensions and densities. Miscuts have also been observed to have a role in size calibration [4]. These variations have been discussed in terms of the growth kinetics and energetics of island formation with varying growth conditions and substrate morphology [5].

Tuning the emission in III-V QDs has also been achieved by thermally induced compositional disordering. The luminescence emission is blue shifted with interdiffussion in InGaAs/GaAs [2], InGaAs/AlGaAs and InAlAs/AlGaAs QDs [6]. Recent studies found that greater blue shifts can be obtained in InGaAs QDs than in quantum wells (QW) for the same value of diffusion length [2], and additionally, inhomogeneously broadened photoluminescence (PL) peaks become significantly narrower with interdiffusion. Here it is shown that the intersublevel spacings from excited states emission in QDs can also be tuned. The effects of interdiffusion on excited state emission from QDs [7] is demonstrated and analyzed. This gives a range in tunability for applications like infrared detectors and lasers based on intrasubband transitions. Numerical simulations for the PL spectra using the rate equations estimated intrinsic carrier relaxation ratios for different values of $\Delta E_{[(i+1)-i]}$. These findings allow experimental determination of the effects of intersublevel spacings on carrier relaxation in semiconductor QDs.

In order to obtain a clearer understanding of the mechanisms in Stranski-Krastanow (S-K) QD formation and the influence of structural features on the optical properties of QDs, the progression of structural and radiative QD evolution has been studied using graded growth rates [8]. Correlating the structural and luminescence results allowed studying the progression in photoluminescence (PL) evolution from initial wetting layer (WL) growth to QD formation, island saturation and the effects of coalescence and incoherent island formation on QD PL emission.

B. EXPERIMENTAL DETAILS

InGaAs/GaAs and AlInAs/AlGaAs structures used in this study were grown by metalorganic chemical vapor deposition (MOCVD) using a horizontal reactor operating at 76 Torr. For the growth of InGaAs islands, $(CH_3)_3Ga$, $(CH_3)_3In$, and AsH_3 were used as precursors. The H_2 carrier flow rate was 17.5 standard liters per minute (slm) which was lowered to 5 slm to obtain a graded deposition. After growth of GaAs buffer layers (BL) at 650°C, the temperature was lowered within the range 490-630°C and nanometer sized InGaAs islands were grown by depositing 5 monolayers (ML) of $In_{0.6}Ga_{0.4}As$. $Al_{0.4}In_{0.6}As/Al_{0.31}Ga_{0.69}As$ QDs were grown in similar fashion, using $(CH_3)_3Al$ and $(CH_3)_3In$ as precursors and growth temperatures ranging from 550 C to 675 C. Growth rates used for both compositions were nominally 0.5 ML per second. GaAs capping layer thicknesses were 100 nm for samples used in PL experiments. Substrates were (100) semi-insulating GaAs with miscut angles (θ_m) ranging from 2° towards (110) to very close to the exact (100). Ternary compositions were determined from PL spectroscopy of thick and thin films grown at various temperatures. Force microscopy (FM) with standard etched silicon nitride tips gave statistical information on island sizes, surface densities, and provided a very accurate technique to measure the miscut angle.

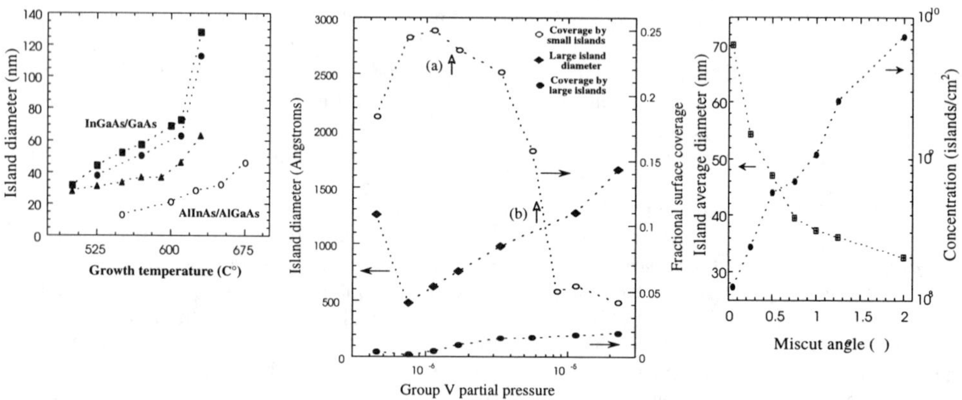

Figure 1. QD diameters with growth temperature.

Figure 2. QD surface coverages with group V partial pressures.

Figure 3. QD densities and diameters vs. substrate miscut.

Plan view transmission electron microscopy and scanning electron microscopy were also used. Low temperature PL spectra were obtained using a 780 nm laser diode for excitation, dispersed with a single grating 0.67 meter monochromator, and the signal collected using Ge detectors and lock-in techniques. Thermal treatments were performed in a N_2 ambient using rapid thermal annealing (RTA) and proximity capping.

C. RESULTS AND DISCUSSION

C.1. ACHIEVING TUNABILITY WITHIN A FIXED TERNARY COMPOSITION
C.1.1. VARIATIONS WITH TEMPERATURE

Dramatic differences in island size are obtained with InGaAs and AlInAs growth using MOCVD. This is illustrated in Figure 1, which shows the results of InGaAs and AlInAs island growth. Significant increases in island average diameters, concentrations and average heights are observed as a result of increasing T_g during formation of InGaAs islands. Similar trends were seen in the growth of AlInAs QDs, for which the curve is shifted towards lower values of island dimensions and higher island densities.

C.1.2. VARIATIONS WITH GROUP V PARTIAL PRESSURE

Changing the AsH_3 partial pressure [pp] can produce large variations in density for equal island sizes. We have found this to be the only parameter that produces such variability in a reproducible and controllable manner. Figure 2 shows the variation in surface coverage from nanometer size InGaAs with Arsine pp. Statistical information for two types of islands is presented: small coherent island (QDs), and larger islands found to coexist with the QDs in different ratios. Both are seen to depend critically on growth conditions.

C.1.3. VARIATIONS WITH MISCUT ANGLE

The miscut angle [θ_m] can be used to tune self-organized QDs. As shown in Figure 3, subtle changes in substrate orientation produce significant differences in average island diameters and island surface densities at the same growth temperature [550 °C].

C.2. EVOLUTION IN ISLAND FORMATION FOR InGaAs/GaAs QUANTUM DOTS

The FM micrographs in Figure 4 show the progression in island density and surface morphology from the onset of island formation to growth past the S-K transformation. The island concentration as a function of coverage are also shown in the lower plot. The onset of island coalescence becomes apparent with the decrease in island concentration with further deposition. The morphology indicated in each FM frame corresponds to the stages indicated by arrows (a) to (e).

The PL plot in Figure 4 shows that the evolution of the luminescence spectra from these structures can be divided into four distinct regions, labeled (a) to (d) in Figure 3: WL emission, simultaneous WL and QD emission, QD saturation, and lastly, the dislocation/coalescence regime. Luminescence emission begins with a thin QW which progressively red shifts (becomes thicker) as InGaAs deposition is increased. This is indicated in the lower portion of the PL spectra in Figure 3. In the next stage, the QD concentration rises until the QD PL signal is strong enough for detection. Once the QD PL peak increases in intensity, the WL peak diminishes rapidly. The evolution of WL to QD luminescence occurs over a broad range in QD concentrations but this corresponds to a very narrow range in InGaAs deposition: from 4.08 ML to 4.14 ML. At the next stage, which occurs over the next ~ 0.5 ML deposition, the PL QD emission does not change significantly. This stage corresponds to island saturation. In the last stage, the PL intensity drops in magnitude to roughly a third of its former intensity, and stays at this lower intensity over the next ~ 2 ML deposition.

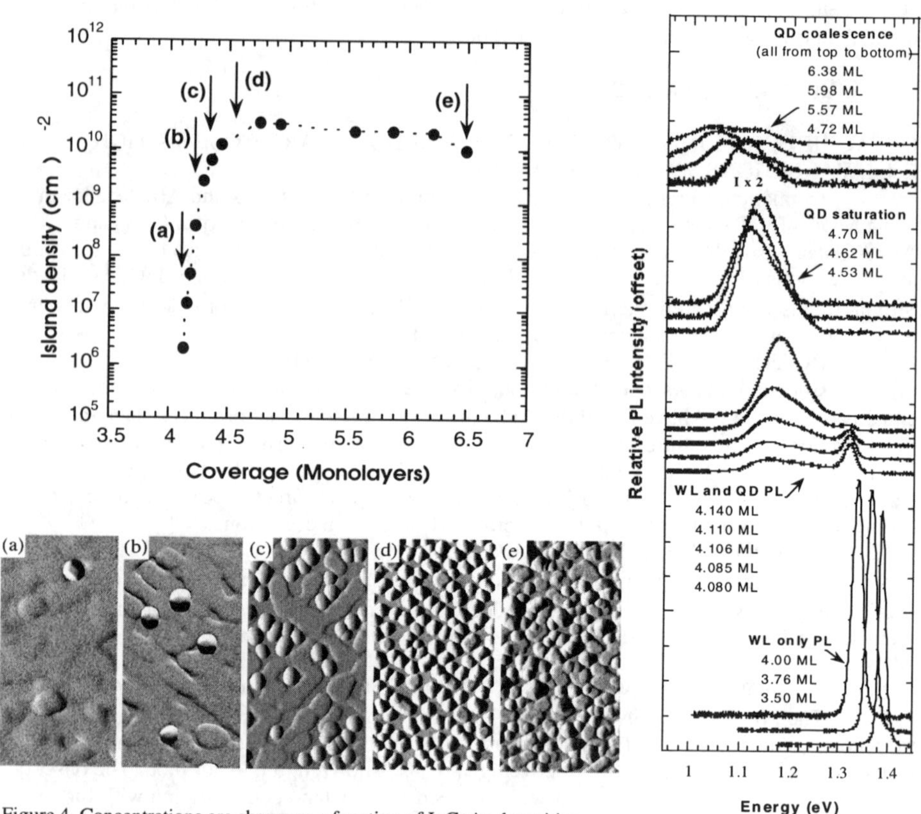

Figure 4. Concentrations are shown as a function of InGaAs deposition with corresponding morphology of QD evolution studied by Force Microscopy. Corresponding PL spectra from WL to island coalescence is shown on the right.

C.3. TUNING INTERSUBLEVEL TRANSITIONS IN QUANTUM DOTS.

Figure 5 (left frame) shows the PL spectra from InGaAs/GaAs QDs as a function of excitation power. Higher energy peaks become more prominent with increased excitation. Excited state emission can be observed even at very low excitation power, with emission from the (i+1) levels before the (ith) level saturates. A small spread in island sizes gives small inhomogeneous broadening, which allows resolving excited states. The spectra seen in Figure 5 thus reflect the excited states emission of a single QD. Figure 5 also shows WL luminescence, which is expected for low QD concentrations. The WL is effectively a thin (4-5 ML) quantum well with PL emission at higher energies. The peak intensity ratio between WL/QD increases with excitation power.

Figure 5 (right frame) shows the results of thermally induced intermixing on the sample producing the spectra shown on the left frame. When excitation power was varied, spectra for all annealed samples showed similar excitation dependent behavior. Figure 5 shows that all PL peaks blue-shift, and the inhomogeneous broadening diminishes. Furthermore, the intersublevel spacings become narrower with increased diffusion lengths. All spectra in Figure 2 have been simulated using solutions to the rate equations [9] calculated for four discrete levels:

$$PL(E) = \sum_{n=1}^{4} \left(f_i N_i / \tau_r\right) \exp\left(-(E-E_i)^2 / 2\Gamma^2\right)$$

Where the filling ratio f_{i-1} of the (i-1) level is proportional to the electronic relaxation rate in the ith level. ($N_i = 2 \times i$) was used for level degeneracies and τ_r is the carrier recombination lifetime. Different values for τ_{0i} were needed to achieve acceptable fits to the experimental spectra. Diffusion lengths can be established from measurements and calculations of diffusivity values in quantum wells of the same

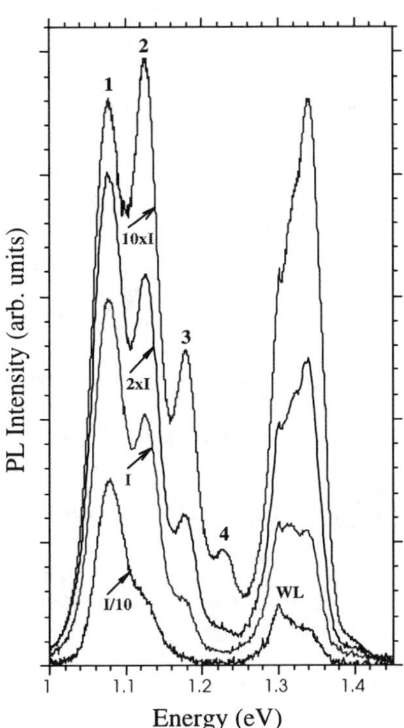

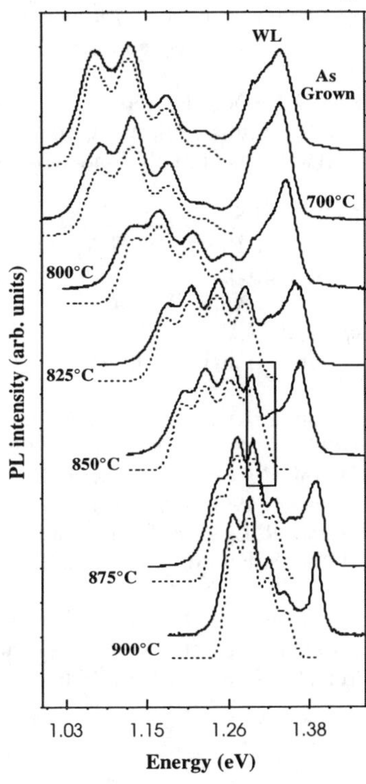

Figure 5. The left frame shows the PL spectra as a function of excitation intensity and the right frame shows how excited state luminescence shifts as a function of anneal temperature after intermixing [7].

ternary composition [2]: D (T) = [0.19] exp(-(3.4eV)/kT).

While some subtle effects seen from these experiments and simulations could be attributed to changes in carrier relaxation from matching $\Delta E_{[(i+1)-i]}$ and LO phonon energies, strong peaks from excited state emission are present at different values of $\Delta E_{[(i+1)-i]}$. This makes this finding very promising for device applications based on QD intraband transitions.

D. CONCLUSIONS

Wide range tunability can be achieved using different approaches: by varying growth temperature, varying the Group V flow during island formation, and using different substrate miscut angles in vicinal (100) GaAs. A study of the structural and radiative evolution in InGaAs QDs indicates that island concentrations rise exponentially after the onset of the S-K transformation. An abrupt drop in PL intensity is observed beyond saturation island densities. This drop in PL corresponds with a sudden increase in the concentration of incoherent islands. It was also found that while the integrated QD PL increases with QD concentration, it does not do so proportionally to concentration, so the intensity per QD drops as the concentration of QDs increases. The effects of intermixing from QD structures with excited state luminescence also show that these can be tuned. All peaks blue-shifted and narrowed and the intersublevel transitions $\Delta E_{[(i+1)-i]}$ were reduced and could be tuned continuously for values of $\Delta E_{[(i+1)-i]}$ greater, similar and lower than the LO phonon energies in InAs and GaAs.

E. AKNOWLEDGMENTS

Part of this work was carried out at the Australian National University (Canberra, Australia) and was partially funded by the Australian Research Council.

F. REFERENCES

[1]. S. Fafard, S. Raymond, G. Wang, R. Leon, D. Leonard, S. Charbonneau, J. L. Merz, P. M. Petroff, and J. E. Bowers, Surf. Sci. 361-362, (1996) 778.

[2]. R. Leon, D. R. M. Williams, J. Krueger, E. R. Weber and M. R. Melloch, *Phys. Rev. B* **56**, R4336 (1997).

[3]. R. Leon, S. Fafard, D. Leonard, J. L. Merz, and P. M. Petroff, *Appl. Phys. Lett.* **67**, (1995) 521.

[4]. R. Leon, T. J. Senden, Yong Kim, C. Jagadish, and A. Clark, *Phys. Rev. Lett* **78**, 4942 (1997) 4942.

[5]. R. Leon, C. Lobo, A. Clark, R. Bozek, A. Wysmolek, A. Kurpiewski, and M. Kaminska, J. Appl. Phys 83 (1998) (in press)

[6]. C. Lobo, R. Leon, S. Fafard and P. G. Piva, Appl. Phys. Lett (1998) (in press).

[7]. R. Leon, S. Fafard, P. G. Piva, S. Ruvimov, and Z. Liliental-Weber, Phys. Rev B (1998) (in press).

[8]. R. Leon and S. Fafard, Phys. Rev. B (1998) (in press).

[9]. K. Mukai, O. Nobuyuki, H. Shoji, and M. Sugawara, *Appl. Phys. Lett* . **68**, 3013 (1996), and *Phys. Rev. B* **54**, R5243 (1996).

InAs/In$_{0.52}$Al$_{0.48}$As quantum wire structure with
the specific layer-ordering orientation on InP (001)

J. Wu, B. Xu, H. X. Li, Q. W. Mo, and Z. G. Wang

Laboratory of Semiconductor Materials Science, Institute of Semiconductors, Chinese Academy of Sciences, P. O. Box 912, Beijing 100083, China

X. M. Zhao and D. Wu

Department of Metal Deformed, Northeast University, Shengyang 110006, P. R. China

Quantum wires were formed in the 6-period InAs/In$_{0.52}$Al$_{0.48}$As structure on InP (001) grown by molecular beam epitaxy. The structure was characterized with transmission electron microscopy. It was found that the lateral periodic compositional modulation in the QWR array was in the $[1\bar{1}0]$ direction and layer-ordered along the specific orientation deviating [001] by about 30 degrees. This deviating angle is consistent with the calculation of the distribution of elastic distortion around quantum wires in the structure using the finite element technique.

Low dimensional semiconductor heterostructures, such as the quantum well (QW), quantum wire (QWR), and quantum dot (QD), are of great interest due to their physical properties potentially applicable in microelectronic and optoelectronic devices. For the growth of these structures of high quality and uniformity with reproducibility, prerequisite for both of investigation and application, it is necessary to study the distribution of strain or elastic distortion inside or around the active regions in a structure.

In a heterostructure, the presence of strain field is an unavoidable consequence of the mismatch between the materials used in the structure. The consideration of strain is of important in low dimensional heterostructures as it has strong effects on the spontaneous formation of QDs or QWRs during growth as well as the influence on band structures. It is well understood now that the quantum well in a heterostructure is pseudomorphic and biaxially strained across the interface and relaxed freely along the growth direction. Whereas the pattern of strain distribution is comparatively complex and highly nonuniform in a self-organized island in the matrix, and, in addition, elastic deformation occurs in the surrounding material to release strain energy to some extent [1]. In comparison to QW and QD, the QWR structure has its own characteristic in symmetry and strain distribution, and many works [2-5] have been performed, both experimentally and theoretically, on the strain distribution associated with QWRs. In this work, we report the formation and characterization of the QWRs in the 6-period of InAs/In$_{0.52}$Al$_{0.48}$As heterostructure on InP (001). Transmission electron microscopy (TEM) reveals that quantum wires in the structure are layer-ordered along a specific orientation deviating the [001] growth direction by about 30 degrees. The specific orientation in layer-ordering is consistent with the result of the calculation of the

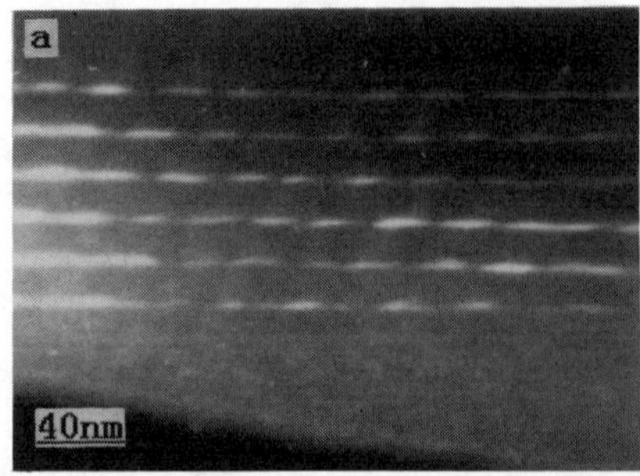

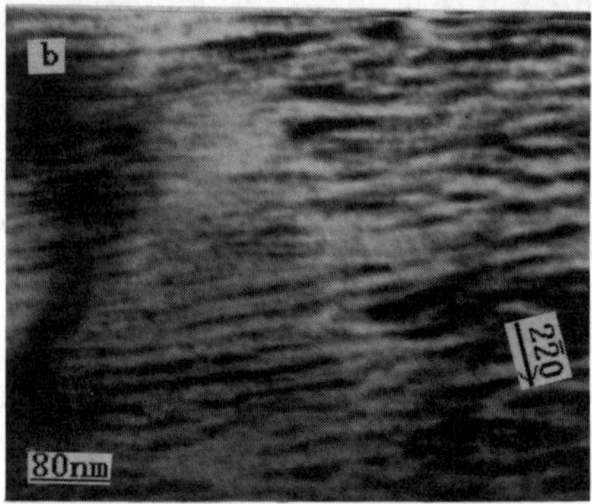

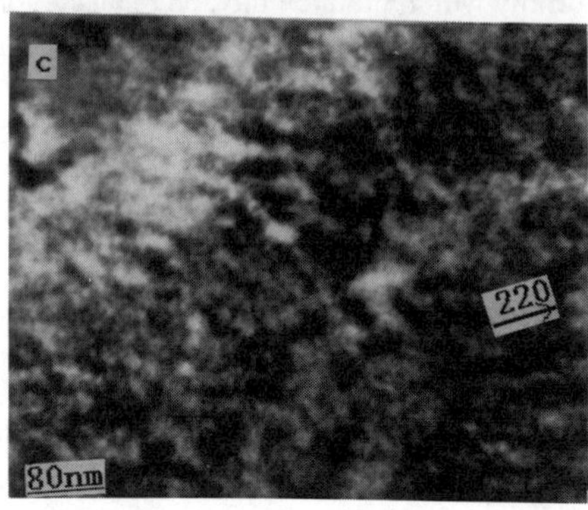

Fig. 1 TEM micrographs a, cross-section 200 dark_field; b and c, plane-view with g=2$\bar{2}$0 and 220, respectively.

elastic distortion in the material around a quantum wire using the finite element technique based on a simplified model.

The material system investigated in this study is the MBE-grown 6-period InAs/In$_{0.52}$Al$_{0.48}$As on (001) InP. The mismatch in lattice parameter in InAs/In$_{0.52}$Al$_{0.48}$As is about 3%. The InAs layers in the structure are 4.5 monolayers (MLs) in thickness and the In$_{0.52}$Al$_{0.48}$As separating layers are 20 nm thick. The In$_{0.52}$Al$_{0.48}$As cap and buffer layers are 80 and 100 nm, respectively. Specimens for TEM analysis, at 200 kV voltage, in this work are prepared in the conventional way.

Fig. 1 are the TEM micrographs of the multiple QWR (MQWR) structure in InAs/In$_{0.52}$Al$_{0.48}$As. Fig. 1 (a) is the 200 dark-field cross-section image viewing along [110] and light spots are observed, while only white stripes are observed along [1$\bar{1}$0], not shown here. These light spots and stripes should correspond to the regions where the composition is mostly of InAs. The figures demonstrate that the InAs layers in the structure are modulated along [1$\bar{1}$0], resulting in the MQWR structure with quantum wire lines along [110]. The lateral composition modulation in the structure should be formed by the strain induced lateral layer ordering (SILO) process [6] occurring spontaneously during growth. Fig.1 (a) shows that QWR lines are layer-ordered along a specific direction deviating [001] by about 30 degrees, in comparison with the vertically alignment of InAs dots in the GaAs matrix [7-9]. Fig. 1 (b) and (c) show the plane-view TEM images of the MQWR structure with $\mathbf{g}=2\bar{2}0$ and $\mathbf{g}=220$

diffraction vectors, respectively. In Fig. 1 (b), quantum wire lines along [110] are observed with $\mathbf{g}=2\bar{2}0$ normal to them. While in Fig. 1 (c), with $\mathbf{g}=220$ parallel to quantum wire lines,

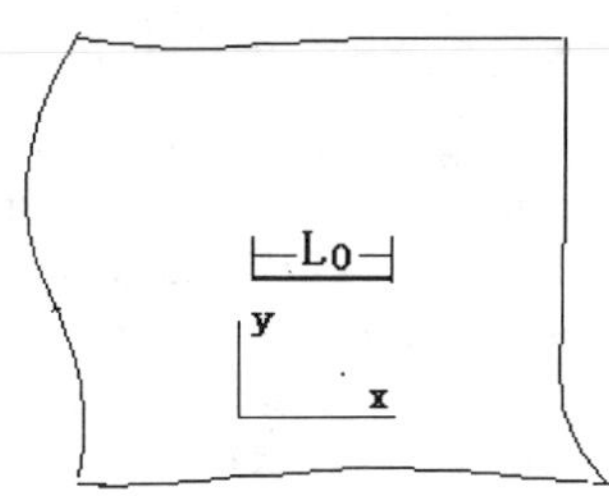

Fig. 2 Schematic model of the cross section of the quantum wire.

wire lines along [110] are observed with $\mathbf{g} = 2\bar{2}0$ normal to them. While in Fig. 1 (c), with $\mathbf{g} = 220$ parallel to quantum wire lines, the contrast showing quantum wire lines disappears. With $\mathbf{g} \cdot \mathbf{b} = 0$ criterion, the two TEM images indicate that the elastic relaxation of strain in the MQWR structure occurs along the directions normal to quantum wire lines, two dimensional in distribution. This observation is consistent with the measurement made by Shen et al [3] using X-ray diffraction demonstrating that quantum wires are pseudomorphic along the line direction and elastic relaxation of strain occurs normal to the wire line and in the growth direction.

The elastic distortion associated with QD may extend "deeply" into the surrounding material and creates "priority region" for the onset of 3D growth to occur in the subsequent strained layer. This leads to the vertically alignment of InAs islands in the InAs/GaAs system [7-9]. As mentioned above, elastic distortion emanating from quantum wires is two-dimensional and normal to the quantum wire line [3], and may have its own characteristic in far-reaching distribution pattern, in comparison with QD. As shown in Fig. 1, QWR lines in the 6-period InAs/In$_{0.52}$Al$_{0.48}$As structure are *skewly* layer ordered, in comparison with the vertical alignment of InAs dots in the multiple InAs/GaAs structure [7-9]. If the *skew* alignment of quantum wires in the InAs/In$_{0.52}$Al$_{0.48}$As structure results from the elastic distortion emanating from themselves, the distribution of distortion should also be *skew*. To confirm this point, we calculate the distortion distribution in the surrounding material using the finite element method based on a simplified model.

It has been clarified by x-ray analysis that the strain inside quantum wires is highly inhomogeneous [2,4], which indicates that the strained quantum wires cannot be treated by a simplified strain model. However, in this work, we are only concerned with the "far-

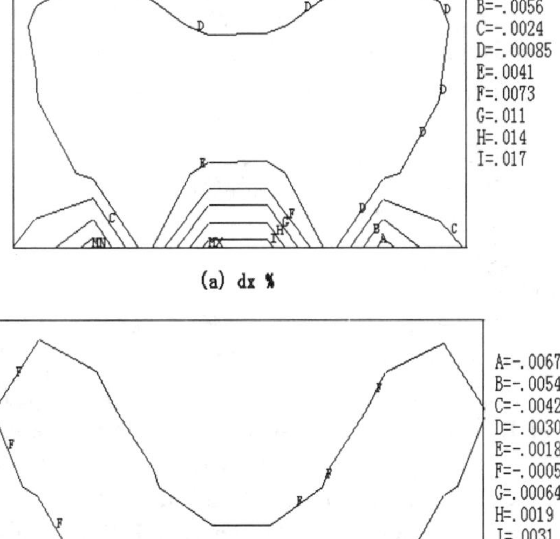

Fig. 3 calculated elastic distortion d around the quantum wire. (a) in the x-direction, (b) y-direction.

reaching" elastic distortion in the surrounding material, and, therefore, the detail in strain distribution inside the quantum wire or nearby has not significant effect on the result. Just like the situation in the calculation of the potential around a charged sphere, the charge is assumed concentrated in a geometry point. In our calculation, the quantum wire in the structure is intuitively simulated with a line of L_0 original length in a two-dimensional matrix of distortion-free, as shown in Fig. 2. The line is homogeneously stretched 3 per cent in length. Therefore, the region surrounding the line is elastically distorted to some extent to be calculated. This simplification is based on the following three facts. No elastic distortion or relaxation occurs along the quantum wire line as the structure is pusedomorphic along this direction and the calculation is two-dimensional; apparently, the cross-sectional aspect ratio of quantum wires is comparatively large; the main characteristic in elastic distortion at the interface is the mismatch of 3% between InAs and $In_{0.52}Al_{0.48}As$. The Poisson's ration is taken to be 0.36 and the Young's modulus 6.07×10^{10} Pa, the same to that of InP [10].

Fig. 3 are the calculated elastic distortion around the line with 3 % extension in the material. The figures show that the most far reaching direction for elastic distortion both in the x and y direction, d_x and d_y, is the one making the angle of about 30 degrees with the normal to the extended line in the material. Apparently, the anisotropy in the calculated distribution of elastic distortion around quantum wires is consistent with the observed layer-ordering orientation of QWR lines in the InAs/ $In_{0.52}Al_{0.48}As$ structure.

In conclusion, the *skew* layer-ordering orientation of quantum wires formed in $InAs/In_{0.52}Al_{0.48}As$ is consistent with the calculation of the elastic distortion in the surrounding material using the finite element technique.

[1]. D. J. Eaglesham and M. Cerullo, Phys. Rev. Lett. **64**, 1944 (1990).

[2]. Y. P. Chen, J. D. Read, W. J. Schaff, and L. F. Eastman, Appl. Phys. Lett. **65**, 2202 (1994).

[3]. Q. Shen, S. W. Kycia, E. S. Tetarelli, W. J. Schaff, and L. F. Eastman, Phys. Rev. **B54**, 1638 (1996).

[4]. M. Notomi, J. Hammersberg, H. Weman, S. Nojima, H. Sugiura, M. Okamoto, and T. Tamamura, Phys. Rev. **B52**, 11147 (1995).

[5]. D. A. Faux, J. R. Downes, and E. D. O'Reolly, J. Appl. Phys. **80**, 2515 (1996).

[6]. K. Y. Cheng, K. C. Hsieh, and J. N. Baillargeon, Appl. Phys. Lett. **60**, 2892 (1992).

[7]. J. Y. Yao, T. G. Anderson, and G. L. Dunlop, J. Appl. Phys. 69, 2224 (1991).

[8]. Q. Xie, A. Madhuka, P. Chen, and N. P. Kobayashi, Phys. Rev. Lett. **75**, 2542 (1995).

[9]. V. A. Shchukin, N. N. Ledentstov, P. S. Kop'ev, and D. B. Bimberg, Phys. Rev. Lett. **75**, 2698 (1995).

[10]. K. Nishi, A. A. Yamaguchi, J. Ahopelto, A. Usui, and H. Sakaki, J. Appl. Phys. **76**, 7437 (1994).

GaAs and SiC Quantum Wire Arrays: Fabrication and Characterisation

V.A. Samuilov, I.A. Bashmakov, I.B. Butylina,
I.M. Grigorieva, V.K. Ksenevich, L.V. Solovjova

Belarus State University, F. Skaryna avenue, 4, 220080 Minsk, Republic of Belarus

Abstract

We report a novel approach for fabrication of mesoscopic ordered arrays (networks) based on self-organized patterning in complex liquids (nitrocellulose solution). In order to prepare semiconductor quantum wire arrays the networks are used as masks for further reactive-ion-beam etching of GaAs surface or as a precursors for synthesis of SiC on the surface of Si. Auger electron spectroscopy carbon peaks of converted networks exhibit a fine structure typical for SiC.

A. Introduction

While quantum well structures are widely used for electronic applications, the structures with lower dimension (quantum wires - 1D and quantum dots - 0D) appear to be more difficult to fabricate. Nevertheless, several methods were proposed for the fabrication of 1D and 0D including a lithographic patterning of 2D structures and a self-organized growth process [1]. Both of the methods involve very expensive and precise technique, like e-beam lithography and ultrathin film epitaxy (MBE) and vapor phase epitaxy (MOVPE).

The lithographic patterning of 2D structures leads to the creation of confining potential in the case of strong potential modulation. When the Fermi energy intersects the attractive potential, quantum dots array is obtained [2] and if the potential peaks are repulsive - antidots array [3]. The method of self-organized growth process gives a possibility to create dense arrays of isolated 0-D with dimensions lower than exciton Borh radius [1,4]. Researchers at different laboratories are very active in this 1-D and 0-D systems, having a narrow size distribution, for the purpose of fabricating quantum wire and dot lasers [1], 2-D photonic crystals [5,6], arrays of mesoscopic electronic devices [7,8], etc.

Here we report an alternative approach for fabrication of arrays of quantum wires based on mesoscopic self-organized patterning in complex liquids (nitrocellulose solution) and further reactive-ion-beam etching (RIE) of GaAs surface through prepared permanent cellulose mesoscopic network as a mask and synthesis of SiC using the network as a precursor.

B. Experimental procedure. Mesoscopic self-organized patterning in complex liquids

The technological approach is believed to be described in the frame of Benard-Marangoni effect due to chemical variations, i.e. self- organization patterning at the liquid - gaseous interface [9,10]. A permanent regular structure is fabricated in a thin layer of double- diffusive system of quickly and slowly evaporating solvents after an initial perturbation of the surface by a vapour of small water droplets, which can undergo a phase transition to symmetry breaking ordered state [11]. A solution of nitrocellulose in amil-acetate as a solvent is used. A network is easily removed from the surface of the water to a solid substrate (Fig. 1a and Fig.2a) for further patterning as a mask or synthesis procedures, structural and electrical investigations.

As a routine we manufacture networks with the size of hexagonal cells about 500 nm. The thickness of bonds is about 50 nm.

Scanning electron microscopy (SEM) observations were performed in S-806 (Hitachi) with the magnification up to 100,000 and resolution of 4 nm.

The scanning Auger electron spectrometer type PHI-660 (Perkin Elmer) with a locality of the electron beam of 100 nm and a sensitivity of atomic concentration 0.1 % was used for Auger electron spectroscopy investigations (AES). The spectrometer was equipped with a standard sputter gun for AES - depth profile investigations and for Ar - RIE through a cellulose mesoscopic mask. The diameter of the sputter spot was in mm-range. The etching rate was determined to be about 100 nm/min for silicon.

C. GaAs quantum wire arrays

GaAs arrays of wires were prepared by Ar - RIE at 3.5 kV acceleration voltage (Fig. 1.b).

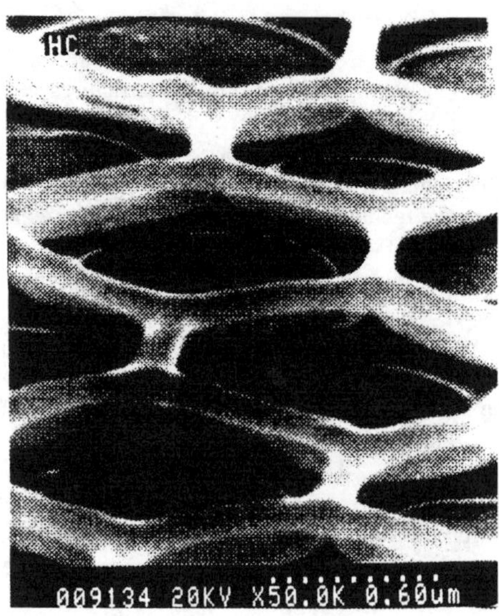

Fig. 1a: SEM image of the initial cellulose mask prepared by mesoscopic self-organized patterning.

Fig. 1b: GaAs honeycomb - shape network prepared by Ar - RIE of GaAs surface through a cellulose mesoscopic mask.

The initial structures for patterning were epitaxial n^+-GaAs, (n(300K) =1.3x10^{17} cm^{-3}, the thickness of the epi-layer d=0.25 μm, the thickness of the buffer layer d=1.0 μm, SI GaAs:Cr substrate, d=400 μm). All the samples were with Van-der-Pauw configuration with contact distance of 300 μm . I-V characteristics were tested at T=300K and 4.2K.

The samples with the thickness of bonds of the networks about 50 nm were completely depleted at 4.2K and could not be turned on [8]. The estimated thickness of a depletion layer for the carrier concentration in used epi-layer (n=1.3x10^{17} cm^{-3}) was about 30 nm from the surface. So, these wires should be depleted. But if the thickness of the bonds was about 100 nm, the resistance was finite and was increased only in about 1 order in value when the temperature was decreased from 300K down to 4.2K

D. SiC quantum wire arrays synthesis

Our preparation of SiC mesoscopic networks involves the solid state reaction of carbon of the cellulose precursor (Fig. 2a) with silicon of a substrate by 4-hour annealing in vacuum at T=1100-1200 °C (Fig. 2b). It was shown [12], that a reaction:

$$[Si] + [C] \rightarrow SiC,$$

is consistent with the formation of silicon carbide in nanoscale 1-D structures at these temperatures.

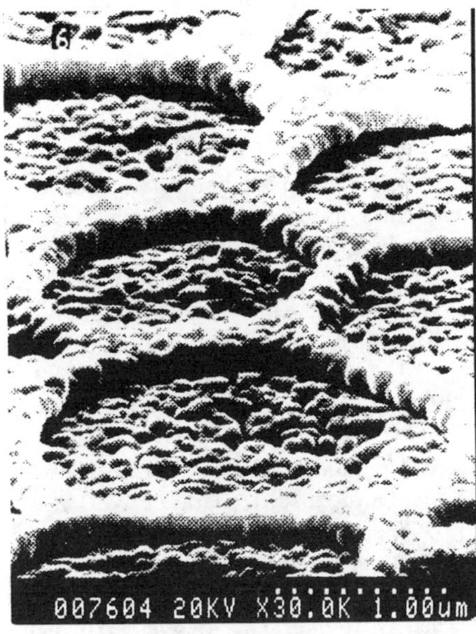

Fig 2a: SEM image of an initial cellulose network, which was used as a precursor	Fig 2b: SEM image of a cellulose network, which was converted to SiC by reaction of C of the precursor with Si of a substrate by 4-hour annealing in vacuum at T=1100°C.

To approve SiC mesoscopic networks formation the AES investigations, including depth profile, were done. Fig. 3 shows AES signals of cellulose network, which was converted to SiC by reaction of C of the precursor with Si - substrate by 4 - hour annealing in vacuum at T=1100°C. The AES signals of SiC specimen and amorphous carbon film were measured for comparison. The AES carbon peak of the cellulose network converted to SiC exhibited a fine structure typical for SiC specimen. The size of the plateau on the AES depth profiles (Fig.4) correlates with the thickness of the nearly stoichiometric SiC network.

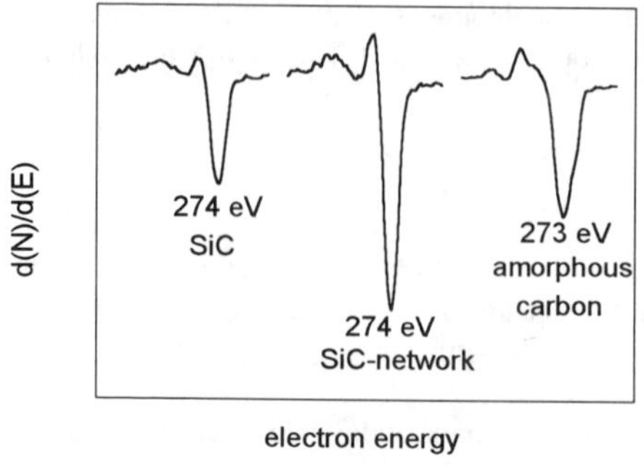

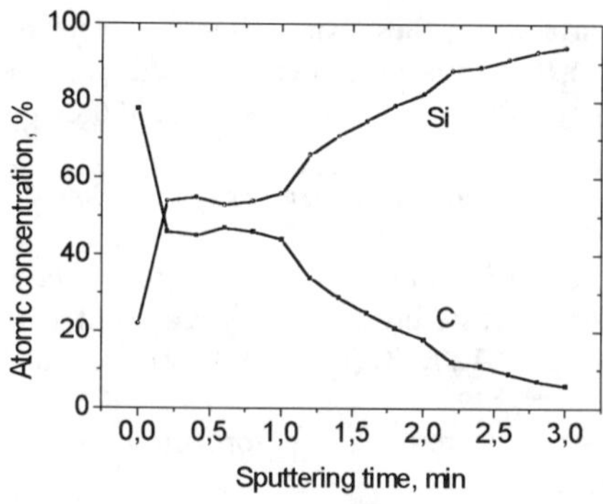

Fig.3: Shape of AES signal of converted to SiC network and SiC specimen and amorphous carbon film.

Fig.4: AES depth (Ar ion etching time) profile of converted to SiC network.

E. Conclusions and prospects

A novel approach for fabrication of mesoscopic ordered networks based on self-organized patterning in complex liquids (nitrocellulose solution) was proposed. In order to prepare semiconductor quantum wire arrays, the networks were used as masks for further reactive-ion-beam etching of GaAs surface or as a precursors for synthesis of SiC on the surface of Si.

Preliminary studies show that these ordered mesoscopic networks share their properties of the bulk materials with the effects of confinement and deduced dimensionality. These arrays of quantum wires might find technological applications in nanostructured composite materials.

References

[1] M. Henini, III-V Review 9 (1996) 64.

[2] C.I. Duruoz, R.M. Clarke, C.M. Marcus, J.S. Harris, Jr., Phys. Rev. Lett. 74 (1995) 3237.

[3] D. Weiss, M.L. Roukes, A. Menschig, P. Grambow, K. von Klitzing, G. Weimann. Phys. Rev. Lett. 66 (1991), 2790.

[4] M. Grundmann, J Christen, N.N. Ledentsov, J. Boehrer, D. Bimberg, S.S. Ruvimov, P. Werner, U. Riehter, U. Goesele, J. Heydenreich, V.M. Ustinov, A.Yu. Egorov, A.E. Zhukov, P.S. Kop'ev, Zh.I. Alferov, Phys. Rev. Lett. 74 (1995) 4043.

[5] T. Krauss, Y.P. Song, S.Thoms, C.D.W. Wilkinson, R.M.DelaRue, Electr. Lett.30 (1994), 1444.

[6] T. Krauss, R.M.DelaRue, S. Brand, Nature 383 (1996) 699.

[7] H. Noge, A. Shuimizu, H. Sakaki, Nanotechnology 3 (1992) 180.

[8] K. Ismail, S. Washburn, K.Y.Lee. Appl. Phys. Lett. 59 (1991) 1998.

[9] A.Thess, S.A. Orszag, Fluid Mech., 283 (1995) 201.

[10] V.A. Samuilov, I.B. Butylina, I.A. Bashmakov, L.V. Govor, I.M. Grigorieva, V.K. Ksenevich, in Proc. of the Int. Conf. on Advanced Semiconductor Devices and Microsystems, October 20-24, 1996, Smolenice, Slovakia, p.p. 313-317.

[11] C. Van den Broeck, J.M.R. Parrondo, R.Toral, Phys. Rev. Lett. 73 (1994) 3995.

[12] H. Dai, E. W. Wong, Y. Z. Lu, S. Fan, C. M. Lieber, Nature 375 (1995) 769.

VI.

Group III-V Oxides

The Technology and Applications of Selective Oxidation of AlGaAs

Kent D. Choquette, K. M. Geib, and H. Q. Hou
Center for Compound Semiconductor Science and Technology
Sandia National Laboratories
Albuquerque, NM 87185-0603

D. Mathes and Robert Hull
University of Virginia
Charlottesville, VA

Wet oxidation of AlGaAs alloys, pioneered at the University of Illinois a decade ago, recently has been used to fabricate high performance vertical-cavity surface emitting lasers (VCSELs). The superior properties of oxide-confined VCSELs has stimulated interest in understanding the fundamentals of wet oxidation. We briefly review the technology of selective oxidation of III-V alloys, including the oxide microstructure and oxidation processing as well as describe its application to selectively oxidized VCSELs.

A. Introduction

Wet oxidation of AlAs was first reported in 1979, although the native (Gibbsite) oxide phase was formed at the low oxidation temperature used (100°C) [1]. Ten years later during studies of atmospheric degradation of AlAs, it was discovered by researchers at the University of Illinois that wet oxidation of Al-containing semiconductors (AlGaAs, AlInAs, AlGaInP, etc.) at temperatures above ≈300 °C produces a mechanically robust oxide with a low refractive index [2]. The utility of this oxide was initially demonstrated in a variety of edge emitting laser structures where the oxide layers provide both index-guiding and buried current apertures [3-5].

Oxidized AlGaAs layers have also recently been introduced within vertical cavity surface emitting lasers (VCSELs) [6,7]. Selectively oxidized VCSELs have exhibited superior laser performance due to the efficient electrical and optical confinement afforded by the buried oxide layers. Other applications of wet oxidized AlGaAs have included high index contrast distributed Bragg reflector mirrors [8], birefringent optical waveguides [9], buried microlenses [10], and metal-insulator-semiconductor transistors [11]. Nevertheless as shown in Fig. 1, research on the oxidation of III-V alloys as measured by journal publications has been driven over the past few years by the significant interest in selectively oxidized VCSELs. Understanding the microstructure and processing of oxides converted from AlGaAs is essential for the development of a robust fabrication technology for optoelectronic devices [12]. We briefly review selective wet oxidation of AlGaAs alloys and its application to VCSEL fabrication.

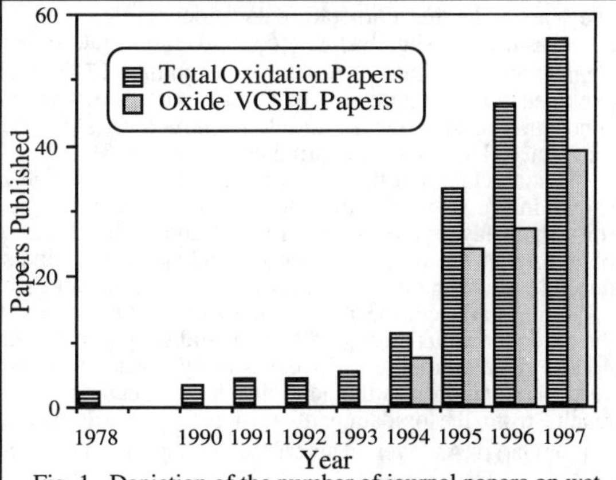

Fig. 1. Depiction of the number of journal papers on wet oxidation of AlGaAs and selectively oxidized VCSELs.

B. Oxide Microstructure

Cross section transmission electron microscopy (TEM) has been applied to characterize the microstructure of buried oxide layers as shown in Fig. 2. Electron diffraction patterns indicate that the oxide converted from AlGaAs is an amorphous solid solution of $(Al_xGa_{1-x})_2O_3$ [13]. The bright field image in Fig. 2 reveals only granular amorphous contrast within the oxide, with an $\approx$17 nm thick interface zone at the oxide/semiconductor interface. No extended defects are

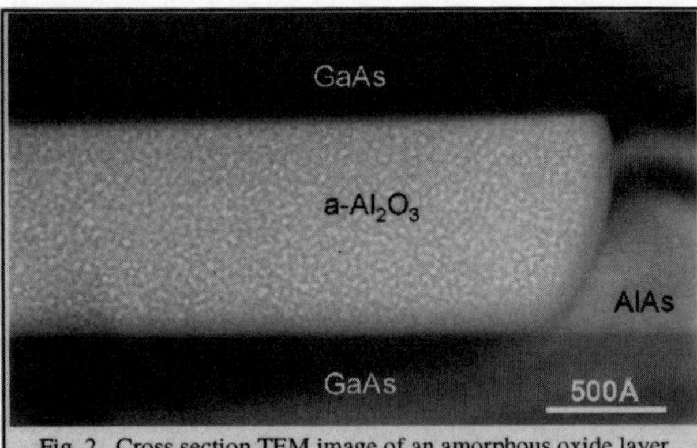

Fig. 2. Cross section TEM image of an amorphous oxide layer converted from 100 nm thick AlAs surrounded by GaAs layers.

apparent along the oxide/semiconductor interfaces and a smooth transition over several lattice constants is apparent between the oxide and semiconductor.

The thickness of an oxide formed at temperatures above 300 °C tends to be reduced compared to the original semiconductor layer [2]. The linear shrinkage is measured to be 12 to 13% in the amorphous oxide layers converted from AlAs [8]. However, shrinkage for oxidized $Al_{0.92}Ga_{0.08}As$ layers is measured to be only 6.7% [14]. The reduced thickness of the oxide is thus dependent upon composition of the converted semiconductor and can result in strain fields at the oxide terminus [15].

C. Oxidation Processing

To develop a manufacturable wet oxidation technique, the influence of process parameters such as gas flow, gas composition, temperature, Al composition, and layer thickness on the rate of oxidation have been examined [12,16]. The lateral oxidation has a linear oxidation rate which obeys an Arrhenius temperature dependence. A strong compositional dependence of the oxidation rates is shown in Fig. 3(a) where the oxidation rate of $Al_xGa_{1-x}As$ for x varying from 1 to 0.84, changes by more than 2 orders of magnitude [7]. It is this oxidation selectivity which can be exploited for the fabrication of buried oxide layers within a VCSEL; with minute changes of Ga concentration, specific or multiple AlGaAs layer(s) can be selected out to form buried oxide layers for electrical and optical confinement (see Fig. 5).

The thickness of the semiconductor layer to be oxidized can also influence its oxidation rate as shown in Fig. 3(b). For thickness $\geq$ 60 nm, a relatively constant lateral oxidation rate is observed; for thinner layers the oxidation rate dramatically decreases. Comparing Figs. 3(a) and 3(b), it is obvious that the oxidation rate dependence on thickness can compensate for the compositional dependence: thin layers of AlAs may oxidize slower than thick layers of AlGaAs.

Finally, the composition of the surrounding layers will affect the oxidation rate and shape of the oxide terminus through diffusion and supply of reactants [17]. For example as shown in Fig. 4, a rapidly oxidizing layer can supply reactants to its surrounding layers such that significant vertical oxidation into the adjoining layers occurs. Fig. 4 shows a tapered oxide terminus which results from the oxidation of a 14 nm thick $Al_{0.98}Ga_{0.02}As$ layer embedded within a thicker $Al_{0.92}Ga_{0.08}As$ layer. The effective lateral oxidation rate of the $Al_{0.92}Ga_{0.08}As$ layer in Fig. 12 is enhanced by a factor of 2 over the rate predicted from Fig. 3(a) due to vertical oxidation originating from exposure to reactants supplied from the thin oxidized $Al_{0.98}Ga_{0.02}As$ layer.

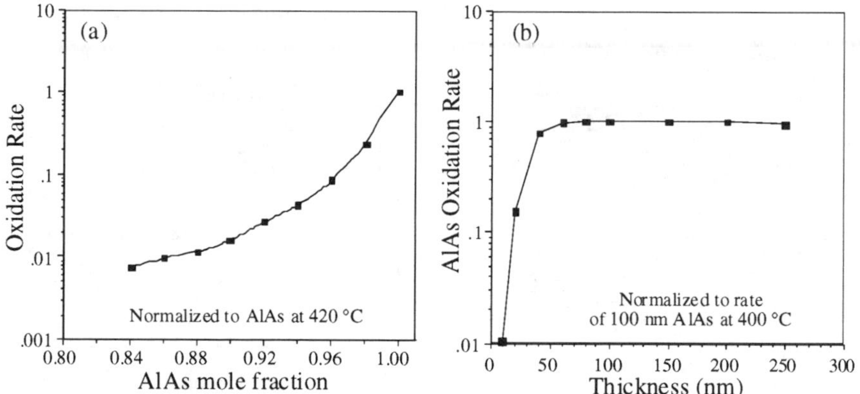

Fig. 3. (a) Normalized $Al_xGa_{1-x}As$ oxidation rate at 420 °C versus composition; (b) normalized oxidation rate of AlAs at 400 °C versus layer thickness.

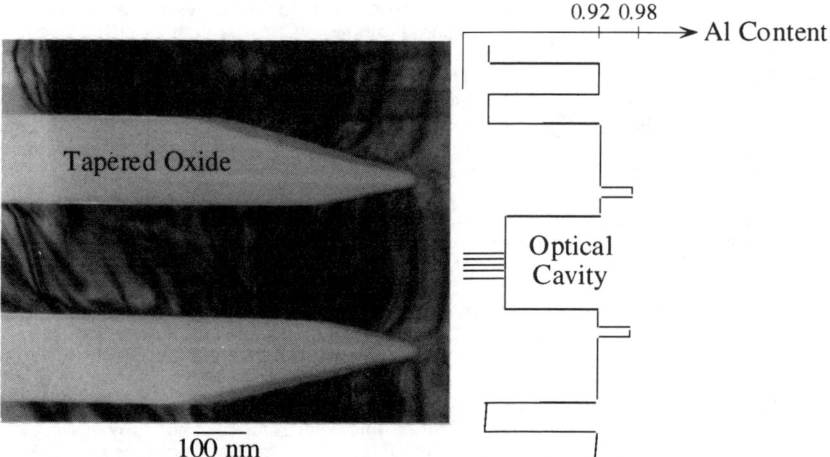

Fig. 4. Cross section TEM image of two tapered oxides (adjacent to optical cavity) produced from 14 nm thick $Al_{0.98}Ga_{0.02}As$ embedded within a 160 nm thick $Al_{0.92}Ga_{0.08}As$ layer.

D. Selectively Oxidized VCSELs

Significant performance advances have been achieved from selectively oxidized VCSELs which, as depicted in Fig. 5, incorporate one or more buried oxide apertures for efficient electrical and optical confinement [7,18]. These devices are fabricated by first growing the material by metalorganic vapor phase epitaxy. A mesa is etched to expose the sidewalls for oxidation as sketched in Fig. 5. The lateral extent of oxidation within the laser mesa can be controlled by the oxidation time [18]. Moreover, through design of the composition of the various layers within the VCSEL, the oxidation profile within the laser cavity can be tailored to affect the optical confinement and/or electrical injection into the laser.

211

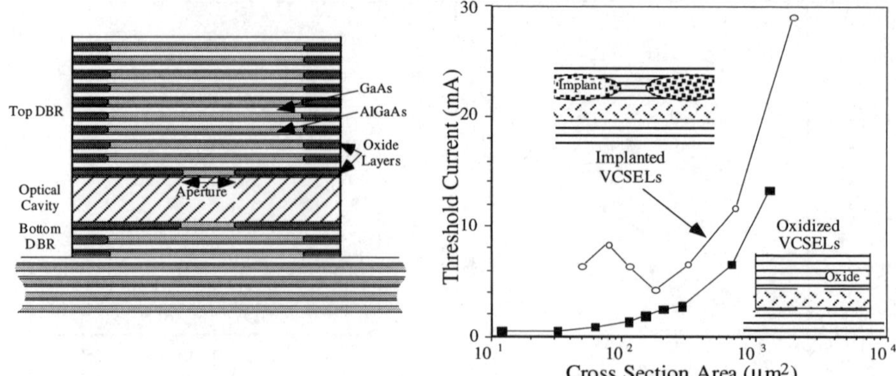

Fig. 5. Selectively oxidized VCSEL composed of GaAs/AlGaAs multilayers and buried oxide layers.

Fig. 6. Comparison of threshold current of implanted and selectively oxidized VCSELs.

The improved electrical confinement arising from the oxide apertures is particularly manifest as reduced threshold current and voltage compared to other VCSEL structures [19]. In Fig. 6 we compare selectively oxidized and ion implanted 850 nm VCSELs with various cavity sizes fabricated from the same epitaxial wafer. The ion implanted VCSELs rely on ion implantation to render a region around the laser nonconductive, thus defining the laser cavity. The reduced threshold current for the selectively oxidized VCSELs in Fig. 6 arises due to the improved electrical confinement and reduced current spreading. Moreover, the refractive index of the buried oxide layer changes from 3.0 for the original AlGaAs layer to ≈1.6 for the oxidized layer. This induces a significant index difference between the laser cavity and the region surrounding the laser thus providing index-guiding optical confinement. The ion implanted VCSELs in Fig. 6 do not exhibit monotonic decreasing threshold current with decreasing size, due to increased diffraction loss. In contrast, the index-guiding in the selectively oxidized VCSELs produces a monotonic decrease of the threshold currents in Fig. 6. Selectively oxidized VCSELs have also demonstrated high power conversion efficiency [20], high output power, and high speed modulation [21].

E.   Conclusion

Further work will certainly be required to develop selective wet oxidation of AlGaAs into a manufacturing technology for VCSELs or other microelectronic/photonic devices. Although selective oxidation has been shown to be a stable fabrication process, in order to scale up to full wafer manufacture, issues pertaining to reproducibility and uniformity obviously remain. Moreover, developing a means to photolithographically predefine the length of the oxide layer within a given device structure would be advantageous. Finally, analogous to process simulators available for Si oxidation, it will be necessary to develop predictive models of the oxidation of AlGaAs. In spite of these remaining challenges, the motivation to develop a wet oxidation technology is clearly driven by the demonstrated high performance of selectively oxidized VCSELs. Therefore, to meet the requirements of emerging applications, selective wet oxidation of AlGaAs may play an important role in the manufacture of high performance VCSELs.

Sandia is a multiprogram laboratory operated by Sandia Corporation for the United States Department of Energy under contract No. DE-AC04-94AL85000.

References

[1] W. T. Tsang, *Appl. Phys. Lett.*, 1978, vol. 33, 426.

[2] J. M. Dallesasse, N. Holonyak, Jr., A. R. Sugg, T. A. Richard, and N. El-Zein, *Appl. Phys. Lett.*, 1990, vol. 57, 2844.

[3] J. M. Dallesasse and N. Holonyak, Jr., *Appl. Phys. Lett.*, 1991, vol. 58, 394.

[4] F. A. Kish, S. J. Caracci, N. Holonyak, Jr., J. M. Dallesasse, K. C. Hsieh, M. J. Ries, S. C. Smith, and R. D. Burnham, *Appl. Phys. Lett.*, 1991, vol. 59, 1755.

[5] S. A. Maranowski, A. R. Sugg, E. I. Chen, and N. Holonyak, Jr., *Appl. Phys. Lett.*, 1993, vol. 63, 1660.

[6] D. L. Huffaker, D. G. Deppe, K. Kumar, and T. J. Rogers, *Appl. Phys. Lett.*, 1994, vol. 65, 97.

[7] K. D. Choquette, R. P. Schneider, Jr., K. L. Lear, and K. M. Geib, *Electron. Lett.*, 1994, vol. 30, 2043.

[8] M. H. MacDougal, H. Zhao, P. D. Dapkus, M. Ziari, and W. H. Steier, *Electron. Lett.*, 1994, vol. 30, 1147.

[9] A. Flore, V. Berger, E. Rosencher, N. Laurent, S. Theilmann, N. Vodjdani, and J. Nagle, *Appl. Phys. Lett.*, 1996, vol. 68, 1320.

[10] O. Blum, K. L. Lear, H. Q. Hou, and M. E. Warren, *Electron. Lett.*, 1996, vol. 32, 1406.

[11] E. I. Chen, N. Holonyak, Jr., and S. A. Maranowski, *Appl. Phys. Lett.*, 1995, vol. 66, 2688.

[12] K. D. Choquette, K. M. Geib, C. I. H. Ashby, R. D. Twesten, O. Blum, H. Q. Hou, D. M. Follstaedt, B. E. Hammons, D. Mathes, and R. Hull, *J. Special Topics Quan. Electron.*, 1997, vol. 3, 916.

[13] R. D. Twesten, D. M. Follstaedt, and K. D. Choquette, *SPIE Proc.*, 1997, vol. 3003, 56.

[14] R. D. Twesten, D. M. Follstaedt, K. D. Choquette, and R. P. Schneider, Jr., *Appl. Phys. Lett.*, 1996, vol. 69, 19.

[15] K. D. Choquette, K. M. Geib, H. C. Chui, B. E. Hammons, H. Q. Hou, T. J. Drummond, and R. Hull, *Applied Physics Letters*, 1996, vol. 69, 1385.

[16] K. Geib, K. D. Choquette, H. Q. Hou, and B. E. Hammons, *SPIE Proc.*, 1997, vol. 3003, 69.

[17] R. L. Naone, E. R. Hegbloom, B. J. Thibeault, and L. A. Coldren, *Electron. Lett.*, 1997, vol. 33, 300.

[18] K. D. Choquette, K. L. Lear, R. P. Schneider, K. M. Geib, J. J. Figiel, and R. Hull, *Photon. Tech. Letters*, 1995, vol. 7, 1237.

[19] K. D. Choquette, K. L. Lear, R. P. Schneider, and K. M. Geib, *Applied Physics Letters*, 1995, vol. 66, 3413.

[20] K. L. Lear, K. D. Choquette, R. P. Schneider, Jr., S. P. Kilcoyne, and K. M. Geib, *Electronics Letters*, 1995, vol. 31, 208.

[21] K. L. Lear, A. Mar, K. D. Choquette, S. P. Kilcoyne, R. P. Schneider, Jr., and K. M. Geib, *Electron. Lett.*, 1996, vol. 32, 457.

Oxidation of GaAlAs Red LED Surface Due to Light Irradiation

Toshitada Shimozaki*, Takahisa Okino** and Masahiro Yamane***

* *The Center for Instrumental Analysis, Kyushu Institute of Technology, Tobata, Kitakyushu. 804-8550. Japan*
** *College of Liberal Arts and Sciences, Nippon Bunri University, Ichigi Oita, 870-0316. Japan*
*** *Department of materials science and engineering, Kyushu Institute of Technology, Tobata, Kitakyushu. 804-8550. Japan*

abstract

Formation of light absorptive layer on the surface of GaAlAs LED due to light irradiation has been studied by irradiating red, green light and infrared ray LED on a pellet of GaAlAs device. The auger electron spectroscopy depth profiles and electron probe micro analysis have revealed that light absorptive layer is formed on the n-site surface by the light irradiation. Photocatalysis reaction may play very important role in the formation of the layer.

A. Introduction

It has been well known that GaAlAs red LEDs used outside degrade faster than those used inside due to the formation of light absorptive thin film on the surface of the GaAlAs. According to the previous work[1] the formation of the thin layer occurs only on the LEDs which are operated in a humid atmosphere but not on the LEDs which are not operated even in the humid atmosphere. From this result, it has been considered that water and an electric current play an important role in the degradation of LED.

It has also been well known that irradiation of high energy light on a semiconductor generates electrons on the n site and positive holes on the p site. These electrons or positive holes contribute to the chemical reactions such as oxidation or reduction on the surface of the semiconductor. This phenomenon is so-called photocatalysis[2]. There is a possibility that the photocatalysis reaction occurs on the red LED surface by the irradiation of sunshine or the red light of LED itself. To confirm the photocatalysis reaction on the GaAlAs LED surface, the red, green and infrared LED lights are irradiated on pellets of GaAlAs device in the humid atmosphere.

In this experiment, degradation of red LED due to operation of it in the humid atmosphere has also been studied as a reference experiment. The results are also reported.

B. Experimental procedure

The GaAlAs device is made on the GaAs substrate by liquid phase epitaxial growth method. LED on the market is made by depositing Au electrode, bonding Au wires on the pellet and covering them with epoxy resin.

The red LEDs having Au wire before epoxy resin are covered were operated in an environmental chamber keeping the temperature at 358K and the humidity at 85%. The current was controlled to be 40 or 60mA. After the operation, the part of the device about 0.3mmx0.3mmx0.1 mm in size was carefully removed from the base by a cutter having sharp edge. The procedure is schematically illustrated as "Experiment 1" in **Fig.1**.

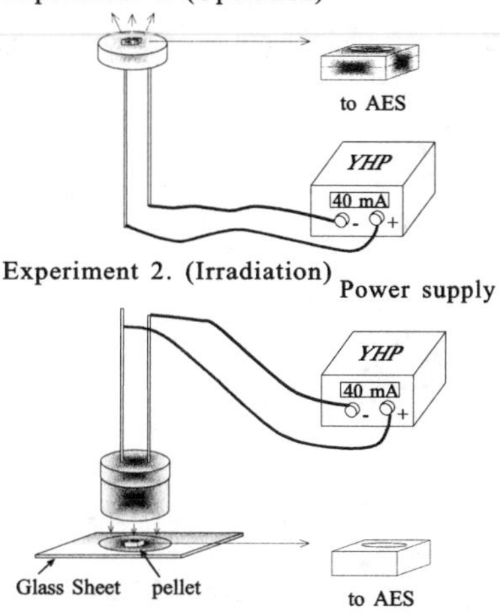

The pellets used for the irradiation experiment are the ones before Au wire and epoxy resin are fabricated. The natural oxide film on the surface of the pellets was dissolved by 0.1N NaOH solution for 60s. A few pellets were put on a glass sheet and were irradiated by red, green or infrared ray LEDs on market, respectively, in the environmental chamber as shown "Experiment 2" in **Fig.1**. The distance between the pellets and the LED is set to be less than 1mm. The current is controlled to be 40mA and 60mA for red LED and 40mA for green LED by a constant current generator.

The surface of the pellet after the operation as well as that after the irradiation was analyzed by the AES (JEOL JAMP-10SX). The energy ranges of the AES analysis are from 460 to 560eV for O, 1050 $\sim$ 1150eV for Ga, 1210 $\sim$ 1310eV for As and 1350 $\sim$ 1450eV for Al. The etching speed was not well known. But the conditions were set as same conditions which enable to etch SiO_2 surface by a speed of 0.556nm/s.

Fig.1   Schematic illustration of Experiment. These experiments are employed in an environment chamber kept at 358 K and humidity of 85%.

C. Results and Discussions

Typical auger depth profiles for the n-site surface of the device which was operated by the current of 40mA for 432 ks are shown in **Fig.2** together with the profiles for the initial one. The peak heights of As and Ga increase gradually from surface to the inside and then saturate. On the other hand, the peak height of Al gradually decreases. This corresponds to the concentration gradient which arises during liquid phase epitaxial growth. Oxygen penetration for the pellet after the operation is deeper than the initial one and that

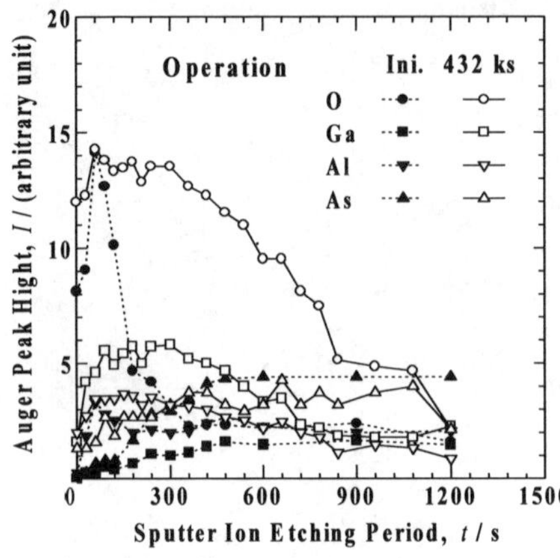

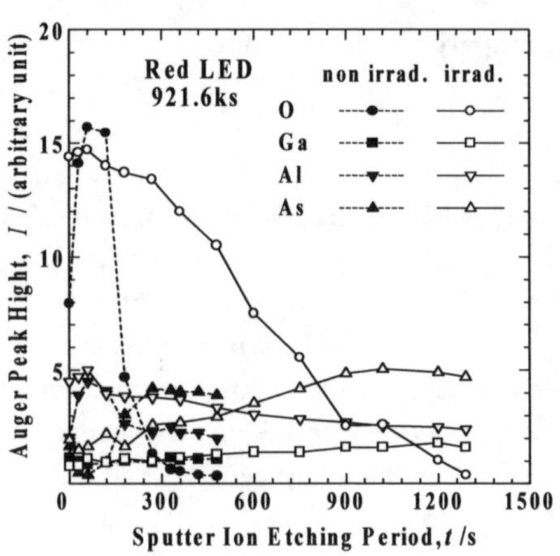

Fig.2   AES depth profiles of GaAlAs initial surface and after aging with operation.

Fig.3   AES depth profiles of GaAlAs surface after aging with and without irradiation.

after aging without operation.

The typical auger depth profiles for the pellet irradiated by red LED light for 921.6 ks at 358K are shown in **Fig.3** together with those which were left in the environment chamber without irradiation. The profiles are very similar to those shown in Fig.2. In general, the penetration of oxygen is deeper the longer the irradiation period or the larger the current density although there are some exceptions.

The depth profiles of O, Ga, As and Al for the n-site surface of the pellet obtained by irradiating green light for various period of aging time at 358K suggested that green light also enhances the oxygen penetration. A similar experiment was also done by using infrared ray LEDs. But penetration of oxygen could not be observed. This result is reasonable because red and green lights have higher energy than the energy gap of the GaAlAs but infrared ray does not.

A typical depth profiles for the p-site of the pellet which was irradiated by green light and the enlargement of near surface are shown in **Fig.4** in comparison with those for the n-site. In the p-site,

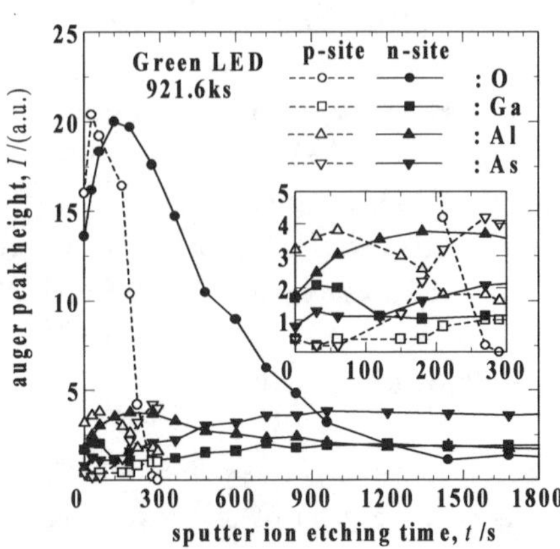

Fig.4   AES depth profiles for n-site and p-site of GaAlAs after aging for 921.6 ks with irradiation by green LED.

oxygen is also detected near surface, however, Ga and As cannot be detected at surface while they can be detected at the n-site surface. Oxygen penetration from surface on the n-site surface is obviously deeper than that on the p-site surface.

These AES depth profiles suggest the formation of light absorptive layer due to irradiation of LED light. In order to confirm the formation of the absorptive layer, the cross section of the GaAlAs pellet which was irradiated by green LED for 1800 ks is analyzed by EPMA. We could find thin layer about 5 μ m in thickness which was formed on the n-site surface as shown in **Fig.5**. The morphology is very similar to the light absorptive layer obtained by operating a red LED in the 358K and 85% humid atmosphere as has been reported by Yamanaka et al[1]. So, we conclude that the thin layer is formed by the light irradiation. On the other side, the formation of oxide layer on the p-site could not be observed by EPMA because the thickness may be too thin to be detected.

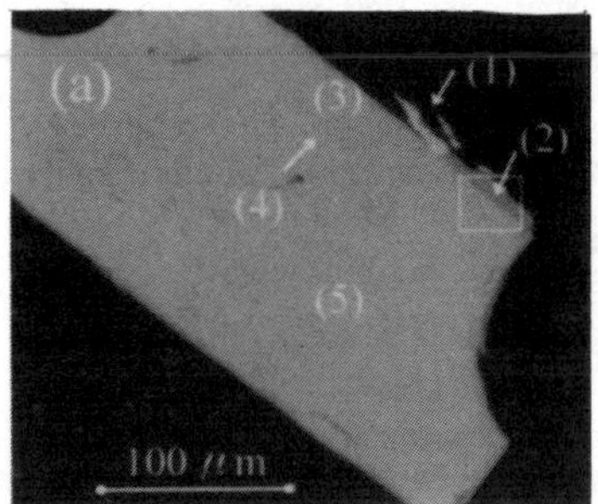
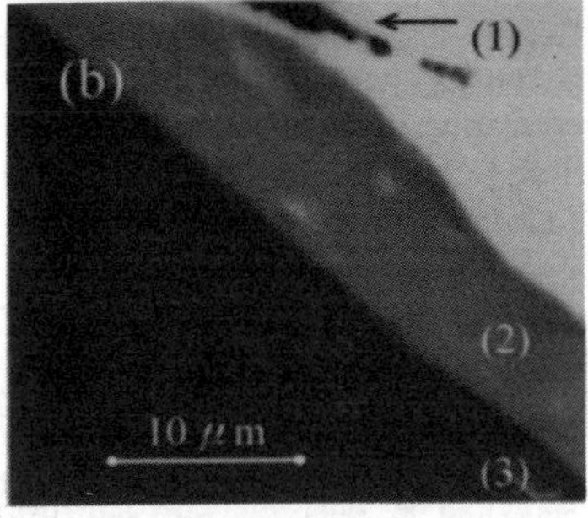

Fig.5　EPMA component image of the light absorptive layer formed on the n-site surface (a) and the enlargement (b). (1)Au electrode, (2) light absorptive layer, (3) n-type GaAlAs (4) emitter, (5) p-type GaAlAs.

In general, electrons on the n-site surface take part in the reduction reaction. But in an atmosphere where oxygen exists, very active OH⁻ or O⁻ are formed and they take part in an oxidation reaction. Therefore, on the n-site surface hydroxide which containes not only Al but also Ga and As may be formed. The hydroxide may change to oxide, although we could not confirm whether the thin layer is oxide or hydroxide.

On the other side, electron holes on p-site surface take part in the selective oxidation of Al. The Al_2O_3 layer thus formed may be compact and may protect GaAlAs surface from further oxidation.

References

[1]　H. Yamanaka and S. Koike : J. Electrochem. Soc., **132**(1985), 381.
[2]　A. Fujishima and K. Honda : Nature, **238**, (1972), 37.

CHANGES IN INTERDIFFUSION ASSOCIATED WITH THERMALLY OXIDIZED GaAs.

R.M. Cohen(1), Gang Li,(2), C. Jagadish,(2), Patrick T. Burke,(3) and Michael Gal(3)

(1) Materials Science and Engineering Dept., Univ. of Utah, Salt Lake City, UT 84112 USA
(2) Electronic Materials Engineering, Research School of Physical Sciences and Engineering,
Australian National University, Canberra, ACT 0200, AUSTRALIA
(3) School of Physics, University of New South Wales, Sydney, NSW 2052, AUSTRALIA

Abstract

Thermal oxidation and subsequent processing steps were found to significantly affect interdiffusion in
AlGaAs/GaAs quantum wells. GaAs layers were grown on top of the quantum wells and oxidized at
450 C, and this was followed with a nonoxidizing ambient during annealing at 950 C. The resulting
oxide, largely composed of Ga_2O_3, affects the native defect concentration and interdiffusion rate. Low
temperature photoluminescence demonstrated that the oxidation plus annealing process significantly
increased the rate of interdiffusion relative to structures which were annealed but not oxidized.
However, when oxide layers covered with Al were annealed at 950 C, the Ga_2O_3 was reduced and Ga
interstitials were forced into the GaAs. Using Al layers of various thickness, different quantities of Ga
interstitials were injected into the epitaxial layer and a consistent reduction in interdiffusion was related
to the Al layer thickness. The results are found to be consistent with previously a proposed model of
interdiffusion controlled by Ga vacancies.

Control of the rate of interdiffusion on the group III sublattice, D_{III}, is important for device
processing technology. Attempts to control D_{III} involve the use of either dopants or encapsulants, i.e.,
so called impurity-induced or impurity-free effects on diffusion rate. Increases or decreases in D_{III} can
be repeatedly obtained by using appropriate donor or acceptor atoms, and a wide range of results have
been shown to be consistent with a diffusion mechanism associated with a Ga vacancy with a charge
of -1.[1] However, it is seldom practical to use dopants to control interdiffusion because the doping
concentration is normally chosen to meet device performance specifications. Thus, one would also like
to control D_{III} without adding dopants. Some recent reports discuss the effect on D_{III} caused by the
encapsulants SiO_x and SiN_x,[1] SrF_2,[2] WN_x,[3] and anodically formed oxides of GaAs.[4,5] In this
paper, we shall discuss changes in the interdiffusion of quantum wells associated with a thermally
grown oxide of GaAs.

It has been shown that a wide range of annealing conditions cause the group III interdiffusion to
occur by a vacancy mechanism associated with the group III sublattice.[1] For such a case, one may
empirically relate the measured interdiffusion, D_{III}, to

$$D_{III} = D_{III}^{Std}(T, P_{As}^{Std}) \frac{[V_{Ga}]}{[V_{Ga}(T, P_{As}^{Std})]^{Std}} \qquad (1)$$

where D_{III}^{Std} is the measured rate of interdiffusion when the actual Ga vacancy concentration equals its
equilibrium value, $[V_{Ga}]=[V_{Ga}]^{Std}$, for the chosen standard conditions, at T and, say, $P_{As}=1$ atm.[6] It
should be clear from eq. 1 that the measured D_{III} may scale either up or down if the actual $[V_{Ga}]$
changes with the choice of processing conditions.

Changes in $[V_{Ga}]$ in an epilayer can occur not only when P_{As} is changed, but may occur under an
encapsulant which cuts off contact with the vapor. In such a case, $[V_{Ga}]$ and the measured D_{III} in the
epilayer may change with time because the substrate acts as a source or sink of vacancies, or because
the metallurgical reactions which generally occur at the GaAs-encapsulant interface generate a source or
sink of vacancies.[7] To minimize the complexity of metallurgical reactions between GaAs and multielement encapsulants, we have studied how a simple encapsulant based upon the oxide of the host
material GaAs affects interdiffusion.

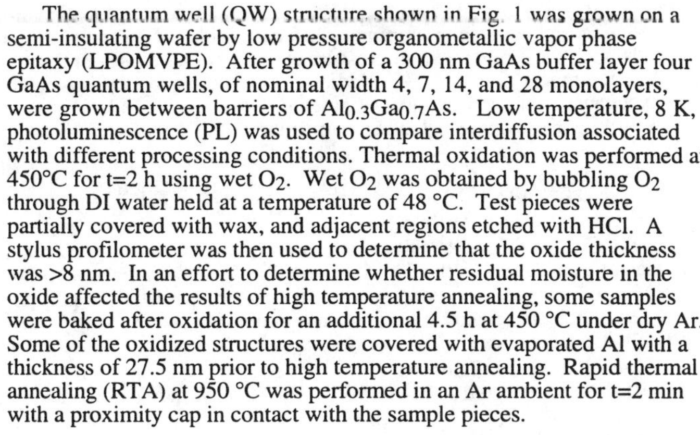

Fig. 1. Samples have 4 QWs within $Al_{0.3}Ga_{0.7}As$ barriers.

The quantum well (QW) structure shown in Fig. 1 was grown on a semi-insulating wafer by low pressure organometallic vapor phase epitaxy (LPOMVPE). After growth of a 300 nm GaAs buffer layer four GaAs quantum wells, of nominal width 4, 7, 14, and 28 monolayers, were grown between barriers of $Al_{0.3}Ga_{0.7}As$. Low temperature, 8 K, photoluminescence (PL) was used to compare interdiffusion associated with different processing conditions. Thermal oxidation was performed at 450°C for t=2 h using wet O_2. Wet O_2 was obtained by bubbling O_2 through DI water held at a temperature of 48 °C. Test pieces were partially covered with wax, and adjacent regions etched with HCl. A stylus profilometer was then used to determine that the oxide thickness was >8 nm. In an effort to determine whether residual moisture in the oxide affected the results of high temperature annealing, some samples were baked after oxidation for an additional 4.5 h at 450 °C under dry Ar. Some of the oxidized structures were covered with evaporated Al with a thickness of 27.5 nm prior to high temperature annealing. Rapid thermal annealing (RTA) at 950 °C was performed in an Ar ambient for t=2 min with a proximity cap in contact with the sample pieces.

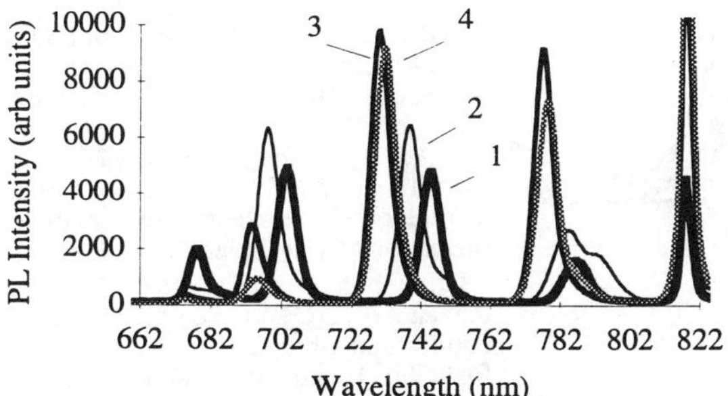

Fig. 2. PL spectra from samples which were (1) as-grown, (2) annealed after growth, (3) oxidized prior to annealing, and (4) baked for an extra 4.5 h after oxidation, prior to annealing.

Fig. 2 compares the PL spectra from four samples. Spectrum 1, from the as-grown control sample, shows five PL peaks originating with the substrate and the four QWs. Spectrum 2 shows PL from a piece of the as-grown (unoxidized) sample after RTA. The PL from the QWs shift to higher energies, and the full width at half maximum of each peak is similar to its counterpart in the control sample, although weak low energy tails have developed for unknown reasons. For annealing without an oxide, $D_{III}\approx3\times10^{-18}$ cm^2/s. Spectra 3 and 4 show PL taken after RTA of samples which had received no additional baking or additional baking with Ar immediately after oxidation,

respectively. The two spectra are nearly the same, although there is a slightly smaller PL energy shift observed for the sample which received additional baking under dry conditions. The presence of the encapsulant clearly speeds up interdiffusion, similar to that reported after room temperature anodic oxidation and RTA of similar structures,[5,8] and $D_{III}\approx3\times10^{-17}$ cm^2/s.

Fig. 3 shows the spectra from six samples. Spectra 1, 2, and 3 are the same as in Fig. 2. Spectra 4, 5, and 6 show the results after RTA of Al-covered samples, and correspond to as-deposited Al film thicknesses of 17, 27.5, and 35 nm, respectively. An expanded region is shown above Fig. 3 for the peaks associated with the 14 ML wide QW. A small amount of diffusion occurs for initial Al film thicknesses of 17 and 35 nm, but so little interdiffusion occurs for the 27.5 nm thick Al film that it is only possible to estimate an upper limit of $D_{III}\leq2\times10^{-19}$ cm^2/s.

Fig. 4 plots D_{III} vs the thickness of the deposited Al film on the thermally grown oxide layer for these QWs. All results have been duplicated at least once. To understand the differences observed, it is worth considering the Ga-As-O ternary phase diagram presented by Thurmond et. al.[9] There is no stable oxide of GaAs but there are several oxides of Ga and As which may be formed on the GaAs.

Except at very low oxygen partial pressure, the stable oxide of Ga is gallia, Ga_2O_3, whose physical properties resemble those of alumina. Different oxides of As may form and the equilibrium phase depends upon the oxygen partial pressure chosen, but all oxides of As are relatively volatile and leave the film at temperatures ≥ 450 °C. Although continuous oxide films can be formed, they tend to be rough because of the gradual loss of arsenic.

A fairly general empirical model can be written to describe the measured D_{III},

$$D_{III} = D_{III}^{Std,V}(T,P_{As}^{Std})\frac{[V_{Ga}]}{[V_{Ga}(T,P_{As}^{Std})]^{Std}} + D_{III}^{Std,I}(T,P_{As}^{Std})\frac{[I_{Ga}]}{[I_{Ga}(T,P_{As}^{Std})]^{Std}} \qquad (2)$$

where the first term describes the contribution of the Ga vacancy mechanism to the interdiffusion, and the second term describes the contribution of the kick-out mechanism to the interdiffusion on the group III sublattice. Based upon dozens of results from many different groups, the rate of interdiffusion is expected to be determined by the concentration of V_{Ga} over a very wide range of annealing conditions, and only in special cases will it be dominated by the interstitial-driven kick-out mechanism, i.e., the second term in eq. 2.[1] Thus, for annealing without an encapsulant, i.e., spectrum 2 in Fig. 3, the interdiffusion is expected to be controlled by a Ga vacancy mechanism. Annealing with a Ga_2O_3 encapsulant is new and the reason for the increase in interdiffusion (spectrum 3, Fig. 3) is not immediately clear. However, the dramatic reduction in interdiffusion when Al is present on the oxide surface allows us to determine the point defect chemistry influencing the diffusion.

At 950 °C, the free energy of formation of Al_2O_3 is approximately 600 kJ/mol lower than the free energy of formation of Ga_2O_3, and even lower relative to the oxides of As.[10] This means that an Al/Ga_2O_3 couple will approach equilibrium when Ga_2O_3 is reduced to form Al_2O_3 and atomic Ga, i.e., Al + $Ga_2O_3 \leftrightarrow Al_2O_3$ + Ga. Atomic As also will be produced by the reduction of any As-oxides present, and the atomic Ga and As produced might reform GaAs. However, Thurmond et. al.[9] (and references therein) have shown that oxide films become strongly Ga-rich during the oxidation process because of the high volatility of arsenic and its oxides. Thus, more Ga than As is expected to be produced by the reduction of the oxide film during RTA. Because of the relatively low melting temperature of Al, $T_m \approx 660$ °C, molten Al is expected to provide a fresh metal surface on top of the oxide film and to rapidly drive the reduction reaction. Much of the atomic Ga produced under the Al_2O_3 is trapped and diffuses into the semiconductor as Ga interstitials, I_{Ga}.

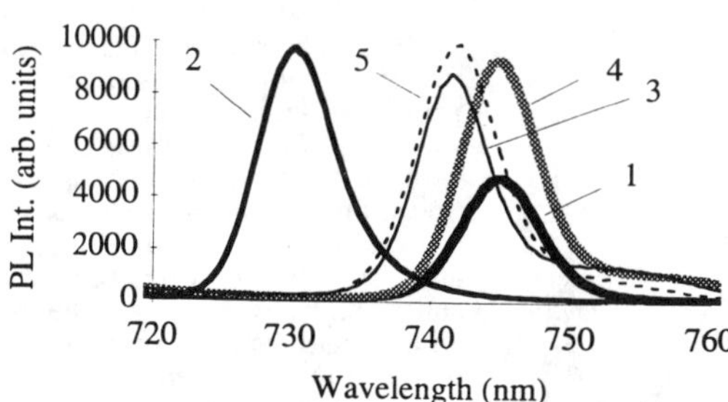

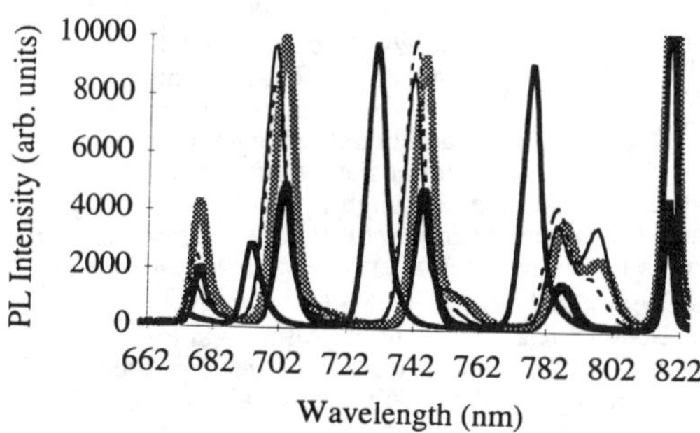

Fig. 3. PL spectra from samples which were (1) as-grown, (2) oxidized prior to annealing, or oxidized and covered with (3) 17 nm, (4) 27.5 nm, or (5) 35 nm of Al prior to annealing. Expanded view of the PL peaks associated with the 14 ML QW is shown above.

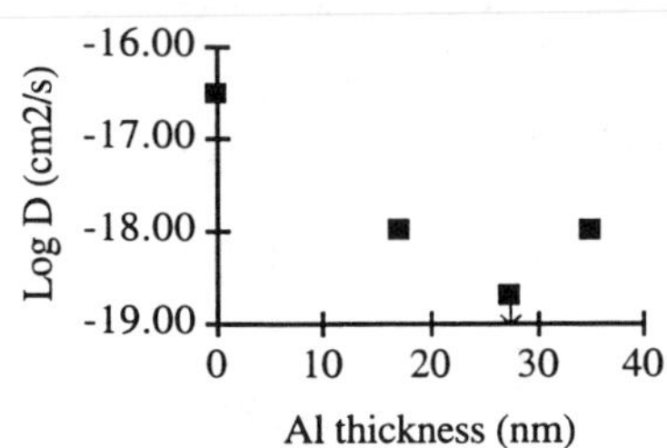

Fig. 4. D_{III} vs thickness of Al layer deposited on oxide. Al induces reduction of Ga_2O_3. Injected interstitials slow the vacancy controlled interdiffusion.

An increase in the Ga interstitial concentration will push the Frenkel reaction, $Ga_{Ga} \leftrightarrow V_{Ga} + I_{Ga}$, to the left, i.e., it will immediately reduce the Ga vacancy concentration and the rate of vacancy-controlled interdiffusion. The reduced interdiffusion is strong evidence that V_{Ga} controls the interdiffusion when annealing with thermally grown oxide films. Since each of the Al films contain more than enough Al atoms to completely reduce the oxide film, the observed thickness dependence implies that all of the Al did not immediately reach the Ga_2O_3. This suggests that the formation of the Al_2O_3 crust slows the interaction of liquid Al with Ga_2O_3. As the Al film thickness increased to 27.5 nm, D_{III} was observed to decrease. However, D_{III} increased for the thickest Al film of 35 nm. This behavior suggests that either more Al reached the oxide, or Al reached the substrate and both Ga and Al atoms entered the crystal as interstitials, and caused an increase in D_{III} via the kick-out mechanism. The precise cause of this increase in D_{III} remains to be determined.

The above results strongly suggest that the rapid D_{III} measured after RTA of oxide covered GaAs (spectra 3 and 4, Fig. 2) was caused by an increase in the concentration of V_{Ga}. We consider it likely that an inflow of vacancies from a reservoir of point defects, i.e., the substrate. However, an increase in [V_{Ga}] in the epilayer might also occur if vacancies are generated at the oxide-GaAs interface during RTA, and their source remains to be determined.

In summary, it has been shown that QW interdiffusion can be either increased or decreased by manipulating the point defect concentrations during high temperature processing. Relative to samples which have not been encapsulated, an order of magnitude increase in D_{III} has been measured as a result of the thermal oxidation of GaAs. More than an order of magnitude decrease in D_{III} has also been measured as a result of the chemical reduction of the thermally grown Ga_2O_3-rich film. Changes in D_{III} are primarily a result of changes in the Ga vacancy concentration, but the kick-out mechanism appears to begin increasing D_{III} when larger amounts of Al are present. These results show that metallurgical reactions may be an effective tool for engineering native defect concentrations and associated diffusivities.

Acknowledgments. Support from the National Science Foundation, Division of Materials Research is gratefully appreciated by RMC. G. Li gratefully acknowledges fellowship support from Australian Research Council. Australian authors gratefully acknowledge Australian Research Council for financial support.

[1] R. M. Cohen, Materials Science and Engineering Reports **R20,** 167 (1997).

[2] I. Gontijo, T. Krauss, J. H. Marsh, and R. M. D. L. Rue, IEEE J. Quantum Electron. **30,** 1189 (1994).

[3] E. L. Allen, C. J. Pass, M. D. Deal, J. D. Plummer, and V. F. K. Chia, Applied Physics Letters **59,** 3252 (1991).

[4] S. Yuan, Y. Kim, C. Jagadish, P. T. Burke, M. Gal, J. Zou, D. Q. Cai, D. J. H. Cockayne, and R. M. Cohen, Appl. Phys. Lett. **70,** 1269 (1997).

[5] S. Yuan, Y. Kim, H. H. Tan, C. Jagadish, P. T. Burke, L. V. Dao, M. Gal, M. C. Y. Chan, E. H. Li, J. Zou, D. Q. Cai, D. J. H. Cockayne, and R. M. Cohen, J. Appl. Phys. **83,** 1305 (1998).

[6] F. A. Kroger, *The Chemistry of Imperfect Crystals*, Vol. 2, 2 ed. (North-Holland, Amsterdam, 1974).

[7] R. M. Cohen, Defects and Diffusion Forum **in press** (1998).

[8] S. Yuan, C. Jagadish, Y. Kim, Y. Chang, H. H. Tan, R. M. Cohen, M. Petravic, L. V. Dao, M. Gal, M. C. Y. Chan, E. H. Li, J. S. O, and P. S. Zory, IEEE Journal of Special Topics in Quantum Electronics, in press (1998).

[9] C. D. Thurmond, G. P. Schwartz, G. W. Kammlott, and B. Schwartz, Journal of the Electrochemical Society **127,** 1366 (1980).

[10] I. Barin, *Thermochemical Data of Pure Substances*, 3 ed. (VCH, New York, 1995).

Structural, electronic and optical properties of AlN-based oxides on Si and GaN substrates

Johnson O. Olowolafe[1], Ingolf Rau[1], James Kolodzey[1], Enam A. Chowdhury[2], Karl M. Unruh[2], Charles P. Swann[3]

1 *Department of Electrical and Computer Engineering, University of Delaware, Newark, DE 19716*
2 *Department of Physics and Astronomy, University of Delaware, Newark, DE 19716*
3 *Bartol Research Institute, Sharp Laboratory, Newark, DE 19716.*

Abstract

The properties of novel high-quality AlN-based oxides suitable for device applications are presented in this paper. The process steps employed in this investigation are similar to those of SiO_2, providing new technology for group III-nitride based electronic devices. The oxides were prepared by thermal oxidation of AlN films or direct deposition of ultra-pure aluminum in a O_2-N_2 ambient on Si or GaN substrates. The structure of the films depends on deposition conditions for oxidized AlN and on composition for the Al films deposited in the N_2-O_2 ambient. The dielectric constant was evaluated to be 12.74 while the effective charge density is about 10^{11} cm^{-2} for fully oxidized AlN. The refractive index of the oxide was about 3.9 using infrared techniques. Our results indicate that the AlN-based oxides provide a wide choice of high-quality insulators for electronic and optoelectronic applications.

A.   Introduction

Device-quality oxides are essential components of electronic devices. So far, SiO_2 has played the dominant role in the fabrication of Si-based field-effect devices and layering of interconnect wiring in integrated circuits (ICs). As the field effect transistors scale down in size and the thickness of the gate oxides shrinks further device reliability issues such as leakage currents and dielectric breakdown become a major issue.[1] Furthermore, the oxides of compound semiconductors, except AlAs,[2, 3] and group III nitrides are unstable for device applications either because they are thermally unstable or are of poor quality.[4,5] In this paper we present the processing and evaluation of AlN-based oxides on Si and GaN substrates.

B.   Experiment

Two experimental procedures were taken in the processing of the AlN-based oxides. The first approach involves the deposition of AlN films by reactive ion sputtering on Si or GaN substrates. We used an ultra-high-vacuum magnetron sputtering system operated between 200 and 400 watts at a vacuum close to 10^{-8} prior to deposition. Oxidation was performed in a furnace at temperatures ranging from 800 to 1100 °C with O_2 flow for times ranging from 1 to 2 hours. In the second deposition procedure Al was sputtered in a O_2-N_2 ambient, with no oxidation required after deposition. The film thickness was measured by stylus profilometry at step edges and the atomic force microscopy (AFM), while film purity was evaluated with secondary ions mass spectromery (SIMS) and Auger electron spectroscopy (AES). The structure of the films was determined by the transmission electron microscopy (TEM) and x-ray diffraction (XRD) techniques. The Rutherford backscattering spectroscopy (RBS) was used to analyze the composition and kinetics of growth of the oxides.[6] The dielectric properties of the oxides were evaluated using the capacitance-voltage (C-V) profiling of MOS structures.

C. Results

C1. Thermally Oxidized AlN

Figure 1 shows the TEM electron diffraction of AlN as-deposited showing that the structure of the film is polycrystalline with small crystal sizes. This microstructure is typical of the films used in this project.

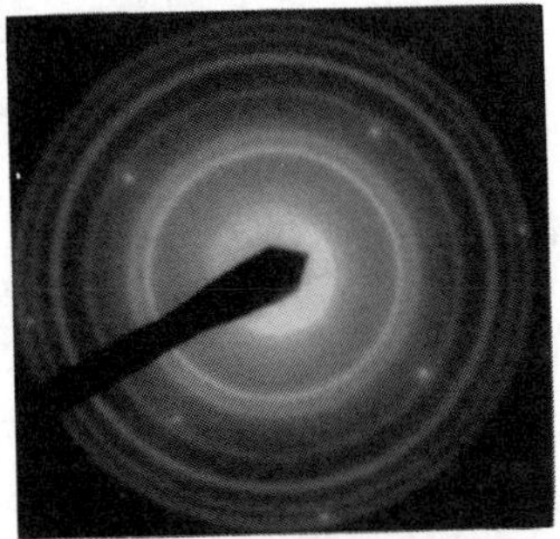

Fig. 1. TEM electron diffraction of an AlN film as-deposited, showing that structure is nanocrystalline

Figure 2 shows the XRD intensity scans of AlN as-deposited and oxidized between 800 and 1100 °C for 1 or 2 hr. The scans cover the range of angles associated with AlN and Al_2O_3. As-deposited and oxidized up to 900 °C small-grain polycrystalline ("nanocrystalline") phases present while Al_2O_3 phases, including sapphire (α-Al_2O_3), were present at higher temperatures.

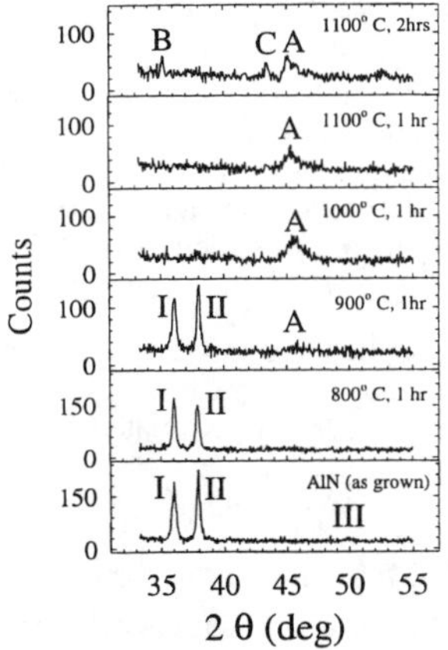

Fig. 2. XRD intensity versus 2θ showing AlN as-deposited and gradual transformation to Al_2O_3 phases with increasing oxidation temperature. (I): AlN (002), (II): AlN (101), (III): AlN (102); (A): θ-Al_2O_3 (B): α-Al_2O_3 (104), (C): α-Al_2O_3 (113)

To evaluate the dielectric properties of the oxides, Al/oxide/Si MOS structures were fabricated. The C-V profiles of multi-frequency measurements performed at room temperature with measurement frequency between 10 KHz and 10 MHz are shown in Fig. 3.

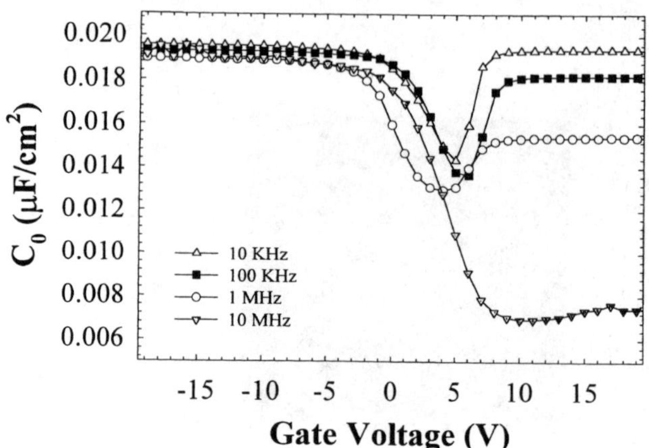

Fig. 3. C-V profiles of Al/ Al$_2$O$_3$:N/p-Si/Al MOS structure

The characteristics indicate that the Si surface was at accumulation at negative voltages, depletion at higher voltages and inversion if the frequency is high enough (2 MHz to 10 MHz). This is similar to the well-known reported behavior of SiO$_2$.[7] To calculate the dielectric properties, the maximum capacitance (accumulation), C_{acc}, and the minimum capacitance (inversion), C_{inv}, were analyzed using the following expressions:

$$C_{acc} = \varepsilon_{ox}/t_{ox}, \quad C_{inv} = (t_{ox}/\varepsilon_{ox} + d_p/\varepsilon_s)^{-1}, \tag{1}$$

where $d_p = (2\varepsilon_s E_g/q^2 N_A)^{1/2}$ is the $_{depletion}$ width in the p-type Si substrate and ε_{ox} and ε_s are the dielectric permittivities of the oxide and Si, respectively; q is the magnitude of the electron charge (1.6 x 10^{-19} C), Eg is the bandgap of Si (1.1 eV) and N_A = 5 x 10^{15} cm^{-3} is the concentration of acceptor ions in the substrate as determined by the four-point probe measurement. The Al$_2$O$_3$ and SiO$_2$ layers are modeled as two dielectric films in series. With a combined thickness, t_{ox} = 625 and ε_{ox} = 7.78 ε_o , $\varepsilon_{AlN:O}$ = 12. 74 for the electric permittivity of Al$_2$O$_3$:N. ε_o is the permittivity of free space and κ = 3.9 was used as the dielectric constant of SiO$_2$.
The flatband voltage of the structures was evaluated from the expression :

$$V_{FB} = -Q_{ox}/C_{ox} + \Phi_{ms} \tag{2}$$

where Qox is the net oxide charge (the first moment of the oxide charge distribution divided by the oxide thickness), Φms is the metal-semiconductor work function difference (about 0.92 V for Al and p-type Si). For the sample annealed at 1000 °C, Q_{ox}/q = 1.4 x 10^{11}, with the flatband and threshold voltages $_{close}$ to zero, indicating few defects and that the semiconductor Fermi level was not pinned at the oxide-semiconductor surface.

C2. Al sputtered in N_2-O_2 ambient

The oxides, $Al_xO_yN_z$, obtained from Al films sputtered in a O_2-N_2 ambient varies in composition from the O-rich $Al_{0.45}O_{0.50}N_{0.05}$ to the N-rich $Al_{0.42}O_{0.23}N_{0.35}$ films. The oxides, whether O-rich or N-rich, are amorphous as-deposited and remain stable at high temperatures. TEM micrographs and diffraction patterns indicate that the films are amorphous as-deposited and annealed. The dielectric properties of the oxide vary with composition as shown in Table 1. Films that are N-rich have the highest dielectric constant (about 15.49-16.53) while sputtered Al_2O_3 has a dielectric constant of about 9.08 close to what is obtained in the literature.

Table 1. Composition and Dielectric Properties of $Al_xO_yN_z$

Film Composition	Dielect. Const.
AlN	8.7
$Al_{0.42}O_{0.23}N_{0.35}$	15.49
$Al_{0.40}O_{0.38}N_{0.32}$	16.53
$Al_{0.45}O_{0.50}N_{0.05}$	8.35
Al_2O_3	9.08

C2.1 GaN substrates

The GaN substrates used in this experiment are epitaxially grown n-type films on sapphire substrates. The samples are about 1/2 inch in diameter with a thin (conducting epi layer (sample # 1994.3) on insulating GaN grown on Al_2O_3 (sapphire). Fig. 4 shows the C-V profile of about 500 nm thick $Al_{0.35}O_{0.59}N_{0.06}$ obtained on the GaN epi-layer. This profile indicates maximum capacitance (accumulation) at positive voltage which decreases (depletion) as the voltage decreases, attaining minimum (inversion) at negative voltages. The non-attainment of a perfect maximum capacitance may be due to the oxide-GaN interface.

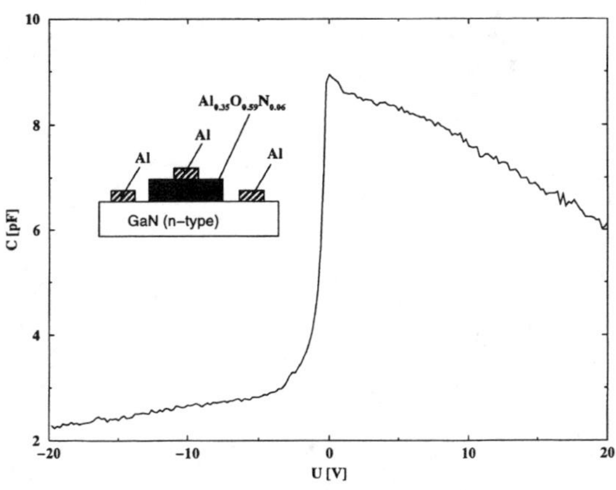

Fig. 4. C-V profile of Al/ $Al_{0.35}O_{0.59}N_{0.06}$/n-GaN MOS structure

Conclusion

We have demonstrated that device-quality AlN-based oxides can produced either by conventional thermal oxidation or by reactive ion sputtering. The oxides have been shown to be amorphous and thermally stable. The oxide films have high dielectric constants, about 12.74 for oxidized AlN, and up to 16.53, and varies with composition, for the reactively sputtered oxide, thus permitting high charge storage capacity and thicker layers without breakdown or leakage problems. The C-V profiles exhibit behaviors typical of high-quality SiO_2-based MOS structures; the capacitance is maximum at accumulation, decreases in depletion and minimum in inversion.

References

[1] C. Hu, Tech. Dig. Int. Electron Devices Meet., (1996) 319.
[2] M. J. Ries, N. Holonyak, Jr., E. I. Chen, S. A. Maranowski, M. R. Islam, A. L. Holmes, and R. D. Dupuis, Appl. Phys. Lett. 67, (1995) 1107.
[3] M. R. Krames, A. D. Minervini, and N. Holonyak, Jr., Appl. Phys. Lett. 67 (1995) 73.
[4] R. J. Molner, R. Singh, and T. D. Moustakas, Appl. Phys. Lett. 66, (1995) 268.
[5] A. Ozgur, W. Kim, Z. Fan, A. Botchkarev, A. Salvador, S.-N. Mohammad, B. Sverdlov, and H. Morkoc, Electron. Lett. 31, (1995) 1389.
[6] L. R. Doolittle, Nucl. Instrum. Methods Phys. Res. B 9, (1985) 344.
[7] E. H. Nicollian and J. R. Brews, MOS Physics and Technology, (John Wiley, Inc., NY, 1982).

Lateral Oxidation of AlAs Thin Films

Z. Liliental-Weber, O. Richter, W. Swider, M. Li,* G.S. Li,* C. Chang-Hasnain,* and
E.R.Weber**

Materials Science Division, Lawrence Berkeley National Laboratory, Berkeley, CA 94720

* Electrical Engineering and Computer Science Department, UC Berkeley, Berkeley, CA 94720
** Material Science Dept., UC Berkeley, Berkeley, CA 94720

Abstract

Transmission electron microscopy was used for characterization of microstructure resulting from wet oxidation of AlAs layers of different thicknesses separated by 20 nm thick GaAs layer. In general, oxidation rate was related to the AlAs layer thickness, but some deviation from this rule was observed for layer thicknesses in the range of 65 to 30 nm. Accumulation of As precipitates at the AlAs/GaAs/AlAs interfaces and in the GaAs separation layers was observed. Build up of stress at these interfaces occasionally caused formation of stacking faults close to the oxidation front. Large differences in interface abruptness was observed for direct (AlAs on GaAs) and the inverted (GaAs on AlAs) interface. The oxidation front was asymmetric with the faster front close to the inverted interface.

A. Introduction

Gallium Arsenide (GaAs) has long been studied for use in high speed electronics due to its high electron mobility. One major problem, however, is the lack of a stable oxide in the GaAs system. The Aluminum Arsenide system (AlAs) on the other hand has a nearly perfect lattice match with GaAs and does form a stable oxide (Al_2O_3) [1-12]. It was shown that the addition of Ga to the AlAs system can have a profound effect on the oxidation rate [13]. Reaction of $Al_xGa_{1-x}As$ with H_2O vapor at elevated temperatures ($\sim$ 400-450° C), produces high quality, stable $-Al_2O_3$ oxides.
The purpose of this current study is to use transmission electron microscopy to observe the microstructure and dependence of oxidation rate on the AlAs layer thickness.

B. Experimental

The AlAs layers studied varied in thickness, while the GaAs spacer layers remained constant. Two sets of samples were grown by molecular beam epitaxy (MBE) on n-type (0010 GaAs substrate. One contained 7 layers of AlAs with thicknesses of 20 nm, 10 nm, 5 nm, 2.5 nm, 5 nm, 10 nm and 20 nm separated by 50 nm GaAs layers with a 500 nm top layer of GaAs (sample A). The second set also contained 7 layers of AlAs with thicknesses of 60 nm, 45 nm, 30 nm, 15 nm, 30 nm, 45 nm, and 60 nm separated by 50 nm GaAs layers with a 50 nm top layer of GaAs (sample B). These samples were patterned into long strips along a <110> direction. The oxidation was carried out in a quartz furnace at 450°C in $H_2O:N_2$ by bubbling nitrogen through de-ionized water at 95°C. This process resulted in lateral oxidation of the AlAs layers. The oxidized AlAs layer has a different refractive index and the oxidized layers become amorphous, therefore, the oxidation terminus can be easily observed.
Cross-section TEM samples were prepared perpendicular to the strip direction. The samples were mechanically polished followed by ion milling. Immediately after obtaining electron transparent samples TEM observation was performed to avoid any damage due to storage in the atmosphere. A Topcon 002B electron microscope with acceleration voltage of 200 KeV and point-

to-point resolution 1.8 Å was used for the studies.

C. Results

Oxidation started from the mesa's walls. High resolution TEM analysis of the as-deposited layers showed perfect crystallinity with near-atomically flat interfaces (Fig. 1). Careful adjustment of the oxidation time allows study of the lateral oxidation front. Upon oxidation the AlAs layer was transformed to γ-Al$_2$O$_3$, as shown by previous TEM studies [5-7, 11]. Selective area diffraction obtained from the oxidized AlAs layers gave ring patterns suggesting that the layer consists of small crystallites with grains not larger than 2-5 nm.

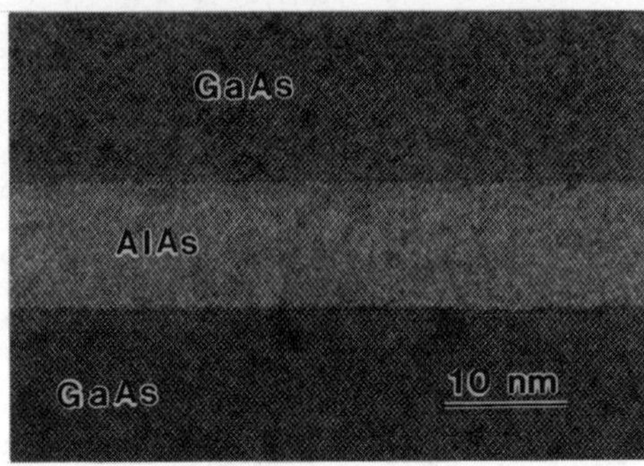

Fig. 1. High resolution micrograph showing as-grown AlAs layer. Note that both interfaces - direct (AlAs grown on GaAs) and inverted (GaAs grown on AlAs) are abrupt.

Clear dependence of oxidation rate on the AlAs layer thickness was observed. For the layers in the sample B the difference in oxidation rate is not linear with thickness (Fig. 2). The highest oxidation rate was observed for the 45 nm thick layer. For layers with thickness ranging from 20 nm to 2.5 nm (sample A) an almost liner dependence was observed.

Fig. 2. Oxidation of AlAs layers with different AlAs layer thicknesses (sample B). Note that layers with greater thickness oxidize at faster rates and that layers of the same thickness closer to the surface oxidize more slowly.

The second observation was that layers closer to the sample surface oxidized slightly slower than layers closer to the substrate independently from the layer thickness (Figs. 2 and 3a). This is most probably an influence of the sample surface. The third observation was that the oxidation reaction results in the production of excess As (reported already earlier [5-7, 11]), that is clearly visible in the form of precipitates near the oxidized layers (Figs. 3 a,b). Accumulation of As and shrinkage of the oxidized layer can lead to stress build up close to the oxidation front which sometimes results in the formation of defects such as stacking faults (Fig. 3a). The released As can diffuse through the surrounding layers towards the surface of the structure which can be deleterious to device applications [11].

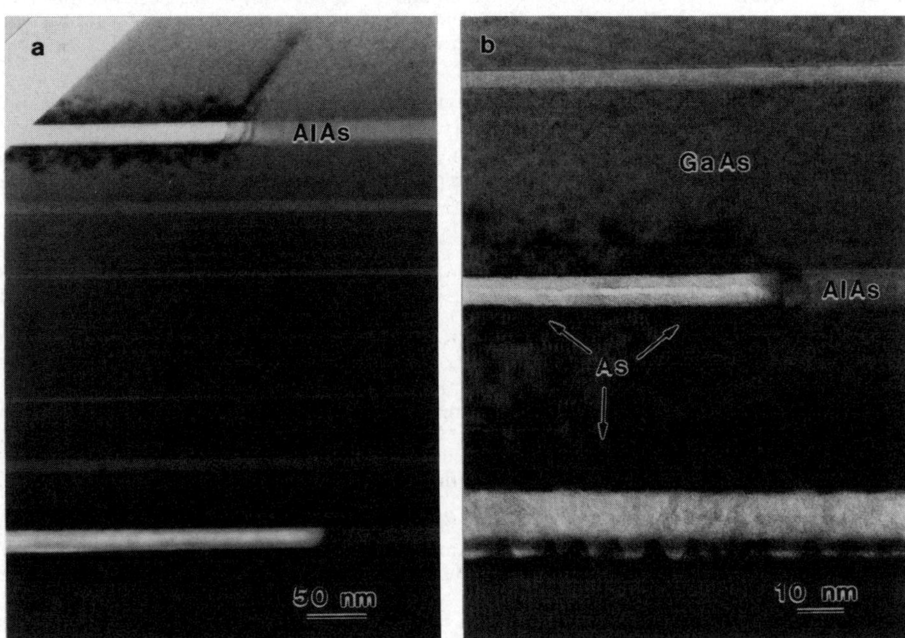

Fig. 3. Oxidation of thin AlAs layers (sample A). Arrangement of layers: 20 nm, 10 nm, 5 nm with 2.5 nm in the center. (a). Formation of a stacking fault at the inverted interface. Note typical shape of the oxidation front and formation of As precipitates in the GaAs close to the oxidized interfaces; (b) Oxidation front of 10 nm thick AlAs layer. Note formation of voids at the direct interface of the 20 nm thick AlAs layer and high density of As precipitates between these two oxidized layers.

The oxidation front has a specific asymmetric shape with the mo· ng front at the inverted (AlAs/GaAs) interface slightly ahead of that at the lower interface. This was true independent of the layer thickness. In areas close to the oxidation front no voids can be observed. However, further back from the oxidation front, voids are formed especially at the lower interface (Fig. 3b). This void formation was observed to occur at the direct interface (GaAs/AlAs) independent of the layer thickness (Fig. 3b). Such findings are important since formation of voids on these interfaces can weaken the structural integrity of devices and cause delamination.

D. Conclusions

Detailed structural characterization of wet oxidation of AlAs layers by transmission electron microscopy showed that dense γ-Al_2O_3 oxide was formed as the result of wet oxidation. A result of the oxidation is accumulation of As at the interfaces and in the GaAs surrounding the oxidized layers. The oxidation front is not symmetric showing faster motion at the inverted interface than at the direct interface. Clear dependence of oxidation rate on the AlAs layer thickness was observed. For a range of layer thicknesses a linear dependence of oxidation rate on thickness was observed. In areas some distance back from the advancing oxidation front voids were formed at the direct interfaces.

Acknowledgment

This work was supported by DARPA/ETO project. Electron microscopy was performed in the National Center for Electron Microscopy supported by the US Department of Energy under Contract No. DE-AC03-76FS00098.

References:

1.   A.R. Sugg, E.I. Cnen, N. Holonyak, Jr., K.C. Hsieh, J.E. Baker, and N. Finnegan, J. Appl. Phys., 74, 3880 (1993).
2.   G. S. Li, S. F. Lim, W. Yuen and C. J. Chang-Hasnain, Electronics Letters., Vol. 31, 23, pp. 2014-5, 1995.
3.   D. L. Huffaker, J. Shin, and D. G. Deppe, Electron. Lett., 30, 1946 (1994).
4.   K.L. Lear, K.D. Choquette, R.P. Schneider, JR., S.P. Kilcoyne, and K.M. Geib, Electron. Lett, 31, 208 (1995).
5.   Z. Liliental-Weber, M. Li, G.S. Li, C. Chang-Hasnain, and E.R. Weber, Proc. of the 9th Conference on Semiconducting and Insulating Materials (SIMC'9) , Toulouse, France, edt. C. Fontaine, (1996), p. 159.
6.   Z. Liliental-Weber, M. Li, G.S. Li, C. Chang-Hasnain, and E.R. Weber, "Proc. of 54th Annual Meeting Microscopy Society of America, edts. G.W. Bailey, J.M. Corbet, R.V.W. Dimlich, J.R. Michael, and N.J. Zaluzec, San Francisco Press, Inc, San Francisco (1996) p. 942
7.   S. Guha, F. Agahi, B. Pezeshi, J.A. Kash, D.W. Kisker, and N.A. Bojarczuk, Appl. Phys. Lett., 68, 906 (1996).
8.   D.L. Huffaker, D.G. Deppe, C. Lei, and L.A. Hodge, Appl. Phys. Lett. 68, 1948 (1996).
9.   C.I.H. Ashby, J.P. Sulivan, P.P. Newcomer, N. A. Missert, H. Q. Hou, B.E. Hammons, M.J. Hafich, and A.G. Baca, Appl. Phys. Lett. 70, 2443 (1997).
10.   D.L. Huffaker and D.G. Deppe, Appl. Phys. Lett. 71, 1449 (1997).
11.   Z. Liliental-Weber, S. Ruvimov, W. Swider, J. Washburn, M. Li, G.S. Li, and C. Chang-Hasnain, and E.R. Weber, "SPIE-The International Society for Optical Engineering Proceedings, **3006**. 15 (1997).
12.   O. Blum, C.I. Ashby, and H.Q. Hou, Appl. Phys. Lett., 70, 2870 (1997).
13.   P.W. Evans and N, Holonyak, Jr., Appl. Phys. Lett. 71, 261 (1997).

VII.

Group III-Nitrides

Doping and segregation effects in AlGaN systems

P. Bogusławski

Instytut Fizyki PAN, 02-668 Warsaw, Poland

J. Bernholc

North Carolina State University, Raleigh, NC 27695

Doping properties of substitutional C, Si, Ge, Be, and Se impurities in wurtzite GaN and AlN are studied by quantum molecular dynamics. We investigate effects that potentially quench doping efficiency, namely the self-compensation and the compensation by native defects. We also predict a strong interfacial segregation of Se_N, C_N, and nitrogen vacancies at the AlN/GaN heterointerface towards the GaN layer.

A. Introduction

Wide band-gap nitrides are of considerable interest due to applications in blue/UV light-emitting diodes and lasers, and in high-temperature electronics [1,2]. To exploit fully the potential of these materials understanding and control of technologically important issues still need to be achieved. In this paper we focus on two of them, compensation of doping and segregation effects. In $Al_xGa_{1-x}N$ alloys, doping efficiency of shallow impurities decreases with the increasing Al content because the shallow states become progressively deeper due to both the increasing effective mass and the decreasing dielectric constant. However, we will show below that there are other processes that may quench the doping efficiency in a more drastic way. The first one is self-compensation, *i. e.*, simultaneous incorporation of a group-IV dopant on both cation and anion sublattices, which will be analyzed here for C, Si, and Ge impurities in the wurtzite GaN and AlN [3-9]. Under appropriate growth conditions self-compensation is blocked by strain effects. However, in competition with the strain effects are processes of electron transfer from donors to acceptors, very efficient due to the wide band gap of nitrides; they enhance both self-compensation and the compensation by native defects. The latter issue was briefly discussed for Se [10]. Due to limited space the formation of nearest-neighbor donor-acceptor pairs contributing to self-compensation [11] will not be discussed here, and neither will the problem of the stability of DX states of Si and Ge, for which conflicting results were obtained [11-13]. In the second part of this paper we consider the effect of interfacial segregation, which leads to accumulation of dopants or defects in the GaN layer of AlN/GaN heterostructures. The driving force for segregation is found to be significantly stronger in nitrides than in SiGe systems. Details of our quantum molecular dynamics calculations are given in Ref. [11].

B. Doping and compensation of dopants

In GaN, both Si_{Ga} and Ge_{Ga} are found to be shallow effective-mass donors. In AlN, Si_{Al} is still shallow, but Ge_{Al} is a deep donor, with the energy level located at about 1 eV below the bottom of the conduction band. The localization of the wave function on Ge is accompanied by a large outward relaxation of the nearest N neighbors with a breathing-mode symmetry. On the other hand, both Si_N and Ge_N are deep acceptors, with the levels at about 1.3 eV in GaN and 2.0 eV in AlN. In both materials the acceptor levels are split by ~0.6 eV by the hexagonal crystal field. Therefore, the incorporation of Si and Ge on the N sites does not result in p-type samples; this situation is very different from that in, $e.g.$, GaAs, where both Si_{As} and Ge_{As} are shallow acceptors.

From the above results it follows that the self-compensation of both Si and Ge can occur. However, the degree of self-compensation largely depends on the conditions of growth. Formally, this dependence is given by the formation energy E_{form} [14]

$$E_{form}(q) = E_{tot}(q) - n_{Ga}\,\mu_{Ga} - n_N\,\mu_N - \mu_{impurity} + q\,E_F ,\qquad (1)$$

where E_{tot} is the total energy of the supercell with the impurity, n_{Ga} and n_N are the numbers of Ga and N atoms in the supercell, μ is the chemical potential, q is the charge state of the impurity, and E_F is the Fermi energy. The chemical potentials depend on the source of atoms involved in the process, and therefore on the actual experimental situation. Calculated formation energies are given in Table 1 for both Ga-rich and N-rich growth conditions.

Table 1. Calculated formation energies (in eV) of neutral substitutional impurities in GaN at Ga-rich and N-rich conditions.

dopant	Ga-rich	N-rich
Si_{Ga}	0.9	1.4
Si_N	3.0	6.9
Ge_{Ga}	2.3	2.2
Ge_N	3.1	6.4
C_{Ga}	5.7	4.0
C_N	1.1	2.8

It follows from Table 1 that both Si and Ge are preferentially incorporated on the Ga sublattice. This can be explained by the presence of residual strain when the mismatch between the host and the impurity atoms is large. This is the case for Si or Ge substituting for N, which leads to a the strain energy of the order of a few eV [15]. We also notice that the difference between the formation energies of X_{Ga} and X_N (X=Si or Ge) is higher in the N-rich limit, since under these conditions the incorporation on the Ga sites is easier. However, electron transfer between the dopants and the Fermi level may overcome the strain effects. In the case of charged impurities, formation energies are in general reduced from the values of Table 1 according to Eq. (1). Since the energy gain upon electron transfer from a donor to an acceptor is of the order of the band gap, the formation of charged defects is of particular importance in wide band gap materials. For the dopants considered here, the electron transfer effects render formation energies of X_{Ga} and X_N close to each other, and self-compensation becomes possible. However, the degree of self-compensation strongly depends on the conditions of growth, and may vary from none to total self-compensation. From Table 1 it follows that in the N-rich conditions the difference of formation energies between X_{Ga} and X_N exceeds the value of the band gap. (This also holds in AlN [11].) Therefore, under these conditions, the energy gain due to charge transfer cannot overcome strain-driven effects, and the self-compensation is negligible. In contrast, for the case of GaN:Ge under Ga-rich conditions, a value of the Fermi level exists at which $E_{form}(Ge_{Ga}^{+}) = E_{form}(Ge_N^{-})$, which should lead to equal concentrations of Ge_{Ga}^{+} and Ge_N^{-}. This implies that doping with Ge should lead to self-compensation and pinning of the Fermi level. For Si in GaN, however, self-compensation effects are not expected except in the very Ga-rich limit.

Thus far, we considered the possibility of self-compensation of a dopant by itself. A second possible mechanism acting to reduce the doping efficiency is compensation by native defects. As was discussed in our previous paper [16], the relatively high formation energies of native defects in the neutral charge state are greatly reduced for charged defects by electron transfer effects. In particular, the efficient doping by both Si and Ge donors predicted for the N-rich conditions of growth is compensated by the dominant native acceptors, which are cation vacancies [16]. We find that the calculated solubility limit of Si at 800 °C exceeds 10^{20} cm^{-3}, but the degree of compensation by V_{Ga} is high, because formation of cation vacancies is an efficient process under N-rich conditions. We also notice that the energy gains induced by electron transfer increase with the increasing band gap, and therefore the degree of compensation by native defects in $Al_xGa_{1-x}N$ alloy should increase with increasing Al content. This effect may be partially responsible for the observed difficulties with *n*-doping of Al-rich AlGaN [6,17].

A high degree of compensation by Ga vacancies in GaN was recently demonstrated for Se_N donors [10]. We note that in this case the amount of compensation, determined by the concentration of V_{Ga}, may be increased by the formation of $Se_N^{+}-V_{Ga}^{-}$ nearest-neighbor complexes. This effect is driven by the very high residual strain energy around Se atom substituting for the much smaller N. According to our calculations [15], the strain energy is about 6 eV. A part of this energy is released after a removal of one of the Ga nearest neighbors, which results in the high value of the calculated binding energy of a $Se_N^{+}-V_{Ga}^{-}$ pair, 1.7 eV. The experimentally determined value of E_{form} of the compensating defect is 0.9 eV, which is close to our theoretical value for V_{Ga}^{3-} , 0.5 eV. Since the compensation by native defects is independent on the doping species, one should expect similar effects for other donors.

We end this Section by considering *p*-type doping. From Table 1 it follows that C preferentially substitutes for N (except for the N-rich limit in AlN, where the formation energies of C_{Al} and C_N are comparable). Under the cation-rich conditions of growth, the incorporation on the cation sublattice is very weak, and thus self-compensation is not expected. However, C_N may be compensated by dominant native donors, which are N vacancies and/or Ga interstitials [16,18]. According to our calculations, C_N is a shallow acceptor, with the level at 0.2 eV above the top of the valence band in GaN, and somewhat deeper in AlN. Thus C_N is slightly more shallow, by about 20 meV, than the commonly used Mg [7]. However, according to our calculations Be should be even even more shallow.

C. Interfacial segregation

Thus far, we have considered GaN and AlN with a uniform spatial distribution of dopants. We now turn to the AlN/GaN semiconductor heterostructures, where the equilibrium concentration of a given dopant in the two constituents is in general different. In such cases one may expect the effect of interfacial segregation, *i.e.*, the diffusion of dopant atoms across the heterointerface, directed towards the constituent where the chemical potential of the dopant atoms is lower. The effect of interfacial segregation was predicted by Hu [19]. For example, in SiGe/Si systems doped with B, the measured concentration of B is about twice higher in the SiGe layer than in Si [20]. This leads to the conclusion that the total energy of B in SiGe is lower by 0.3 eV than in Si. Comparable concentration differences are measured for P and As [21].

In Hu's approach [19], the chemical potential of a dopant is given by two contributions; the first one corresponds to the insertion of a neutral impurity into the crystal, and the second one is due to the ionization of impurities and redistribution of carriers in the heterostructure, which minimizes the free energy. The first contribution is given by

$$\mu_{imp} = E_{form} + k_BT \ln(N_{imp}/N_{site}) + E_{strain} , \qquad (2)$$

where the first term is the formation energy defined in Eq. 1, the second term is the configurational entropy expressed by N_{imp} and N_{site}, the concentrations of impurities and of the lattice sites accessible to the impurity (*e. g.*, the concentration of Ga sites in the case of Si_{Ga} in GaN), respectively, and E_{strain} is the strain energy induced by the dopants. In the case of a high doping level, and under the assumption of the local charge neutrality, the electronic contribution is

$$\mu_{el} = E_c - E_{imp} + k_BT \ln(N_{imp}/N_c) , \qquad (3)$$

where E_c and E_{imp} are the energies of the bottom of the conduction band and of the impurity, and N_c is the density of states in the conduction band, respectively. At equilibrium, the concentrations on both sides of the interface are determined by the condition of equal chemical potential $\mu = \mu_{imp} + \mu_{el}$.

In this work, we analyze the interfacial segregation in a short-period $(AlN)_2(GaN)_2$ superlattice for three defects, the nitrogen vacancy, C_N acceptor, and the Se_N donor. To this end, we have performed calculations of the total energy of the system placing the defect either in the GaN or in the AlN layer. The calculated energy difference, which is the segregation energy, is a very good approximation of the difference of the chemical potentials $\Delta\mu = \mu(GaN) - \mu(AlN)$. This can be justified by the following arguments. First, the difference between the number of sites in AlN and GaN is very small and leads to negligible contributions at typical growth temperatures. Second, to minimize the role of strain we assume a free-standing superlattice, with the lateral lattice constant a equal to the average of

a(AlN) and a(GaN). The neglected strain effects provide minor corrections of about 0.1 eV to the obtained values. Third, the calculations include automatically the electronic contribution (Eq. 3) in a self-consistent way, accounting for charge transfer and screening effects. Therefore, in the case under consideration the driving force for the segregation is the difference of formation energies. It depends mainly on the nearest neighbors of the defect, and is reliably calculated in spite of the very short period of the superlattice.

We find that Se_N, C_N, and the nitrogen vacancy should segregate towards the GaN layer, and the calculated segregation energies are 0.95, 1.2, and 2.4 eV, respectively. These values are much larger than that of B in the SiGe/Si system. Therefore, the equilibrium concentrations of these defects at typical growth temperatures should be significantly lower in AlN layers than in GaN layers in AlN/GaN heterostructures. However, interfacial segregation will only occur if the diffusion rate of vacancies in the nitrides is sufficiently fast. In this context one should note the recent work of Ambacher *et al.* [22], which demonstrated that considerable self-diffusion of nitrogen in GaN occurs at temperatures of 780-980 °C. Since diffusion in semiconductors is usually mediated by native defects, it is likely that the observed self-diffusion of N is mediated by nitrogen vacancies. The measured diffusion lengths are sufficient to enable interface segregation of vacancies in short-period superlattices. Finally, we stress that the considered dopants, as well as N vacancies (being shallow donors in GaN [16,23] and deep ones in AlN [24]) are electrically active, and therefore their redistribution can drastically affect the electronic properties of heterostructures.

D. Summary

In summary, we have investigated the doping properties of C, Si, Ge, Se, and Be in wurtzite GaN and AlN by quantum molecular dynamics. Incorporation of group-IV atoms on both cation and N sublattices was considered. We have analyzed effects that could potentially limit the doping efficiency, namely self-compensation and compensation by native defects. The strain-driven preference to substitute for the host atom with a similar size may be overcome by the energy gain due to electron transfer between a donor and an acceptor. In $Al_xGa_{1-x}N$ alloys with high Ga content and grown in the N-rich conditions, both Si and Ge are excellent effective-mass donors, and neither self-compensation nor the formation of nearest neighbor pairs is expected. C_N is an excellent acceptor, as it is somewhat more shallow than the commonly used Mg [1,7,16], and it should have a significantly greater solubility than atomic Mg. Be may be even more shallow than C_N. With the increasing Al content the compensation by native defects becomes more important. We have also considered the interfacial segregation at the AlN/GaN heterointerface; after a reformulaiton of the work of Hu in modern terms we have performed the calculations of the segregation energies for two donors, the nitrogen vacancy and Se_N. The values obtained for AlN/GaN are much higher than these measured for SiGe/Si systems, we thus expect that segregation may play an important role in the nitrides.

Acknowledgements This work was supported in part by Grants ONR N00014-92-J-1477, NSF DMR 9408437, and KBN 2 P03B 093 14.

REFERENCES

[1]. R. F. Davis, Physica B185 (1993), 1.

[2]. H. Morkoc *et al.*, J. Appl. Phys. 76 (1994), 1363.

[3] S. Nakamura, T. Mukai, and M. Senoh, Jpn. J. Appl. Phys. 31 (1992), 2883.

[4] L. B. Rowland, K. Doverspike, D. K. Gaskill, Appl. Phys. Lett. 66 (1996), 1495.

[5] X. Zhang, P. Kung, A. Saxler, D. Walker, T.C. Wang, and M. Razeghi, Appl. Phys. Lett. 67 (1995), 1745.

[6] M. Bremser *et al.* , MRS Internet J. Nitride Semicond. Res. 1 (1996), 8.

[7] S. Fisher, C. Wetzel, E. E. Haller, and B. K. Meyer, Appl. Phys. Lett. 67, 1298 (1995).

[8] C. R. Abernathy, J. D. MacKenzie, S. J. Parton, and W. S. Hobson, Appl. Phys. Lett. 66, 1969 (1995).

[9] G.-C. Yi and B. Wessels, Appl. Phys. Lett. 70 (1997), 357.

[10] G.-C. Yi and B. Wessels, Appl. Phys. Lett. 69 (1996), 3028.

[11] P. Bogusławski and J. Bernholc, Phys. Rev. B 56 (1997), 9496.

[12] C. H. Park and D. J. Chadi, Phys. Rev. B 55 (1997), 12995.

[13] C. Van de Walle, Phys. Rev. B 57 (1998), R2033.

[14] C. G. Van De Walle, D.B. Laks, G. F. Neumark, and S. T. Pantelides, Phys. Rev. B 47 (1993), 9425.

[15] In cases when the mismatch between the atomic radii of impurity and the host atom is large (Si_N,Se_N, etc.), the atomic relaxation from the ideal substitutional configuration to the final one induces energy gains of about 5 eV, changes the bond lengths by about 15%, and shifts deep impurity-induced levels by about 1 eV. We observe that even after the relaxation, *i. e.*, at equilibrium, bonds around an impurity remain strained, which induces a residual strain energy. Simple arguments show that the strain energy is comparable to the relaxation energy [11]. Put somewhat differently, only half of the initial strain is released during the relaxation.

[16] P. Bogusławski, E. L. Briggs, and J. Bernholc, Phys. Rev. B 51 (1995), 17255.

[17] C. Stampfl and C. Van de Walle, Appl. Phys. Lett. 72, 459 (1998).

[18] P. Perlin, T. Suski, H. Teisseyre, M. Leszczynski, I. Grzegory, J. Jun, S. Porowski, P. Bogusławski, J. Bernholc, J. C. Chervin, A. Polian, and T. D. Moustakas, Phys. Rev. Lett. 75, 296 (1995).

[19] S. M. Hu, Phys. Rev. Lett. 63 (1989), 2492; Phys. Rev. B 45 (1992), 4489.

[20] N. Moriya, L. C. Feldman, S. W. Downey, S. A. King, and A. B. Emerson, Phys. Rev. Lett. 75 (1995), 1981.

[21] S. M. Hu, D. C. Ahlgren, P. A. Ronsheim, and J. O. Chu, Phys. Rev. Lett. 67 (1991), 1450.

[22] O. Ambacher, F. Freudenberg, R. Dimitrov, H. Angerer, and M. Stutzmann, Jpn. J. Appl. Phys. 1998, in print.

[23] D. C. Look, D. C. Reynolds, J. W. Hemsky, F. R. Sizelove, R. J. Jones, R. J. Molnar, Phys. Rev. Lett. 79, 2273 (1997).

[24] I. Gorczyca, N. E. Christensen, and A. Svane, Acta Phys. Polonica A92 (1997), 785.

Optical Properties of GaInN/GaN Heterostructures and Quantum Wells

C. Wetzel, T. Takeuchi, S. Nitta, S. Yamaguchi, H. Amano, and I. Akasaki
High-Tech Research Center & Dept. of Electrical and Electronic Engineering,
Meijo University, Nagoya, Japan

Photoreflection and photoluminescence spectroscopies have been used to identify details of the electronic bandstructure in GaInN/GaN strained heterostructures and multiple quantum well structures. Franz-Keldysh oscillations in the ternary layers are identified in both systems revealing large piezoelectric fields of 240 kV/cm (x=0.079, thin film) and 0.65 MV/cm (x=0.187). From spatially resolved luminescence a very narrow distribution ΔE =28 meV of the bandgap energy is derived (x=0.187). From the variation of the field with strain a piezoelectric coefficient $dP_z/d\varepsilon_{zz}$=0.3 C/m^2 is obtained corresponding to e_{14} = 0.1 C/m^2 and an equilibrium polarization of GaN of P_{eq} =43 mCm^2 is extrapolated in qualitative agreement with recent calculations.

A. Introduction

Recent dramatic progress in alloys of group-III nitrides has led to high performance new devices such as short wave light emitting devices and high temperature and highest frequency field effect transistors [1]. Moreover, further progress is expected once the properties controlling the electronic bandstructure in thin films and quantum well structures is identified. The alloy of $Al_yGa_{1-x-y}In_xN$ covers both a record in variation of the bandgap energy and in lattice constants. This advantage in the capability for bandgap engineering should therefore be coupled to challenges in stress design on heterostructures. Here we present the results of our recent studies in GaInN/GaN single and quantum well heterostructures.

B. Experimental

Samples in this study were prepared by metal organic vapor phase epitaxy on (0001) sapphire. Single 40 nm $Ga_{1-x}In_xN$/GaN heterostructures at various composition x ($0<x<0.2$) and multiple quantum well structures consisting of five 3 nm $Ga_{1-x}In_xN$ wells embedded in 6 nm GaN barriers were studied by photoluminescence (PL) and photoreflection (PR) spectroscopy (T=300 K). Composition and stress conditions were analyzed by high resolution x-ray diffraction mapping of a and c lattice constants. A Xe-lamp white light source was used for PR measurements. Photomodulation was performed by a HeCd 325 nm laser. The same system was used for PL. Further details can be found in Ref. [2] .

C. Single GaInN/GaN heterostructures

C.1. Stress and composition

The growth mode of this strongly lattice mismatched system GaInN/GaN was identified by x-ray mapping revealing pseudomorphic growth of the ternary layer at the a lattice constant of GaN. Despite the large thickness (40 nm) and the large lattice mismatch ($\Delta a/a$=1.1% for x=0.1) films were homogeneously strained under biaxial compression [3]. This result can not be explained by the current models of critical layer thicknesses in strained systems. This unexpected behavior strongly affects the interpretation of the In composition x from x-ray data. Assuming Vegard's law for the interpolation of the stress-free lattice

    239

constant and linearly interpolated bulk moduli we find a relation between experimental c-lattice constant and x in the pseudomorphic system:

$$x = 1.165\,(c/\text{Å} - 5.184)\,[1 - 0.16\,(c/\text{Å} - 5.184)] - 0.01;\quad a = 3.182\,\text{Å}. \tag{1}$$

Herein, lattice constants $c = 5.184$ Å (5.705 Å), $a = 3.188$ Å (3.540 Å) and elastic constants (in 10^{11} dyn/cm^2) $c_{13}=11.4$ (9.4), $c_{33}=38.1$ (20.0) in GaN (InN), respectively are used [4]. The constant offset originates in the stressed condition of the GaN reference layer. Consequently compositions derived without such a consideration overestimate the actual composition by roughly a factor of two. This may have contributed to discrepancies in the interpretation of results in the literature [5].

C.2. Luminescence bandgap in GaInN

Photoluminescence together with PR of a GaInN/GaN single heterostructure (x=0.18) is presented in Fig. 1. Luminescence was excited with a spatial resolution of 1 μm^2 and mapped over an area of 50x50 μm^2. The spatial distribution of the peak center energy if given in Fig. 1.a). We find a center energy of 2.612 eV and a distribution with a halfwidth of only 25 meV. Weighing each peak energy position with the luminescence intensity we obtain the spectrum in Fig. 1.b) identifying the dominant emission energy at 2.620 eV at a distribution of 29 meV. The very good coincidence of both distributions indicates a very homogeneous distribution of the effective luminescence bandgap on the length scale relevant for optical emission. At the same time the PR spectrum (Fig. 1.c) shows a very broad oscillation in energy without any narrow (10-20 meV) features typically indicating the presence of excitons. In this data we furthermore find no evidence for a significant variation of the In composition at x=0.18. It has been proposed that apparent discrepancies in the expected transition energies are caused by the formation of In-rich islands or dots [6].

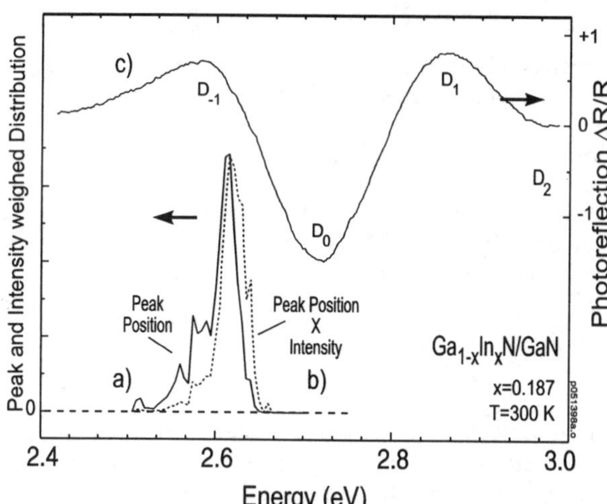

Figure 1. Distribution functions of the maximum in spatially resolved photoluminescence. (a) Distribution of peak maximum. (b) Distribution of peak maximum weighed by its intensity. (c) Photoreflectance for comparison showing only Franz-Keldysh oscillations.

C.3. Piezoelectric field in GaInN/GaN

We assign the wide oscillation in the PR signal to Franz-Keldysh oscillations (FKO's) caused by an electric field across the ternary layer [7,8]. Another PR spectrum for x=0.079 is given in Fig. 2.a). A clear minimum D_0 is seen in the vicinity of the GaInN bandgap. We recently used this to study the composition dependence of the bandgap energy [9]. Characteristic subsidiary oscillations (D_1... D_3) can be identified on the high energy side (see also Fig. 1.c), x=0.18). An interpretation of the period according to

Aspnes [10] leads to a field value of 260 kV/cm (x=0.079) and a very high field of 0.65 MV/cm for x=0.187. We propose that this field is caused by piezoelectricity and strain in the non-centro-symmetric wurtzite structure. The polarization shows a trend to increase with composition or strain at an averaged rate of $dP_z/d\varepsilon_{zz}$=0.3 C/m^2 corresponding to e_{14} = 0.1 C/m^2 interpreting a set of 13 samples [11]. Considerable spread of the data indicates the presence of an effective field relaxation mechanism. Besides screening by mobile carriers domains of inverted crystal polarity may play a role. An extrapolation of the data indicates that the piezoelectric polarization should not vanish for vanishing strain but exhibit an offset corresponding to $|P_{eq}|$=43 mC/m^2 of opposite direction. Such a behavior in GaN has recently been predicted in first principles calculations [12]. Values of $dP_z/d\varepsilon_{zz}$ and P_{eq} found here however are only about 10% of the values predicted. The PR signal of the GaN layer underneath the ternary film does only reveal a third-derivative-like structure of the excitonic nature of the bandgap. We find no indications of the presence of an electric field across the GaN layer. This may be the result of an effective overall screening of the net field across the sample by charges and reconstruction on the surfaces to maintain the thermodynamic equilibrium. In turn no features indicating the dominance of excitons are observed in the PR signal of the GaInN layer. Applying the mechanism of Frenkel-Poole ionization excitons should be dissociated already at fields of 100 kV/cm.

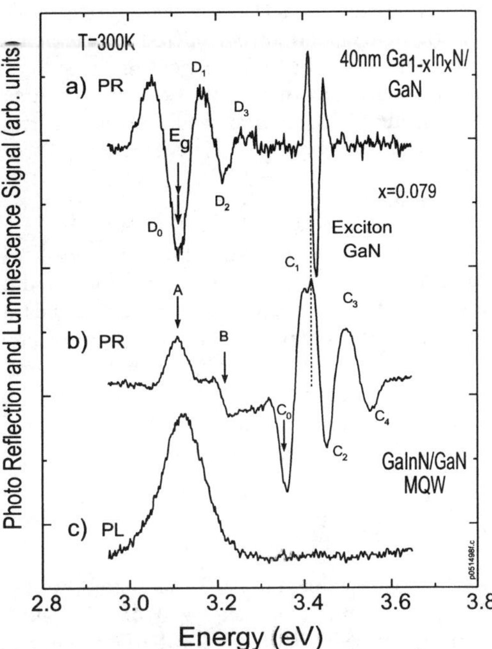

Figure 2. Photoreflectance a),b) and photoluminescence c) spectra of a GaInN/GaN single heterostructure a) and a quantum well structure b),c). Clear Franz-Keldysh oscillations are seen in both structures a): D_0 ... D_3, b) C_0 ... C_4. Luminescence in corresponds to a clear feature in the PR spectrum.

D. GaInN/GaN multiple quantum well structures

PR of a multiple quantum well structure is shown in Fig. 2.b) [2]. We again identify excitonic contributions in the vicinity of the GaN excitonic bandgap of the buffer layers. In addition, however, there is an oscillatory contribution (peaks C_0 ... C_4) in the energy range above and below the GaN bandgap starting sharply at E = 3.33 eV in this structure. There is a strong resemblance with FKO's in the non-quantized thin films. An equivalent interpretation of the period results in an electric field value of 240 kV/cm. All extrema values excluding the ones associated with excitons in GaN are very well described using the approximate description or the electrooptical functions as given by Aspnes [10]. The mere superposition of the GaN signal rather than incorporation of it indicates that this field again does not apply to either the GaN epilayer, the barrier layers or both. Instead in these higher states of the GaInN well the effect of FKO's dominate over the two-dimensional system. As a consequence the electric field can not be considered a weak perturbation in these states. At lower energies two more transitions A and B are identified (see Fig. 1.b). Level A very closely corresponds to the luminescence maximum (Fig. 1.c)) and is tentatively assigned to the lowest quantized state in the quantum well system, while level B corresponds to the first excited state. A detailed identification of both levels is underway.

E. Summary

In summary we have performed a spectroscopic analysis of the electronic bandstructure in $Ga_{1-x}In_xN/GaN$ thin films and multiple quantum well structures. A spatially resolved photoluminescence mapping reveals a very homogenous distribution of the InN fraction on the length scale of 1 to 50 μm resulting in a luminescence peak distribution of 25 meV (x=0.18). Franz-Keldysh oscillations were observed in both structure types identifying electric fields in the range of 260 kV/cm (x=0.078) and similar values in the quantum well case. In the case of a thin film sample of x=0.187 a maximum field of 0.65 MV/cm is observed. An average strain derivative $dP_z/d\varepsilon_{zz}$=0.3 C/m^2 corresponding to $e_{14} = 0.1$ C/m^2 is observed.

Acknowledgement

This work was partly supported by the Ministry of Education, Science, Sports and Culture of Japan (contract nos.09450133 and 09875083, and High-Tech Research Center Project) and JSPS Research for the Future Program in the Area of Atomic Scale Surface and Interface Dynamics under the project of Dynamic Process and Control of the Buffer Layer at the Interface in a Highly-Mismatched System.

References

[1]*For a recent review see* I. Akasaki and H. Amano, Jpn. J. Appl. Phys. **36** (1997) 5393.
[2]C. Wetzel, T. Takeuchi, H. Amano, and I. Akasaki, J. of Crystal Growth August 1, (1998).
[3]H. Amano, T. Takeuchi, S. Sota, H. Sakai, and I. Akasaki, in "III-V Nitrides" vol. 449 (Mater. Res. Soc. 1997).
[4]T. Takeuchi, S. Sota, M. Katsuragawa, M. Komori, H. Takeuchi, H. Amano, and I. Akasaki, Jpn. J. Appl. Phys. **36** (1997) L 382.
[5]S. Nakamura, N. Iwasa, and S. Nagahama, Jpn. J. Appl. Phys. **32** (1993) L 338; S. Nakamura, J. Vac. Sci. Technol. **A13** (1995) 705.
[6] S. Chichibu, T. Azuhata, T. Sota, and S. Nakamura, Appl. Phys. Lett. **70** (1997) 2822.
[7]C. Wetzel, H. Amano, I. Akasaki, T. Suski, J.W. Ager, E.R. Weber, E.E. Haller, and B.K. Meyer, in "Nitride Semiconductors", vol. 482 (Mater. Res. Soc. 1998). *in print*
[8]C. Wetzel, T. Takeuchi, H. Amano, and I. Akasaki, *submitted to* Appl. Phys. Lett. 1998.
[9] C. Wetzel, T. Takeuchi, S. Yamaguchi, H. Katoh, H. Amano, and I. Akasaki, *submitted to* Appl. Phys. Lett. (1998).
[10] D.E. Aspnes, Phys. Rev. B **10** (1974) 4228; Phys. Rev. **153** (1967) 972.
[11] C. Wetzel, T. Takeuchi, H. Amano, and I. Akasaki, in "Wide-Bandgap Semiconductors for High Power, High Frequency and High Temperature" (Mater. Res. Soc. 1998). *in print*
[12] F. Bernardini, V. Fiorentini, and D. Vanderbilt, Phys. Rev. Lett. **79** (1997) 3958.

Optical processes in In$_x$Ga$_{1-x}$N epitaxial films grown by metalorganic chemical vapor deposition

W. Shan, J.W. Ager III, W. Walukiewicz, and E.E. Haller

Materials Sciences Division, Lawrence Berkeley National Laboratory, Berkeley, CA 94720

Abstract

We have studied the properties of optical transitions in In$_x$Ga$_{1-x}$N epitaxial films (0<x<0.2) grown by metalorganic chemical vapor deposition (MOCVD). The fundamental band gap energies of the In$_x$Ga$_{1-x}$N alloys were determined using photomodulation spectroscopy measurements and the variation of the fundamental band gap was measured as a function of temperature. The effects of pressure on the photoluminescence (PL) emission lines were studied. Our results show that PL originates from effective-mass conduction-band states. The anomalous temperature dependence of the PL peak shift and linewidth and of the Stokes shift between the photoreflectance (PR) and PL lines is explained by composition fluctuations in as-grown InGaN alloys.

A. Introduction

The In$_x$Ga$_{1-x}$N alloy system has recently attracted much attention because of its importance in both scientific and technological aspects. In the course of studying the bulk optical properties of In$_x$Ga$_{1-x}$N films, an anomalous temperature dependence of the PL peak shift and linewidth and a very large Stokes shift between PR and PL lines have often been observed in this material system. However, apparently there is no model that can account for all the spectral characteristics of the luminescence emission from InGaN alloys. In this work, we show that the observed peculiarities of the optical spectra of InGaN alloys can be explained consistently by large fluctuations of the In composition without the need to invoke highly localized states or strain induced piezoelectric effects.

B. Experimental

Samples used in this work were nominally undoped single crystal epitaxial films grown by MOCVD on thick GaN layers (~1-2μm) using sapphire substrates. The thickness of the In$_x$Ga$_{1-x}$N epitaxial layers ranges from several hundred to a few thousand Å. The alloy compositions were determined by X-ray diffraction (XRD) and Rutherford Backscattering Spectrometry (RBS). PL and PR measurements were performed over a temperature range from 3 K up to 295 K. Pressure-dependent measurements were carried out using gasketed diamond anvil cells.

C. Results and discussion

Fig.1 shows the optical transition energies measured by PR and PL for the samples used in this work. as a function of the InN mole fraction. The inset of Fig.1 is a comparison of PR and PL spectra taken at 295 K from an In$_{0.08}$Ga$_{0.92}$N sample. The dashed line in the figure is a least-squares fit to PR data using the band-gap energy equation for an alloy

$$E(x)=E(0)+ax+bx^2. \tag{1}$$

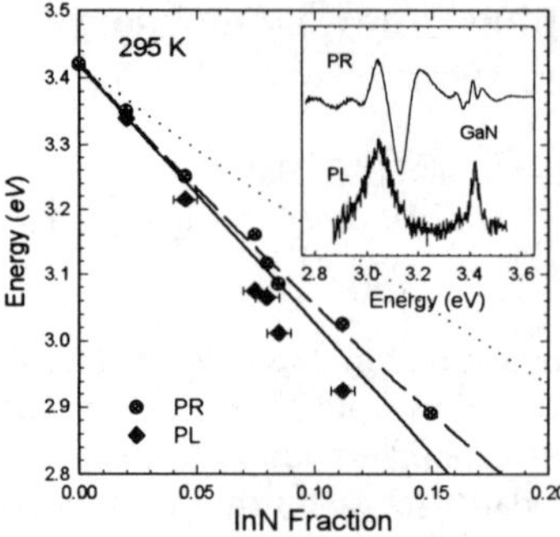

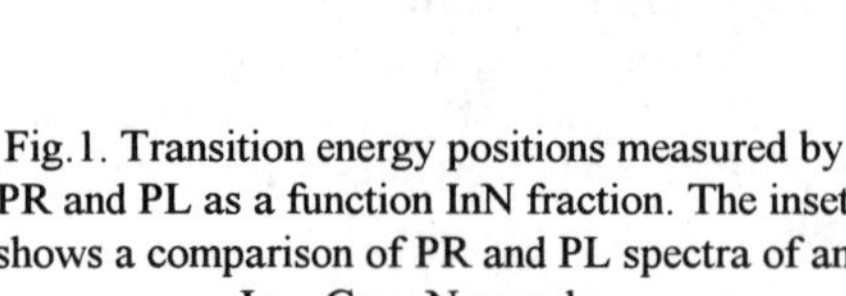

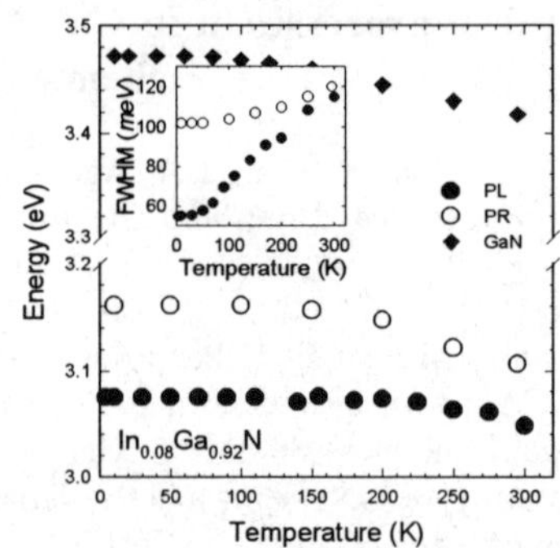

Fig.1. Transition energy positions measured by PR and PL as a function InN fraction. The inset shows a comparison of PR and PL spectra of an In$_{0.08}$Ga$_{0.92}$N sample.

Fig.2. The change of transition energies in the In$_{0.08}$Ga$_{0.92}$N measured by PR and PL with temperature. The PL results of GaN are shown for comparison. The inset shows the variation of PR and PL linewidths for the sample.

A best fit yields a = -3.91 and b = 2.39 eV. A linear relation between band gap and In concentration for pseudomorphically strained In$_x$Ga$_{1-x}$N ($x \leq 0.12$) recently derived by McCluskey *et al.*[1] (solid line) and the theoretical predication by Wright and Nelson[2] (dotted line) are also shown in the figure. The band-gap bowing parameter for the InGaN alloy system is much larger than the theoretically predicted value (0.9 eV)[2].

The most striking features of the optical spectra of InGaN alloys, as shown in the inset in Fig.1, are very large Stokes shifts between PL and PR transition energies and the large linewidths. The temperature dependencies of the energy positions of PL line and PR transition for the In$_{0.08}$Ga$_{0.92}$N sample are shown in Fig. 2. The position of the PL peak is practically independent of temperature from 10 to 200 K. By contrast, the PR line shows the typical behavior of the temperature dependence of the energy gap. The inset of the figure shows that there are also quite significant differences in the temperature dependencies of the PL and PR spectral linewidths. The lineshape of the PL spectrum is strongly temperature dependent. We also note a very distinct correlation between the temperature dependence of the PL linewidth and the PL intensity. At temperatures below 50 K, both the linewidth and the intensity are almost constant. At higher temperatures the gradually increasing linewidth is accompanied by a dramatic decrease in the intensity.

It has been proposed recently that piezoelectric fields induced by built-in strain are responsible for large PL linewidths observed in InGaN/GaN QW structures and in In$_x$Ga$_{1-x}$N epilayers[3]. To examine the importance of the piezoelectric fields in our samples we have measured the dependence of PL spectra on the excitation intensity. It was expected that the screening of the internal piezoelectric fields by the photoexcited carriers would result in a blue shift of the PL peak. We have found that changing of the excitation intensity by three orders of magnitude does not produce any noticeable shift of the PL line. This result demonstrates that the piezoelectric effects do not play any significant role in our samples.

The very weak temperature dependence of the PL peak position was also reported by Chichibu *et al*[4]. They assigned the observed luminescence emission to bound, localized states in their InGaN samples. It is well known that localized states have pressure dependencies which are smaller, in most instances, than the band edge states. However, our pressure-dependent PL results, as shown in Fig.3, demonstrate that the shift of the luminescence peak positions with pressure follow the direct band gap of the $In_xGa_{1-x}N$ epilayers[5]. This indicates that the electronic states involved in the recombination processes contributing to the PL emissions are associated with the conduction and valence band edges and can be described within the framework of the effective mass approximation. Therefore, no highly localized states are involved in the PL process.

Several recent studies have suggested that a strong immiscibility of InN in a nitride ternary alloy leads to a phase separation resulting in much larger than random compositional inhomogeneities[6,7]. The existence of such inhomogeneities can indeed provide a basis for a qualitative explanation of the optical spectra in InGaN alloys. Positions of the conduction and the valence band edges

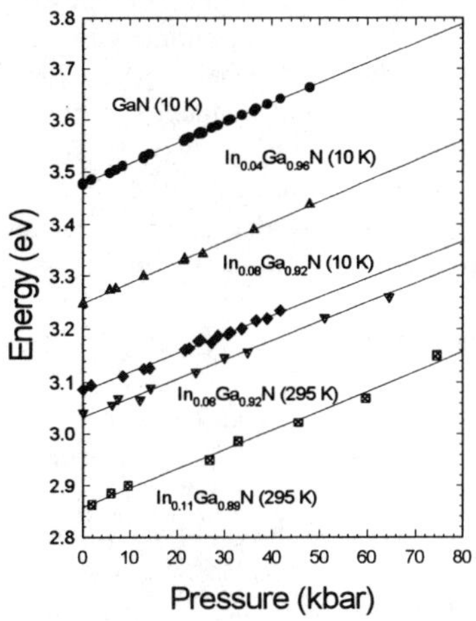

Fig. 3. Shift of PL peak positions for $In_xGa_{1-x}N$ samples and GaN with pressure.

in InGaN with spatial variations of the alloy composition are schematically shown in Fig. 4. Because of the large, $\Delta E_C/\Delta E_V = 5$ ratio of the conduction to the valence band offsets, most of the change in the band gap is accommodated by the variation in the conduction band edge energy[8]. Also, since as-grown InGaN is always n-type, we expect a significant density of states below the conduction band edge. The effective mass shallow donor states are delocalized and therefore they closely follow the conduction band edge. Within this picture we can qualitatively understand the origin of the large Stokes shift of the PL line. In a PL process photoexcited electrons relax by diffusing to the lowest available energy state. Then they recombine with holes in the valence band. As is shown in Fig. 4 we assume that the low energy states are shallow impurity states associated with local minima in the conduction band. By definition the local minima are mostly located below the average conduction band edge that corresponds to the transition energy for the PR spectral feature. Therefore one expects that the PL linewidth will be approximately equal to half of the PR linewidth.

So far we have assumed that the nonradiative lifetime is long enough so that electrons can diffuse to the local minima before recombining with holes. This is true at low temperatures when the PL lifetimes exceeding a few hundreds of picoseconds (ps) are characteristic for these materials[9,10]. However, shortening of the nonradiative lifetime with increasing temperature changes the kinetics of the PL process. The diffusion length of photoexcited carriers is reduced and they are no longer able to diffuse to a local minimum. Thus more and more electrons with energies exceeding the average conduction band edge contribute to the PL signal. This obviously leads to a broadening and a shift to higher energy of the PL spectrum. In the limit of very short lifetime, the PL linewidth should approach that of PR spectrum. The results in Fig. 2 show that this is exactly what is observed in our samples.

It is now quite easy to understand the differences in the temperature dependencies of the PR and PL energy positions shown in Fig. 2. The PR energy position is defined by the average energy gap that has a very similar temperature dependence to an energy gap in a homogenous material such as GaN. On the other hand, as we have argued above, the PL peak position shifts to higher energy with increasing temperature. In the temperature range T<200K this shift compensates for a downward shift of the band gap resulting in a temperature independence of the PL peak position. A small downward shift of the PL peak position is observed at T> 200 K when the band gap becomes more strongly dependent on temperature.

D. Conclusions

Pressure and temperature dependencies of the optical properties of $In_xGa_{1-x}N$ alloys have been studied using

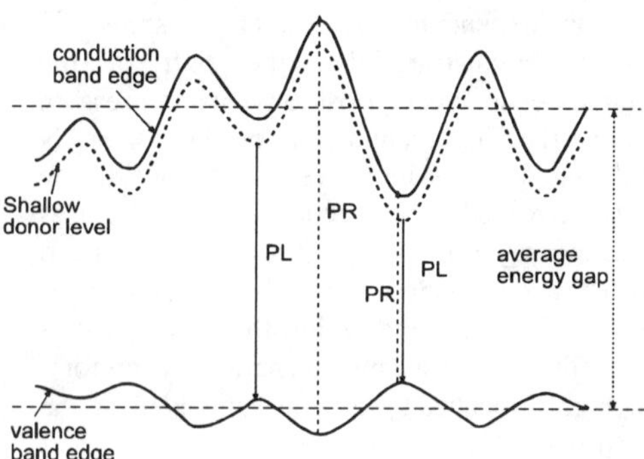

Fig. 4. Schematic representation of PL and PR in inhomogeneous InGaN alloys.

photoluminescence measurements and photomodulation spectroscopy. The highly sensitive photomodulation spectroscopy allowed us to determine the average band-gap energy as a function of InN content. The pressure-dependent PL measurements indicate that the low-energy states contributing to the low temperature PL signal follow the band edges and therefore can be treated within the effective mass approximation. The anomalous temperature dependence of the PL peak shifts and the linewidth can be qualitatively explained in terms of the spatial fluctuations of In content. The fluctuations are much larger than predicted for ideally random alloys indicating a phase separation in as grown InGaN alloys.

Acknowledgments

The authors are pleased to thank K.M. Yu for the RBS measurements and gratefully acknowledge helpful discussions with P. Perlin. This work was performed under a Cooperative Research and Development Agreement (CRADA) with Hewlett Packard laboratories and was supported by the Director, Office of Energy Research, Office of Computational and Technology Research, of the U.S. Department of Energy under Contract No.DE-AC03-76SF00098.

References

[1] M.D. McCluskey, C.G. Van de Walle, C.P. Master, L.T. Romano, and N.M. Johnson, Appl. Phys. Lett. **72**, 2725(1998).

[2] A.F. Wright and J.S. Nelson, Appl. Phys. Lett. **66**, 3051(1995).

[3] K.P. O'Donnell, T. Breitkopf, H. Kalt, W. Van der Stricht, I. Moerman, P. Demeester, and P.G. Middleton, Appl. Phys. Lett. **70**, 1843(1997).

[4] S. Chichibu, T. Azuhata, T. Sota, and S. Nakamura, Appl. Phys. Lett. **70**, 2822(1997).

[5] W. Shan, J.J. Song, Z.C. Feng, M. Schurman, R.A. Shall, Appl. Phys. Lett. **71**, 2433(1997).

[6] A. Wakahara, T. Tokuda, X.Z. Dang, S. Noda, and A. Sasaki, Appl. Phys. Lett. **71**, 906(1997).

[7] S. Chichibu, K. Wada, and S. Nakamura, Appl. Phys. Lett. **71**, 2346(1997).

[8] C.G. Van de Walle and J. Neugebauer, Appl. Phys. Lett. **70**, 2577(1997).

[9] M. Smith, G.D. Chen, J.Y. Lin, H.X. Jiang, M. Asif Khan, and Q. Chen, Appl. Phys. Lett. **69**, 2837(1996).

[10] W. Shan, B.D. Little, J.J. Song, Z.C. Feng, M. Schurman, and R.A. Stall, Appl. Phys. Lett. **69**, 3315(1996).

TEM/HREM of Ti/Al and WNi/Ti/Al Ohmic Contacts for n-AlGaN

S. Ruvimov,[1] Z. Liliental-Weber and J. Washburn,
Lawrence Berkeley National Laboratory, Berkeley, CA 94720,

D. Qiao, Q.Z. Lui, S.S. Lau,
University of California, San Diego, La Jolla, CA 92093

High resolution and analytical transmission electron microscopy was applied to characterize the microstructure of Al/Ti and W/Al/Ti ohmic contacts to AlGaN /GaN HFET structures. Formation of a 15-25 nm thick interfacial AlTi$_2$N layer appears to be essential for Ohmic contact behavior indicating a tunneling contact mechanism. Contact resistance was found to depend on the structure and composition of the metal and AlGaN layers, and on atomic structure of the interface. Contact resistivity increases with Al fraction in the AlGaN layer.

A. Introduction

GaN and related compounds are promising for electronic applications [1]. Low resistance ohmic contacts to III-nitrides are critical to achieve desirable parameters of electronic devices, e.g. a high transconductance and a high saturation current, for example in AlGaN/GaN heterostructure field effect transistors (HFET) [2]. For many practical applications contact resistance needs to be about 1 Ohm-mm or less. Several metallization schemes for Ohmic contacts to *n*-GaN [3-6] and *n*-AlGaN [7] with low contact resistivities have recently been reported. However, the mechanisms of Ohmic behavior are still under discussion [6-9]. Two models, low barrier Schottky or tunneling contact, have been proposed to explain Ohmic behavior of metal contacts to n-type III-nitrides [5,8]. The second mechanism involves outdiffusion of nitrogen toward the metal layer and formation of a heavily doped (by N vacancies) material under it which provides the configuration for tunneling. Because this mechanism, to some extent, is independent of the particular metallization scheme it could be applicable to various contacts [5,6,8]. The contact structure can be different even for similar metalization schemes [6,8,9]. Thus, further detailed structural studies are necessary for better understanding of Ohmic contact behavior.

Here we report the results of high resolution and analytical electron microscopy studies of Al/Ti and W/Al/Ti Ohmic contacts to n-AlGaN /GaN HFETs

B. Experimental

Typical AlGaN/GaN HFET structures grown on (0001) sapphire substrates have a 1 μm thick insulating GaN layer followed by a 3 nm thick undoped Al$_{0.15}$Ga$_{0.15}$N layer and a 30 nm thick doped Al$_{0.15}$Ga$_{0.15}$N layer. Two types of metal contacts, Al/Ti and W$_{0.96}$Ni$_{0.04}$/Al/Ti were studied (see Table 1).

Details of the metallization schemes and surface treatment prior the metal deposition have been reported elsewhere [7]. Contact resistance was measured by the transmission line method (TLM). Al/Ti contacts were compared for two AlGaN/GaN HFETs (samples #1 and #2 in Table 1). The AlGaN/GaN structures were grown in the same process, but slightly differ in sheet charge mobility product (nsμ). Despite of nominally identical material, contacts show significant difference in resistances after annealing. Effect of the Al/Ti ratio on contact structure and electrical properties was studied for W/Al/Ti (sample #4 Table 1).

---

[1] On leave from A.F. Ioffe Physical-Technical Institute, S.Petersburg 194021, Russia

TEM studies were carried out on Topcon 002 and JEOL 200CX microscopes operated at 200 kV and on an ARM microscope operated at 800 kV. Cross-sectional specimens were prepared for TEM study by dimpling followed by ion milling.

Table 1. Contact and sheet resistance and $n_s\mu$ product for HFET samples

	#1	#2	#3	#4
Contact composition	Al/Ti (71/30 nm)	Al/Ti (71/30 nm)	$W_{0.96}Ni_{0.04}$/Al/Ti (30/71/30 nm)	$W_{0.96}Ni_{0.04}$/Al/Ti (30/10/50 nm)
$n_s\mu$ product, 1/Vs	0.94^x10^{16}	1.09^x10^{16}	0.94^x10^{16}	0.94^x10^{16}
R_{min}, Ω mm	0.50+0.17	4.00+0.20	0.57+0.20	0.86+0.20
Annealing T, C/time, s	950/80	950/80	950/80	950/80
R_s, Ω cm^2	4^x10^{-6}	2^x10^{-4}	4^x10^{-6}	$8x10^{-6}$

R_{min} is a minimum contact resistance, R_s is a specific contact resistivity

C. Results and discussion

Figs. 2 shows typical cross-sectional electron microscopy images of Ti/Al contacts (#1 and 2) to AlGaN/GaN HFET structures after high temperature annealing. Each contact contains two layers of different contrast due to different Al/Ti ratio in the layers. Energy dispersive x-ray (EDX) analysis (not shown) evidences that the composition of the top layer is close to Al_3Ti in a good agreement with previous results [7,8], while the interfacial layer (with darker contrast) is rich in Ti (with Ti/Al ~ 2).

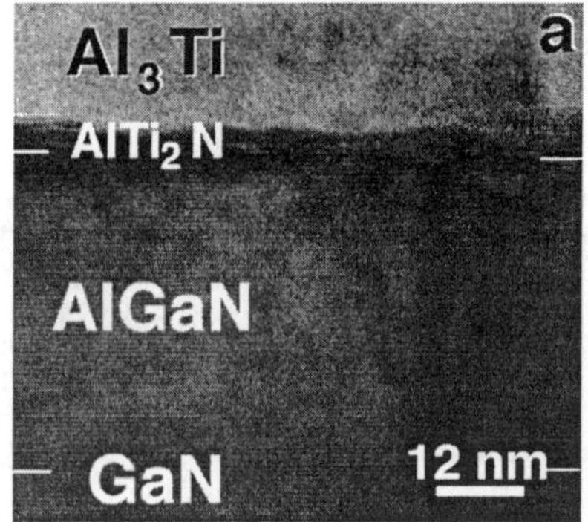

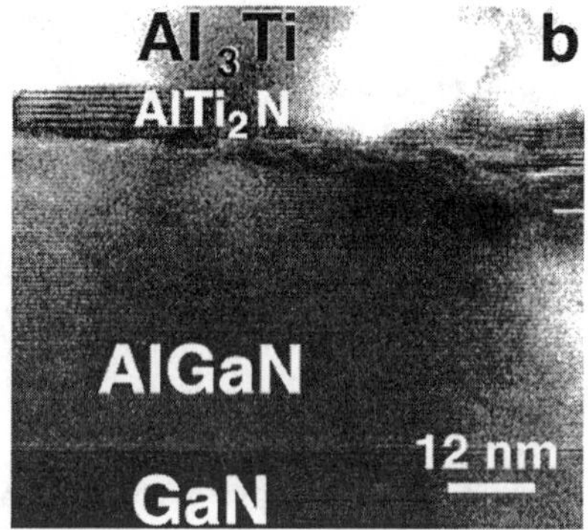

Fig. 1, a-b. Bright-field TEM cross-sectional images of Al/Ti metal layers on AlGaN/GaN HFET heterostructures (samples #1(a) and #2 (b)) after annealing at 950 C for 80 s in flowing N_2.

All contact layers in Fig. 1 are polycrystalline. The thicknesses of the interfacial layers are similar for both contacts, but the contacts differ in interface roughness and electrical resistivity. The low resistance contact (Fig. 1a) has a uniform interfacial layer over all the visible area and an abrupt an interface with AlGaN (interface roughness is about 2nm), while the interfacial layer in the high resistance contact (sample #2 in Table 1) is often discontinious and has a rough interface (about 6 nm) with AlGaN (Fig. 1b).

A high resolution electron microscopy image of the interfacial layer is shown in Fig. 2. Lattice parameters of the layer evaluated from the electron diffraction pattern (not shown) and HREM image of Fig.2 (a= 0.296±0.01 nm, c= 1.43±0.03 nm) are close to those taken from crystallographic tables [10] for hexagonal the Ti_2AlN phase (a=0.299 nm, c=1.361 nm). Electron energy loss spectroscopy confirms the presence of nitrogen in the interfacial layer. The orientation relationship between the $AlTi_2N$ and the AlGaN was found to be: $(0001)AlTi_2N$ //

(0001)AlGaN, (11-20)AlTi$_2$N//(11-20)AlGaN. This differs from previous results reported for Al/Ti contacts to n-GaN [6,8,9].

Formation of this Ti$_2$AlN interfacial layer indicates nitrogen outdiffusion from the AlGaN toward the metal contact during annealing with the formation of heavily doped material under the contact. Because the Ti$_2$AlN layers are rather thick (15-25 nm) their formation might be associated with partial decompopsition of the AlGaN at high temperatures (above 850 C). This results in outdiffusion of Ga and, sometime, in the formation of small pockets at the interface (not shown) which were observe in several samples. The presence of Ga in the metal layer after annealing was detected by both EDX and SIMS measurements.

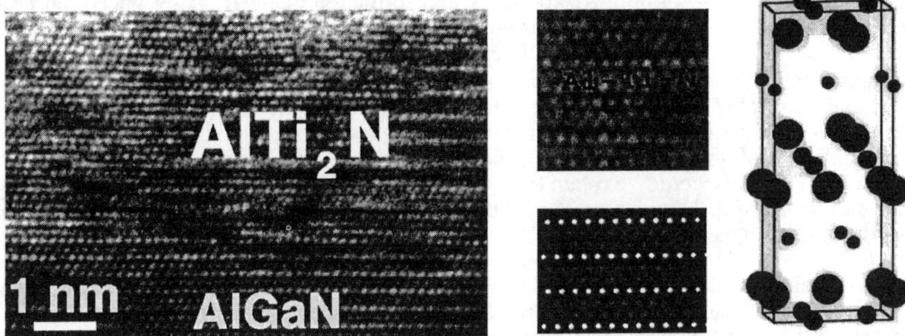

Fig. 2. Cross-sectional HREM image of the Ti$_2$AlN interfacial layer on AlGaN.

Similar results were obtained also for W/Al/Ti contacts. Fig. 3 shows cross-sectional TEM images of W/Ti/Al contacts (#3 and #4) to AlGaN/GaN HFET structures after high temperature annealing. The contact #3 contains three layers with W (top), Al$_3$Ti and Ti$_2$AlN phases: W remains on the top even after high temperature annealing with some interdiffusion in the Al$_3$Ti layer (EDX data are not shown). Except W, its structure is simular to that of sample #1 and contact resistivity is also low. In sample #4, excess of Ti (compared to other samples) results in heavy reactions with the AlGaN which sometimes was completely consumed. This observation confirms our conclusion of partial decomposition of AlGaN due to reaction with Ti during high temperature annealing.

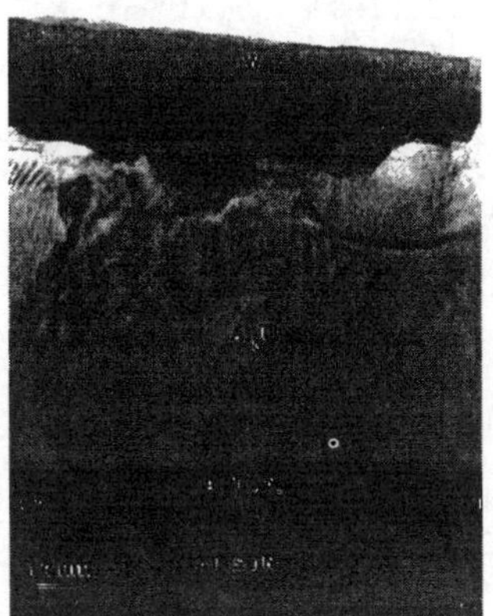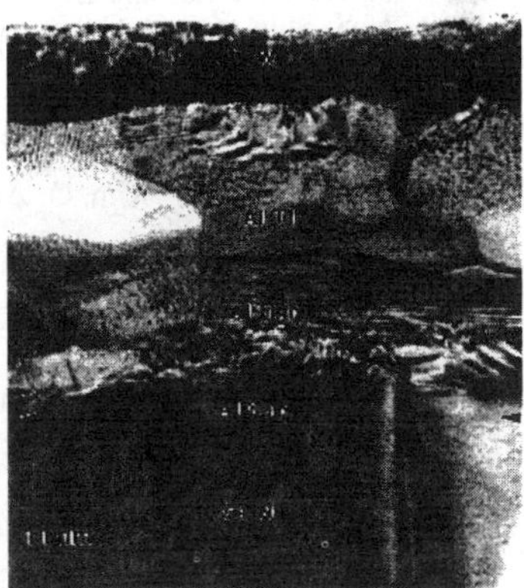

Fig. 3, a-b. Bright-field TEM cross-sectional images of W/Al/Ti metal layers on AlGaN/GaN HFET heterostructures (samples #3 (a) and #4 (b)) after annealing at 950 C for 80 s in flowing N$_2$.

Based on these observations, the following scenario for solid state reactions in Al/Ti contacts to AlGaN during annealing is suggested. The reactions first start between Al and Ti at rather low temperatures (in the range of 250-300 C) with formation presumably of the Al_3Ti phase as expected according to the binary Al-Ti phase diagram. Its formation was also confirmed by RBS measurements [7]. Formation of an Al_3Ti layer from individual Al and Ti layers requires the Al/Ti thickness ratio to be equal to 2.82 taking into account the densities of Ti and Al metals. This ratio was slightly less (about in 2.37) for samples in the present study suggesting excess of Ti (about 5 nm) reacting with the AlGaN. Reaction of Ti with AlGaN which starts probably above 400 C results, first, in dissolution of the native oxide from the AlGaN surface and, then, in outdiffusion of N and formation of the Ti-Al-N interfacial phase. Decomposition of the AlGaN that happens probably at higher temperatures (above 850 C) leads to formation of the Ti_2AlN phase at the interface. Because AlN appears to be more stable at high temperatures (above 850 C) than GaN, stability of AlGaN increases with Al content. This suggests that Al might create a barrier for outdiffusion of N and, hence, for the reaction between Ti and AlGaN. Fluctuations of Al composition over the layer and/or the effect of defects would lead to roughening of the reaction front, epecially at higher Al concentrations, during annealing of Al/Ti contacts. This could explain the difference in interface roughness observed for samples #1 and #2 (Fig. 1) which were slightly different in Al composition of the AlGaN layer according to SIMS data (not shown).

In conclusion, the formation of Ohmic contacts appears to be a complex process that includes solid state reactions at the interface with the substrate. These reactions have been shown to result in a nitrogen-containing interfacial layer and in formation of a heavily doped layer (probably rich in N-vacancies) under it providing the configuration for tunneling. Formation of a 15-25 nm thick interfacial $AlTi_2N$ layer appears to be essential for Ohmic contact behavior suggesting the tunneling contact mechanism. Contact resistance was found to depend on structure and composition of the metal and AlGaN layers, and on atomic structure of their interface. Contact resistivity increases with Al fraction in the AlGaN layer.

This study was supported by BMDO (Dr. K. Wu) monitored by the the US Army Space and Strategic Defence Command. The study at LBNL was partly supported by the Director, Office of Energy Research, U.S. Department of Energy under Contract No.DE-AC03-76F00098. The use of the facilities at National Center for Electron Microscopy and assistance of W.Swider for sample preparation are gratefully appreciated.

1.   S.N. Mohammad, A. Salvador, and H. Morkoç, Proc. IEEE **83**, 1306 (1995).
2.   M. Asif Khan, M.S. Shur, and Q. Chen, Appl. Phys. Lett. **68**, 3022 (1996).
3.   M.E. Lin, Z. Ma, F.Y. Huang, Z.F. Fan, L.H. Allen, and H. Morkoç, Appl. Phys. Lett. **64**, 1003 (1994).
4.   J.D. Guo, C.I. Lin, M.S. Feng, F.M. Pan, G.C. Chi, C.T. Lee, Appl. Phys. Lett. **68**, 235 (1996).
5.   Z.-F. Fan, S.N. Mohammad, W. Kim, O. Aktas, A.E. Botchkarev, and H. Morkoç, Appl. Phys. Lett. **68** 1672 (1996).
6.   B.P. Luther, S.E. Mohney, T.N. Jackson, M. Asif Khan, Q. Chen, J.W. Yang, Appl. Phys. Lett. **70**, 57 (1997)
7.   Q.Z. Lui, L.S. Yu, F. Deng, S.S. Lau, Q. Chen, J.W. Yang, M. A. Khan, Appl. Phys. Lett. **71**, 1658 (1997
8.   S. Ruvimov, Z. Liliental-Weber, J. Washburn, K.J. Duxstad, E.E. Haller, Z.-F. Fan, S.N. Mohammad, W. Kim, A.E. Botchkarev, and H. Morkoç, Appl.Phys. Lett. **69**, 1556 (1996)
9.   B.P. Luther, J.M. DeLucca, S.E. Mohney, R.F. Karlicek, Jr., Appl. Phys. Lett. **71**, 3859 (1997)
10.   P.Villars and L.D. Calvert, Pearson's Handbook of Crystallographic Data for Intermatallic Phases, (Second Ed), Materials Park, ASM

MECHANISMS OF GENERATION AND ATOMIC STRUCTURE OF DEFECTS IN III-NITRIDES EPITAXIAL SYSTEMS FOR DEVICE APPLICATIONS

S. Ruvimov, Z. Liliental-Weber, J. Washburn,
Lawrence Berkeley National Laboratory, MS 63-203, Berkeley, CA 94720,
H. Amano, and I. Akasaki,
Meijo University, Tempakuku-ku, Nagoya 468, Japan

High resolution electron microscopy was applied to study the atomic structure of defects (dislocations, stacking faults, grain boundaries and nanopipes) and the mechanisms of their generation in III-nitrides epitaxial layers grown by MOVPE on sapphire. Formation of vertical boundaries in epitaxial layers was often associated with specific defects at the interface with a substrate. Dislocation generation and annihilation were shown to be mainly growth-related processes and, hence, can be controlled by the growth conditions, especially during the first growth stages.

I. Introduction

GaN and related compounds are promising for electronic applications in the short wavelength range [1-4]. Visible light-emitting diodes (LEDs) [1-4], blue lasers [3,4] and metal-semiconductor field-effect transistors [5] have been successfully fabricated based on GaN. Good optical characteristics of LEDs and lasers have been achieved despite a high dislocation density in epitaxial layers [4]. The typical dislocation density reported [6-12] for GaN-based heterostructures ranges from 10^8 to 10^{10} cm^{-2} and results from a high mismatch in the lattice parameters and thermal expansion coefficients of the GaN layers and the sapphire substrate. Besides dislocations, major defects are stacking faults (SFs) [13], subgrain boundaries [10,11] and nanopipes [7,12]. Defect formation has been shown to depend on growth conditions, level of doping and impurities, substrate and buffer layer. Knowledge about atomic structure of these defects and mechanisms of their generation in GaN is of importance for device applications because it is a key for the reduction of defect density and optimization of GaN technology and processing. For example, performance of devices, in particular, high power devices such as HFETs could be limited by defect-induced centers in the band-gap of the active region.

Here we report the results of an electron microscopy study of defects in GaN epitaxial layers grown on sapphire and discuss the mechanisms of their generation and annihilation.

II. Experiment

GaN epitaxial layers (see Table 2) were grown by MOVPE on (0001) sapphire substrates with an AlN buffer layer. Growth details are described elsewhere [14]. The AlN buffer layers were grown at 400 °C. GaN layers were either undoped or doped with Si (up to 4×10^{18} cm^{-3}) and Mg. Mg-doped samples were annealed at 800°C for 30 min.

TEM studies were carried out on Topcon 002 and JEOL 200CX microscopes operated at 200 kV and on an ARM microscope operated at 800 kV. Cross-sectional specimens were prepared for TEM study by dimpling followed by ion milling.

III. Results and discussion

According to recent reports [15], the dislocation mobility in GaN and other III-nitrides is very low even at growth temperatures of about 1000 °C. Therefore, the mechanisms of defect generation and annihilation typical for other III-V epitaxial systems might not play an important role in III-nitrides. While some dislocations in the GaN layer can move under the high stress developed during the cooling due to the high mismatch in thermal expansion between AlN, GaN and sapphire, most defects do not move after the growth. Thus, dislocation generation and annihilation appear to be mainly growth-related processes and, hence, can be controlled by the growth conditions, especially during the first growth stages.

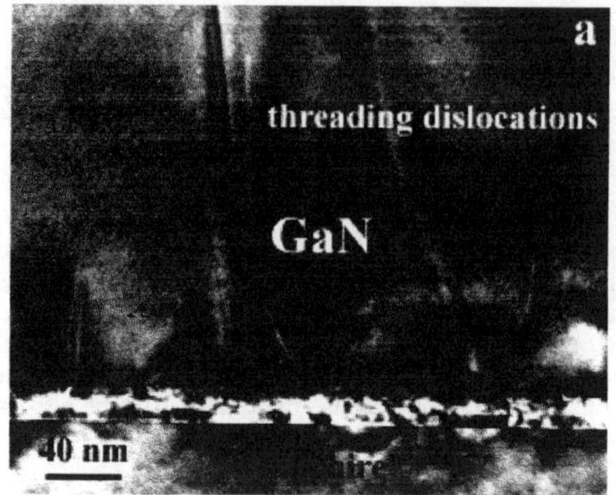

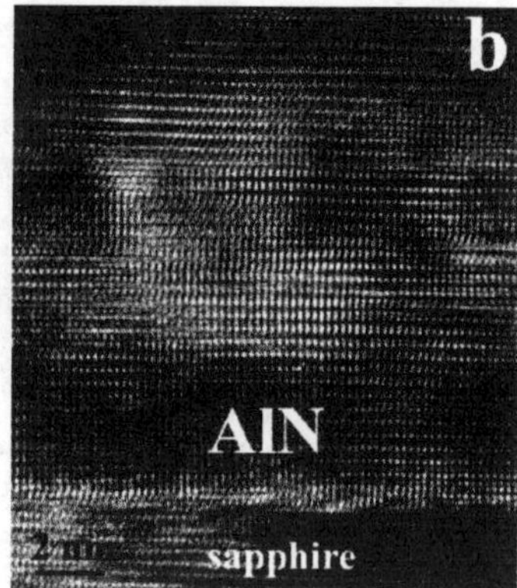

Fig. 1, a-b. Bright-field TEM cross-sectional images of an epitaxial GaN layer (a) and AlN buffer layer (b). Arrows show periodic contrast oscillations at the AlN/sapphire interface.

The dislocation distribution in the GaN layer is, to a large extent, frozen after the layer growth and, hence, reflects the growth process itself. Fig. 1 (a) shows a typical cross-sectional TEM image of undoped epitaxial GaN layers grown on sapphire with AlN buffer layer. Fig. 1 (b) is a magnified image of the AlN buffer layer. The dislocation density is high in the AlN buffer layer and at the GaN/AlN interface, then drastically decreases over 0.2 μm toward the GaN top surface. This dislocation distribution being typical for other GaN samples indicates that most of threading dislocations originate and annihilate during the early growth stages of the AlN and GaN layers.

Initial growth of both GaN and AlN layers proceed in three dimensional (3D) fashion due to the high mismatch in lattice parameters between AlN and sapphire (~12.46%), and between GaN and AlN (~2%). Both AlN/sapphire and GaN/AlN interfaces contain high densities of misfit dislocations which accommodate almost all the misfit between their crystalline lattices during the first growth stages. The orientation relationship between AlN and sapphire, $(0001)_{AlN}//(0001)_{Al2O3}$, $[1\bar{1}00]_{AlN}//[11\bar{2}0]_{Al2O3}$, provides the best match between their crystalline lattices, but the mismatch still remains high. The orientation relationship results in a 6 to 7 "magic" ratio between crystalline lattices of AlN and sapphire: every 6 planes ($[1\bar{1}00]$ or $[11\bar{2}0]$) of AlN fits to 7 planes ($[11\bar{2}0]$ or $[1\bar{1}00]$, respectively) of sapphire with an error of about 2 %. The period of 1.65 nm for the contrast oscillations observed at the AlN/sapphire interface (Fig. 1b) fits well to this "magic" value. The coincidence site lattice for adjacent Al layers (two-dimensional lattices of Al atoms) in crystalline AlN and sapphire (Fig. 2) has a period of 3.3 nm in the $[1\bar{1}00]_{AlN}$ direction. The areas of "good" match between these two-dimensional lattices of Al atoms are divided by the areas of poor match which are, in fact, misfit dislocations with delocalized cores. Periodic arrangement of misfit dislocations at the AlN interface corresponds to the periodical oscillation contrast so that any disturbance in the observed oscillations reflects disturbance in the arrangement of misfit dislocations. The atomic structure of this AlN/sapphire interface suggests formation of a high density of small coherent hexagonal AlN nuclei as the first stage of the growth and, then, their coalesce into larger incoherent AlN islands with misfit dislocations at the AlN/sapphire interface .

Threading dislocations in AlN appear at points where the misfit dislocation network is disturbed, for example, by the presence of atomic steps at the interface or at merging points of adjacent islands. Threading dislocations in the GaN layer are also generated at the interface with AlN during coalescence of 3D island, but some propagate from the AlN buffer layer as well. Both AlN buffer and GaN layers have a columnar structure resulting from 3D growth.

Crystalline subgrains in both layers are slightly misoriented around the c-axis, but almost perfectly oriented in the c-plane. This misorientation is accommodated by the threading dislocations formed at vertical merging boundaries between the adjacent subgrains .

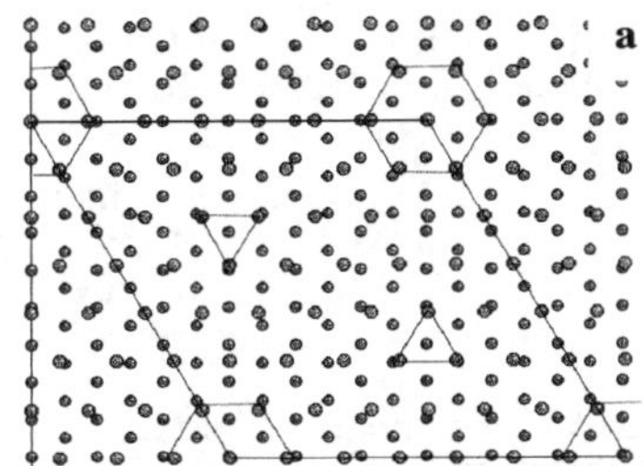
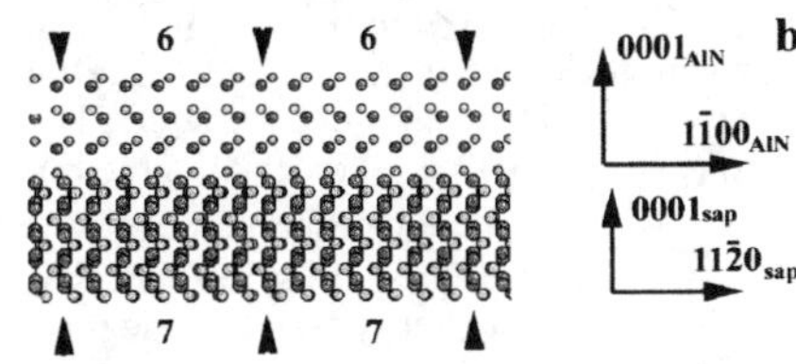

Fig. 2. Coincidence site lattice (a) for adjacent Al layers (two-dimensional lattices of Al atoms) in crystalline lattices of AlN and sapphire, and AlN/sapphire interface (b).

Fig. 1 shows that the subgrain size increases in the growth direction: the GaN subgrains are much larger than the AlN subgrains, and the AlN subgrains at the GaN/AlN interface are larger then those at the AlN/sapphire interface. Such evolution of the subgrain size is caused by lateral overgrowth of some of the AlN and GaN subgrains. The growth conditions for various grains are locally different because the islands differ in their initial size and strain distribution within each island. Therefore, larger islands with lower strain will grow faster than others. As a result, these islands will laterally overgrow the others leaving them buried near the interface. Some of the AlN subgrains in Fig. 1 (b) have buried smaller subgrains under them. Threading dislocations that accommodate the misorientaion between adjacent AlN grains are forced by this lateral overgrowth to bend into basal planes, interact with other dislocations and often annihilate. Schematically it is shown in Fig. 3. This process results in the formation of many half loops in the AlN buffer and in the first 0.2 μm of the GaN observed in Fig.1 and in the decrease of overall dislocation density in the growing GaN layer.

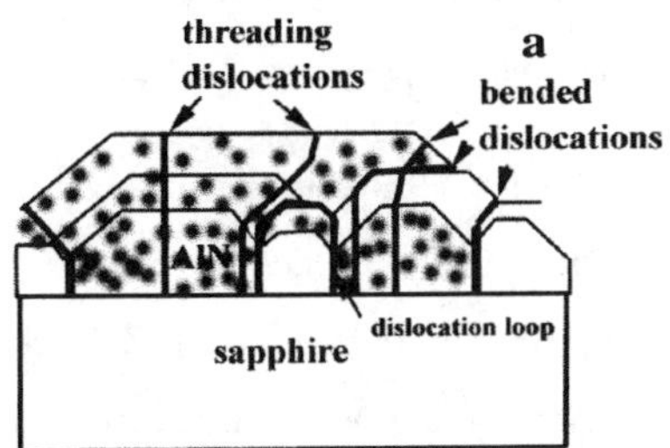

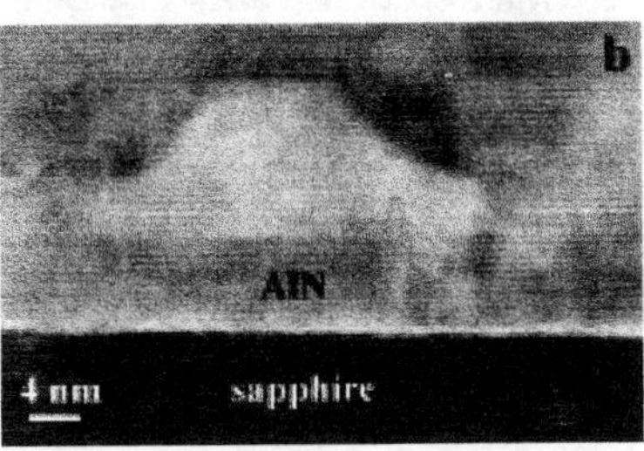

Fig. 3. Lateral overgrowth of islands during the early growth stages (scheme (a) and cross-sectional TEM image (b) of the AlN buffer layer).

The efficiency of this process depends on both thickness and structure of the buffer layer, and on the growth conditions. The AlN buffer layer reduces possible deterioration of sapphire and the resulting interface roughness and, hence, reduces the number of threading dislocations and effects their arrangement in the GaN layer. Formation of subgrain boundaries in the AlN buffer and generation of threading dislocations in the GaN layer were often observed above sites of interface deterioration. The deterioration is associated with oxygen outdiffusion from the sapphire that may cause defect generation in the GaN layer. Threading dislocations are usually arranged in small angle boundaries with both tilt and twist components. TEM diffraction analysis shows the presence of edge, screw and mixed threading dislocations of $1/3{<}11\bar{2}0{>}$, $1/2{<}0001{>}$ and $1/3{<}11\bar{2}3{>}$ Burgers vectors with a majority of edge and mixed dislocations.

Doping also affects the density and the distribution of the dislocations in GaN layers. Si-doping was found to reduce the dislocation density while Mg doping seems to increase it. A high density of horizontal dislocation segments in the Si-doped GaN layers [11] increases the probability of dislocation interaction and annihilation in agreement with our earlier observation. Formation of such horizontal segments suggests a higher probability of the lateral overgrowth for Si-doped GaN compared to that of Mg-doped layers.

Formation of SFs in GaN results in a wurtzite-sphalerite transition within a few basal plane atomic layers. There are three types (I_1, I_2 and E) of basal stacking faults [16] observed in GaN with one (b), two (c) and three (d) cubic bilayers, respectively [13]. The energy of SFs in III-nitrides, γ, is proportional to the number of cubic bilayers: $\gamma_{I1} = 1/2\gamma_{I2} = 1/3\gamma_E$, similar to that for hexagonal metals [16]. SFs in GaN have lower energy than in AlN. Basal stacking fault, I_1, with the lowest energy was the one most frequently found in the MOVPE grown GaN (Fig. 1). SFs in the AlN buffer and in the GaN epitaxial layers were found to result from both growth mistakes and movement of partial dislocations.

Pinholes at GaN layer surfaces were observed in almost all samples while nanopipes were mostly observed in GaN:Mg layers. The density of pinholes was found to increase with doping and impurity concentration in the layer (in particular, oxygen) [18]. Both pinholes and nanopipes were often associated with threading dislocations, but some were formed inside the GaN layer, far from dislocations. This observation suggests that both types of defects are related to a local instability of growth possibly caused by impurity segregation on the growth front. Such segregations are often associated with dislocations which explains the frequent attachment of nanopipes to dislocations. The density of pinholes was also found to be much lower in samples grown on SiC compared with the samples grown on sapphire. This observation supports an idea that pinhole density correlates with impurity level (in particular with oxygen concentration) in the GaN layer.

In conclusion, defect generation and annihilation were shown to be mainly growth-related processes and, hence, can be controlled by the growth conditions, especially during the first growth stages.

This study was supported by the Director, Office of Energy Research, U.S. Department of Energy under Contract No.DE-AC03-76F00098. The use of the facilities at National Center for Electron Microscopy and assistance of W. Swider for sample preparation are appreciated.

1. H. Amano, M. Kito, X. Hiramatsu, and I. Akasaki, Jpn. J. Appl. Phys. **28**, L2112 (1989)
2. S. Nakamura, T. Mukai, and M. Senoh, Jpn. J. Appl. Phys. **30**, L1998 (1991)
3. S. Nakamura, M. Senoh, S. Nagahama, et al, Appl. Phys. **69**, 4056 (1996)
4. F.A. Ponce, D.P. Bour, Nature **386**, 351 (1997)
5. S.N. Mohammad, A. Salvador, and H. Morkoç, Proc. IEEE **83**, 1306 (1995)
6. F.A. Ponce, MRS Bulletin **22**, 51 (1997)
7. W. Qian, G.S. Rohrer, M. Skowronski, K. Doverspike, L.B. Rowland, and D.K. Gaskill, Appl. Phys. Lett. **67**, 2284 (1995)
8. Z. Liliental-Weber, H. Sohn, N. Newman, and J. Washburn, J. Vac. Sci. Technol **B 13**, 1578 (1995)
9. L.T. Romano, B.S. Krusor, R.J. Molnar, Appl. Phys. Lett. **71**, 2283 (1997)
10. X.H. Wu, L.M. Brown, D. Kapolnek, S. Keller, et al, J. Appl. Phys. **80**, 3228 (1996)
11. S. Ruvimov, Z. Liliental-Weber, T. Suski, J.W. Ager, J. Washburn, J. Krueger, C. Kisielowski, E.R. Weber, H. Amano, and I. Akasaki, Appl. Phys. Lett. **69**, 1454 (1996)
12. Z. Liliental-Weber, S. Ruvimov, T. Suski, J.w. Ager III, W. Swider, J. Washburn, H. Amano, I. Akasaki, W. Imler, Mat. Res. Soc. Symp. Proc. **423**, 487 (1996)
13. Z. Liliental-Weber, C. Kisielowski, S. Ruvimov, Y. Chen, and J. Washburn, J. Electron. Mat. **25**, 1545 (1996)
14. I.Akasaki, et al. J. Crystal Growth **98** (1989) 209
15. L. Sugiura, J. Appl. Phys. 81, 1633 (1997)
16. D. Hull and D.J. Bacon, Introduction to Dislocations, Pegamon Press, 1984
17. C. Stampfl, C. Van de Walle, Phys. Rev. B, 57, R15052 (1998)
18. Z. Liliental-Weber, Y. Chen, S. Ruvimov, and J. Washburn, Phys. Rev. Lett. **79**, 2835 (1997)

Electrical characterisation of Mg-related energy levels and compensation mechanism in Mg-doped GaN

D. Seghier and H.P. Gislason

Science Institute, University of Iceland, Dunhagi 3, IS-107 Reykjavik, Iceland

Using admittance spectroscopy and optical deep-level transient spectroscopy (ODLTS) we investigate activation of acceptors in GaN:Mg samples induced by annealing. Conductance measurements reveal two peaks, H1 and H2, with activation energies 130 and 170 meV, respectively, from the valence band. The concentration of H1 is proportional to the acceptor concentration in the samples. Capacitance measurements show that H1, which we attribute to a Mg-related acceptor level, is the shallowest level in our samples. ODLTS spectra exhibit two electron traps at 0.28 and 0.58 eV from the conduction band. Their concentrations are too weak to influence the free carrier concentration. We conclude that compensation occurs through passivation of Mg acceptors by hydrogen, rather than self-compensation by new donor levels.

A. Introduction

Because of their direct wide band gap, III-V nitrides are interesting for applications in short-wavelength optical devices as well as high-power and high-frequency electronic devices. The successful development of short-wavelength light emitting diodes and the recent realisation of nitride semiconductor lasers are promising for the future application of this material for blue and ultraviolet optoelectronic devices. Hole conductivity in GaN is the key for producing optoelectronic devices. So far Mg has been the only dopant in GaN which generates useful p-type conduction in a reproducible manner. For reasons of inadvertent compensation, treatments such as electron-beam irradiation or thermal annealing [1,2] are usually required in order to activate the p-type conduction in layers grown by metal-organic chemical vapor deposition (MOCVD). It has been suggested that the Mg acceptors are passivated by hydrogen in the as-grown state of the material [1,2]. The electrical properties of Mg in GaN are, however, not well understood. Several energy levels and activation energies have been detected by various characterisation techniques and attributed to Mg [3-5]. In this work we examined GaN:Mg using admittance spectroscopy and optical deep-level transient spectroscopy (ODLTS). We traced the electronic states in the band gap of GaN from samples of lower to higher conductivity. We show that the shallowest acceptor level which controls the conductivity in our samples is related to Mg and located 130 meV above the valence band. We suggest that the compensation mechanism in GaN:Mg before annealing is mainly due to direct passivation of Mg, probably by hydrogen.

B. Experimental

The measurements were conducted on Schottky diodes made by evaporating Au dots on 2 μm thick layers of GaN:Mg, grown by MOCVD on c-plane sapphire substrates. Four pieces, denoted #1, #2, #3, and #4, were cut from the same as-grown wafer. Samples #1 to #3 were annealed at 500 °C under flowing N_2 for time duration 1, 3 and 6 minutes, respectively. Sample

Table 1: Effect of annealing of as-grown GaN:Mg on active acceptor concentration.

Sample	Annealing at 500 °C (minutes)	Acceptor concentration (cm^{-3})
#1	1	3.0×10^{16}
#2	3	4.5×10^{16}
#3	6	7.0×10^{16}
#4	20 at 800 °C	1.0×10^{17}

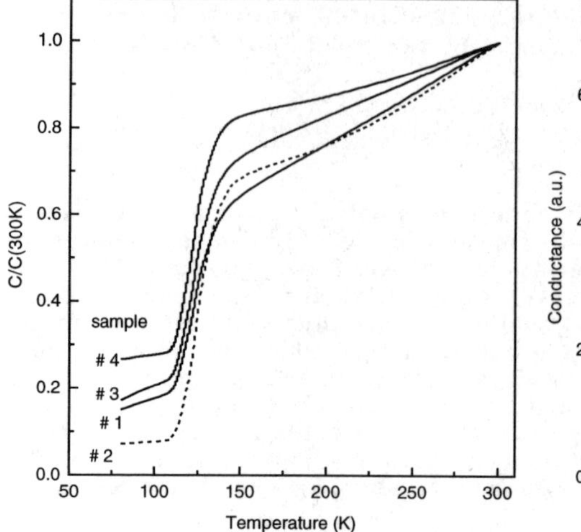 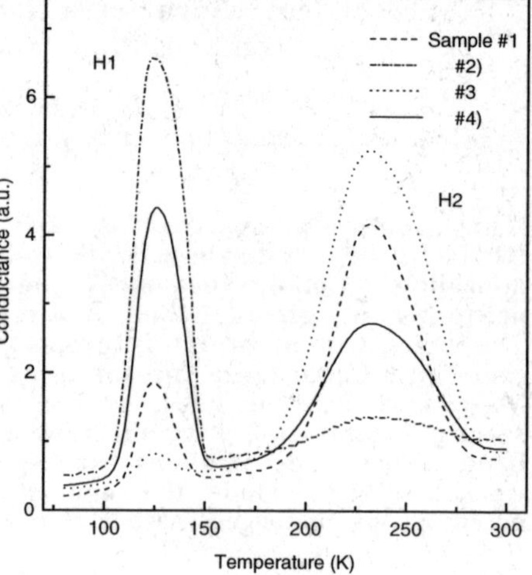

Fig. 1. Temperature variation of the capacitance at a.c. frequency of 10 kHz. The capacitance is normalised by its value at 300K for each sample.

Fig. 2. Variation of the conductance with temperature at a.c. frequency of 10 kHz.

#4 was annealed at 800 °C under flowing N_2 for 20 minutes. The aim of this study was to obtain gradual electrical activation of the Mg acceptors. Ohmic contacts were made by melting In on the front face of the GaN layer. Admittance measurements were performed in the dark using a Hewlett-Packard 4284A LCR meter, at an a.c.-modulation level of 10 mV and frequencies ranging from 10 Hz to 1 MHz. The conductance and the capacitance were monitored in the temperature range 80-300 K. Standard set-up was used for the ODLTS technique which enables detection of minority carrier traps [6].

C.   Results

We performed capacitance-voltage (CV) measurements at room temperature in order to determine the acceptor concentration in the samples. The as-grown samples were too highly resistive to be measured by Hall- and CV techniques. Annealing causes a gradual activation of the acceptors, as Table 1 shows. We also measured the temperature dependence of the capacitance at different a.c. frequencies. The capacitance behaves in a similar way in all samples. Fig. 1 shows the capacitance as a function of temperature at 10 kHz and 0 V bias. After a smooth decrease when cooling down from room temperature it drops drastically around 100 K and remains low at lower temperatures. This result agrees well with the CV measurements according to which the samples are completely depleted at these low temperatures. Hence, we attribute the abrupt capacitance drop to a freeze-out of free carriers. These results also show that the shallowest acceptor level has similar ionisation energies in all samples. The variation of conductance with temperature is shown in Fig. 2. The measurements were performed at 10 kHz under reverse bias of 0.5 V. Two peaks, labelled H1 at lower and H2 at higher temperature, are observed in all samples. Measurements at different frequencies show that these peaks are related to traps in the GaN epilayers, since the corresponding plots of $\log(\omega/T^2)$ versus T^{-1} give straight lines, where ω is the angular frequency of the a.c. signal and T the temperature. H1 and H2 are hole traps since usually only majority carrier traps are observed in admittance spectroscopy. The apparent activation energies calculated from the slopes of the $\log(\omega/T^2)$ versus T^{-1} curves are 130 and 170 meV for H1 and H2, respectively.

The relative height of a conductance peak is proportional to the concentration of the corresponding trap [7]. Fig. 3 shows the height of the H1 peak as a function of the acceptor

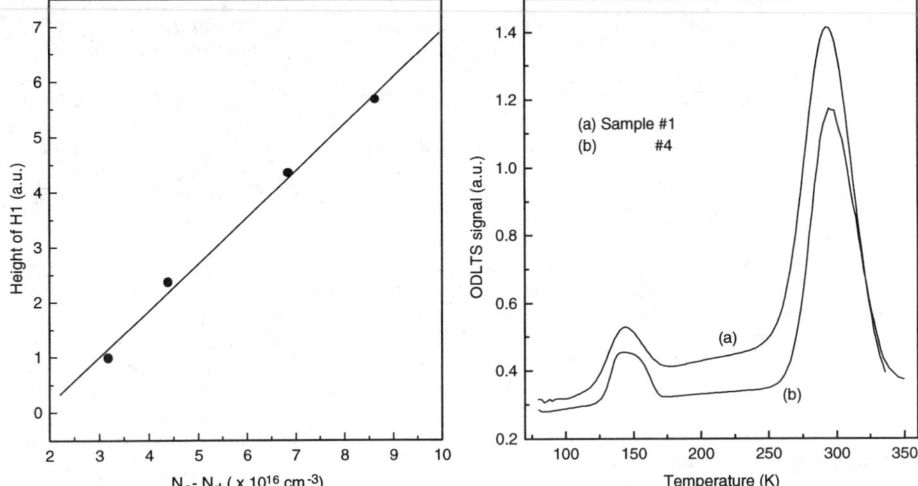

Fig. 3. Variation of the height of the conductance peak H1 with acceptor concentration.

Fig. 4. ODLTS spectra of samples #1 and #4.
The rate window was 50 s^{-1}.

concentration in the samples. The linear relation demonstrates that the concentration of H1 is directly proportional to the acceptor concentration of the samples which increases with the annealing. For this reason we attribute H1 to a Mg-related acceptor. The temperature at which this peak is observed coincides with the inflection point of the capacitance versus temperature curves for the same frequency, 10 kHz (see Fig. 1). Therefore, we deduce that H1 is the shallowest acceptor level in the GaN layers. The height of H2 seems to decrease with the acceptor concentration, in a sublinear way. From the absence of a clear corresponding step in the capacitance curves in Fig. 1 we deduce that the concentration of H2 is too weak to affect the carrier concentration in our samples significantly. Fig. 4 shows ODLTS spectra obtained from samples #1 and #4. The ODLTS technique makes it possible to investigate the upper half of the energy gap in p-type samples. Two positive peaks related to electron traps are observed around 150 and 300 K. Their activation energies are 0.28 and 0.58 eV, respectively. This is in agreement with other work, where similar electron trap levels have been reported [8,9]. The ODLTS spectra are similar in all our samples. We estimate the concentrations of the two defects as 5×10^{13} and 2×10^{14} cm^{-3}, respectively. Therefore, these electron traps have negligible effect on the carrier concentration in the material. We conclude that the electrical compensation in GaN:Mg cannot be caused by the presence of donor levels in the upper part of the band gap. If this were the case, one would observe electron traps with significant and very different concentration in samples with low conductivity and ones with high conductivity.

D.   Discussion

Tanaka *et al.* [4] observed in Hall effect measurements two activation energies in samples annealed at 1140 °C, 125 meV in samples annealed for 60 s and 175 meV in samples annealed for 15 s. Huang *et al.* [5] detected acceptor levels at 136 meV in samples highly doped with Mg, but 124 and 160 meV in samples with lower density of Mg. All our samples have the same Mg concentration, their difference being the portion of Mg acceptors activated by the annealing. By tracing the main energy levels which may play a role in the conductivity of the material we conclude that Mg creates an acceptor level at 130 meV above the valence band, the shallowest one in our samples. The binding energy of this acceptor level may depend on the Mg concentration [5]. In the as-grown samples only a few of the Mg atoms contribute to this energy level. The

257

remaining fraction is electrically inactive, presumably due to passivation by hydrogen, as suggested by several authors [1,2]. The linear dependence of the H1 concentration on N_a - N_d supports the hypothesis that hydrogen is being released from the magnesium acceptors during the heat treatment.

The obvious alternative reason for charge compensation of Mg acceptors is self-compensation during growth caused by the creation of Mg-related donor levels. In this case activation of the acceptors would occur through dissociation of these donors which increases the net acceptor concentration N_a - N_d. Since the height of the conductance peak H1 depends on the total concentration of N_a a non-linear variation is expected in this case. Further, our ODLTS results give no evidence for self-compensation by the creation of donor levels since we only observe donor levels with very low concentrations.

Hole traps with scattered energies close to that of H2 have been reported in the literature [4,5,10].   If they have similar origin one can attribute the discrepancy in binding energies to different concentrations of Mg atoms in the samples. In our samples H2 has no significant influence on the conductivity. Indeed, from the C(T) measurements its concentration can be estimated to be much lower than that of H1, the dominating acceptor. It is possible that in samples with lower carrier concentration than ours the conductivity may be dominated by H2. For example, some workers report conductivity with an activation energy around 170 meV [4,5,10]. This observation is not in disagreement with our findings since there is an evident anticorrelation between the heights of levels H1 and H2 in Fig. 2.

E.   Conclusion

GaN:Mg samples with acceptor concentrations ranging 3×10^{16} to 1×10^{17} cm^{-3} were obtained by annealing several pieces cut from the same as-grown wafer for different duration of time. Using admittance spectroscopy we observed two acceptor levels H1 and H2 with activation energies 130 and 170 meV above the valence band, respectively. We attribute H1 to a Mg-related acceptor level which controls the conductivity of the samples. We also show that H1 is the shallowest level in our samples. ODLTS spectra exhibit two electron traps at 0.28 and 0.58 eV from the conductance band, with concentrations too weak to influence the conductivity of the samples. These results suggest that charge compensation in as-grown GaN:Mg is caused by passivation of Mg acceptors, probably by hydrogen, rather than self-compensation through creation of new donor levels.

Acknowledgements

This research was supported by the Icelandic Research Council and the University Research Fund. We are grateful to Dr. O. Dulac , EMCORE, for providing the GaN material.

References

[1]   S. Nakamura, T. Mukai, M. Senoh and N. Iwasa, Jpn. J. Appl. Phys. **31**, L139 (1992).
[2]   S. Nakamura, N. Iwasa, M. Senoh and T. Mukai, Jpn. J. Appl. Phys. **31**, 1258 (1992).
[3]   S. Fischer, C. Wetzel, E.E. Haller and B.K. Meyer, Appl. Phys. Lett. **67**, 1298 (1995).
[4]   T. Tanaka, A. Watanabe, H. Amano, Y. Kobayashi, I. Akasaki, S. Yamazaki
       and M. Koike, Appl. Phys. Lett. **65**, 593 (1994).
[5]   J.W. Huang and T.F. Kuech, Appl. Phys. Lett. **68**, 2392 (1996).
[6]   A. Mitonneau, G.M. Martin and A. Mircea, IOP Conf. Ser. No. 33a,
       Bristol: IOP, 1977, pp. 73-83.
[7]   D.L. Loose, J. Appl. Phys. **46**, 2204 (1975)
[8]   P. Hacke, H. Nakayama, T. Detchprohm, K. Hiramatsu and N. Sawaki,
       Appl. Phys. Lett. **68**, 1362 (1996).
[9]   W. Götz, N.M. Johnson and D.P. Bour, Appl. Phys. Lett. **68**, 3470 (1996).
[10]   W. Götz, N.M. Johnson, J. Walker, D.P. Bour and R.A. Street,
       Appl. Phys. Lett. **68**, 667(1996).

Degradation Mechanisms in GaN/AlGaN/InGaN LEDs and LDs

Daniel L. Barton and Marek Osiński*

Sandia National Laboratories, MS 1081, Albuquerque, New Mexico 87185-1081
Center for High Technology Materials, University of New Mexico, Albuquerque, New Mexico 87131-6081

Abstract

Our studies of device lifetime and the main degradation mechanisms in Nichia blue LEDs date back to spring 1994. Following the initial studies of rapid failures under high current electrical pulses, where metal migration was identified as the cause of degradation, we have placed a number of Nichia NLPB-500 LEDs on a series of life tests. The first test ran for 1000 hours under normal operating conditions (20 mA at 23 °C). As no noticeable degradation was observed, a second 23 °C test was performed with the same devices but with a range of currents between 20 and 70 mA. For subsequent tests, the temperature was increased by 10 °C in 500 h intervals up to a final temperature of 95 °C using the same currents applied in the second test. This work presents a thermal degradation mechanism that dominates degradation at high ambient temperatures.

A. Introduction

In order to investigate the lifetime of the Nichia LEDs, 20 devices were mounted inside a large environmental chamber that could be maintained at a constant temperature [1]. The light output of each LED was sampled by an optical fiber that was connected to its own photovoltaic detector located outside the chamber. The LED-to-fiber connection was both mechanically stable and light tight, eliminating intensity variations due to mechanical misalignments and ambient light. The system used a switching device to select a single detector's current, which was fed to a meter for automated reading of each LED's output.

A comparison between the life test data for the early stages of the life test (at low ambient temperatures) and the data shown in Figure 1 for the eleventh segment of the test (at 95 °C) showed a distinct difference in the degradation rates for the same LEDs. The obvious explanation is that the increase in the degradation rate in the higher temperature test is an acceleration of the same degradation mechanism because of the increase in temperature. The degradation at lower ambient temperatures has been shown to be related to the changes in the plastic packaging material with exposure to the short wavelength emission from the LED [2]. The problem with this degradation mechanism being responsible for the degradation at high ambient temperatures was identified in Figure 2. Figure 2 shows how the light output is reduced at higher temperatures which would cause degradation mechanisms related to light output to be reduced as the temperature is raised. Thus, a different mechanism must be responsible.

B. Thermal Degradation of Package Transparency

In order to identify the high temperature mechanism, the degradation rates or the slopes of the intensity versus time curves shown in Figure 1 were extracted form the data. The data, as shown in Figure 3, indicates that there is a dramatic increase in the degradation rate for currents greater than 30 mA at an ambient temperature of 95 °C.

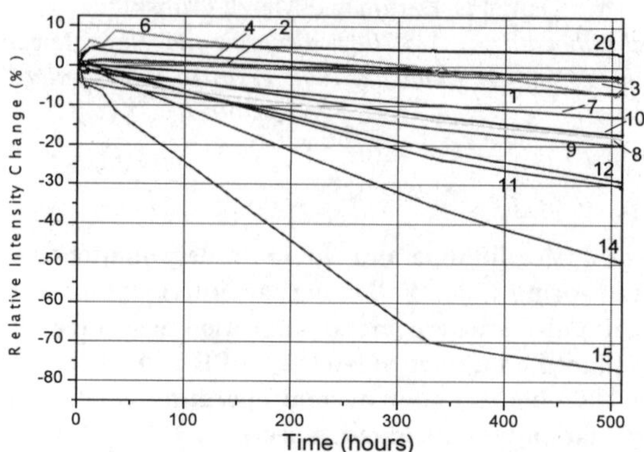

Figure 1 - Relative change in intensity of remaining Nichia blue LEDs at 95 °C for the final 510 hours.

Since the life test data suggests that the degradation may be related to a combination of ambient temperature and LED self-heating, the changes in the plastic with temperature were investigated. Figure 4 shows three sections of plastic removed from an Nichia LED control sample. The piece on the left has not been exposed to more than normal room temperature and has a clear appearance. The sample in the middle was placed in a burn-in oven at 150 °C in air for a 133 hours. This sample has turned to a deep brown color. The sample in the right was placed in the same oven at 200 °C for 130 minutes and shows a similar change to the sample exposed to 150 °C only to a lesser degree. This test indicated that at temperatures of around 150 °C (or slightly lower) the plastic can change by a purely thermal effect in such a manner that the light output of the LED could be attenuated. While the LEDs used in the life test did not show plastic that had browned that was visible on the outside of the package, the plastic in contact with the LED may have. This evidence showed that a more accurate junction temperature measurement was needed.

The only way to accurately measure the temperature at the diode is by measuring the junction voltage. To make this measurement, a control LED was placed in a burn-in oven and allowed to come to thermal equilibrium with the oven temperature. The junction voltage was measured when the LED was first turned on to minimize changes in the voltage across the junction due to self-heating. An HP 4145 semiconductor parameter analyzer was used for the measurement allowing the voltage to be accurately measured within 10 ms after the current was applied. This process was repeated at currents from 10 mA to 70 mA (coinciding with the range of currents used in the life tests) and at temperatures from 30 °C to 150 °C. The data showed that there was a linear relationship between the voltage measured across the junction and the temperature of the junction. From this data, a better measure of the temperature that the plastic in contact with the diode was at could be made.

To calculate the diode temperatures experienced by the LEDs in test 11 (95 °C), a similar LED was placed in an oven at 95 °C and allowed to come to thermal equilibrium. The same forward currents that were used in the life tests were applied and allowed to stabilize meaning that the final temperature resulting from the ambient temperature and self-heating had been reached. This point was determined by observing the junction voltage and waiting for it to stop decreasing. Using this data plus the linear fit parameters from the data collected on junction voltage versus temperature at turn on, the junction temperatures were calculated for each forward current. The results are shown in Figure 5. The calculations showed that for forward currents of 40 mA and above, the junction temperature was in excess of 145 °C which is very close to the 150 °C temperature where the plastic was shown to discolor in Figure 4. For currents smaller than 40 mA, the temperature is less than 135 °C where, according to the data in Figure 3, the degradation rate is significantly lower. While the exact temperature at which the plastic begins to discolor has not been determined, the temperature where the degradation begins to affect this material in a time frame important to LED lifetimes has been shown to be between 135 and 145 °C.

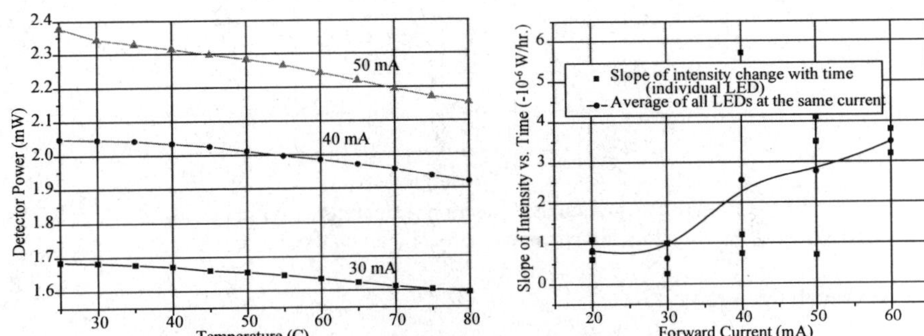

Figure 2 (left) - LED output power at various temperatures.
Figure 3 (right) - Slopes extracted from test 11 data showing a sharp change in degradation rate at 95 °C for currents around 40 mA.

Figure 4 - Sections of Nichia LED package material: (left) unstressed, (middle) 133 hours at 150 °C, (right) 130 minutes at 200 °C.

C. Conclusions

The life tests of Nichia blue LEDs completed have not produced significant degradation on any of the devices in a manner consistent with a degradation mechanism attributable to the diode. These results indicate that Nichia devices enjoy a remarkable longevity in spite of their high density of defects. An analysis of the degradation characteristics of the plastic packaging material demonstrated that a degradation in light output could indeed be caused by the ultraviolet light emitted by the LED at lower temperatures [2]. At higher life test temperatures, the degradation mechanism changed from a light intensity related mechanism to a pure thermal mechanism. By calculating the junction temperature of the diode at the life test temperatures, excellent correlation was made between the observed degradation rate and the temperatures achieved through the combination of ambient temperature and diode self-heating. This analysis was supported by the observation of plastic discoloration by exposure to temperatures consistent with those calculated from the junction temperature measurements.

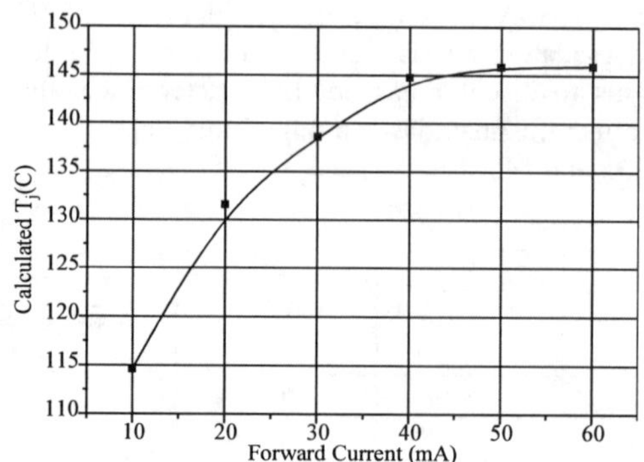

Figure 5 - Calculated junction temperature (T_j) at various forward currents (ambient temperature 95 °C).

D. Acknowledgements

The support of the Defense Advanced Research Projects Agency (DARPA) and the New Energy and Industrial Technology Development Organization (NEDO) of Japan is gratefully acknowledged. This work was performed at Sandia National Laboratories and is supported by the U.S. Department of Energy under contract DE-AC04-94AL85000. Sandia is a multiprogram laboratory operated by Sandia Corporation, a Lockheed Martin Company, for the United States Department of Energy.

E. References

1. C. J. Helms, N. H. Berg, D. L. Barton, B. S. Phillips, and M. Osin´ski in <u>Fabrication, Testing, and Reliability of Semiconductor Lasers</u>, edited by M. Fallahi and S. C. Wang (SPIE International Symposium on Lasers and Integrated Optoelectronics OE/LASE '96, SPIE Proc. **2683**, San Jose, CA, 1996) pp. 74-80.

2. D. L. Barton, M. Osin´ski, P. Perlin, C. J. Helms, and N. H. Berg, (Proceedings of 5th Annual IEEE International Reliability Physics Symposium, Denver, CO, 1997) pp. 276-281.

Time-Resolved Laser Spectroscopy of Wide Band Gap Semiconductors

Frederick H. Long[1], Milan Pophristic
Department of Chemistry, Rutgers University
Piscataway, NJ 08855-8087

Chuong Tran, Robert F. Kalicek Jr., Zhe Chuan Feng[2], Ian Ferguson
EMCORE Corporation
Somerset, NJ 08873

Abstract

We present time-resolved photoluminescence measurements of InGaN MQWs, films, LEDs and SiC.

A. Introduction

In recent years there has been considerable progress in the use of nitride semiconductors for photonic applications including blue LEDs, lasers, and solar blind photodetectors. In spite of the impressive results obtained by Nakamura[1], the optimization of laser performance requires that the fundamental mechanisms upon which these devices operate must be better understood.

Time-resolved spectroscopy provides a method for the direct measurement of exciton and carrier lifetimes in semiconductors and semiconductor structures. Lifetimes are critical to device efficiency and therefore overall device performance. An understanding of the fundamental physics underpinning such phenomena is essential for the improvement of nitride semiconductor devices.

Silicon carbide (SiC) is currently being revisited as a material for the next generation of high-power and high-temperature electronics. New improved noncontact, in situ diagnostics for SiC will help accelerate the development of low cost devices. The time-resolved PL of SiC is largely unexplored at room temperature.

B. Experimental

The laser used was an amplified Ti-sapphire laser from Coherent Corporation operating at 250 kHz. The time-resolved photoluminescence (PL) measurements were done with a Hamamatsu streak camera model C5680. The excitation pulse was at 400 nm (3.10 eV). The excitation power was adjusted by using calibrated neutral density filters. The scattered laser light was filtered with a 420 nm band pass filter. The typical response time was 60 picoseconds and was determined by electrical jitter in the triggering electronics.

The $In_xGa_{1-x}N$ multiple quantum well samples were grown by metal organic chemical vapor deposition at EMCORE Corporation. The average indium mole fraction was 22 %. The whole structure was on c-plane sapphire with 3 microns of unintentionally doped (n-type

---

[1] To whom correspondence should be addressed, fhlong@rutchem.rutgers.edu
[2] Institute of Materials Research & Engineering, S7, NUS, Singapore 119260

$5x10^{16}/cm^3$) GaN as a substrate.[2] The multiple quantum well consists of five layers of $In_xGa_{1-x}N$ 35 Å thick and four layers of GaN each 45 Å thick.

The 4H- and 6H-SiC used were from a variety of sources. The wafers were n-type (nitrogen) doped with nominal nitrogen concentrations of 10^{18} to 10^{19} cm^{-3}, depending on the specific sample of interest. The carrier concentration of the SiC wafers was verified by Raman spectroscopy.

C. Results and Discussion

C.1 InGaN Multiple Quantum Wells

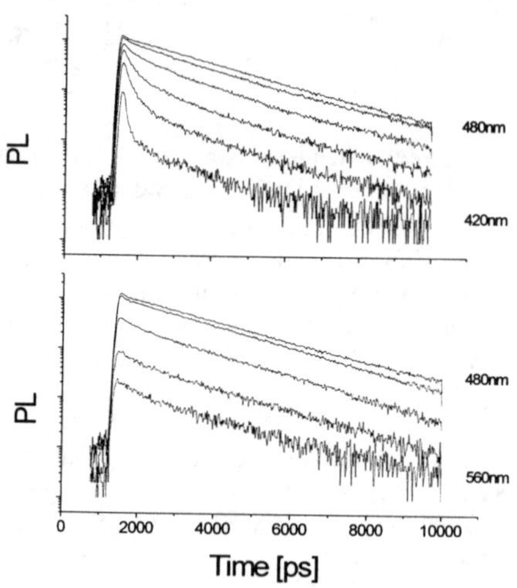

Figure 1:   Wavelength dependence of the emission kinetics for MQW C. Top λ=420, 430, 440, 450, 460, 480 nm. Bottom λ=480, 500, 520, 540, 560 nm.

The wavelength dependence of the emission kinetics for a MQW are shown in figure 1. A dramatic wavelength dependence is observed. For wavelengths longer than 480 nm (2.58 eV), near single exponential decays are observed, which are nearly independent of wavelength, figure 1 bottom.(τ =1.87 ± 0.02 ns, at 480 nm, and 9 mW) As the emission wavelength is moved to shorter wavelengths, the lifetime is observed to shorten and the kinetics are clearly no longer single exponential, figure 1 top. Analagous results have been reported in low temperature studies of InGaN MQWs.[3] These changes in lifetime can be attributed to a tail in the density of states. According to Fermi's golden rule, the transition rate will be proportional to the density of states; therefore, if the frequency dependence of the electronic matrix element is neglected, an exponential tail in the density of states, will be reflected in the energy dependence of the emission lifetime. This threshold like behavior is similar to a mobility edge, often observed in

264

disordered media.

C.2 InGaN films

Typical lifetime data for an InGaN film at room temperature is given in figure two. It is important to note that there is relatively little change in the emission lifetime with emission wavelength in the film. We attribute this to a dimensionality effect. In the film, the excited states are free to migrate in three dimensions, thereby increasing the rate of spectral diffusion. Therefore the PL lifetime is nearly independent of energy at room temperature.

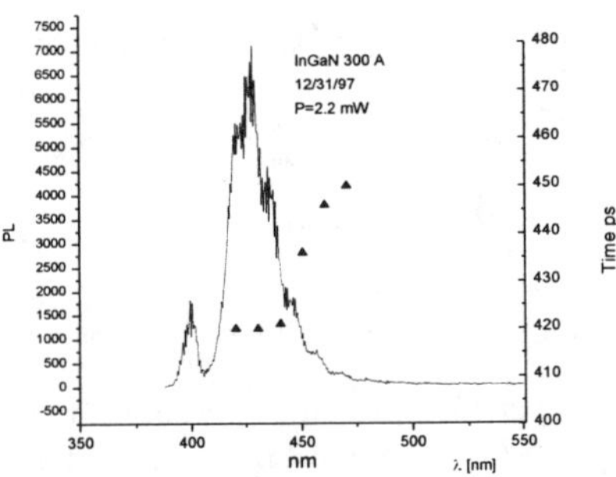

Figure 2: Time-resolved PL from an 300 Å InGaN film.

C.3 Light-Emitting Diodes

The PL decays for this bright LED are clearly stretched exponentials, figure 3. This is

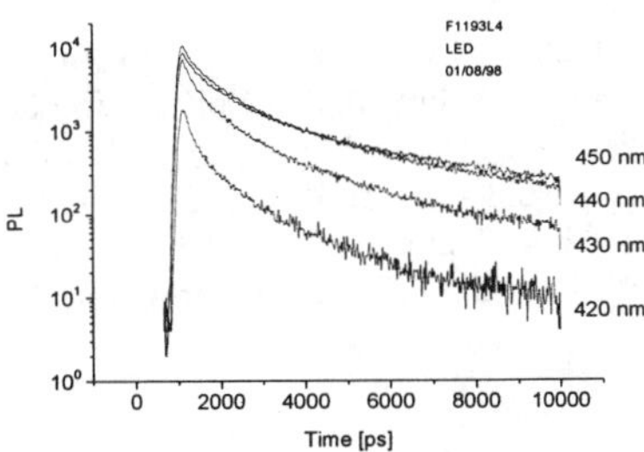

Figure 3: Time-resolved emission from a bright LED.

seen most clearly for 420 nm (2.95 eV), when the decay appears to be exponential on a logarithmic vertical scale. For MQWs, biexponential or near single exponential decays are observed in the room temperature PL. It is well known that stretched exponential kinetics are associated with disorder. It is most likely that the molecular origin of the disorder is the fluctuations in relative indium concentration. In contrast for the dim LED structure studied, we observed that the PL lifetime is much shorter and the decays do not appear to be sensitive to the emission wavelength.

C.4 Silicon Carbide

We have examined the TRPL from both 4H- and 6H-SiC wafers, typical results are shown

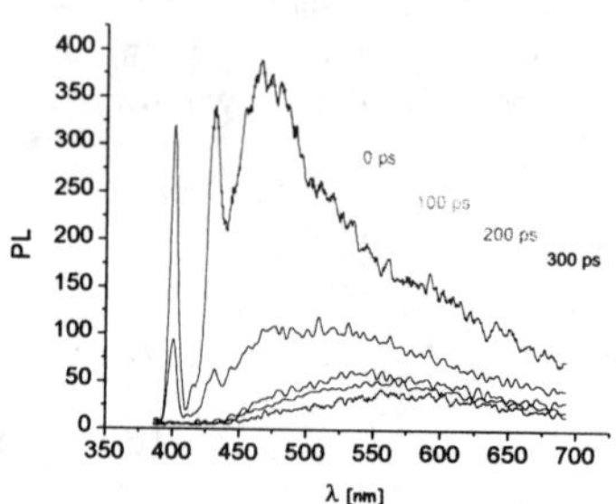

Figure 4: Time-resolved PL spectra from 6H-SiC

in Figure 4. The fast component at 430 nm is due to excitons bound to nitrogen donors. The broad PL bands between 450 and 700 nm is most likely due to transitions involving Aluminum and Boron.

D. Summary

In summary we have used time-resolved spectroscopy to investigate SiC, InGaN films, MQWs and LEDs.

FHL would like to thank the Rutgers Research Council for financial support. The authors would also like to thank Bill Cieslik and Hamamatsu Corporation for the loan of the streak camera. Acknowledgement is also made to the donors of the Petroleum Research Fund, administered by the ACS, for the partial support of this research. ZCF acknowledges the support of Prof. C.F. Shih and Prof. S. J. Chua. The work at EMCORE was supported in part by ONR contract number N00014-97-C-0210 and monitored by Dr. Colin Wood.

References

[1] S. Nakamura, M. Senoh, S. Nagahama, N. Iwasa, T. Yamada, T. Matsushita, Y. Sugimoto, and H. Kiyoku, Applied Physics Letters **70**, 868-870 (1997).
[2] C. A. Tran, R. F. K. Jr., M. Schurman, T. Salagaj, A. Thompson, and R. Stall, in *Phase Separation in Bulk InGaN and Quantum Wells Grown by Low Pressure MOCVD*, IEEE Meeting on GaN Materials and Devices, Montreal, Canada, 1997.
[3] Y. Narukawa, Y. Kawakami, S. Fujita, S. Fujita, and S. Nakamura, Physical Review B **55**, R1938-R1941 (1997).

VIII.

Silicon Carbide

Step-Controlled Epitaxial Growth of SiC and Its Conductivity Control

Hiroyuki Matsunami and Tsunenobu Kimoto

Department of Electronic Science and Engineering, Kyoto University, Japan

Abstract
Polytype-controlled epitaxial growth of SiC is achieved by utilizing step-flow growth on off-oriented SiC{0001} substrates. High-quality of SiC epilayers has been elucidated through photoluminescence, Hall effect, and deep level analyses. The background doping level of undoped epilayers can be reduced down to $0.5\sim2\times10^{14}cm^{-3}$. *In-situ* n- and p-type impurity doping from the 10^{15} to $10^{19}cm^{-3}$ range has been realized. In ion implantation of nitrogen donors and aluminum/boron acceptors, high-temperature annealing is a key issue to obtain nearly perfect electrical activation. Vanadium ion implantation can successfully be applied to form semi-insulating SiC layers.

A. Introduction

Silicon carbide (SiC) has received increasing attention as a vital candidate for high-power, high-frequency, and high-temperature device applications. Recent breakthroughs in crystal growth technology have led to the availability of high-quality SiC material, with which outstanding potential of SiC has been demonstrated in prototype devices such as high-voltage diodes, vertical MOSFETs, high-power MESFETs, and ICs operating at high temperature [1].

Table 1 lists up major physical properties of Si, GaAs, SiC(4H-SiC) and GaN. In addition to distinguished properties, SiC is an exceptional wide bandgap semiconductor in a sense that both n- and p-type conductions can easily be obtained by either *in-situ* doping or ion implantation, though a success in p-type doping of GaN up to the $10^{17}cm^{-3}$ range has been realized. SiC is the only compound semiconductor which forms high-quality SiO_2 by thermal oxidation like in Si. Besides, both low-resistivity [2] and semi-insulating SiC wafers [3] are available, eventually enabling the fabrication of vertical high-power devices and lateral high-frequency devices, respectively. In this paper, recent progress in fundamental issues on SiC epitaxial growth and conductivity control is discussed.

Table 1 Physical properties of Si, GaAs, SiC(4H-SiC) and GaN.

	Si	GaAs	SiC(4H)	GaN
bandgap (eV)	1.12	1.43	3.26	3.39
electron mobility (cm^2/Vs)	1350	8000	1000	1000
breakdown field (MV/cm)	0.3	0.4	3.0	3.0
saturation velocity (cm/s)	1×10^7	1×10^7	2×10^7	2×10^7
thermal conductivity (W/cmK)	1.5	0.5	4.9	1.3
p, n control	◯	◯	◯	△
conducting wafer	◯	◯	◯	×
insulating wafer	×	△	◯	◯ (sapphire)

◯: excellent, △: fair, ×: difficult

B. High-Quality Homoepitaxial Growth

SiC is a representative material which exhibits "polytypism": One-dimensional stacking variations of a Si-C pair along the *c* axis result in more than 200 different SiC polytypes. In homoepitaxial growth of hexagonal SiC, polytype mixing, especially cubic SiC (3C-SiC) inclusions, had been observed, which was a severe obstacle for electronic device applications. The authors' group has found that the surface steps existing on off-oriented SiC{0001} substrates serve as a template which forces the replication of stacking sequence of the substrate polytype in the epilayer (*step-controlled epitaxy*) [4].

Epitaxial growth was performed by atmospheric-pressure chemical vapor deposition (CVD) in a SiH_4-C_3H_8-H_2 system [4]. A typical growth temperature is 1500°C at which a growth rate of 2.5~4.0μm/h is obtained. To realize polytype replication by step-controlled epitaxy, hexagonal (4H or 6H) SiC{0001} with 3.5~8° off-angles toward <11$\bar{2}$0> was employed as substrates.

Figure 1 represents the surface morphology of a 15μm-thick 4H-SiC epilayer observed with (a) a Nomarski microscope and (b) an atomic force microscope (AFM). In the low magnification (Fig.1(a)), the epilayer surface is flat with a low morphological defect density of $10^3 cm^{-2}$. The AFM image revealed macrostep formation caused by step bunching in step-flow growth [4,5]. Bunched steps have unique heights of either the half or full unit cell of a SiC polytype being grown: For example, two or four Si-C bilayer heights are dominant on a 4H-SiC surface.

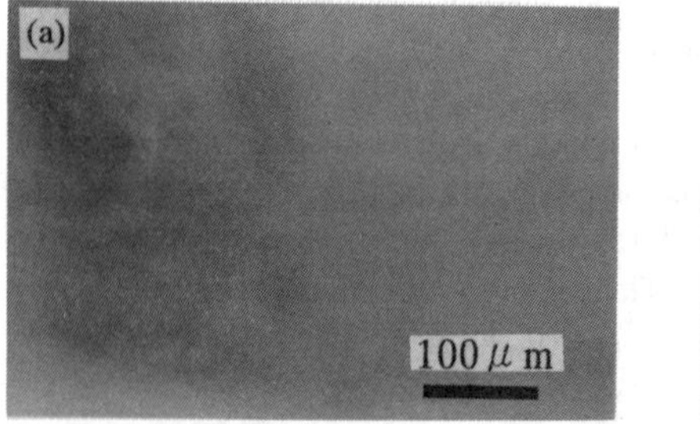
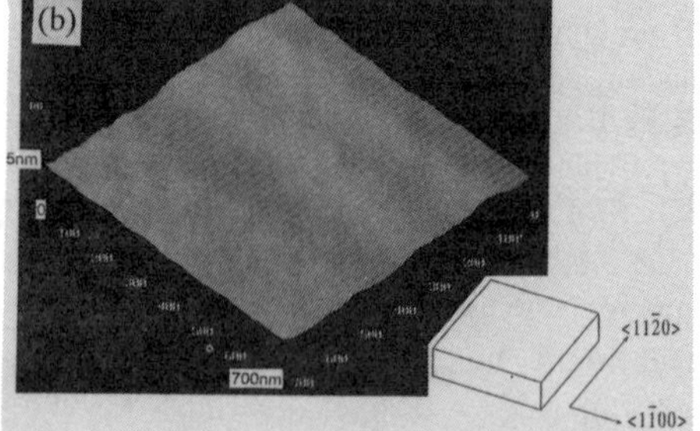

Fig.1 Surface morphology of a 15μm-thick 4H-SiC epilayer observed with (a) a Nomarski microscope and (b) an AFM.

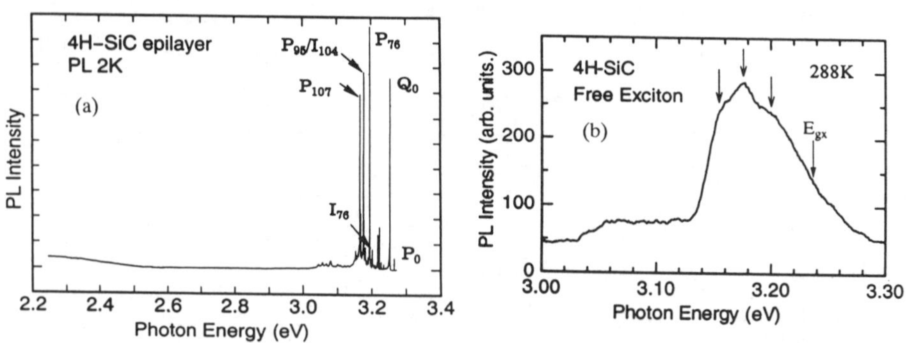

Fig.2 (a) Low temperature and (b) room temperature PL spectra taken from a 10μm-thick n-type 4H-SiC epilayer.

A low-temperature photoluminescence (PL) spectrum taken from a 10μm-thick n-type 4H-SiC epilayer is shown in Fig.2(a), demonstrating sharp exciton-related peaks near the band edge (exciton energy gap=3.265eV at 2K). These peaks can be attributed to free excitons, excitons bound to neutral nitrogen (N) donors, and their phonon replicas. The relative intensity of free exciton peaks increases with decreasing the donor concentration of epilayers. The PL spectrum indicates very little contamination of aluminum (Al) acceptors or titanium (Ti) isoelectronic traps. A small bump in the region from 2.2 to 2.4eV is ascribed to the N donor-boron (B) acceptor pair band. This PL band may come from a substrate, because the penetration depth of excitation light (He-Cd laser: 325nm) exceeds a few tens μm, larger than the epilayer thickness. Above 50K, N bound exciton peaks are quenched, and free exciton peaks dominate. Even at room temperature (288K), clear free exciton peaks could be observed in spite of the indirect band structure of SiC, as shown in Fig.2(b). This result demonstrates high purity and high quality of SiC epilayers, since free excitons in SiC easily suffer from impurity- or defect-related non-radiative recombination channels due to the band structure.

In Hall effect measurements, an n-type 4H-SiC epilayer with a donor concentration of $4\times10^{16}cm^{-3}$ (N doped) exhibits high electron mobilities of $724cm^2/Vs$ at 292K and $11,000cm^2/Vs$ at 77K [6]. This significant increase in mobility at low temperature reflects low impurity compensation in the epilayer. The compensating acceptor concentration was estimated to be as low as $10^{12}\sim10^{13}cm^{-3}$ from the fitting of the electron concentration vs. temperature curve with an electric neutrality equation.

Figure 3 shows the deep level transient spectroscopy (DLTS) spectra obtained from an as-grown n-type 4H-SiC epilayer. Although one DLTS peak, which is located at 0.63~0.68eV below the conduction band edge ("Z_1 center"), is observed at around 350K, the trap concentration is very low, $4\times10^{12}cm^{-3}$. Detailed investigation elucidated the nature of this defect center as an acceptor-like complex containing intrinsic defects such as divacancy [7].

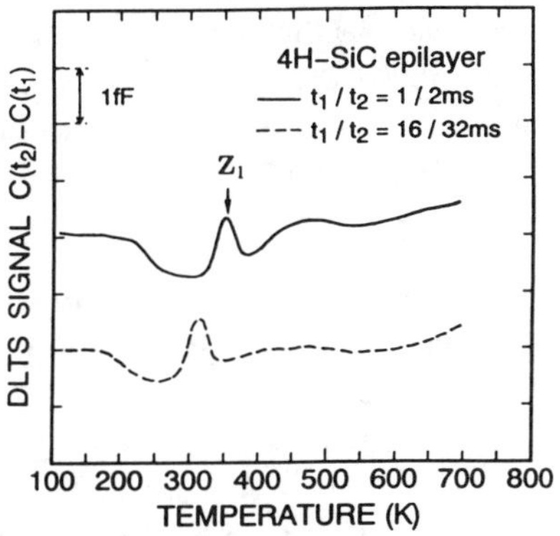

Fig.3 DLTS spectra obtained from an as-grown n-type 4H-SiC epilayer.

C. Conductivity Control

C.1 In-situ impurity doping

The control of doping concentration by changing the C/Si ratio during CVD has provided a path to grow high-purity materials (site-competition epitaxy) [8]. Figure 4 represents our result on the C/Si ratio dependence of background doping level of unintentionally doped 4H-SiC epilayers grown at 1500°C. On a SiC(0001) Si face, the donor concentration can drastically be reduced down to $0.5\sim2\times10^{14}cm^{-3}$ by increasing the C/Si ratio. Since N atoms, major residual donors, substitute at the C sub-lattice in SiC, high C coverage under C rich condition outcompetes N atoms, and thereby lowly-doped epilayers can be obtained. On a SiC(000$\bar{1}$) C face, however, the donor concentration is not sensitive to the C/Si ratio at this temperature, being

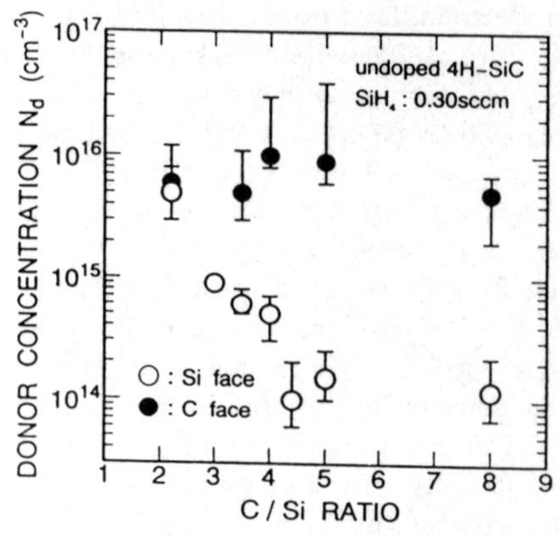

Fig.4 C/Si ratio dependence of donor concentration for unintentionally doped 4H-SiC epilayers.

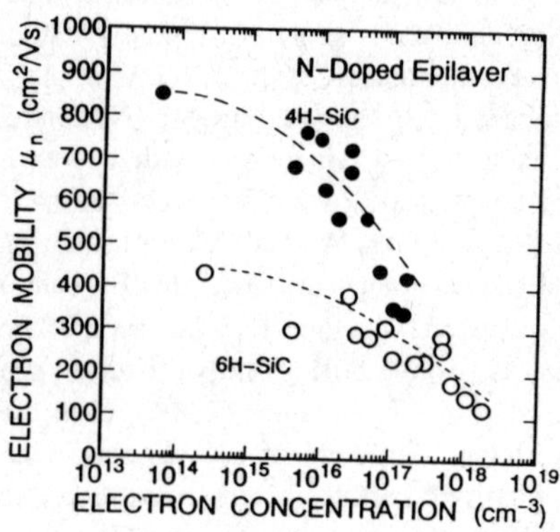

Fig.5 Electron mobility vs. electron concentration at RT for 4H- and 6H-SiC epilayers.

a mid 10^{15}cm^{-3}.

In-situ n-type doping from the 10^{15} to 10^{19}cm^{-3} range can easily be achieved by the introduction of N_2. The electron mobility vs. free electron concentration at room temperature (RT) for 4H- and 6H-SiC epilayers is plotted in Fig.5. For a lowly-doped 4H-SiC epilayer, a high electron mobility of $851\text{cm}^2/\text{Vs}$ was obtained. The authors have mapped the doping concentration of epilayers grown on one-inch wafers by capacitance-voltage (C-V) measurements on Schottky structure. With an exception of a few mm near the wafer edge, the doping inhomogeneity was very small, less than 7% in the $10^{14}\sim10^{15}\text{cm}^{-3}$ range, and less than 4% in the $10^{17}\sim10^{18}\text{cm}^{-3}$ range.

Even in Al-doped p-type materials, 4H-SiC exhibits higher mobility than 6H-SiC, though the mobility value itself ($10\sim80\text{cm}^2/\text{Vs}$) is smaller than electron mobility. The large ionization energy of Al acceptors ($190\sim240\text{meV}$) causes the activation ratio p/N_a (p: hole concentration, N_a: acceptor concentration) as low as $0.01\sim0.1$ at RT. Nevertheless, a high hole concentration over 10^{19}cm^{-3} is achievable in highly-doped epilayers, resulting in the lowest p-type resistivity of $0.025\Omega\text{cm}$ for 4H-SiC. Concerning the growth of high-resistivity SiC layers, vanadium doping during CVD has recently been reported [9].

C.2 Ion implantation

Ion implantation is a key technique for selective doping in SiC, because the low diffusion coefficients of impurities in SiC make the diffusion process impractical in real device fabrication. N^+ ions were implanted into Al- or B-doped p-type SiC epilayers with a net doping concentration of $3\sim9\times10^{16}\text{cm}^{-3}$ [10]. The multiple implantation was performed to obtain a box profile, junction depth of which is approximately $0.45\mu\text{m}$. During implantation, samples were kept at RT, or elevated temperatures of 500°C or 800°C. Post-implantation annealing was carried out at 1500°C for 30min in Ar ambience. Figure 6 shows the dependence of the sheet resistance on the total implanted dose [10]. In the case of RT implantation, the sheet resistance takes a minimum value of $770\Omega/\square$ at a dose of $8\times10^{14}\text{cm}^{-2}$. The higher–dose implantation resulted in the higher sheet resistance with lower activation ratios, due to residual damage which cannot be removed by the annealing. Although the effects of hot implantation are small in the low-dose region, the sheet resistance can be reduced at high dose, a low sheet resistance of 542Ω/□ being obtained at a 4x10^{15}cm^{-2} dose. The electrical activation ratio of implanted N atoms is estimated to be higher than 80%. Hot implantation is effective to suppress leakage current of pn diodes formed by N$^+$ implantation [10,11].

Al$^+$ and B$^+$ implantations have also successfully been applied for selective p-type doping [12,13]. The authors have investigated RT-implantation of Al$^+$ or B$^+$ ions into n-type 6H-SiC epilayers doped with 3~4x10^{15}cm^{-3}. The total dose was varied in the range from 1x10^{14} to 5x10^{15}cm^{-2}, and the junction depth was estimated to be 0.6~1μm. The acceptor concentration of the implanted layers was determined by C-V measurements on Schottky structure. Figure 7 denotes the annealing temperature dependence of activation ratio for Al$^+$ and B$^+$ implantations (implanted dose: 1x10^{14}cm^{-2}), indicating that high-temperature annealing is a key issue to activate implanted ions. Lower sheet resistance (22kΩ/□ at a 2x10^{15}cm^{-2} dose) could be obtained by Al$^+$ implantation compared to B$^+$ implantation (250kΩ/□) [12]. An even lower sheet resistance below 10kΩ/□ has been reported by utilizing hot implantation of Al$^+$ [14]. More recently, a success of co-implantation of Al$^+$/B$^+$ and carbon ions (C$^+$) has been demonstrated [15]: The C$^+$ co-implantation enhances the substitution of Al/B acceptors at the Si sub-lattice, leading to the higher activation ratio with lower resistivity.

Vanadium (V) atoms act as a deep amphoteric impurity, being a V(3d^1)^0/V(3d^2)$^-$ acceptor in n-type SiC and a V(3d^1)^0/V(3d^0)$^+$ donor in p-type SiC [16]. V$^+$ implantation into either n-type or p-type 6H-SiC epilayers was performed at RT followed by annealing at 1500°C [17]. A very low implanted dose required (typically the 10^{12}cm^{-2} range) leads to small lattice damage in implanted layers. The resistivity at RT exceeds 10^{13}Ωcm for p-type and 10^{7}Ωcm for n-type 6H-SiC [17]. Semi-insulating SiC layers formed by this technique have potential for device isolation [18], edge termination, reduction of parasitic impedance in high-frequency devices, and the realization of SiC ICs on insulators.

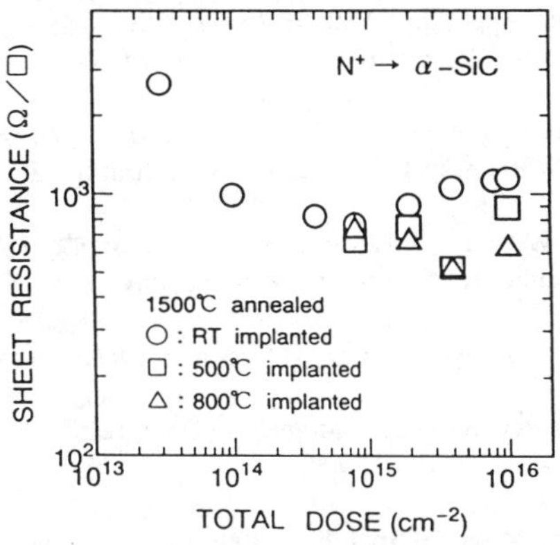

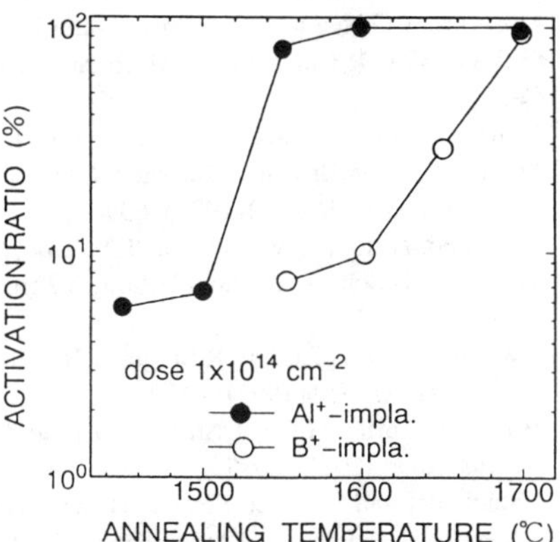

Fig.6 Dependence of the sheet resistance on the total dose and the implantation temperature of N$^+$-implanted layers annealed at 1500°C.

Fig.7 Annealing temperature dependence of activation ratio for Al$^+$ and B$^+$ implantations.

D. Conclusions

Recent progress in homoepitaxial growth and conductivity control of SiC has been reviewed. Step-controlled epitaxy provides the perfect polytype replication in epitaxial growth of SiC. Free exciton peaks in RT-PL, high electron mobility, and low deep level concentration demonstrate device-quality of SiC epilayers. The lowest doping level of unintentionally doped epilayers has been reduced down to $0.5\sim2\times10^{14}cm^{-3}$. *In-situ* n- and p-type impurity doping from the 10^{15} to $10^{19}cm^{-3}$ range has been realized with good uniformity. In N^+, Al^+, and B^+ implantations, high electrical activation ratios over 80% could be achieved by employing high-temperature annealing at 1500~1700°C. V^+ implantation can successfully be applied to obtain semi-insulating properties. These results show that the resistivity of SiC can be controlled over the widest range (from 10^{-3} to $>10^{13}\Omega cm$ at RT) among semiconductor materials.

Acknowledgments

The authors would like to express deep gratitude to Prof.W.J.Choyke at University of Pittsburgh and Dr.G.Pensl at University of Erlangen-Nürnberg for collaboration on characterization of SiC epilayers and fruitful discussion on ion implantation. They thank the late Dr.A.Itoh, N.Inoue, O.Takemura, T.Nakashima, and T.Yamamoto for their contribution to the present works. A part of this work was supported by a Grant-in-Aid for Specially Promoted Research from the Ministry of Education, Science, and Culture of Japan. Some of experiments were performed at Venture Business Laboratory of Kyoto University.

References

[1] *Silicon Carbide, III-Nitrides and Related Materials* (Trans Tech Publications, 1998), Chapt.7.

[2] V.F.Tsvetkov, S.T.Allen, H.S.Kong, and C.H.Carter, Jr., *Silicon Carbide and Related Materials 1995* (IOP, 1996), p.17.

[3] H.M.Hobgood, R.C.Glass, G.Augustine, R.H.Hopkins, J.Jenny, M.Skowronski, W.C.Mitchel, and M.Roth, Appl. Phys. Lett. **66** (1996), 1364.

[4] *for review, see* H.Matsunami and T.Kimoto, Mat. Sci. & Eng. **R20**(1997), 125.

[5] T.Kimoto, A.Itoh, and H.Matsunami, Appl. Phys. Lett. **66**(1995), 3645.

[6] A.Itoh, H.Akita, T.Kimoto, and H.Matsunami, Appl. Phys. Lett. **65**(1994), 1400.

[7] T.Dalibor, C.Peppermuller, G.Pensl, S.Sridhara, R.P.Devaty, W.J.Choyke, A.Itoh, T.Kimoto, and H.Matsunami, *Silicon Carbide and Related Materials 1995* (IOP, 1996), p.517.

[8] D.J.Larkin, P.G.Neudeck, J.A.Powell, and L.G.Matus, Appl. Phys. Lett. **65**(1994), 1659.

[9] B.E.Landini, G.R.Brandes, and M.Vollaro, *presented at 1997 Electronic Materials Conference* (Colorado, 1997), S2.

[10] N.Inoue, A.Itoh, T.Kimoto, H.Matsunami, T.Nakata, and M.Inoue, J. Electron. Mat. **26**(1997), 165.

[11] M.Ghezzo, D.M.Brown, E.Downey, J.Kretchmer, W.Hennesy, D.L.Polla, and H.Bakhru, IEEE Electron Device Lett. **12**(1992), 639.

[12] T.Kimoto, A.Itoh, H.Matsunami, T.Nakata, and M.Watanabe, J. Electron. Mat. **25**(1996), 879.

[13] T.Troffer, M.Schadt, T.Frank, H.Itoh, G.Pensl, J.Heindl, H.P.Strunk, and M.Maier, phys. stat. sol. **162**(1997), 277.

[14] J.W.Palmour, L.A.Lipkin, R.Singh, D.B.Slater, Jr., A.V.Suvorov, and C.H.Carter, Jr., Diamond and Related Materials **6**(1997), 1400.

[15] H.Itoh, T.Troffer, G.Pensl, *Silicon Carbide, III-Nitrides and Related Materials* (Trans Tech Publications, 1998), p.685.

[16] J.Schneider and K.Maier, Physica B **185**(1993), 199.

[17] T.Kimoto, T.Nakajima, H.Matsunami, T.Nakata, and M.Inoue, Appl. Phys. Lett. **69**(1996), 1113.

[18] M.P.Lam, K.T.Kornegay, J.A.Cooper, Jr., and M.R.Melloch, IEEE Electron Devices, **44**(1997), 907.

Stresses in 3C-SiC films grown on Si substrates

Chacko Jacob and Pirouz Pirouz
Dept. of Materials Science and Eng., Case Western Reserve University, Cleveland, OH 44106

Abstract

3C-SiC films were grown epitaxially on Si(001) substrates by an atmospheric pressure chemical vapor deposition method. The stresses in the films were determined by Raman spectroscopy and compared to data from load-deflection measurements on similar films. The films are tensile as grown and have a stress of 0.3 GPa, which is lower than the reported values for similar films. A possible explanation for the lower stresses as well as other observed trends is suggested.

A. Introduction

In order for 3C-SiC films epitaxially grown on Si substrates to be used in devices, it is important that the stresses in the films be well understood and controlled, as these stresses determine the flatness and usability of these wafers, as well as other subsequent device characteristics. These stresses do not arise from the lattice mismatch of 20%, but rather from the thermal mismatch of 8% and other factors. The lattice mismatch is accommodated within the first monolayer by the formation of misfit dislocations. However, the thermal coefficient mismatch cannot be avoided and the stresses develop as the film-substrate system cools down after growth. While a number of explanations have been offered in the past for these stresses in chemical vapor deposition (CVD) grown films [1], there is no clear quantitative explanation that explains all the data.

B. Experimental Results

The films were grown in an atmospheric pressure CVD (APCVD) reactor and the details of the growth have been previously reported [2]. The stresses were measured by micro-Raman spectroscopy and compared with data on films grown in the same reactor and measured by a load-deflection technique [3]. The calculations of stress are based on the work of Feng *et al.* [4]. It is well known that stresses cause a shift in the Raman peaks. These stresses can be estimated from the shifts in the LO (longitudinal optical) and TO (transverse optical) phonon positions. The peak positions for the films are tabulated in Table 1. From these values, the stress was determined.

A stress of 0.3±0.2 GPa was measured in a 0.3 μm thick film grown in the APCVD reactor at 1360 °C. This value of stress is lower than the values on similar films obtained by the load-deflection technique [3]. These films were all grown in the same reactor under identical conditions. The resolution of the micro-Raman measurements was lower than that of the load deflection technique and that could explain the higher stress value measured by Raman spectroscopy. It has already been established in the SiC/Si system, that there are no stresses due to the film-substrate lattice mismatch, because of the presence of misfit dislocations with the appropriate spacing at the interface. Thus, any residual stresses are due to the thermal mismatch and other intrinsic stresses.

The thermal mismatch stresses (σ_f) can be calculated for each film orientation, using

        275

$$\sigma_f = \Delta\alpha\Delta T\left(\frac{E}{1-v}\right) \tag{7}$$

where $\Delta\alpha$ is the difference in the coefficient of thermal expansion between the film and the substrate, ΔT is the difference between the growth temperature and room temperature, E is Young's modulus of the film and v is Poisson's ratio for the film.

Table 1 Raman peak positions used in the calculation of stress

	3C-SiC films
Ω_o^{LO} (cm^{-1})	965.56
Ω_o^{TO} (cm^{-1})	789.3
Ω^{LO} (cm^{-1})	965.56
Ω^{TO} (cm^{-1})	787.87
Stress (GPa)	0.3±0.2

The biaxial modulus $\left(E/1-v\right)$ is orientation dependent [5] and is given by

$$\left(\frac{E}{1-v}\right)_{100} = \frac{1}{S_{11}+S_{12}} \tag{8}$$

where S_{ij}s are the elastic compliances and they can easily be converted to elastic stiffnesses, C_{ij}s. A search of the existing data for elastic constants in the literature revealed that the old data published in the book edited by Marshall, Jr., Faust and Ryan [6] is of questionable value[7]. Feng et al. [7] suggest that Slack's data [8] for the elastic constants are possibly the most reliable. However, Lambrecht et al. [9] made some theoretical estimates of the elastic constants and arrived at values considerably different from those obtained by Slack. They also pointed out that Slack's data contained an obvious mistake and, after correction, this data is much closer to the theoretically calculated values. A more recent theoretical treatment of the problem of elasticity in cubic SiC [10] states that while three sets of "experimental" values are reported in literature, there are no directly measured experimental values of the elastic constants. They, too, calculated the elastic constants for 3C-SiC and their values along with those of Lambrecht et al.[9], and the corrected values of Slack [from Ref. 9], are listed in Table 2.

Table 2 Elastic constants of 3C-SiC

	C_{11}	C_{12}	C_{44}	S_{11}	S_{12}	S_{44}
Marshall, Faust and Ryan	289	234	55.4	12.6	-5.6	18.1
Slack (uncorrected)	540	180	250	2.22	-0.56	4
Slack (corrected)	440	148	207	2.74	-0.69	4.83
Lambrecht et al.	420	126	287	2.76	-0.64	3.48
Mirgorodsky et al.	428.2	165.4	246.2	2.98	-0.83	4.06

Units: C_{ij}: 10^9 Pa S_{ij}: 10^{-12} Pa^{-1}

From these values, the biaxial modulus values as well as the expected thermal stress for growth at a temperature of 1360 $^\circ$C is calculated and the results are shown in Table 3. Possibly a reliable estimate would be the average of the values from Lambrecht et al. and Mirgorodsky et al. [10]. This yields a value of 0.6 GPa for films grown on (100) Si substrates at 1360 $^\circ$C. Also, the ratio between the stresses on (111) and (100) substrates, $\sigma_{111}/\sigma_{100}$, is considerably smaller than 2.

Table 3 Calculated values of biaxial modulus and thermal stresses in 3C-SiC grown at 1360 $^\circ$C

	Biaxial modulus		Thermal stress		
	(111)	(100)	(111)	(100)	$\sigma_{111}/\sigma_{100}$
Marshall, Faust and Ryan	2.57	1.44	0.34	0.19	1.78
Slack (uncorrected)	7.11	6	0.94	0.79	1.18
Slack (corrected)	5.84	4.88	0.77	0.65	1.19
Lambrecht et al.	6.36	4.7	0.84	0.62	1.35
Mirgorodsky et al.	6.43	4.66	0.85	0.62	1.38

Units: Biaxial modulus: 10^2 GPa Thermal stress: GPa

C. Discussions

In the light of these calculations, the values of stress that we have measured on films grown in the APCVD reactor turns out to be lower than what could be generated by thermal stresses alone. To understand this relaxation, we need to take into account the additional causes of stress in the films. As mentioned earlier, internal stresses (alternately referred to as CVD stresses), are also present. In polycrystalline films, the formation of grain boundaries leads to the generation of additional tensile stresses [1]. However, for epitaxial films, when epitaxially oriented islands coalesce, they form domain walls. Koch [11] have shown that in this case, compressive stresses are generated as opposed to the tensile stresses generated by grain boundary formation in the case of polycrystalline films. This could explain the stress relaxation that occurs in the SiC films that we have grown. The compressive stresses generated by the coalescing of islands relaxes the thermal mismatch stresses to some extent, leaving a net tensile stress in the films. Since the films appear to be continuing to grow by island formation (Volmer-Weber growth) [12], this could explain the decreasing stress level with increasing thickness observed by Veprek et al.[13]. The thermal stress is not a function of thickness. Therefore, as the film thickness increases, the increasing compressive stresses generated due to the continuing coalescence of islands provides stress relief. This was confirmed over a limited film thickness range by Chandra [3], who reported values of 0.44 GPa, 0.34 GPa and 0.33 GPa for epitaxial films 1, 2 and 3 microns respectively, grown at 1360 $^\circ$C.

This model suggests that polycrystalline films of SiC should have a higher residual stress than epitaxial films as the tensile stresses generated by grain coalescence adds to the tensile stresses generated by thermal mismatch. This was also seen in the data reported by Chandra [3] where for a given film thickness, the polycrystalline film had a higher tensile stress. Finally, as the major stress component is due to the thermal mismatch, if it is possible to lower the temperature of epitaxial growth in our systems, we should expect to see lower stresses in the epitaxially grown films.

D. Conclusions

The stresses in the 3C-SiC films epitaxially grown on Si substrates are primarily due to the thermal mismatch between the film and the substrate. However, the measured stresses are lower than calculations based on elastic constants. An additional source of stress (compressive) as suggested by Koch [11] is due to the coalescence of islands and this is thickness dependent as the film grows by a Volmer-Weber mechanism. This is used to explain the lower measured stresses as well as the trend of decreasing stresses with increasing film thickness. Additionally, it is demonstrated that polycrystalline films should have a higher stress than single crystal films of the same thickness and that films grown at a lower temperature should have lower stresses.

Acknowledgments

This work was performed at Case Western Reserve University, Cleveland, Ohio under Defense Advanced Research Projects Agency (DARPA) Contract No. DABT63-95-C-0070. The authors would like to acknowledge Aaron Fleischman and Dr. Christian Zorman for growth of SiC and Chien-Hung Wu for useful discussions.

References

[1] R. W. Hoffman, in "Physics of Thin Films" vol. 3, ed. G. Hass and R. E. Thun (Academic: New York,1966) 211
[2] C A Zorman, A J Fleischman, A S Dewa, M Mehregany, C Jacob, S Nishino and P Pirouz, J. Appl. Phys. 78(8) (1995) 5136
[3] K. Chandra, "Characterization of Residual Stress and Elastic Modulus of Silicon Carbide Films Grown on Silicon by APCVD", M.S. Thesis, Case Western Reserve University (1997)
[4] Z.C. Feng, W.J. Choyke and J.A. Powell, J. App. Phys. 64(12) (1988) 6827
[5] D. Olego, M. Cardona and P. Vogl, Phys. Rev. B. 25(6) (1982) 3878
[6] R.C. Marshall, J. W. Faust and C. E. Ryan, Eds. Silicon Carbide-1973 (University of South Carolina, Columbia, SC, 1974) 668
[7] Z.C. Feng, A.J. Mascarenhas, W.J. Choyke and J.A. Powell 64(6) (1988) 3176
[8] G. A. Slack, J. Appl. Phys. 35 (1964) 3460
[9] W.R.L. Lambrecht, B. Segall, M. Methfessel and M. van Schilfgaarde, Phys. Rev. B 44 (8) (1991) 3685
[10] A. P. Mirgorodsky, M. B. Smirnov, E. Abdelmounim, T. Merle and P. E. Quintard, Phys. Rev. B 52(6) (1995) 3993
[11] R. Koch, J. Phys.: Condens. Matter 6 (1994) 9519
[12] C. Jacob, "Growth and Characterization of Epitaxial 3C-SiC Films on Silicon for Electronic Applications", Ph.D. Thesis, Case Western Reserve University (1997)
[13] S. Veprek, Th. Kunstmann, D. Volm and B. K. Meyer, J. Vac. Sci. Technol. A 15(1) (1997) 10

Structural and electronic properties of clean and defected Si-SiC(001) surfaces

Giulia Galli[*], Francois Gygi[*] and Alessandra Catellani[**]

[*] Lawrence Livermore National Laboratory, P.O.Box 808, Livermore, CA 94551, USA.
[**] CNR-MASPEC, Via Chiavari, 18/A, I-43100 Parma, Italy.

Abstract

We have studied the reconstructions and electronic properties of both clean and defected Si-terminated (001) surfaces of cubic SiC, by performing first principles computations within density functional theory. We find that the unstrained bulk exhibits a stable p(2x1) reconstruction, whereas a bulk under tensile stress shows a c(4x2) reconstruction. Furthermore our calculations indicate that ad-dimers are common defects on the Si-terminated SiC(001) surface. These results permit the interpretation of recent STM and X-ray-photoemission experimental data.

A Introduction

The characterization of SiC surfaces is an essential prerequisite to understanding the growth of silicon carbide, as well as its applications as a semiconductor for high-power, high-temperature and high-radiation environments [1].

Here we focus on the Si-terminated (001) surface of the cubic polytype of SiC (β-SiC), which is one of the most studied surfaces, although not yet well characterized [2]. There are two main issues which need to be addressed in order to understand the physical properties of SiC surfaces: (i) the influence of stress on the surface reconstruction and (ii) the modifications introduced by defects on the surface structure and electronic properties. Stress effects on cubic SiC surfaces represent a key issue since cubic SiC films are presently prepared by chemical vapor deposition on Si(001) substrates. The lattice mismatch between Si and SiC is almost 20 %, and thus SiC samples grown on Si are expected to be strained. Furthemore stoichiometric Si-terminated SiC(001) surfaces are prepared by evaporating excess-Si, and the presence of defects such as missing- and ad-dimers and missing- and ad-atoms are expected on the surface. The presence of defects on Si-SiC(001) is clearly visible from STM images [3] although the type of defects has not yet been identified.

In this paper we report the results of theoretical studies of the influence of stress (section B) and defects (section C) on the properties of Si-SiC(001). Our investigations have been carried out by performing a series of ab-initio molecular dynamics calculations within the local density functional approximation, using pseudopotentials and plane wave basis sets [4,5].

B Clean Si-terminated surface

Experimentally both p(2x1) [2] and c(4x2) [2,3] patterns have been observed on clean Si-SiC(001), with p(2x1) reconstructions being often seen in areas of missing dimers [3] and low coverage. The results of our calculations show that an unstrained bulk exhibits a p(2x1) reconstruction, which is under tensile stress. In agreement with Sabisch et al. [6], we have found that the p(2x1) reconstruction is characterized by dimer rows, with dimers much longer (2.6 Å) than those of Si(001) ($\simeq$ 2.3Å). These dimers are weakly bonded, with no important hybridization involved. The presence of weak bonds on the surface has not been confirmed experimentally, and the only fit to LEED data available in the literature points at shorter dimers [7]. However very recent ARUPS data [8] are consistent with models implying a weak bonding of the Si dimers. In our calculations we found that the p(2x1) reconstructed surface is non metallic at least up to 400 K. The surface has a gap between π^*-like antibonding states and σ bonding states. This is different from the electronic structure of the Si(001) and C(001) surfaces where the reconstruction opens a band gap between π and π^* surface states.

When applying small stresses to the cubic SiC bulk, we have observed a symmetry breaking of the surface reconstruction, leading to a c(4x2) pattern [4]. In our calculation, the c(4x2) surface geometry is characterized by alternating unbuckled short and long dimers, the short dimers having a component perpendicular to the surface smaller than the long ones. The dimer bond lengths are 2.54 and 2.62 Å in the case, e.g., of a 3 % strained bulk. This surface geometry is in agreement with the alternating-up-and-down-dimer (AUDD) model proposed on the basis of STM experiments [3]. Similarly to the p(2x1) reconstruction, the c(4x2) geometry exhibits a gap between antibonding occupied and bonding empty surface states.

Calculated STM images for both p(2x1) and c(4x2) reconstructed surfaces are displayed in Fig. 1. The top panels display the derivative of the tunneling current with respect to applied voltage: these images clearly show the π-like bonding states on the dimers. Bright spots appear on all dimers of the p(2x1) dimer rows (left); on the contrary only the up dimers are visible on the c(4x2) rows (right). Constant current plots at V=-1.5 eV (lower panels of Fig. 1) show instead surface states having large components *between* dimers. These are bonding and antibonding π-like states. On the p(2x1) surface bright spots are identical on all dimers, while on the c(4x2) surface they clearly show the difference in height between up and down dimers. When lowering the voltage from -1.5 to -3 eV in constant current plots, we found spots localized also on dimers, showing the difference in height between up and down dimers for the c(4x2) reconstruction. As expected, lowering the voltage makes the

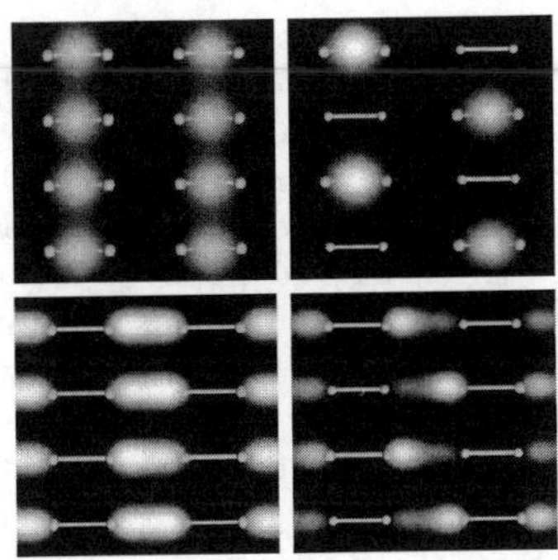

Figure 1: Computed STM images of the p(2x1) (left panel) and c(4x2) (right panel) reconstructions of Si-SiC (001): the upper and lower panels show a plot of $\frac{\partial I(x,y,z_0;V)}{\partial V}$ $\propto$ $\Sigma_i |\psi_i(x,y,z_0)|^2 f'(\epsilon_i + eV)$ (see text) and $I(x,y,z;V) = I_0$, respectively; in both cases V=-1.5 eV. Here $I(x,y,z;V)$ is the tunneling current computed within the Tersoff-Hamman approximation [10] at a given applied voltage V; the plot $I = I_0$ is drawn by considering all z values between the surface and the top of the slab such that $I = I_0$.
. From Ref. [4].

bonding π states have a larger contribution to the tunneling current. The image at V= -3 eV is in satisfactory agreement with measured [3] images at constant current, at the same voltage. However the measured images show larger components on dimers, indicating that in our calculation of $I(x,y,z;V)$, antibonding components of surface states are weighted more than experimentally. This difference between theory and experiment may be related to the tip induced electric field, which could change the energy separation between π and π^* surface states.

B Defected Si-terminated surfaces

In order to make contacts with several experimental results indicating the presence of defects on Si-terminated surfaces [2], we have studied the structural modifications induced by point defects on the surface structure.

Since p(2x1) reconstructions of Si-SiC(001) have been often seen in areas of missing dimers [3], we first investigated a p(2x1) reconstructed surface with missing dimers. We found that the removal of a dimer relieves surface stress and induces the formation of stronger bonds in four dimers surrounding the missing unit [4]. Nevertheless a dimer removal does not constitute a long ranged perturbation on the p(2x1) reconstruction, whose symmetry and dimer bond lengths are basically unchanged. The computed surface core level shifts (SCLS) [5] of atoms close to the missing dimer are similar to those of surface atoms on clean substrates

(0.9 eV, in our calculation), and vary between 0.8 and 1.1 eV.

We then considered an ad-atom and optimized two different surface geometries, with an extra atom between and on top of dimer rows, respectively. The configuration of minimum energy corresponds to the adatom sitting between rows, forming four long (2.50 Å), equivalent backbonds with the surface atoms. The SCLS of the adatom — computed in the configuration of minimum energy — is very low ($\simeq 0.2$ eV), compared to that of surface atoms.

Finally we considered ad-dimers on a p(2x1) terminated surface, and were inspired by recent investigations [9] of surfaces with excess Si atoms in order to determine a stable ad-dimer geometry. We have considered an ad-dimer *between* rows and optimized the total energy for geometries parallel and perpendicular to the p(2x1) dimer rows. We have found that the perpendicular ad-dimer has a total energy about 0.6 eV lower than the parallel one, and we have computed the SCLS for the configuration of minimum energy. The chemical shifts of atoms belonging to an ad-dimer are larger than those of surface atoms, i.e. $\simeq 1.4$ eV higher in energy than the bulk value. We note that the ad-dimer bond length is much smaller than those of surface dimers, 2.28 Å, and thus the chemical bond of the ad-dimer is expected to be different from that of the weak, much longer surface dimers.

Si-2p spectra [10] obtained in X-ray photoemission experiments show the presence of two peaks, a main peak (S) and a much less intense feature (S') centered at 0.5-0.7 eV and 1.2-1.7 eV above the bulk contribution, respectively. In view of the results discussed above for SCLS, we have suggested that while surface atoms are responsible for the main peak (S) observed experimentally in X-ray photoemission spectra, ad-dimers are responsible for the less intense S' feature. Our results [5] point at ad-dimers as common defects on Si-SiC(001) surfaces.

Work performed by the Lawrence Livermore National Laboratory under the auspices of the U. S. Department of Energy, Office of Basic Energy Sciences, Division of Materials Science, Contract No. W–7405–ENG–48.

[1] See, e.g., MRS bulletin **22**, (1997).

[2] For a review see V. Bermudez, Phys. Stat. Sol. A **201**, 1997.

[3] P. Soukiassian, F. Semond, L. Douillard, A. Mayne, G. Dujardin, L. Pizzagalli, C. Joachim, Phys. Rev. Lett. **78**, 907 (1997).

[4] A. Catellani, G. Galli, F. Gygi and F. Pellacini, Phys. Rev. B **57**, 12255 (1998).

[5] A. Catellani, G. Galli, F. Gygi, Appl. Phys. Lett. **72**, 1902 (1998).

[6] M. Sabisch, P. Krûger, A. Mazur, M. Rohling and J. Pollmann, Phys. Rev. B **53**, 13121 (1996).

[7] J. Powers, A. Wander, P.J. Rous, M.A. Van Hove and G.A. Somorjai Phys. Rev. B **44**,11159 (1991).

[8] P. Käckell, F. Bechstedt, H. Hüsken, B. Schröter and W. Richter, Surf. Sci. **391**, L1183 (1997).

[9] F. Semond, P. Soukiassian, A. Mayne, G. Dujardin, L. Douillard, and C. Jaussau, Phys. Rev. Lett. **77**, 2013 (1996); P. Soukiassian, F. Semond, A. Mayne, G. Dujardin, Phys. Rev. Lett. **79**, 2498 (1997).

[10] V. Bermudez and J. Long, Appl. Phys. Lett. **66**, 475 (1995).

[11] J. Tersoff and D. Hamann, Phys. Rev. B **31**, 805 (1985).

The 1.1 eV Deep Level in 4H-SiC

W. C. Mitchel(1), R. Perrin(1), J. Goldstein(1), M. Roth(1)*, S. R. Smith(2),
J. S. Solomon(2), A. O. Evwaraye(2), H. M. Hobgood(3)**, G. Augustine(3), and V. Balakrishna(3)

*(1)Materials Directorate, Air Force Research Laboratory, AFRL/MLPO, W-PAFB, OH
45433, USA*
(2)University of Dayton, 3000 College Park, Dayton, OH 45469, USA
(3)Nortrop Grumman Corp., Science and Technology Center, Pittsburgh, PA 15235, USA
*(Present Address: *Sterling Semiconductor, Reston, VA,*
***Cree Research, Inc., Durham, NC)*

Abstract: Temperature dependent Hall effect and optical absorption measurements of vanadium doped semi-insulating 4H-SiC samples are reported along with optical admittance spectroscopy measurements of related material. In addition to the deep donor and acceptor levels of substitutional vanadium, E_C-1.6 eV and E_C-0.7 eV respectively, we report an additional level at 1.1 eV. This level has been seen in undoped 4H-SiC by optical admittance spectroscopy which has also detected a similar level at E_C-1.0 eV in 6H-SiC, and by temperature dependent Hall effect in vanadium doped material. Optical absorption measurements of the vanadium intra-center absorption show that this level is not either of the two substitutional vanadium levels. SIMS measurements support the hypothesis that the 1.1 eV level is a complex of vanadium and another impurity, possibly titanium.

A. Introduction: Semi-insulating silicon carbide (SiC) is required for a variety of applications, including microwave FET's.[1] Doping with vanadium results in compensation of residual acceptors or donors and produces material that is nearly insulating at room temperature.[2,3] Vanadium substitutes for silicon in SiC and produces both a donor and an acceptor level. The acceptor level, however, is located only 0.7 eV below the conduction band. This leads to thermal excitation of trapped electron from the ionized vanadium centers to the conduction band and reduces the resistivity at elevated temperatures[4] so the donor level, located near mid-gap, is preferred.

In addition to the beneficial aspects of semi-insulating material, deep levels also limit mobility and act as lifetime limiting defects. Son et al. have reported a level in 6H-SiC at E_C-1.1 eV that is the dominant lifetime limiting defect in both the bulk and epitaxial material they studied.[5] Other deep levels have also been reported.[6,7] We have previously reported a level at 1.18 eV in 4H-SiC and a related level at 1.0 eV in 6H-SiC.[8,9] We report here further studies of this defect, including temperature dependent Hall effect measurements up to 1000K and optical absorption measurements on intentionally vanadium doped material.

B. Experimental Details: The samples were all fabricated from boules grown by the sublimation vapor transport technique. All samples in this study were grown along the c-axis and cut into wafers perpendicular to that direction. Square samples approximately 0.75 cm on a side were cut for van der Pauw measurements. Contacts for the semi-insulating samples were gold on tantalum annealed at 925°C in forming gas for two minutes. Transport measurements on the semi-insulating samples were made in a fully guarded DC system at temperatures from 300 to 1000K. The samples were immersed in nitrogen gas during the experiment. Schottky contacts for the optical admittance spectroscopy experiment diodes were unannealed aluminum. Ohmics contacts were made before the deposition of the Al by annealing Ni at 950°C for ten minutes in forming gas. Optical absorption measurements were made with a Cary model 5E spectrophotometer.

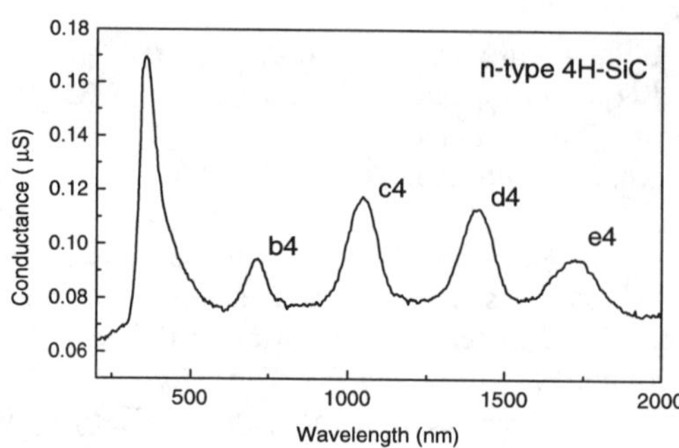

Fig. 1: Optical Admittance spectra for n-type 4H-SiC sample. b4 = 1.73 eV, c4 = 1.18 eV, d4 = 0.87 eV, and e4 = 0.72 eV.

C. Results: Figure 1 shows optical admittance spectroscopy data on an n-type 4H-SiC sample. The peak at 364 nm is due to the band gap. The peak at 716 nm (1.73 eV) has been shown to be due to the vanadium donor level, $V^{4+}(3d^1)/V^{5+}(3d^0)$. The peak at 1050 nm (1.18 eV) is the subject of this report. The peaks at 0.87 and 0.72 are unidentified but the 0.87 eV level might be the vanadium acceptor level, $V^{3+}(3d^2)/V^{4+}(3d^1)$. The 1.73 and 1.18 eV levels are seen in all bulk samples we have studied.

Temperature dependent Hall effect measurements up to 1000K were made on vanadium doped samples. We have observed two activation energies as seen in fig. 2. The carrier concentration versus inverse temperature data are straight lines suggesting that one level dominates each sample. The activation energies of these levels were determined by fitting the data to equation 1:

$$\log(nT^{-1.5}) = A - (E_a/kT) \qquad \text{eq. 1}$$

The activation energies are E_a = 1.0±0.1 eV and E_a = 1.6±0.1 eV. A total of three samples from two wafers were measured for the first sample with an average activation energy of 1.1 eV. These energies are similar to the value for the two deepest levels seen in optical admittance spectroscopy experiments discussed above. Furthermore, OAS experiments were run on a sample of the 1.1 eV material. The band edge and peaks b4 and c4 were seen but none of the shallower levels, indicating that they

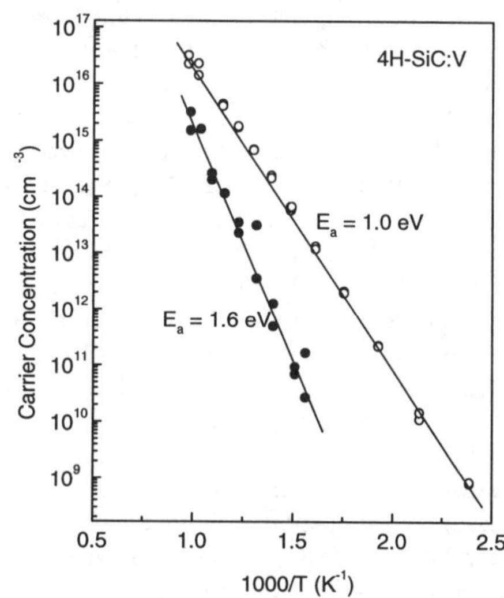

Fig. 2: Carrier concentration versus inverse temperature for samples from two different 4H-SiC:V boules showing two different activation energies.

were compensated and that the 1.1 eV TDH level corresponds to the 1.18 eV OAS level.

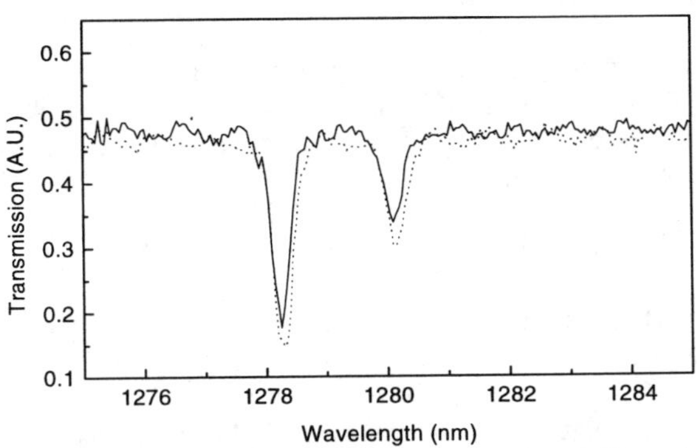

Fig. 3: Vanadium intracenter absorption spectra for 1.1 eV sample. Solid line: spectra before bandgap illumination. Dashed line: Spectra after band gap illumination.

In fig. 3 we show the results of absorption measurements at 10 K of the neutral vanadium intra-center transition, $3d^1E_2 \rightarrow {}^2T_2$, at 0.970 eV (1280 nm)[10] for the 1.1 eV sample from fig. 2. We compare the spectra before and after illumination with white light used to fill traps. There is no significant change in the spectra after illumination.

D. Discussion: There have been several report of deep levels in SiC around 1 eV. Achtziger and Witthuhn[6] have recently reported a level 0.97 eV below the conduction band in 4H-SiC that was doped with vanadium, titanium and chromium by neutron transmutation. They have correlated the 0.97 eV level with vanadium and attribute it to the acceptor level of substitutional, isolated vanadium. Hemmingsson et al.[11] have reported a level at 1.13 eV in electron irradiated 4H-SiC. They suggest it is a relatively stable complex but could not identify the components. Finally, Dalibor et al. [12] have reported several deep levels in oxygen implanted 4H-SiC near 1 eV and speculate that these levels are oxygen complexed with either vacancies, interstitials or both.

The agreement between the activation energies for the 1.1 eV level and the vanadium donor level as measured by Hall effect and optical admittance spectroscopy (OAS) suggests that Hall effect and OAS are looking at the same defects. Our absorption results show that the 1.1 eV level reported here can not be the vanadium acceptor level, since we would have seen an increase in the $3d^1$ absorption after illumination for the Fermi level being pinned at either the donor or acceptor level. For the Fermi level pinned between the two isolated vanadium levels, however, the donor level would be completely occupied, resulting in maximum possible $3d^1$ absorption. We have previously reported increases in the $3d^1$ absorption after illumination for 4H-SiC:V samples with the Fermi level pinned at both the donor and acceptor levels.[13]

Secondary Ion Mass Spectrometry (SIMS) was used to investigate the purity of the samples. Vanadium and titanium were found in all samples. The concentrations for the 1.1 eV sample are 1.9×10^{17} and 4.7×10^{17} cm^{-3} respectively, showing that the titanium concentration exceeded the vanadium concentration in this vanadium doped sample. The two concentrations in the 1.6 eV sample are 4.8×10^{17} and 4.7×10^{17} cm^{-3} respectively. A Cs ion source was used to detect oxygen but this element was below the detection limit, which is below 10^{18} cm^{-3}, in both samples. At present we cannot correlate the 1.1 eV level with any of the impurities or dopants. Isolated substitutional Ti can be ruled out, however, because its level structure is well known and it does not produce deep levels near 1 eV.[14] SIMS measurements of oxygen are inconclusive due to the high background. Due to it's near universal presence in bulk material, we suggest that it is the 4H equivalent of the lifetime limiting defect observed in 6H material by Son et al.[5]

Acknowledgments: The authors are pleased to acknowledge the assistance of Mr. Robert Leese, Mr. G. Landis, and Mr. Robert Culp. AOE was supported in part by a National Research Council Senior Resident Associateship. SRS and JSS were supported in part by Air Force Contract F33615-95-C-5445.

References:

[1] S. Sriram, R. C. Clarke, A. A. Burk, H. M. Hobgood, P. G. McMullen, P. A. Orphanos, R. R. Siergiej, T. J. Smith, C. D. Brandt, M. C. Driver, and R. H. Hopkins, IEEE Electron Device Letters, 15 (1994) 458.

[2] H. M. Hobgood, R. C. Glass, G. Augustine, R. H. Hopkins, J. Jenny, M. Skowronski, W. C. Mitchel and M. Roth, Appl. Phys. Lett. 66 (1995) 1364.

[3] J. R. Jenny, M. Skowronski, W. C. Mitchel, H. M. Hobgood, R. C. Glass, G. Augustine, and R. H. Hopkins, J. Appl. Phys. 78 (1995) 3839.

[4] W. C. Mitchel, R. Perrin, J. Goldstein, M. Roth, M. Ahoujja, S. R. Smith, A. O. Evwaraye, J. S. Solomon, G. Landis, J. Jenny, M. McD. Hobgood, G. Augustine, and V. Balakrishna, *Silicon Carbide, III-Nitrides and Related Materials*, edited by G. Pensl, H. Morkoç, B. Monemar, and E. Janzén, Mat'ls Sci. Forum 264-268 (1998) 545.

[5] N. T. Son, E. Sörman, W. M. Chen, O. Kordina, B. Monemar, and E. Jansén, Appl. Phys. Lett. 65 (1994) 2687.

[6] N. Achtziger and W. Witthuhn, Appl. Phys. Lett. 71 (1997) 110.

[7] T. Dalibor and G. Pensl, Phys. Rev. 55 (1997) 13618.

[8] A. O. Evwaraye, S. R. Smith, and W. C. Mitchel, J. Appl. Phys. 79 (1996) 7726.

[9] A. O. Evwaraye, S. R. Smith, and W. C. Mitchel, J. Appl. Phys. 79 (1996) 253.

[10] J. Schneider, H. D. Muller, K. Maier, W. Wilkening, F. Fuchs, A. Dörnen, S. Leibenzeder, and R. Stein, Appl. Phys. Lett. 56 (1990) 1184.

[11] C. Hemmingsson, N. T. Son, O. Kordina, J. P. Bergman, E. Janzén, J. L. Lindström, S. Savage, and N. Nordell, J. Appl. Phys. 81 (1997) 6155.

[12] T. Dalibor, G. Pensl, T. Yamamoto, T. Kimoto, H. Matsunami, S. G. Sridhara, D. G. Nizhner, R. P. Devaty, and W. J. Choyke, *Silicon Carbide, III-Nitrides and Related Materials*, edited by G. Pensl, H. Morkoç, B. Monemar, and E. Janzén, Mat'ls Sci. Forum 264-268 (1998) 553.

[13] J. Jenny, J. Skowronski, W. C. Mitchel, H. M. Hobgood, R. C. Glass, G. Augustine, and R. H. Hopkins, Appl. Phys. Lett. 68 (1996) 1963.

[14] T. Dalibor, G. Pensl, N. Nordell, and A. Schöner, Phys. Rev. B 55 (1997) 13618.

A NEW TECHNIQUE FOR THE DETECTION OF STRUCTURAL DEFECTS IN SiC BULK CRYSTALS

T.S.Argunova*, J.Baruchel**, J.Härtwig**

*Ioffe Physico-Technical Institute, St.Petersburg, Russia; e-mail: argunova@tania.ioffe.rssi.ru
**European Synchrotron Radiation Facility, Grenoble, France; e-mail: haertwig@esrf.fr

Abstract. **Structural defects in sublimation sandwich grown SiC wafers of a large area have been studied with diffraction, absorption and phase contrast imaging techniques by using highly parallel X-ray flux. The contrast appeared due to either tilt, strain or density gradients between a matrix and a defect. Coherence properties of hard X-ray beams of third-generation synchrotron radiation sources such as the ESRF, were utilized to produce phase contrast in a free-space propagation mode. The nature of the observed defects has been confirmed by scanning electron and optical microscopy data.**

A. Introduction

One of the key problems of nowadays production of large-area SiC single crystals is to get rid of structural defects. Characterization techniques to assist in this effort should be nondestructive and sensitive to different types of defects. Up to now, X-ray diffraction methods have been the basic analytical tools for these purposes [1]. The most significant problem in growing large-area SiC single crystals is the appearance of micropipes. However, it occurs that a high sensitivity of diffraction imaging to lattice distortions make the observation of micropipes generally difficult. It has been shown that the detection of pipes is possible in absorption or radiographic mode [2]. The method works well when a soft radiation impinges upon a sample with rapid density gradients. For low absorbing matter, the shorter are the wavelengths, the weaker is the absorption contrast.

A possibility to improve radiographic imaging is to make use of phase contrast. When the coherence properties of X-ray beams from a third-generation synchrotron source with very low emittance are utilized, the phase contrast could be observed in a free-space propagation mode described in [3-4]. This imaging process, presentable as Fresnel diffraction, transforms the phase shifts appearing due to density gradients inside a transparent object into an amplitude modulation at a sufficiently large distance from the object. Interference pattern arises from the superposition of refractively deflected and non-deflected waves. Unique capabilities of the Synchrotron Radiation (SR) source allowed us to detect phase contrast by a simplest way without the use of a multi crystal set up [5-7] performed elsewhere than at European Synchrotron Radiation Facility (ESRF), when less coherent or larger X-ray sources are utilized.

B. Samples and Techniques.

The studied samples were produced by the Sublimation Sandwich Method (SSM) first proposed in 1970 [8]. SSM growth of bulk SiC crystals is described in [9]. Sample diameters could be varied from 30 up to 65 mm. The samples were 4H polytype N doped in the range $3\ 10^{17}$-$5\ 10^{18}$.

The X-ray experiments were performed at ID-19 Beamline at (ESRF), Grenoble, France, providing a highly parallel coherent X-ray beam in the energy range 8-100 keV. A small source dimension-0.15x0.03 mm^2 (HxV)-and a large source-to-object distance equal to 150 m make the angular size of the source be not more than 1x0.2 μrad^2 (HxV). Diffraction imaging was realized in the transmission geometry with the white beam as well as the mono-chromatic mode. The wavelength used in the monochromatic set up was 0.2 Å. The scheme for obtaining phase contrast in a free-space propagation mode is presented in Fig. 1. The sample is out of the Bragg diffraction position. The distance between the sample and the film is 2 m. In an absorption mode, this distance is reduced to few cm. The film used to record the images provided an image resolution of about 1 μm.

C. Results and Discussion.

For studying typical crystalline defects the samples with a low structural quality were specially chosen. Diffraction images of these samples showed a high density of as-grown dislocations as well as long range distortion fields. Because of the relative misorientation between the different parts of a crystal it was difficult to detect the images of micropipes. Applying the phase contrast (PC) method, the situation changed radically. The PC mode is not sensitive to misorientation and strain. An image appeared due to absorption and refraction only, i.e. due to density variations in the crystal.

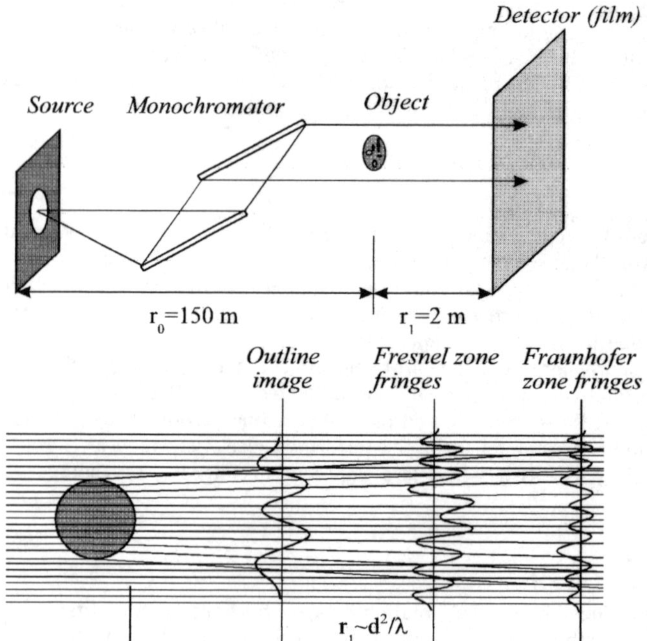

Fig. 1.

Schematic display of the set up used in SR phase contrast imaging.

Principle of X-ray phase contrast imaging in a free space propagation mode. Different degrees of refraction result from different areas of the wave front at inhomogeneities of the sample.

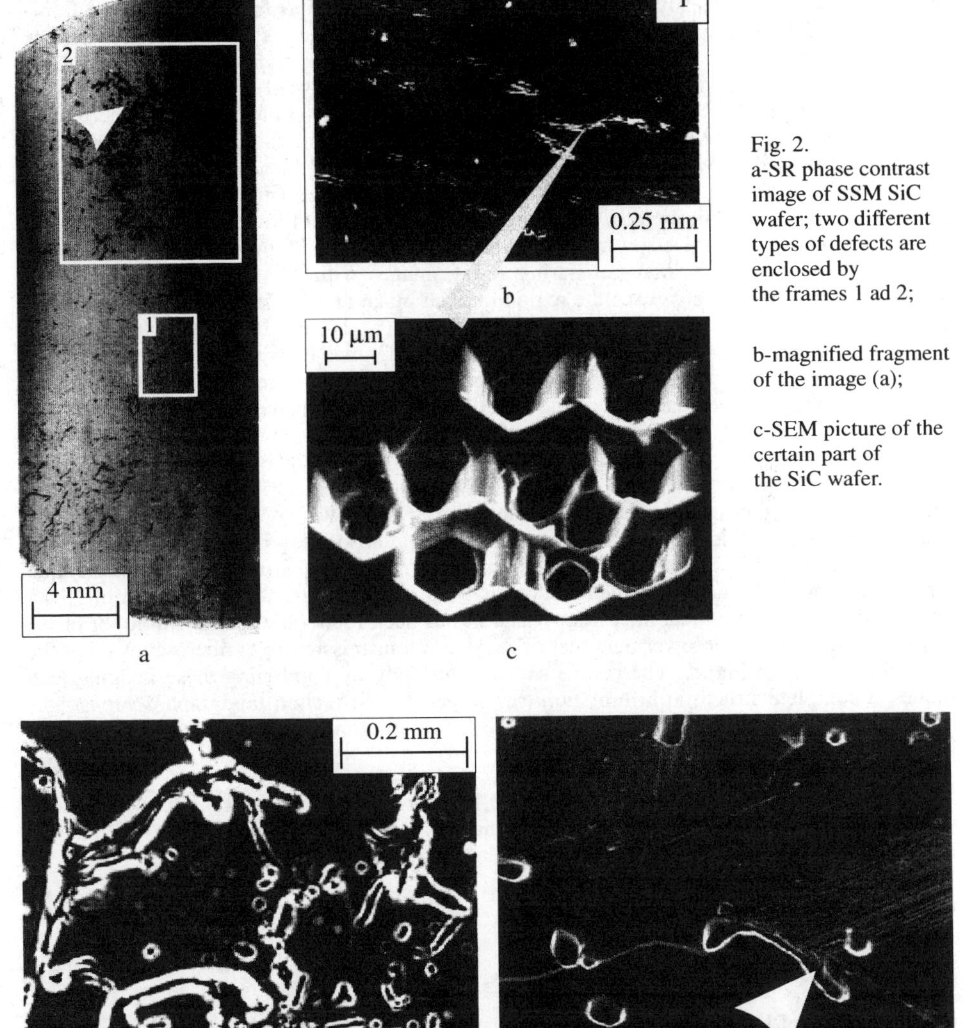

Fig. 2.
a-SR phase contrast image of SSM SiC wafer; two different types of defects are enclosed by the frames 1 ad 2;

b-magnified fragment of the image (a);

c-SEM picture of the certain part of the SiC wafer.

Fig. 3. a-magnified fragment of the PC image in Fig. 2 a; the picture corresponds to the place enclosed by the frame 2; b-SEM image of the same place of SSM SiC wafer.

In Fig. 2 a one can see a whole view of SR PC image. On the image the frames 1 and 2 enclose the maximum density of defects which can be described as follows. Inside the frame 1 (see in Fig. 2 b a magnified fragment) small tubular shaped defects are seen. These defects extend along a line of sample surface miscut. The frame 2 envelopes thick irregular lines forming sophisticated figures disorderly distributed over the crystal area. On an absorption image registered at 10 cm detector-to-sample distance, small tubular shaped defects were totally absent. The visibility of the macro defects was still satisfied. To reveal the nature of the defects, we used SEM technique to investigate the surface of the samples. It appeared that the small tubular shaped defects were associated with micropipes (Fig. 2 c). The conclusion that these defects were micropipes, that is the dislocations with hollow cores, followed from a distinct visibility of spiral steps inside the tubes. The nature of the sophisticated shaped figures also became clear after comparing X-ray images with SEM and light microscopy results. In Fig. 3 (a-b) one can see the region marked by an arrowhead in Fig. 2. A large size inclusion of an other polytype seen on the SEM image (b), is invisible on the X-ray picture (a) which is understandable due to vanishing density difference between polytypes. Over that polytype inclusion several irregular shaped cavities are seen. Micropipes are also present. This damaged region is explainable due to a sublimation growth on the seed holder.

D. Conclusions.

We have reported the results of the investigation of SSM grown SiC wafers by X-ray phase contrast imaging combined with diffraction imaging and absorption radiography. Micropipes in large-area SiC wafers have been nondestructively imaged by the phase contrast technique for the first time. The nature of the defects has been confirmed by the comparison of X-ray images with SEM and optical microscopy data. It has been found that the sensitivity of phase contrast imaging was not lower than that of SEM and transmission light microscopys, but the resolution was even higher. The results showed that only by combining these imaging techniques, a complete structural information can be gained: diffraction topograph demonstrates misorientation and strain, while the PC pictures show micropipes and cavities in the crystal interior.

Acknowledgements. The authors acknowledge Russian Fund of Basic Research (Grants No 97 02 18331 and No 98 02 18309) for the financial support of the presented work and St. Petersburg Joint Research Centre for the assistance in doing measurements.

References

[1] M. Dudley, W. Si, S. Wang, C. Carter, R. Glass, and V. Tsvetkpv, II Nuovo Cimento 19D, 153 (1996).

[2] D. Black and L. Robbins, Inst. Phys. Conf. Ser.No. 137, 337 (1993).

[3] A. Snigirev, I. Snigireva, A. Suvorov, M. Kocsis, and V. Kohn, ESRF Newsletter No. 24, 5 (1995).

[4] P. Cloetens, R. Barrett, J. Baruchel, J.P. Guigay, and M. Schlenker, J. Phys. D: Appl. Phys. 29, 133 (1996).

[5] V. Ingal and E. Beliaevskaya, J. Phys. D: Appl. Phys. 28, 2314 (1995).

[6] D. Gao, T.J. Davis, and S. Wilkins , J. Phys. 48, 103 (1995).

[7] A. Momose, Abstract Booklet Int. Conf. "Highlights in X-ray Synchrotron Radiation Research, Grenoble, France (1997).

[8] Y.A. Vodakov and E.N. Mokhov, Pat. USSR N 403275 (1970); Pat. USA B4147575 (1979).

[9] E.N. Mokhov, M.G. Ramm, A.D. Roenkov and Yu.A. Vodakov, Mater. Sci. Eng. B46, 317 (1997).

XPS study of dependence of Au/6H-SiC Schottky barrier height on carrier concentration.

N.A.Suvorova*, A.V.Shchukarev*, I.O.Usov**, A.V.Suvorov***
*   Regional Analytical Centre Analyt, Surface Diagnostic Laboratory, St.Petersburg, Russia
**   Ioffe Physical Technical Institute, St.Petersburg, Russia
***   Cree Research Inc. Durham, NC, USA

SiC is a wide band gap semiconductor attractive for use in high-frequency and high-power devices (FET's and IMPATT's). To realize such devices, the Schottky contact is preferable to p-n junctions because this enables low-temperature processing and facilitates further improvement of device performance. Since the performance of Schottky-based devices depends on barrier height, the optimum barrier height leading to high performance should be investigated in detail. A large number of studies on Schottky contacts to SiC were reported [1] (and references therein). It was established that barrier height depends on metal work function and SiC polytype, orientation, crystallinity and the surface treatment process before the contacts formation. To the authors knowledge, however, no data have been reported for barrier heights dependence on carrier concentration. In the present work we report a Au/n-6H-SiC contacts Schottky barriers height dependence on uncompensated donor concentration (N_d-N_a). The barrier heights were determined by capacitance-voltage (C-V) and X-ray photoelectron spectroscopy (XPS) techniques.

The Au contacts were prepared on n-type doped with nitrogen Si-face 6H-SiC epitaxial layers obtained from CREE Research Inc. The wafers were treated by using wet chemical etching in 40% solution of NH_4F:HF followed by cleaning in deionized water. After the treatment procedure Au was deposited onto room-temperature surfaces by resistance evaporation in vacuum. Thick Au circular contacts with diameter of 9×10^{-2}cm were formed for C-V measurements and thin film of Au with thickness ~20Å was deposited for XPS analysis. The ohmic contacts were made by Ni on the back side of the wafers for C-V measurements. The donor concentration of 6H-SiC wafers are given in Table I.

Table I. Schottky barrier heights ϕ_b^{XPS} of Au/6H-SiC measured by XPS method and calculated values of barrier heights decreasing caused by image forces $\Delta\phi_{bi}$ and tunneling effect $\Delta\phi_t$ at different N_d-N_a.

N_d-N_a , cm^{-3}	E_{C1s}^{Au} , eV	ϕ_b^{XPS} , eV	$\Delta\phi_{bi}$, eV	$\Delta\phi_t$, eV
2.0×10^{15}	282.64	1.52	0.025	0.0069
1.0×10^{16}	282.76	1.40	0.038	0.0118
3.1×10^{16}	282.76	1.40	0.051	0.017
1.03×10^{17}	282.94	1.22	0.069	0.018
9.2×10^{18}	283.06	1.10	0.21	0.115
2.5×10^{19}	283.07	1.09	0.27	0.16

C-V measurements were made with a capacitance meter operating at 10kHz and reverse bias from 0 to -2V. The C-V data were analyzed according to the conventional model [2] for an abrupt interface in which the Schottky barrier height ϕ_b^{C-V} is directly related, by additive constants, to the intercept on the voltage axis of a $1/C^2$ vs V plot. The estimated ϕ_b^{C-V} is independent of the donor concentration and averaged value of ϕ_b^{C-V}=1.62eV was obtained.

XPS analysis was carried out with electron spectrometer PHI 5400 (Perkin Elmer,USA) $h\nu=1253.6$eV Mg K_α x-ray source. XPS C1s, O1s, Si2p core levels and valence band spectra 6H-SiC and Au 4f core level for Au/6H-SiC were measured. A core level binding energy was referenced to the Au $4f_{7/2}$ peak at 84.0eV.

The calculation of Schottky barrier height ϕ_b^{XPS} was made using the technique described in [3]. The inset of Fig.1 shows the relationship between the C1s binding energy E_{C1s} and E_F^i and ϕ_b^{XPS}. As seen in Fig.1, the interface Fermi energy $E_F^i = E_{1Cs}^{Au}-(E_{1Cs}-E_V)$, were $(E_{1Cs}-E_V)$ is the C1s to valence-band maximum binding energy difference in 6H-SiC, a bulk material constant. A value of $(E_{1Cs}-E_V) = 281.3\pm0.1$ eV was measured by using XPS valence-band data obtained from a surface before metal deposition, which agrees well the work of J.R.Waldrop et.al. [3]. The position of E_V was found from a linear extrapolation of the leading edge of the valence and density of states to zero intensity. The XPS measured Schottky barrier height is thus $\phi_b^{XPS} = 2.86 - E_F^i$, where 2.86 is the 6H-SiC band gap.

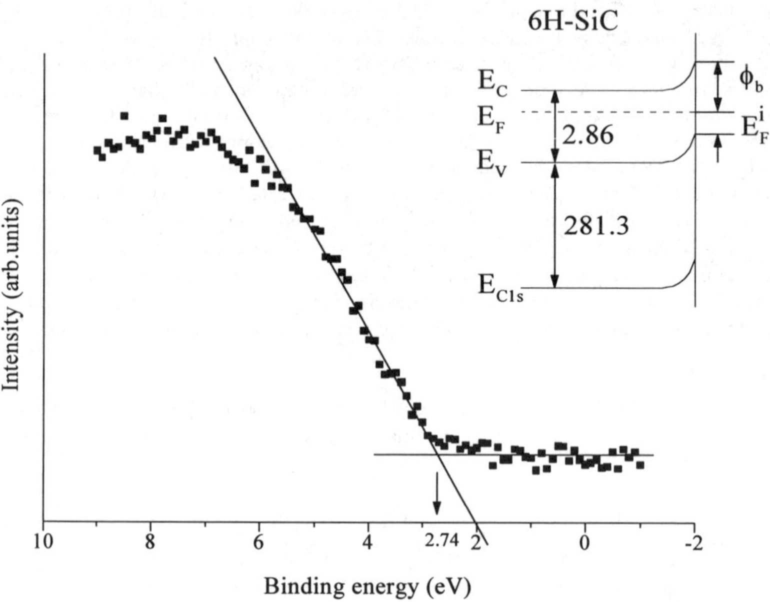

Figure 1. Linear extrapolation of the leading edge of the 6H-SiC valence band measured by XPS before Au deposition. Inset shows the relationship between ϕ_b and C1s binding energy associated with SiC at the Au/SiC interface.

The valence band, C1s and Si2p spectra of 6H-SiC are shown in Fig.2a, b and c, respectively. Fig.2d shows the C1s core level spectrum Au/6H-SiC. (The low-energy C1s peak corresponds to Si-C chemical bond, the high-energy peak to hydrocarbon contaminations.)

The values of Schottky barriers ϕ_b^{XPS} derived from XPS spectra are given in Table I. Fig.3 shows ϕ_b^{XPS} dependence on donor concentration. As seen in Fig.3, the barrier height reduces when the N_d-N_a increasing. In the concentration range from 2×10^{15} to 2.5×10^{19} cm^{-3}, the value of ϕ_b^{XPS} went down to 0.43eV.

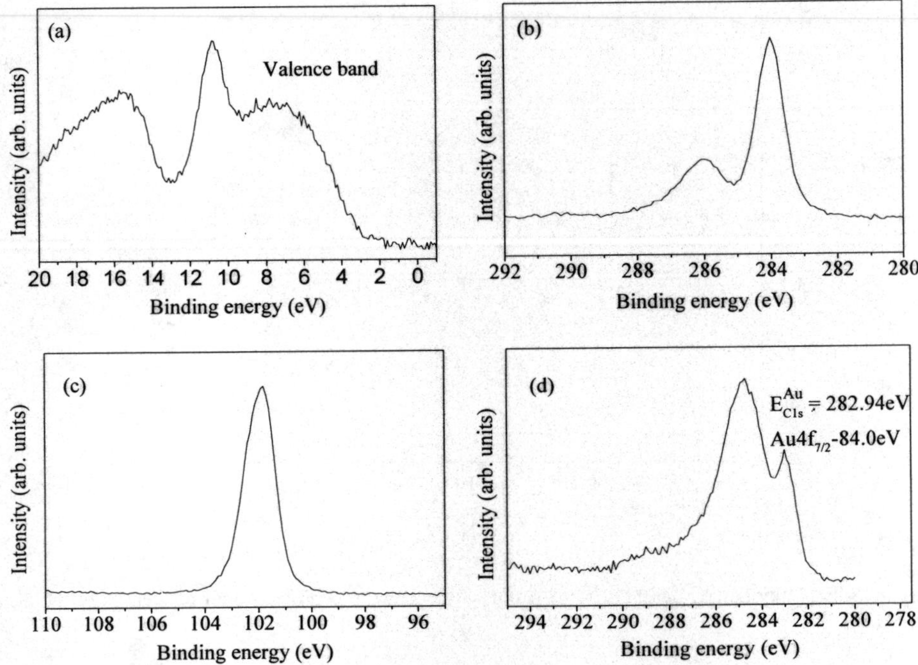

Figure 2.   XPS spectra of (a) valence levels and (b) C1s, (c) Si2p peaks of 6H-SiC with N_d-N_a=1.03×10^{17} cm^{-3} before Au deposition and (d) C1s peak after Au deposition.

In according to Bardin-Schottky-Mott theory the barrier height ϕ_b on metal/semiconductor contact depends on a number of parameters. One of such parameters is $\Delta\phi_{bi}$ - the value of reduction of barrier height caused by image forces [4]:

$$\Delta\phi_{bi} = \left[\frac{q^3(N_d - N_a)\left(\phi_b - \xi - \dfrac{kT}{q}\right)}{8\pi^2\varepsilon_d^2\varepsilon_s\varepsilon_0^3} \right]^{\frac{1}{4}} \tag{1}$$

where q - electron charge, ε_s and ε_d - relative statically and high frequency dielectric constants of semiconductor, respectively, ε_0 - dielectric constant of vacuum; ξ - energy difference between bottom of conduction band and Fermi level.

Another parameter influencing on $\Delta\phi_b$, is barrier height decrease due to electron tunneling through barrier $\Delta\phi_t$ [4]:

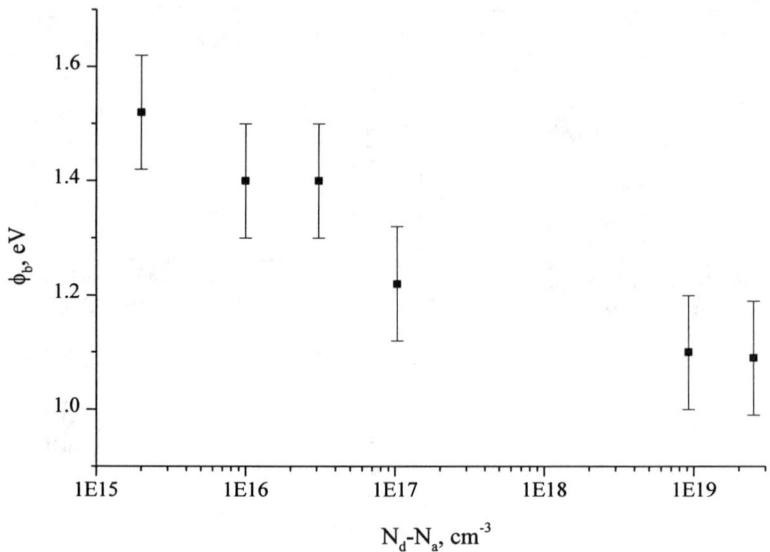

Figure 3. Schottky barrier height ϕ_b^{XPS} of Au/6H-SiC vs uncompensated donor concentration N_d-N_a of 6H-SiC.

$$\Delta\phi_t \approx 1.31 \times \left[\frac{\hbar}{2} \times \left(\frac{N_d - N_a}{m_{d,e}^* \varepsilon_s \varepsilon_0} \right)^{\frac{1}{2}} \right]^{\frac{2}{3}} (V_d)^{1/3} \qquad (2)$$

where V_d - diffusion potential and $m_{d,e}^*$ - density of state effective mass of the conduction band.

The calculations of $\Delta\phi_{bi}$ and $\Delta\phi_t$ was carried out for studied range of donor concentration and given in Table I. It is seen that $\Delta\phi_{bi}$ and $\Delta\phi_t$ magnitudes increase with N_d-N_a increasing and leads to barrier height ϕ_b lowering. It should be pointed out, that the barrier height values calculated from C-V measurements are not affected by the $\Delta\phi_{bi}$ and $\Delta\phi_t$ [4] and in consequence ϕ_b^{C-V} is usually greater than ϕ_b^{XPS} [3]. Also this difference is remained at low donor concentration, when the barrier lowering due to $\Delta\phi_{bi}$ and $\Delta\phi_t$ is negligibly small. This result is caused by that the barrier height obtained from XPS is less sensitive to the problems characteristic for electrical (current-voltage and capacitance-voltage) measurements. The electrical measurements for determining the height of Schottky barrier at a metal-semiconductor contact is based upon a number of assumptions, which may not be valid in practice[2].

References

1. A.Itoh and H.Matsunami Phys.Stat.Sol. (a) 162, 389 (1997)
2. A.M.Goodman J.App.Phys. 34(2),329 (1963)
3. J.R.Waldrop et.al. J.Appl.Phys. 72(10), 4757 (1992)
4. E.H.Rhoderic Metall-Semiconductor Contacts (Clarendon Press, Oxford, 1978).

IX.

Wide Gap II-IV Compounds

Residual strain in a zinc blende Mn$_{1-x}$Mg$_x$Te layers grown by molecular beam epitaxy

E. Dynowska, J. Bąk-Misiuk, E. Janik, J.Trela, T. Wojtowicz

Institute of Physics Polish Academy of Sciences,
al. Lotników 32/46, 02-668 Warsaw, Poland.
e-mail: dynow@ifpan.edu.pl

Abstract

The residual strain in Mn$_{1-x}$Mg$_x$Te layers grown by Molecular Beam Epitaxy (MBE) on CdTe/GaAs(001) substrates has been investigated by X-ray diffraction methods. We confirmed the existence of a tensile strain in the case of layers covered by an additional CdTe cap while in uncapped layers we found a compressive strain. The thermal expansion measurements between 300 - 520 K revealed an anisotropy of the in-plane and out-of-plane thermal expansion coefficients. This helps to understand occurrence of the compressive strain of uncapped layers in terms of a tentative model of relaxation process in both types of layers.

A. Introduction

Molecular beam epitaxy (MBE) has been previously successfully used to grow new materials otherwise not available by traditional equilibrium growth techniques. Diluted magnetic semiconductors (DMS) with very high magnetic ion concentrations, such as Cd$_{1-x}$Mn$_x$Te or Zn$_{1-x}$Mn$_x$Te with x>0.7, are examples of such possibilities of the MBE technique. Moreover, MBE can be used for fabricating materials with crystal structures different from those occurring normally in nature, such as cubic versions of CdSe, MnTe or MgTe [1-5]. Recently, the growth, structural, optical and magnetic properties of a novel DMS system grown by MBE Mn$_{1-x}$Mg$_x$Te with $0 \leq x \leq 1$ have been reported [6]. This new system was expected to be a material well suited for magnetic barriers in quantum heterostructures.

The preliminary structural studies described in [6] revealed that Mn$_{1-x}$Mg$_x$Te layers possessed cubic structure with the unit cell tetragonally distorted at room temperature. This distortion indicated that the layers were residually strained. The knowledge of the strain state of the layers is very important for a quantitative explanation of some physical properties of the heterostructures. In particular, it is necessary for a determination of the composition of the layers consisting of ternary compounds. Our initial measurements yielded an unexpected result: for the layers with $0 \leq x \leq 0.8$ a residual *compressive* strain was observed, while in MgTe layers - a *tensile* strain was found. This finding was interesting because, based on the lattice mismatch alone, only a *tensile* strain could be expected. We speculated that the origin of the *compressive* strain could be in the different thermal expansion coefficients of CdTe and Mn$_{1-x}$Mg$_x$Te. Because thermal expansion coefficients were not known for the cubic modifications of Mn$_{1-x}$Mg$_x$Te and MgTe, suitable temperature measurements had to be done before our conjecture could be verified. The results of such measurements are reported in this paper.

The presence of a *tensile* strain was confirmed in the MgTe layers which were grown on a thick (2 $\square$m) CdTe buffer and covered by a thick (2 $\square$m) CdTe cap layer in contrast to

$Mn_{1-x}Mg_xTe$ layers grown on thin (1000 Å) CdTe buffers without capping. It was not clear if the tensile strain was due to the thickness of the buffer or the presence of the cap or both.

B. Experimental

$Mn_{1-x}Mg_xTe$ epilayers were grown on (001)-oriented GaAs ($2°$ off toward [110]) covered by a thin (~1000Å) CdTe buffer. The growth temperature was about 600 K. The thickness of the films ranged from 3 to 5 $\Box$m with x varying from 0 to 1. Layers with Mg molar fraction exceeding 0.8 decomposed by hydration within several days, but in the case of pure MgTe, the „life time" was several minutes. To prevent hydration, the MgTe films were covered by 2 $\Box$m thick CdTe cap layers. Under such protection the layers have been stable for several weeks. Three samples, two MgTe/CdTe(thick buffer) and one $Mn_{1-x}Mg_xTe$ /CdTe(thin buffer) were additionally covered by 2 $\Box$m thick CdTe cap layers.

A conventional X-ray powder diffractometer (phase analysis) as well as a high resolution Philips Material Research Diffractometer (HRXD) in double and triple crystal configuration were used for structural characterisation. The crystalline quality of the layers was checked by measurements of the full width at half maximum of the 004 rocking curves [6] and reciprocal space maps. The lattice parameter in the layer plane, $a_{\Box\Box}$, and along the growth direction, $a_{\Box}$, were calculated from direct measurements of the doubled Bragg angle positions [7] for symmetrical 004 and asymmetrical 335, 117 and 115 reflections. Because of a lack of knowledge of the elastic constants in the investigated material, the relaxed lattice parameters were calculated assuming that the volume of the tetragonally distorted unit cell is the same as that of the relaxed cubic cell. The relaxed lattice parameter was used to determine the molar fraction in $Mn_{1-x}Mg_xTe$ with the help of Vegard's rule $x=11.905(a_{MnMgTe}-a_{MnTe})$. The thermal expansion coefficients were calculated from the lattice parameter measured by the Bond method [8] in the temperature range 300 - 520 K. A decrease of the lattice parameter and a change of the colour of the layers observed near the upper limit of the temperature range indicated the proximity of a structural phase transition.

C. Results and discussion.

The results of our measurements in all studied samples are listed in Table 1. The most interesting results are the values of in-plane ($a_{\Box\Box}$) and out-of-plane ($a_{\Box}$) lattice parameters because they allow to determine the strain state of the layers. The strain parallel to the interface was calculated according to the formula: $\Box_{\Box\Box} = - (a_{\Box\Box} - a_{rel})/a_{rel}$. As it is seen the layers grown without cap layers exhibit a residual *compressive* strain ($a_{\Box\Box} < a_{\Box}$) while the layers under caps - a *tensile* strain ($a_{\Box\Box} > a_{\Box}$).

Generally, the residual strain should be a result of the lattice mismatch as well as the differences in the thermal expansion coefficients of the layer and the buffer (and/or cap). These two sources of the strain can either act in one direction or in opposite directions and lead to a compensation of the strain. Assuming that the lattice mismatch between the CdTe buffer and the $Mn_{1-x}Mg_xTe$ layer is the sole source of the strain we can expect only a *tensile* strain ($a_{CdTe} > a_{MnMgTe}$). We supposed then the observed compressive strain was due to the thermal mismatch occurring during the post-growth cooling process. If the thermal expansion coefficient of $Mn_{1-x}Mg_xTe$ is smaller than the in-plane thermal expansion coefficient of the CdTe/GaAs substrate ($\Box_{\Box\Box CdTe} = \Box_{GaAs} = 6.7 \cdot 10^{-6}K^{-1}$) and if $Mn_{1-x}Mg_xTe$ layer is almost fully relaxed at the growth temperature, then the observed compressive strain can be justified. To test this possibility HRXD thermal expansion measurements were carried out.

Table 1.

The results of the lattice parameter measurements of $Mn_{1-x}Mg_xTe$ films epitaxially grown on CdTe/GaAs substrates for the directions parallel and perpendicular to the film plane. Also included are information about samples and the calculated values of: relaxed lattice parameter(a_{relax}), composition (x) and strain parallel to the interface ($\varepsilon_{//}$).

SAMPLE number	CdTe buffer	CdTe cap layer	$a_\perp$ [Å]	$a_\parallel$ [Å]	a_{relax} [Å]	x	$\varepsilon_\parallel$ ú10^4
101796A	~1000Å	----	6.341	6.336	6.338	0.0	+3.1
011797B	~1000Å	----	6.349	6.344	6.346	0.09	+3.1
011697C	~1000Å	----	6.357	6.348	6.351	0.15	+4.7
112596A	~1000Å	----	6.365	6.357	6.360	0.26	+4.7
073197A	~1000Å	----	6.370	6.358	6.362	0.29	+6.2
022597B	~1000Å	----	6.382	6.377	6.379	0.49	+3.1
080197A	~1000Å	----	6.392	6.382	6.385	0.56	+4.7
022697A	~1000Å	----	6.397	6.392	6.394	0.67	+3.1
022797A	~1000Å	----	6.410	6.402	6.405	0.80	+4.7
050798A	~1000Å	~2 μm	6.380	6.394	6.389	0.61	-7.8
031197A	~2 μm	~2 μm	6.418	6.422	6.421	1.0	-1.6
030497A	~2 μm	~2 μm	6.406	6.430	6.422	1.0	-12.5

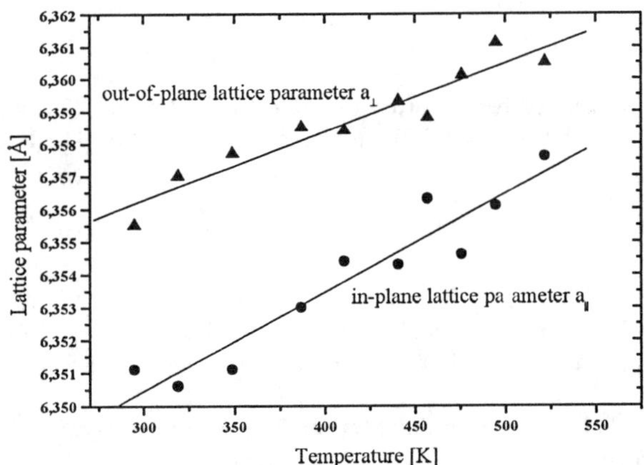

Fig.1 The temperature dependence of the in-plane and out-of-plane lattice parameters for the uncapped $Mn_{0.83}Mg_{0.17}Te$ layer.

The results obtained for one of the samples ($x=0.17$) are shown in Fig.1. On the basis of these measurements the in-plane thermal expansion coefficient $\square_{\square\square}$ and that along the growth direction, $\square_\square$, were found to be $\square_{\square\square}=4.704 \cdot 10^{-6}K^{-1}$ and $\square_\square=3.3 \cdot 10^{-6}K^{-1}$. From these values the relaxed thermal expansion coefficient for $Mn_{1-x}Mg_xTe$ is estimated: $\square_{rel}\square4.2 \cdot 10^{-6}K^{-1}$. This value is, in fact, smaller than that in the CdTe buffer. The difference between the thermal expansion coefficients $\square_{\square\square}$ and $\square_\square$ explains, then, the *compressive* strain of the $Mn_{1-x}Mg_xTe$

films. However, these results do not allow to establish if the observed residual strain arises is „purely" thermal strain. For this to be true, the in-plane thermal expansion coefficients of $Mn_{1-x}Mg_xTe$ and CdTe/GaAs substrate should have been the same.

To explain the tensile residual strain observed in layers covered by CdTe caps we think that at the growth temperature such a $Mn_{1-x}Mg_xTe$ layer is considerably more strained that in the previous case. A very strong tensile strain caused by lattice mismatch can not be completely compensated by the thermal mismatch during the cooling. The reason for such strong tensile strain in the layers covered by CdTe must be a special feature of a relaxation process which must occur in both layers: the increasing of the in-plane lattice parameter of the CdTe involves the increasing of the in-plane lattice parameter in the $Mn_{1-x}Mg_xTe$ layer. A similar model was proposed to explain anomalous interface stress between ZnSe and GaAs by Fujiwara in [9] .

D. Conclusions

The residual strain of the $Mn_{1-x}Mg_xTe$ layers grown by MBE on CdTe/GaAs [001] substrates were systematically investigated by X-ray diffraction methods. We found that at room temperature all the layers were residually strained and we noticed that the sign of this strain depended on whether or not the layers were covered by a CdTe cap. The temperature measurements of $a_{\square\square}$ and $a_{\square}$ allowed to determine the in-plane and out-of-plane thermal expansion coefficients. Their values help to understand the presence of *compressive* strain observed in the uncapped layers. These layers at the growth temperature must be almost fully relaxed. On the contrary, the layers covered by CdTe are more strained at the growth temperature, because the cooling after the growth does not change the strain sign.

Acknowledgements

This work was partially supported by State Committee for Scientific Research under grant PBZ 28.11. The Philips MRD diffractometer was founded by The Foundation for Polish Science (program SEZAM 94).

References

1. S.M.Durbin, J.Han Sungki, O.M.Kobayashi, D.R.Menke, R.L.Gunshor, Q.Fu, N.Pelekanos, A.V.Nurmikko, D.Li, J.Gosvales and N.Otsuka, Appl.Phys.Lett. **55** (1989) 2087.
2. W.Faschinger, F.Hauzenberger, P.Juza, A.Pesek and H.Sitter, J. Electron. Mater. **22** (1993) 497
3. J.R.Bushert, F.C.Peiris, N.Samarath, H.Luo and J.K.Furdyna, Phys.Rev. B **49** (1994) 4619.
4. E.Janik, E.Dynowska, J.Bak-Misiuk, M.Leszczynski, W.Szuszkiewicz, T.Wojtowicz, G.Karczewski, A.K.Zakrzewski, J.Kossut, Thin Solid Films **267** (1995) 74.
5. E.Dynowska, E.Janik, J.Bak-Misiuk, J.Domagala, T.Wojtowicz and J.Kossut, to be published
6. E.Janik, E.Dynowska, J.Bak-Misiuk, T.Wojtowicz, G.Karczewski, J.Kossut, A.Stachow-Wojcik, A.Twardowski, W.Mac, K.Ando, J.Cryst.Growth **184** (1998) 976.
7. P.F.Fewster, N.L.Andrew, J. Appl. Cryst. **28** (1995) 451
8. W.L.Bond, Acta Cryst. **13** (1960) 814
9. Y.Fujiwara,S.Shirakata, T.Nishino, Y. Hamakawa and S.Fujita, Jpn. J. Appl. Phys. **25** (1986) 1628.

Electrical and optical properties of as-grown and electron-irradiated ZnO

D. C. Look, D. C. Reynolds, J. W. Hemsky
University Research Center, Wright State University, Dayton, OH 45435
J. R. Sizelove, R. L. Jones, C. W. Litton, T. Wille
Air Force Research Laboratory, Wright-Patterson Air Force Base, OH 45433
G. Cantwell and W. C Harsch
Eagle-Picher Industries, Inc., 200 B. J. Tunnell Blvd., Miami, OK 74354

Large-diameter (up to 3-inch), n-type ZnO boules grown by a new vapor-phase transport method were studied by temperature-dependent Hall-effect (TDH) and photoluminescence (PL) measurements. From fits to the TDH data, the dominant donor has a concentration N_D of about 1 x 10^{17} cm^{-3} and an energy of about 60 meV, close to the expected hydrogenic value, whereas the total acceptor concentration N_A is much lower, about 2 x 10^{15} cm^{-3}. The 2-K PL data include a series of at least seven sharp lines over the energy range 3.356 - 3.366 eV, and from polarization, magnetic-field, and annealing (up to 1000 °C) experiments, these lines are interpreted as donor-bound excitons associated with donor complexes having different spacings between the atomic species. Electron-irradiation experiments show an electrical and optical threshold, attributed to Zn displacements, at an electron bombardment energy of about 1.3 MeV. The irradiation creates acceptors and destroys shallow donors, but these effects are produced at much lower rates than those found in most other common semiconductor materials, such as Si, GaAs, and GaN.

A. Introduction

The need for blue/uv light emitters, and for high-power/high-temperature transistors has accelerated research and development in wide-bandgap semiconductor materials, especially SiC and GaN [1,2]. Progress in the latter, mostly since 1988, has led to blue lasers with projected life times in excess of 10,000 hours [1], and modulation-doped field-effect transistors with output powers of 2.6 W/mm at 10 GHz [3]. However, a third widegap semiconductor material, with properties very similar to those of GaN, is ZnO [4]. Although ZnO has found commercial application in many different areas (phosphors, piezoelectric transducers, paints, conductive transparent coatings, etc.) it has not yet been explored as an electroluminescent emitter. This is somewhat surprising in light of its strong *photoluminescent* emission in the green and uv regions, but can perhaps be explained by the difficulty in producing p-type ZnO, which is thought necessary for strong *electroluminescence*. Several very recent events will undoubedly produce increased research activity in the future: (1) large-area ingot growth [5]; (2) the demonstration of optical lasing [6]; and (3) the demonstration of p-type material [7]. In fact, it is noteworthy that ZnO has several advantages over the current "hot" material, GaN: (1) a lattice-matched substrate; (2) a 60-meV (vs. 20-25-meV) exciton binding energy; and (3) easier processing with wet chemical etches. In light of these facts, we have undertaken a detailed study of the electrical and optical properties of bulk ZnO, grown by a seeded vapor transport technique. We have also investigated point-defect properties by employing 1-2 MeV electron irradiation. As we will show below, ZnO has an additional advantage in that it is quite resistant to displacement damage from high-energy electrons.

B. Electrical Properties

Two-and three-inch ZnO ingots have been grown by a seeded vapor transport technique developed by Eagle-Picher, Inc.. Typical values of the carrier concentration n and mobility μ at 300 K are 5-10 x 10^{16} cm^{-3} and 200-215 cm^2/V s, repectively[5]. Peak mobilities (at about 50 K), often exceed 2000 cm^2/V s, which is comparable to the best values ever reported for ZnO. Temperature-dependent mobility data, μ vs. T, are shown in Fig. 1 (solid line). These results, along with n vs. T data (not shown) are simultaneously fitted to the Boltzmann transport equation, and change-balance equation, respectively, and yield the following fitting parameters: donor concentration $N_D = 1$ x 10^{17} cm^{-3}; acceptor concentration $N_A = 2$ x 10^{15} cm^{-3}; and dominant donor activation energy $E_D \approx 60$ meV. We note that, for a hydrogenic system, $E_D \approx 13.6m^*/\varepsilon_0^2 \approx$ 66 meV, which is close to the experimental value. However, we don't yet know whether this donor is an impurity, such as Al$_{Zn}$, or a native defect, such as an O vacancy or Zn interstitial. A very surprising result of this study is the low acceptor concentration, only about 2 x 10^{15} cm^{-3}, which leads to a very low compensation ratio, $N_A/N_D \approx 0.02$. Again a firm identification of the acceptor is not possible at this time, although irradiation experiments, discussed below, give some insight.

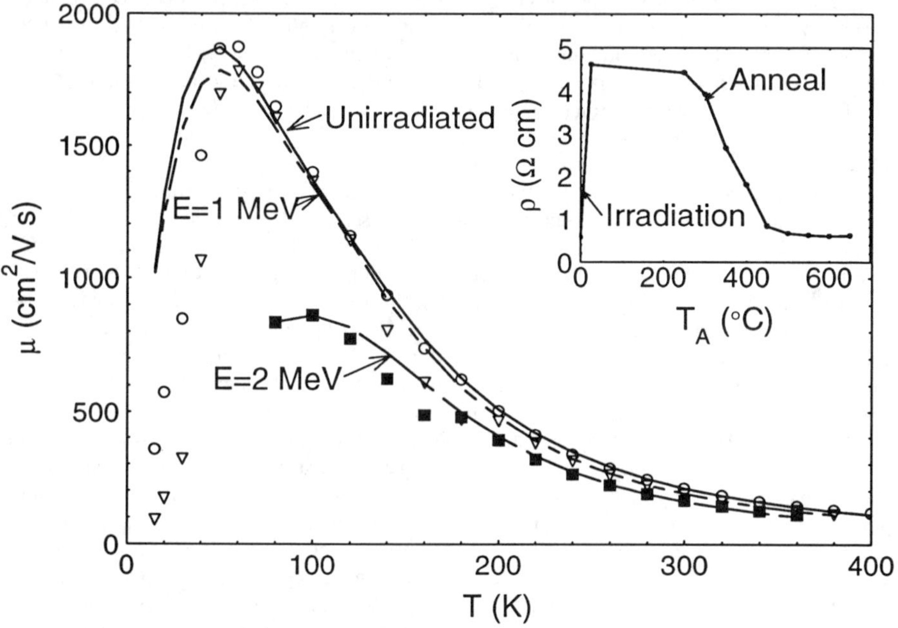

Fig. 1. Mobility vs. temperature plots for bulk ZnO irradiated with electrons of different energies.

C. Optical Properties

Photoluminescence (PL) spectra, such as those shown in Fig. 2, display sharp lines in the near-band-edge (NBE) region, around 3.358-3.366 eV. The difference between these energies and the bandgap, which is about 3.435 eV at 2 K, represents the internal exciton binding energy (≈ 0.06 eV) plus the exciton-donor binding energy (≈ 0.01 eV). Thus, the PL lines shown in Fig 2 are assigned to neutral-donor-bound exciton (D^0X) transitions [4]. Besides the D^0X lines, typical spectra also show the usual assortment of other spectral features, due to free-exciton (FE), neutral-acceptor-bound-exciton (A^0X), free-electron and -hole (e-A and h-D, respectively), and donor-acceptor-pair (D-A) transitions. However, one of the best-known features is a broad band near 2.4 eV, the "green" band, which makes ZnO useful as a phosphor material. In fact, most ZnO samples, including ours, will exhibit a greenish glow when illuminated with a uv lamp or laser, even though the uv emission (invisible) is typically thousands of times stronger.

Returning to the D^0X transitions, we see from Fig. 2 that an anneal at 800 °C eliminates most of the high-energy lines in favor of the lowest, at 3.359 eV. This fact prompts us to assign the various D^0X lines to impurity and/or defect pairs, at different separations. These PL lines are polarized, which also is consistent with a pair model. The pairs may be formed while defects and impurities are congregating into strained regions during crystal cooling; conversely, there may be a coulomb attraction which draws the entities together. In any case, the deeper the donor, the lower the bound-exciton transition energy, by Haynes' Rule, and the deeper donors are probably associated with closer pairs. Then, an 800-°C anneal must allow the elements of the pair to move closer together and enhance the lowest-energy D^0X lines.

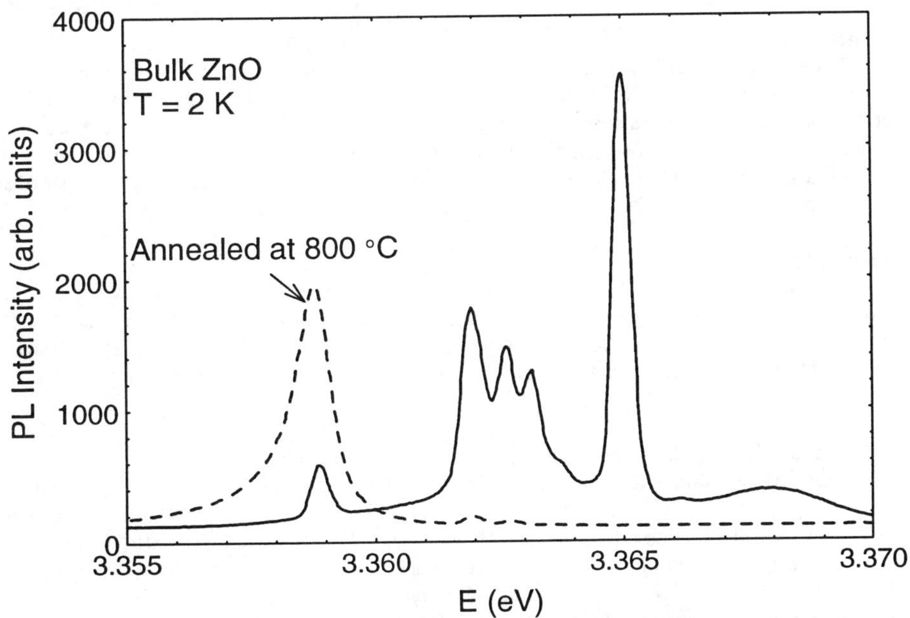

Fig. 2. Photoluminescence spectra in the near-band-edge region for as-grown and annealed ZnO.

D. Irradiation Effects

Irradiation with high-energy electrons decreases the mobility and carrier concentration, increases the resistivity (see Fig. 1), and also enhances the 3.359-eV PL line (not shown). Such irradiation is expected to produce Zn and O Frenkel pairs, i.e., V_{Zn}-Zn_I and V_O-O_I, respectively. Since V_{Zn} is a strong acceptor, and since irradiation on the Zn face produces acceptors at a higher rate than irradiation on the O face, we believe that the V_{Zn}-Zn_I pairs probably dominate. The damage anneals out at about 350 °C, as shown in the inset of Fig. 1, and such a low annealing temperature is also consistent with the recombination of Frenkel pairs. The 3.359-eV PL line, then, likely results from an exciton bound to a Zn_I donor, with a V_{Zn} acceptor nearby. The fact that this line anneals out at 350 °C in irradiated material, but is stable at 800 °C in as-grown material (see Fig. 2), is probably due to strain in the latter case. That is, defects may be attracted to strained regions during the growth process, and such strain may stabilize even close pairs, which would not be stable if created in unstrained regions, as would occur during electron irradiation. However, confirmation of this model must await further experimentation.

The effects of the irradiation on donor and acceptor concentrations can be determined from μ vs. T and n vs. T data. From fits to the data of Fig 1, we can show that the acceptor production rate, $\Delta N_A/\Delta F$, is less than 0.1 cm^{-1} at E =1 MeV, and only about 0.4 cm^{-1} at 2 MeV. These production rates are much lower than those in Si, GaAs[8], or even GaN[9], and suggest that ZnO might be very attractive for certain space applications. As mentioned earlier, since the rate is higher for Zn-face irradiation than for O-face irradiation, we believe that the dominant acceptor being created is V_{Zn}.

E. Conclusions

Excellent, bulk ZnO can be grown by a seeded vapor-phase process, with values of N_D and N_A only about 1×10^{17} and 2×10^{15} cm^{-3}, respectively. The near-band-edge photoluminescence is dominated by several neutral-donor-bound-exciton lines, which are probably associated with defect pairs of various spacings. High-energy electron irradiation is consistent with such a pair model, although the damage produced by the electrons is much less than that measured in Si, GaAs, or GaN.

Acknowledgments

We wish to thank T. A. Cooper for Hall-effect measurements, W. Rice for photoluminescence measurements, and D. Scales for manuscript preparation. D.C.L. and D.C.R. were supported under U. S. Air Force Contract F33615-95-C-1619, and DARPA Contract F33615-95-C-1751; also, G.C. and W.C.H. were supported under the latter contract.

References

1. S. Nakamura, M. Senoh and S. Nagahama et al., Appl. Phys. Lett. 72, (1998) 211.
2. R. J. Trew, M. W. Shin and V. Gatto, Solid-State Electronics 41 (1997) 1561.
3. Y-F. Wu, B. P. Keller and P. Fini et al., IEEE Electron Device Lett. 19 (1998) 50.
4. D. C. Reynolds, D. C. Look, B. Jogai and H. Morkoç, Solid State Commun. 101 (1997) 643.
5. D. C. Look, D. C. Reynolds and J. R. Sizelove et al., Solid State Commun. 105 (1998) 399.
6. D. C. Reynolds, D. C. Look and B. Jogai, Solid State Commun. 99 (1996) 873.
7. K. Minegishi, Y. Koiwai and Y. Kikuchi et al., Jpn. J. Appl. Phys. 36 (1997) L1453.
8. D. C. Look, Appl. Phys. Lett. 51 (1987) 843.
9. D. C. Look, D. C. Reynolds and J. W. Hemsky et al., Phys. Rev. Lett. 79 (1997) 2273.

MBE growth, structural studies and a new optical phenomenon in CdF$_2$-CaF$_2$:Eu superlattices on Si(111)

N.S.Sokolov, S.V.Gastev, A.Yu.Khilko, R.N.Kyutt, L.M.Sorokin, S.M.Suturin, D.B.Vcherashnii, P.D.Brown*, C.J.Humphryes*

Ioffe Physico-Technical Institute, 194021 St.Petersburg, Russia
**University of Cambridge, Cambridge CB 3QZ, UK*

New CdF$_2$-CaF$_2$:Eu superlattices (SLs) were grown from molecular beams and structurally characterized by high resolution X-ray diffractometry and high resolution transmission electron microscopy (HRTEM). A new optical phenomenon in luminescence of Eu^{2+} ions in the SLs was revealed. It was found that the Eu^{2+} luminescence intensity under continious optical excitation drastically decreases in a few seconds after the beginning of the excitation. Possible application of the phenomenon in optical memory cells is discussed.

A. Introduction

It is well-known that considerable efforts have been undertaken for growing, structural characterization and optical studies of superlattices (SLs) based on such semiconductors as Ge, Si, III-V and II-VI compounds, and they resulted in numerous exciting findings. On the other hand, only few studies of structures involving ionic wide bandgap materials are known [1,2]. Though blue shifts of self-trapped exciton absorption lines were observed, their interpretation as due to quantum confinement was shown to be doubtful [3]. Cadmium and calcium fluorides have the same cubic crystal structure and a lattice constant mismatch of opposite signs relative to silicon (-0.78% and 0.6% respectively). This circumstance and the remarkable electronic properties of CdF$_2$ [4] make the growth and studies of CdF$_2$-CaF$_2$ SLs very attractive. Epitaxial growth of such SLs has recently been demonstrated [5]. Their structural and luminescence properties were studied later by the same group [6]. New results on the growth, structural characterization and optical studies of CdF$_2$-CaF$_2$:Eu SLs are presented in this paper.

B. Growth from molecular beams

The superlattices were grown in a small research MBE-system equipped with reflection high energy electron difractometer (RHEED) for *in situ* structural characterization of grown layers. The layer thickness was measured *in situ* with a laser interferometer. Before the growth, thoroughly oriented Si(111) substrates (with miscut less than 5 arc min) were chemically cleaned. Then a thin oxide layer was evaporated in the growth chamber at 1250°C. A distinct 7x7 Si(111) superstucture in RHEED pattern indicated the atomically clean surface of the substrate. The fluoride layers were grown from molecular beams at a rate of 2-3 nm/min. Doped layers were grown from the effusion cell loaded with CaF$_2$:Eu. The RHEED patterns that were observed during the growth showed a good crystalline film quality. Further *ex situ* characterization of the surface morphology was performed by an atomic force microscope produced by NT-MDT (Zelenograd, Russia). Contact mode measurements showed that flat areas dominated on the SL surface. The SLs studied in this work consisted of a periodic sequence of alternating CaF$_2$ and CdF$_2$ layers. The period ranged from 2 to 40 nm and their total thickness was 100-350 nm. CaF_2 layers of most SLs were doped with Eu in 0.1 mol % concentration. Selectively doped SLs were also grown with 10 ML thick spacers of pure CaF_2 introduced at each CdF_2-CaF_2:Eu interface.

C. Structural characterization

The structural SL parameters have been studied by high resolution X-ray diffractometry and high resolution transmission electron microscopy (HRTEM). A large difference between the Cd and Ca scattering abilities (Z_{Ca}=20, Z_{Cd}=48) provides a distinct sattelite pattern on the diffraction curves even for short-period SLs. X-ray measurements have been carried out in a wide range of diffraction conditions including symmetrical and asymmetrical Bragg and Laue [7] geometries. By using the semikinematical approach, a simulation of the rocking curves has been performed and the SL parameters such as thickness, strains and structural factors of individual layers as well as the interface abruptness, have been determined. The structural perfection of the SLs was analysed considering the sattelite shape and broadening. It was shown that the short period SLs have high perfection with the ω-curve width of the sattelites close to the theoretical values. For example, one of the SLs with a 6 nm period showed an additional broadenning of the ω-curve induced by a few structural defects as low as 4 arc. sec. Though SLs with a period larger than 10 nm demonstrated broadened sattelite peaks, most of the SLs studied were pseudomorphic, i.e. all the layers were coherent to each other and to the Si substrate.

Electron microscopy measurements have been performed using a JEM 4000 EXII microscope and sample cross-sections prepared by a standard technique including ion milling. Fig. 1 shows a HRTEM (110) image of CdF_2-CaF_2 SL cross-section with 28 periods. Each period includes 6 monolayers of CdF_2 and 6 monolayers of CaF_2. The fluoride layers can be easily distinguished owing to the higher absorption by cadmium atoms.

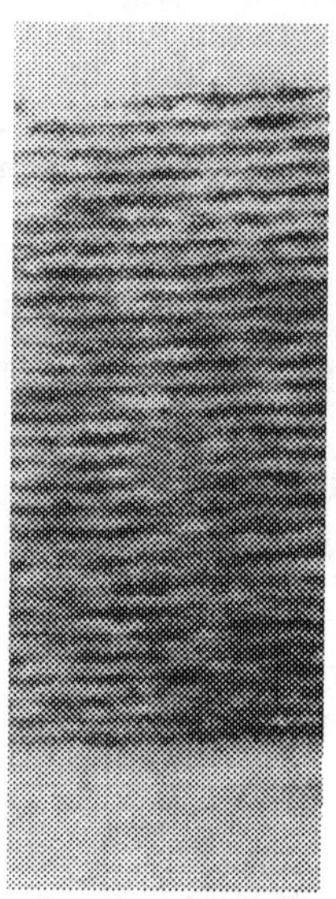

Fig. 1 HRTEM image of $[CaF_2$-6ML-CdF_2-6ML$]_{28}$ SL

This micrograph shows that the interfaces in this SL are flat and abrupt. Their average roughness can be estimated as 1-2 monolayers.

D. Eu^{2+} ion photoluminescence behavior in CdF_2-CaF_2:Eu SLs

Photoluminescence (PL) measurements were carried out at room temperature using Ar^+ and N_2 UV lasers and a grating spectrometer. It was found that after the beginning of the excitation of a previously SL unexposed area, the PL intensity caused by 5d4f6 -4f7 transition is drastically (3-5 times) decreased for a few seconds. The effect becomes stronger with decreasing CaF_2:Eu layer thickness (fig. 2a).Because this phenomenon was not observed in simple CaF_2:Eu layers, it is natural to suppose that it was due to the presence of the CdF_2 layers in the SLs. The decrease in the Eu^{2+} PL intensity can be accounted for by the reduction of the total number of Eu^{2+} ions because their ionization resulting from the transfer of 5d electrons to the neibouring CdF_2 layers.

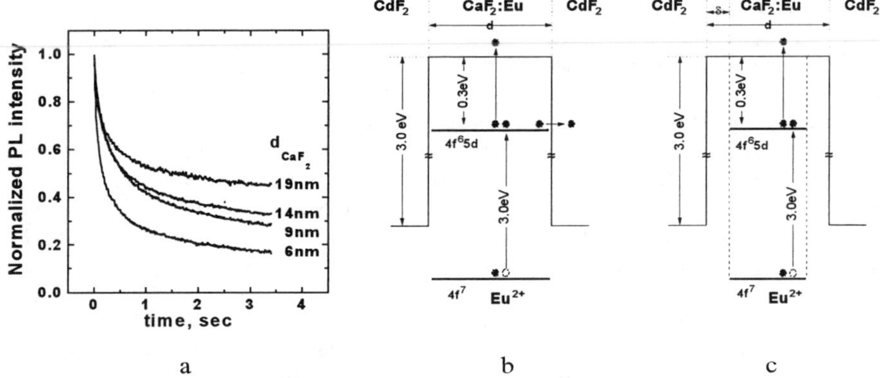

Fig.2. a- Eu²⁺ PL intensity vs time. Energy levels of Eu²⁺ in CdF₂/CaF₂:Eu superlattice: homogeniously doped SL (b) and selectively doped SL (c)

Two possible mechanisms of such ionization can be expected from the energy band diagram (fig. 2b). Taking into account that the $4f^6 5d$ energy level of Eu^{2+} coincides with the CdF_2 conduction band, one can expect electron tunneling from the excited state of Eu^{2+} ions that are close to the adjacent CdF_2 layer. Moreover, the photoionization could take place in two steps through the $4f^6 5d$ intermidiate state into the CaF_2 conduction band with a possible escape of some electrons to the CdF_2 layer. One could distinguish the two mechanisms by studying the dependence of the above effect on the CaF_2 layer thickness of as well as on the optical excitation intensity. Tunneling assisted ionization is expected to increase its rate linearly with increasing excitation power. Moreover, as the probability of the tunneling exponentially drops with the distance from the CdF_2/CaF_2:Eu interface, this mechanism should be important only for those Eu ions that are located in the 3 nm layer in the vicinity of the interface. The thinner the CaF_2:Eu layer, the more part of Eu^{2+} ions can be ionized by electron tunneling. This accounts for higher rates of the PL intensity decrease in the homogeniously doped SLs with short periods, as well as for the larger degree of PL decrease (fig. 2a). In these SLs, we observed a linear increase of the rate of the PL intensity drop with optical excitation power assumed for the single-photon process and a logarithmic dependence of the PL intensity on the time elapsed from the beginning of optical excitation expected for the tunneling mechanism [8].

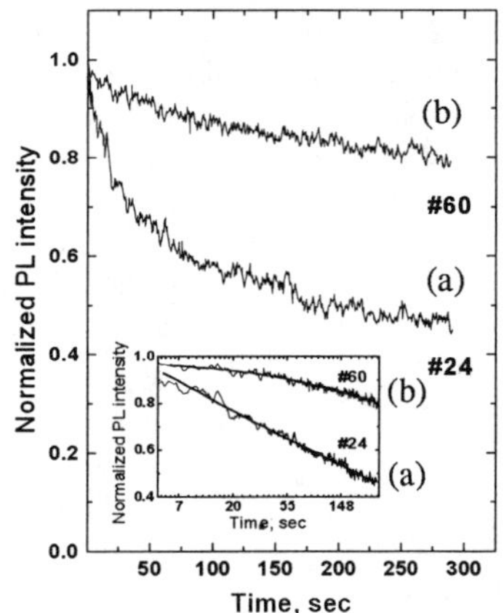

Fig.3. PL behavior in SLs with homogenious (a) and selective (b) doping. Inset represents the same dependences in a logarithmic scale

A less efficient two-step photoionization mechanism was observed in the selectively doped SLs, where the tunneling rate is relatively low because there are no Eu ions in the vicinity of the interface. The non-logarithmic behavior of the PL intensity (fig. 3, #60) and the relatively low rate of the decrease support mechanism.

E. Optical detection of the UV exposed SL spots

The new phenomenon discussed in this paper can be used for recording optical information. UV-exposed spots on the SL can, in principle, be detected via local decrease of Eu^{2+} PL intensity by scanning with a weak probing laser beam. However, during the scanning the existing contrast between the exposed and unexposed SL areas will be inevitably reduced. For this reason, we carried out thorough reflectivity measurements before and during the UV exposure. These measurements were performed using a He-Ne laser. To provide the maximum sensitivity, the incidence angle was chosen close to the Bruster angle, and the reflected light intensity was normalized to the intensity of the incident beam. This provided the measurement accuracy of about 0.1%.

It is clear from fig. 4 that during UV irradiation exciting Eu^{2+} PL and inducing electron transfer from Eu^{2+} ions to CdF_2 layers one can detect a well pronounced decrease in the reflectivity. The decrease was found to be about 2.5%, which considerably exceeds the experimental error. A convincing evidence for the relation between the charge transfer through the CaF_2/CdF_2 interface and the changes in the SL reflectivity is the synchronous character of the changes. Though the detailed mechanism resulting in the change of reflectivity needs further study, one can see that the reflectivity measurements allow differentiation of UV-exposed and unexposed SL areas, thus providing nonviolatile reading of recorded optical information. In order to apply these structures in optical memory cells, it would be very desirable to find a way to erase the recorded information and to restore the cell ability to record new data. Preliminary experiments have shown that this aim can be achieved by exposing the structure to high energy (10-12 eV) optical irradiation.

The authors appreciate stimulating discussions with J.M.Langer and S.A.Basun. The work was supported by grants of the Russian Foundation for Basic Research and the Russian Ministry of Science.

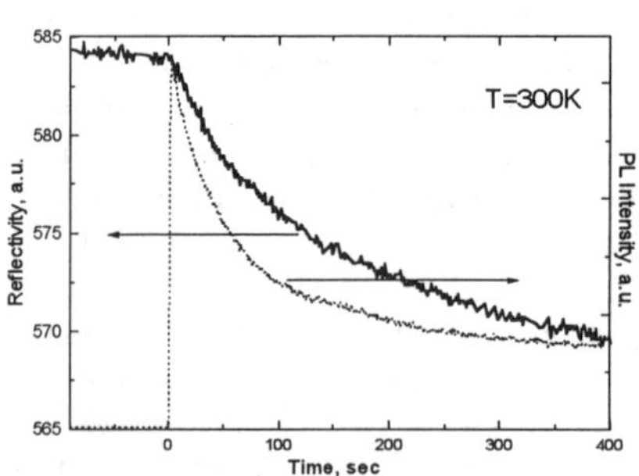

Fig. 4. Reflectivity and PL behavior under continious UV excitation

References

[1] M.H.Yang and C.P.Flynn, Phys.Rev. B 41(1990) 8500

[2] A.Ejiri, K.Nakagawa and A.Hatano, Physica Scripta 41 (1990) 95

[3] K.Edamatsu, S.Hirota and M.Hirai, J.Phys.Soc.Jap. 61 (1992) 2543

[4] In "Crystals with the fluorite structure", Ed by W.Hayes, Clarendon Press, Oxford, 1974

[5] S.V.Novikov, N.N.Faleev, A.Izumi, A.Yu.Khilko, N.S.Sokolov, et al- Abstracts of Intern. Symp. "Nanostructures: Physics and Technology", St.Petersburg, Russia, 1995, 387 - 390

[6] A.Yu Khilko, S.V.Gastev, R.N.Kyutt, M.V.Zamoryanskaya, N.S.Sokolov, Appl.Surf.Sci. 123/124 (1998) 595

[7] R.N.Kyutt, A.Yu.Khilko, N.S.Sokolov, Appl.Phys.Lett. 70 (1997) 1563

[8] N.S.Sokolov, S.V.Gastev, A.Yu.Khilko, S.M.Suturin, I.N.Yassievich, J.M.Langer, A.Kozanecki, to be published

Metastable defects in CdMnTe : Ga.

J.SZATKOWSKI, E.PLACZEK-POPKO, K.SIERANSKI ,B.BIEG[1]
Institute of Physics, Wroclaw University of Technology,
Wybrzeze Wyspianskiego 27, 50-370 Wroclaw, Poland.
[1]*Institute of Physics, Maritime Academy, Waly Chrobrego1,70-500 Szczecin, Poland*

Abstract

Deep levels were studied in bulk Ga doped $Cd_{0.99}Mn_{0.01}Te$ and $Cd_{0.95}Mn_{0.05}Te$ by DLTS method. Some of observed electron traps had strong temperature dependent capture cross-sections. Activation energies for these levels obtained from Arrhenius plots were equal 0.24eV for both types of samples and 0.53eV in $Cd_{0.95}Mn_{0.05}Te$. Energetic barriers for capture processes determined from the temperature dependence of capture cross sections for these traps were found to be equal 0.2eV for the level in $Cd_{0.99}Mn_{0.01}Te$ and 0.15eV and 0.23eV for the two levels in $Cd_{0.95}Mn_{0.05}Te$. These traps are possible candidates responsible for persistent photoconductivity observed by us in both materials.

A.Introduction

In $Cd_xMn_{1-x}Te$ gallium is used as a shallow donor dopant. However it was found that Ga results in strong photomemory effect in the material [1,2]. This effect is related to the bistability of gallium dopant in II-VI compounds. Depending on its charge state, gallium behaves as a shallow or as a deep level, the so-called DX-like center. The name DX-like center has its origin in the DX centers found in GaAs and related ternary and multiternary compounds, responsible for the persistent photoconductivity observed in the materials. So far there were reports on the DX centers in gallium doped $Cd_xMn_{1-x}Te$ of manganese mole fraction equal 0.03 [3,4] and 0.01[5,6]. In the papers [3,4] it has been concluded that both traps observed in $Cd_{0.97}Mn_{0.03}Te$ come from the same defect of two charge states: 0.43eV level was attributed to an acceptor-like two-electron ground state of the defect, and the 0.25eV level to the excited, donor-like one-electron state. The latter one-electron level was the one with thermally activated capture cross section with energetic barrier E_B=0.11eV. In present paper we report on the results of our investigations of deep level defects in bulk Ga doped $Cd_{0.99}Mn_{0.01}Te$ and $Cd_{0.95}Mn_{0.05}Te$ mixed crystals by deep level transient spectroscopy (DLTS) method. The main goal was to find if any of them were DX-like centers in order to explain persistent photoconductivity (PPC).

B.Experiment

N-type Ga doped $Cd_{0.99}Mn_{0.01}Te$ and $Cd_{0.95}Mn_{0.05}Te$ mixed crystals were grown by the Bridgman method. In the former material of lower manganese content donor concentration was about $10^{16}cm^{-3}$ whereas in the latter one the concentration was lower and equal $10^{15}cm^{-3}$. Schottky diodes were realised by evaporation of thick Au layers of 1mm^2 area deposited in vacuum of 10^{-6} Torr on chemically prepared surface. For ohmic contacts dots of indium were soldered in temperature of about 500 K for several seconds. Capacitance-voltage and Deep Level Transient Spectroscopy (DLTS) measurements were performed with the help of the DLS-82E system made by Semitrap, Hungary. This system is a lock-in type DLTS spectrometer. Temperature range for our measurements was within 77K- 400K.

C.Results

In both $Cd_{0.99}Mn_{0.01}Te$ (samples K1) and $Cd_{0.95}Mn_{0.05}Te$ (samples K5) five electron traps were observed. For some of the traps (labelled by us as E2K1, E2K5 and E3K5) thermally activated capture cross sections were observed. Arrhenius plots corresponding to these traps are shown in Fig.1. The data for the trap E3K1 were also included in spite of the fact that for this trap capture process was not thermally activated. The traps observed in $Cd_{0.99}Mn_{0.01}Te$ samples labelled by us as E2K1 and E3K1 are marked as circles and the ones corresponding to the $Cd_{0.95}Mn_{0.05}Te$ material, labelled as E2K5 and E3K5- as crosses. Dotted lines are the best least-square fit to the experimental data. In the figure also data taken from the paper by Semaltianos et al.[4] are included. The solid lines were calculated from the thermal activation energies and capture cross sections for the two levels observed by them: the 0.25eV level of capture cross sections $\sigma_n = 7.7\,10^{-15}cm^2$ and 0.43eV level of $\sigma_n = 5.6\,10^{-14}cm^2$.

Fig.1. Arrhenius plots for the traps observed in gallium doped $Cd_{0.99}Mn_{0.01}Te$ (samples K1) and $Cd_{0.95}Mn_{0.05}Te$ (samples K5). The traps in the K1 samples are labelled as E2K1 and E3K1 (circles) and the traps in the K5 samples are labelled as E2K5 and E3K5 (crosses).

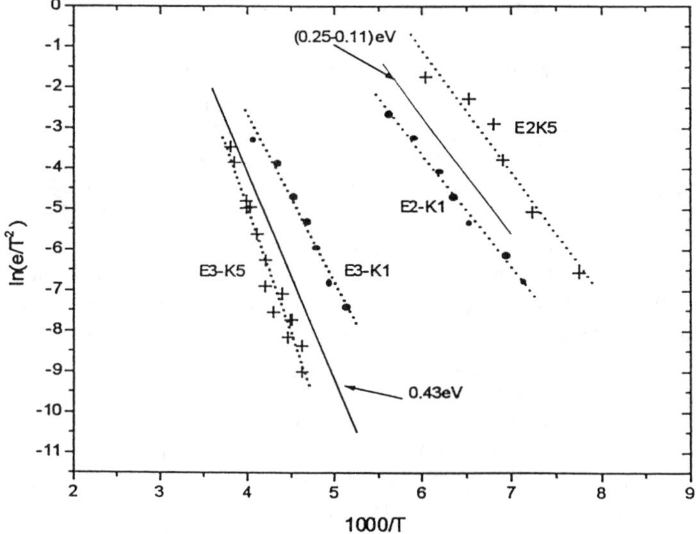

Dotted lines represent a least –square fit to the experimental data. For comparison the Arrhenius plot data corresponding to the 0.25eV and 0.43eV traps observed by Semaltianos et al. in $Cd_{0.97}Mn_{0.03}Te$ [4] are also included as solid lines.

Activation energies and apparent capture cross sections obtained from Arrhenius plots for our experimental data are collected in Table I. For the traps of thermally activated capture cross sections energetic barriers for capture processes (obtained from DLTS data by the method invented by Henry and Lang [7]) are also included in Table I.

Table I. Activation energies, apparent capture cross sections and energetic barriers for capture processes for the traps detected in the samples K1 and K5.

sample ♦	$Cd_{0.99}Mn_{0.01}Te$ (K1)			$Cd_{0.95}Mn_{0.05}Te$ (K5).		
Trap label	Energy activation [eV]	Barrier [eV]	Capture cross section [cm²]	Energy activation [eV]	Barrier [eV]	Capture cross section [cm²]
E2	0.24	0.2	$1.1*10^{-15}$	0.26	0.15	$6*10^{-14}$
E3	0.36	-	$2.2*10^{-15}$	0.53	0.23	$8*10^{-13}$

D.Discussion.

Let us focus our attention on the traps with thermally activated capture cross sections. There are three of them : the one labelled by us as E2K1 in $Cd_{0.99}Mn_{0.01}Te$ and two others labelled as E2K5 and E3K5 in $Cd_{0.95}Mn_{0.05}Te$. In Fig.1. the data collected for these traps seem to be splitted into two sets of data of close position. For the low temperature group activation energy is equal 0.25 ± 0.01eV independent of a sample. However in the case of $Cd_{0.99}Mn_{0.01}Te$ energetic barrier for electron capture (0.2eV) is bigger than in the case of the trap observed in $Cd_{0.95}Mn_{0.05}Te$ (0.15eV). Emission rates from these levels do not depend on electric field, which infers their acceptor-like character. In the case of the high temperature set of data in Fig.1. energetic barrier for capture process is observed only in the K5 samples for the E3K5 trap. The trap E3K1 was completely saturated with filling pulses widths of the order of a few microseconds and this is the reason that there was no possibility to measure a barrier for capture. This trap E3K1 differs from the E3K5 also with respect to the electric field influence as in this case electric field enhanced emission rate was found [6] and explained in the terms of Frenkel-Poole model [8].

The existence of energetic barriers for the levels E2K1, E2K5 and E3K5 can be explained in the terms of large lattice relaxation (LLR) model. According to the model conduction electrons are metastable with respect to the acceptor-like levels in both materials. These results are in accordance with the theoretical model of DX centers in II-VI compounds introduced by Chadi [9]. In this model the DX center acts as an acceptor and is negatively charged when occupied by electron. The presence of fairly high electron capture barriers is accomplished in a very small capture cross sections at low temperatures for the levels E2K1, E2K5 and E3K5. Small capture cross sections indicate that the photomemory effect in both materials originates from these traps.

In the Arrhenius plot we also included for comparison the data concerning $Cd_{0.97}Mn_{0.03}Te$ taken from the paper by Semaltianos et al.[4]. It is seen that the 0.25eV level observed by them, with energetic barrier equal 0.11eV fits in with our low temperature set of data as regards both activation energy and capture cross section which is only slightly different from the ones for the levels E2K1 and E2K5. The 0.43eV level taken from the paper [4] is located in the middle of the high temperature set of our data for E3K1 and E3K5. For this level as in the case of our E3K1 samples complete filling was observed for 2 μs pulses widths. The charge states of these traps were assumed by Semaltianos et al. to be donor-like in the case of the 0.25eV level and acceptor-like in the case of 0.43eV level[4].

In spite of the discrepancies among the properties of observed traps in each set of data it is tempting to assume that related defects have the same origin with the same activation energies of deep levels in the case of the low temperature set and with activation energies increasing with increasing manganese content for the high temperature set.

E.Conclusions

Studies of deep levels in gallium doped ternary semiconducting compounds $Cd_xMn_{1-x}Te$ of manganium contents x=0.01 and x=0.05 yield presence of three traps of thermally activated capture cross sections: E2K1, E2K5 and E3K5 of activation energies E_T equal 0.24eV for the first two and 0.53 for the third one. The results obtained by us are in accordance with the data obtained by Semaltianos et al. for $Cd_xMn_{1-x}Te$ of x=0.03. Observed by them the 0.25eV trap of energetic barrier E_B equal 0.11eV is close to the E2K1 (E_B = 0.20eV) and E2K5 (E_B = 0.15eV) traps found by us. Also the data for the 0.43eV trap observed in [4] are located in Arrhenius plots in the middle of our data for the E3K1 and E3K5 traps. Furthermore we also observe energetic barriers for capture only for the K5 samples of lower net donor concentration ($10^{15}cm^3$) whereas for the samples K1 of net donor concentration of the same order as in the samples in[4] ($10^{16}cm^3$) complete filling of the traps is observed again as in [4] yet for a few microsecond filling pulses. Thus we can say that activation energy of this trap increases with increasing manganese content x in $Cd_xMn_{1-x}Te$. The difference observed by us is in the charge state character of observed traps: in our samples they are acceptor-like whereas in [4] the 0.25eV state is donor-like.

Acknowledgements: We are very indebted to Prof. Giriat from Venezuela for supplying us with the samples.

[1] I.Terry, T.Penney, S.v Molnar , J.M. Rigotty, P.Becla, Solid State Communications **84**, 235 (1992).

[2] N.G.Semaltianos, G.Karczewski, T.Wojtowicz, J.K.Furdyna, Phys. Rev.*B* **47**,12540 (1993).

[3] T.Wojtowicz, G.Karczewski, N.G.Semaltianos, S.Kolesnik, I.Miotkowski, M.Dobrowolska, J.K.Furdyna Mat.Sci.Forum **143-147**,1203(1994)

[4] N.G.Semaltianos, G.Karczewski, B.Hu, T.Wojtowicz, J.K.Furdyna Phys. Rev.B **51** , 17499 (1995).

[5] J.Szatkowski, E.Placzek-Popko, K.Sieranski, P.Fialkowski, A.Hajdusianek, B.Bieg Mat.Sci.Forum **258-263**, 1413, (1997).

[6] J.Szatkowski, E.Placzek-Popko, K.Sieranski, A.Hajdusianek, B.Bieg Inst.Phys.Conf.Ser. **152**, 793 (1998).

[7] C.H.Henry, D.V.Lang Phys.Rev. B **15** , 989 (1977).

[8] J.Frenkel, Phys.Rev. **54**, 647 (1938).

[9] D.J.Chadi, Phys.Rev.Lett. **72**, 534 (1994).

Persistent photoconductivity in nitrogen-doped p-type Zn(S)Se/GaAs heterojunctions

D. Seghier and H.P. Gislason

Science Institute, University of Iceland, Dunhagi 3, IS-107 Reykjavik, Iceland

Nitrogen-doped p-type ZnSe grown by molecular beam epitaxy on p-type GaAs shows persistent photocurrent up to room temperature. A typical decay consists of an initial stretched-exponential transient with a thermally activated decay constant, and a subsequent long transient. We attribute the effect to metastable centers in the ZnSe near the interface to the GaAs substrate, on the one hand, and tunneling of photo-excited holes trapped in a two-dimensional quantum well at the heterojunction through the barrier to the ZnSe, on the other hand.

A. Introduction

Despite much research on the ZnSe/GaAs heterojunction, the nature of the defects which control its electrical characteristics is in some cases incompletely understood. In the literature there are only a few reports on the photoconductivity of p-ZnSe/GaAs heterojunctions [1,2]. Below, we present results of photocurrent measurements on such structures grown by molecular beam epitaxy (MBE) on p-GaAs. When exposed to light these heterojunctions show enhanced carrier concentration after the illumination is terminated, referred to as persistent photocurrent (PPC). Slow relaxation of excess conductivity is caused by energy barriers which cannot be overcome by the carriers at low temperatures. In $Al_xGa_{1-x}As$ persistent photoconductivity is attributed to defect centers which undergo a large lattice relaxation, the well known DX centers [3-5].

B. Experimental procedure

The ZnSe layers were provided by Heriot-Watt University. They were grown at 300 °C by MBE on highly conductive p-type GaAs with a growth rate of 0.5 µm per hour. Solid sources of zinc and selenium were used. The nitrogen was incorporated using an Oxford Applied Instruments plasma source. The doped zinc selenide was grown directly onto the substrate without a buffer layer to a thickness of approximately 1 µm. We made Schottky contacts by evaporating gold contacts with diameter 1 mm and thickness of about 600 Å onto the ZnSe layer and Ohmic contacts by evaporating Au-Zn alloy or using In-Ga on the GaAs substrate. The current-voltage (I-V) and capacitance-voltage (C-V) characteristics of this device are typical of a Au/ZnSe Schottky diode and a ZnSe/GaAs heterojunction diode connected back-to-back [6]. By reverse-bias conditions we refer to the heterojunction. We studied five samples with similar general behavior. Their acceptor concentration ranges from 2×10^{16} to 5×10^{17} cm^{-3}. All curves presented below were obtained from the same sample with the strongest PPC signal. A 100 W Xe lamp was used as a light source. Samples were illuminated on the gold contact. Same initial conditions for each set of data were obtained by warming the system up to 350 K after each measurement, then cooling it down in the dark to the desired temperature. Filters were used for selective spectral illumination of samples. A standard set-up was used for the deep-level transient spectroscopy (DLTS) measurements.

C. Results

Fig. 1 shows an Arrhenius plot of the dark current $I_d(T)$ as a function of temperature, obtained by applying a reverse bias of 5 V to the heterojunction diode. Curve *a* represents the sample cooling down in the dark. At 80 K the sample is illuminated for 5 minutes during which the current rises until saturation. After the illumination the photocurrent is persistent at the saturation value. The sample is then heated up to 400 K in the dark and $I_d(T)$ measured during the heating (curve *b*). In the low temperature range $I_d(T)$ increases slowly with temperature and is

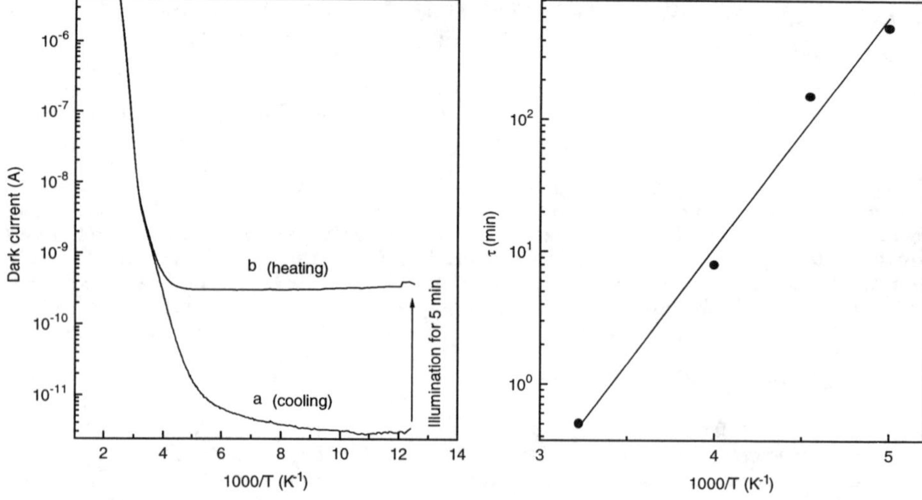

Fig. 1: Arrhenius plot of the dark current as a function of temperature under reverse bias of 5 V. Measured while (a) cooling the sample, (b) warming after illumination at 80 K.

Fig. 2: Arrhenius plot of the relaxation time constant of the persistent photo-current.

higher than the current measured during cooling. The warming curve joins the cooling curve at around 310 K. Hence, there is a PPC effect in the sample below 310 K [3, 4]. At temperatures above 380 K, $I_d(T)$ is thermally activated as described by the formula:

$$I_d(T) = I_\infty e^{-\Delta E_\sigma / kT}. \tag{1}$$

Here I_∞ denotes the high temperature limit of $I_d(T)$ and ΔE_σ the activation energy. Other symbols have their usual meaning. Eq. (1) yields a value which is found to be slightly dependent on the applied bias, $\Delta E_\sigma = 0.8$ eV. It can be taken as a lower limit for the barrier at the hetero-junction, given the high temperature at which the slope is measured, and the possible bias dependence of the barrier. A slow decay of the persistent photocurrent is observed for all temperatures up to room temperature. We define the time-dependent persistent photocurrent as:

$$I_{PPC}(t) = [I_d(t) - I_{d0}]/[I_{PC} - I_{d0}]. \tag{2}$$

Here, the initial dark current level I_{d0} at a given temperature has been subtracted from $I_d(t)$ and the decay curve has been normalized to the value of the current at the moment when the illumination is terminated, I_{PC}. The decay rate increases with temperature. It is non-exponential and can be decomposed into an initial decay followed by a subsequent longer transient. The first transient can be described by a stretched-exponential function of the form:

$$I_{PPC}(t) = I_{PPC}(0) \exp[-(t / \tau)^\beta] \tag{3}$$

as often reported for persistent photocurrent observed in crystalline semiconductors. Here, $I_{PPC}(0)$ is defined as the buildup level when the light excitation is removed, τ a decay time constant, and β a decay exponent. As shown in Fig. 2 the decay time constant τ is thermally

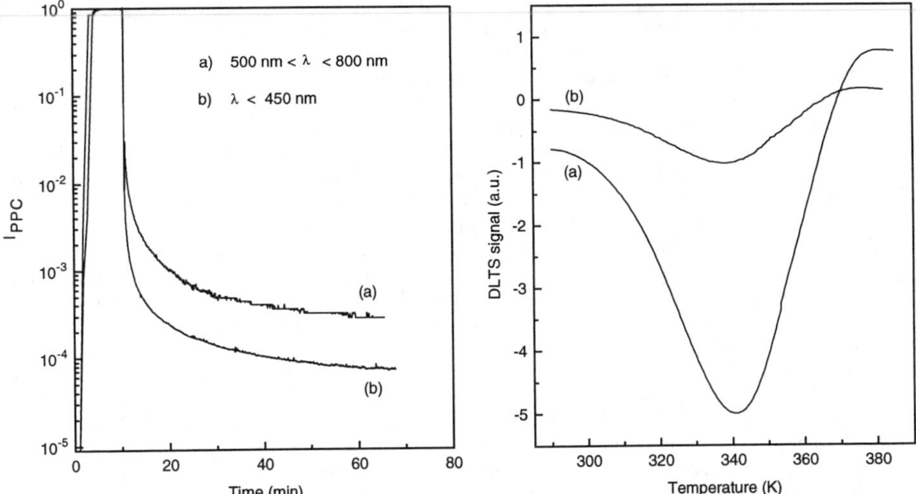

Fig. 3: Decay of the persistent photocurrent at 200 K with excitation light of different wavelength.

Fig. 4: DLTS spectra measured (a) before and (b) after annealing the sample at 150 °C for 30 minutes. The rate window is 10 s^{-1}.

activated. It is given by the expression:

$$\tau = \tau_0 \exp(\Delta E/kT) \qquad (4)$$

with the activation energy $\Delta E = 0.35$ eV. Fig. 3 shows the decay of the PPC at 200 K using different excitation wavelengths. For wavelengths between 500 and 800 nm, approximately, the persistent photocurrent builds up and decays in the way described above. For wavelengths below 450 the PPC effect is much weaker. Since this photon energy is above the bandgap of ZnSe the light is absorbed near the surface. Hence, the centers responsible for the PPC are not present in the surface region but rather likely to be found at the interface between ZnSe and GaAs.

Annealing of the samples at 150 °C for 10 to 30 minutes increases the decay rate of the PPC considerably. The same treatment reduces the height of a DLTS signal around 350 K which was observed in previous work and attributed to interface states between ZnSe and GaAs [6]. Fig. 4 shows the DLTS spectra before and after annealing the sample. There is a direct correlation between the increasing decay rate and the decreasing DLTS signal with annealing time. These observations suggest that the annealing transforms the centers responsible for the persistent photocurrent and can also be taken as an additional argument for attributing some of the PPC effect to interface states.

D.   Discussion

Microscopic energy barriers may form around certain defect centers when they are photoionized [7]. At low temperatures such barriers prevent carriers from being recaptured, which generates persistent photocurrent. An example of this type of metastable defect are the DX centers in GaAlAs [3-5]. The wavelength dependence of the persistent photoconductivity in our samples suggests that it originates from the ZnSe/GaAs interface region. If the centers were evenly distributed in the p-type ZnSe epilayer, illumination above the bandgap would give rise to stronger persistent photocurrent. The effect is unlikely to arise from the GaAs substrate, however, since persistent photocurrent is not observed in p-type GaAs. We therefore propose that the metastable behavior arises from the ZnSe in a region close to the interface. The correlation between the DLTS signal in the interface region and the PPC upon annealing supports this hypothesis. The DLTS measurements show that annealing passivates a fraction of these centers, at the same time as the decay rate of the PPC increases. Hence, the persistent photocurrent is at least partially caused by the defects detected by DLTS. We suggest that they are acceptors with a metastable ionized state. Therefore, free holes have to overcome an energy barrier in order to be recaptured by the photoionized acceptors and the PPC consists of photo-created holes which cannot be recaptured by the ionized acceptors at low temperatures. We conclude that the activation energy of this thermally activated capture is $\Delta E = 0.35$ eV as extracted from Eq. (4) and attribute the initial stretched exponential decay of $I_{PPC}(t)$ to DX-like interface centers.

When exciting only in the surface region with above-bandgap light we find a second, weaker component with a long transient in the decay of $I_{PPC}(t)$. This component is also present when exciting below the ZnSe bandgap, but in part overlapping the above mentioned initial transient. We attribute this effect to tunneling through the barrier at the ZnSe/GaAs interface [8]. Band offsets at the ZnSe/GaAs interface are known to depend strongly on the properties of the interface, as reported by several workers. By controlling the Se to Zn ratio in the first few atomic layers near the interface the valence band offset varies between 0.6 and 1.2 eV [9]. Evidently there is an energy barrier at the heterojunction in our samples high enough to give rise to a two-dimensional quantum well on the GaAs side of the interface. From the $I_d(T)$ measurements we estimate this barrier to be 0.8 eV. We conclude that the persistent photocurrent in our samples arises both from metastable DX-like point defects at the interface and tunneling of carriers through the energy barrier at the ZnSe/GaAs heterojunction.

E. Conclusion

Persistent photocurrent in MBE-grown p-ZnSe/GaAs was observed up to room temperature with time evolution scales ranging from several minutes to hours. Its decay occurs in two steps, a first temperature activated step characterized by a stretched exponential, and a second longer transient. We conclude that the persistent photocurrent has two independent components. The first one is caused by metastable centers at the interface, similar to the DX centers in GaAlAs, which give the thermally activated decay of the excess carriers. The second one is attributed to tunneling of trapped holes from a two-dimensional quantum well at the heterojunction through the interface barrier.

Acknowledgments

This research was supported by the Icelandic Research Council and the University Research Fund. We are grateful to I.S. Hauksson for discussions, B.C. Cavenett and K.A. Prior for providing the samples used in this study.

References

[1] D.J. Olego, J. Vac. Sci. Technol. **B6**, 1193 (1988).
[2] S.G. Ayyar et al. J. Appl. Phys **68**, 5226 (1990)
[3] D.V. Lang and R.A. Logan, Phys. Rev. Lett. **39**, 635 (1977)
[4] P.M. Mooney, J. Appl. Phys. **67**, R1(1990).
[5] D. J. Chadi and K.J. Chang, Phys. Rev. Lett. **57**, 873 (1988).
[6] D. Seghier, and H.P. Gislason, Appl. Phys. Lett. **71**, 2295 (1997).
[7] N.G. Semaltianos, G. Karczewski, T. Wojtwicz, and J.K. Furdyna, Phys. Rev. **B47**, 12540 (1993).
[8] L.X. He, K.P. Martin, and R.J. Higgins, Phys. Rev. B **39**, 13276(1989)
[9] R. Nicollini, L. Vanzetti, G. Mula, G. Bratina, L. Sorba, A. Franchiosi, M. Peressi, S. Baroni, R. Resta, A. Baldereschi, J.E. Angelo and W.W. Gerberish, Phys. Rev. Lett. **72**, 294 (1994).

X.

Devices

Low-Threshold Quantum Dot Injection Laser Emitting at 1.9 μm

A. E. Zhukov, V. M. Ustinov, A. Yu. Egorov, A. R. Kovsh, S. V. Zaitsev, N. Yu. Gordeev,
V. I. Kopchatov, N. N. Ledentsov, A. F. Tsatsul'nikov, B. V. Volovik, P. S. Kop'ev,
and Zh. I. Alferov

A.F. Ioffe Physico-Technical Institute, 194021, Politekhnicheskaya 26, St. Petersburg, Russia

Self-organized InAs quantum dots inserted in an (In,Ga)As matrix lattice matched to an InP substrate were used as an active region of an injection laser. Laser action was observed up to 200 K. Low threshold (11.4 A/cm^2) lasing at 1.894 μm (77 K) via the quantum dot states was realized. The ground-to-excited state transition with increasing threshold gain was observed. The quantum dot material gain of the order of 10^4 cm^{-1} was estimated.

A. Introduction

Extremely low threshold current density (J_{th}), improved thermal stability, and increase in the material gain [1,2] have been predicted for a quantum dot (QD) injection laser. QD injection lasers have been recently realized on the basis of self-organized islands in (In,Ga)As/(Al)GaAs system [3-6]. The J_{th} as low as 60 A/cm^2 has been reported [7]. However, QD lasers of this kind usually emit around 1 μm, which considerably restricts possible important long-wavelength applications. We have previously shown that the QD emission range can be extended up to 2 μm by embedding the InAs QDs into an InGaAs matrix grown on an InP substrate [8,9]. In this work we study threshold and gain characteristics of the InAs/InGaAs/InP QD injection laser.

B. Experimental

The structure studied (Fig. 1) was grown by solid source molecular beam epitaxy (MBE). The active region consisted of the three InAs QD planes separated by the 5 nm InGaAs spacers. The effective thickness of InAs deposited in each island sheet was 7 ML. The active region was inserted into the middle of the 0.6 μm InGaAs waveguiding layer confined by the n$^+$-InP(100) substrate and the 1.5 μm InAlAs:Be as bottom- and top-side cladding layers, respectively. The structure was terminated by the 0.6 μm InGaAs:Be contact layer. The InGaAs and InAlAs epitaxial layers were nominally lattice matched to InP substrate. The growth temperature was 500°C for the whole structure. Laser diodes were fabricated in a 100 μm-wide shallow mesa geometry. The laser characteristics were studied under pulsed excitation (3 μs pulse duration, 5 kHz repetition frequency) in the 77÷300 K temperature interval. Light signal was detected using a cooled InSb photodiode.

C. Results and discussion

QD array is characterized by a certain finite surface density. Therefore, only a finite number of carriers can contribute to lasing. The QD surface density in the InAs/InGaAs/InP

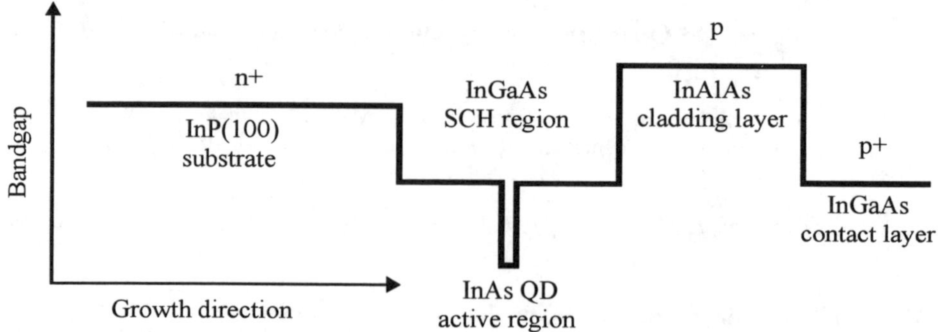

Fig. 1. Schematic bandgap diagram of the laser structure based on InAs QDs.

system was estimated as $7\text{-}8\times10^9$ cm^{-2} [8]. This basically should lead to low values of the transparency current density. On the other hand, one can expect that the effect of gain saturation should be pronounced for the structure under investigation.

To minimize mirror loss we first studied the samples with four cleaved facets. Electro- (EL) and photoluminescence (PL) spectra are given in Fig. 2. It has been shown previously [9] that this 1.9-μm PL line is associated with the recombination via the ground state of InAs QDs. Full width at half maximum of the PL line is as low as 27 meV which is among the best results reported for QD arrays in any material system indicating good size uniformity of the QD array. Lasing line appears in the vicinity of the PL line maximum at 1.894 μm at the threshold current density as low as 11.4 A/cm^2. To our knowledge, this is the lowest J_{th} and the longest emission wavelength ever reported for any kind of the QD laser.

The modal gain g_{mod} as a function of the current density was calculated from the experimental dependence of J_{th} on the cavity length L, using the threshold condition:

$$g_{mod}(J_{th}) = \alpha_m + \alpha_i \tag{1}$$

where α_m is the cavity length dependent mirror loss and α_i is the internal loss. We calculated

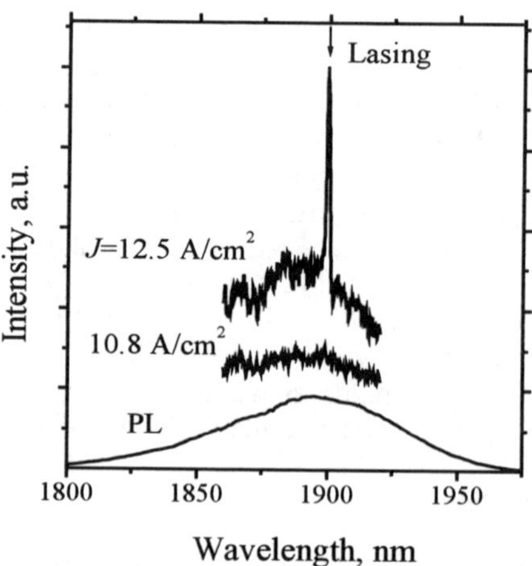

Fig. 2. EL and PL spectra at 77 K of the InAs-QD laser. Lasing line is denoted by the arrow.

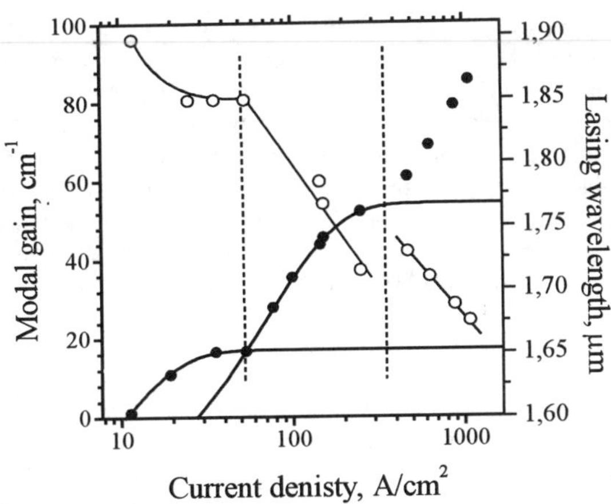

Fig. 3. Modal gain (solid circles) and lasing wavelength (open circles) of the InAs-QD laser (77 K) as a function of the current density.

α_i from the dependence of the external differential quantum efficiency on L. The internal quantum efficiency η_i and α_i were estimated to be 40% and 1.1 cm^{-1}, respectively.

The resultant dependence of the modal gain on the current density is shown in Fig. 3 (solid circles). One can see that there are two pronounced knees in the experimental g-J curve at the current densities of about 53 and 350 A/cm^2 separating three regions of different gain behavior. It should be noted that the g-J curve of the laser under investigation deviates strongly from the shape expected for the QW active region, which is a linear function in semi-logarithmic scale. The main difference is that the gain in the first two regions saturates rapidly with increasing the current density. This fact indicates directly that there is some finite number of states involved into lasing which is characteristic for a QD ensemble. The current density dependence of the gain in these regions can be fitted by the same empirical equation:

$$g = g^{sat}\left(1 - \exp\left(-\gamma\,\frac{J - J_0}{J_0}\right)\right) \qquad (2)$$

Here g^{sat} is the saturated gain and J_0 is the transparency current density. The γ-factor is an additional gain parameter, which is related to the linear gain regime.

The current density dependence of the lasing wavelength is given in Fig. 3 (open circles). Three well-pronounced jumps of the wavelength corresponding to the changes in the g-J behavior are observed at 53 and 350 A/cm^2. The similar multi-region dependence of the gain and the wavelength on the current density has been previously reported for InAs/GaAs QD laser [6] and was attributed to the transition to lasing via the QD excited states. It has been shown theoretically that the excited level can be really localized in a self-organized QD [10,11]. We conclude that the switching to lasing through the QD excited states is responsible for the transition from the first to the second regions in g-J and λ-J curves of this low thrsehold laser in InAs/InGaAs/InP QD system. On the other hand, the appearance of the third region seems to be associated with the transition to lasing via the wetting layer states.

Taking into account the typical size and density of QDs [8,9] we estimated the optical confinement factor Γ_{QD} to be as low as 3.6×10^{-3}. The modal gain sufficient for lasing is reached due to the drastically increased material gain g_{QD} in a QD system. We estimate g_{QD} to be 3×10^3 and 10^4 cm^{-1} for the ground and excited states, respectively. These values agree reasonably well with that of 5000 cm^{-1} calculated for the InGaAs/InP quantum boxes [12].

It should be noted, that the lasing was observed up to 200 K. The laser under investigation demonstrates strong temperature dependence of J_{th} with characteristic temperature T_0 of about 25 K. The similar behavior was observed previously in lasers based on InGaAs QDs in a AlGaAs matrix and was attributed to thermal evaporation of carriers out of QDs into wetting layer and matrix states [7]. We believe, the evaporation can be partly suppressed by a wider bangap InGaAlAs alloy as a matrix material.

D. Conclusion

We have demonstrated the injection laser based on InAs quantum dots in an (In,Ga)As matrix grown on an InP substrate. The threshold current density of 11.4 A/cm^2 and the lasing wavelength of 1.894 μm at 77 K were achieved. Transition from the ground state to the excited state and the wetting layer lasing were observed. High material gain of 10^4 cm^{-1} was determined for the QDs.

Acknowledgments

This work was supported by Russian Foundation for Basic Research (RFBR grant 96-02-17824) and Ministry of Science and Technology of RF, Program "Solid-state nanostructures" (grant 98-1069).

References

[1] Y. Arakawa and H. Sakaki, Appl. Phys. Lett. 40 (1982) 939.
[2] M. Asada, Y. Miyamoto, and Y. Suematsu, J. Quantum Electron. QE-22 (1986) 1915.
[3] N. Kirstaedter, N. N. Ledentsov, M. Grundmann, D. Bimberg, V. M. Ustinov, S. S. Ruvimov, M. V. Maximov, P. S. Kop'ev, Zh. I. Alferov, U. Richter, P. Werner, U. Gosele, and J. Heydenreich, Electron. Lett. 30 (1994) 1416.
[4] K. Kamath, P. Bhattacharya, T. Sosnowski, T. Norris, and J. Phillips, Electron. Lett. 32 (1996) 1374.
[5] R. Mirin, A. Gossard, and J. Bowers, Electron. Lett. 32 (1996) 1732.
[6] H. Shoji, Y. Nakata, K. Mukai, Y. Sugiyama, M. Sugawara, N. Yokoyama, and H. Ishikawa, Electron. Lett. 32 (1996) 2023.
[7] V. M. Ustinov, A. Yu. Egorov, A. R. Kovsh, A. E. Zhukov, M. V. Maksimov, A. F. Tsatsul`nikov, N. Yu. Gordeev, S. V. Zaitsev, Yu. M. Shernyakov, N. A. Bert, P. S. Kop`ev, Zh. I. Alferov, N. N. Ledentsov, J. Bohrer, D. Bimberg, A. O. Kosogov, P. Werner, and U. Gosele, J. Cryst. Growth 175/176 (1997) 689.
[8] V. M. Ustinov, E. R. Weber, S. Ruvimov, Z. Liliental-Weber, A. E. Zhukov, A. Yu. Egorov, A. R. Kovsh, A. F. Tsatsul'nikov, and P. S. Kop'ev, Appl. Phys. Lett. 72 (1998) 362.
[9] V. M. Ustinov, A. E. Zhukov, A. F. Tsatsul'nikov, A. Yu. Egorov, A. R. Kovsh, M. V. Maximov, A. A. Suvorova, N. A. Bert, and P. S. Kop`ev, Semiconductors 31 (1997) 1080.
[10] M. Grundmann, O. Stier and D. Bimberg, Phys. Rev. B52 (1995) 11969.
[11] G. Medeiros-Ribeiro, D. Leonard, and P. M. Petroff, Appl. Phys. Lett. 64 (1995) 1767.
[12] Y. Miyake, H. Hirayama, K. Kudo, S. Tamura, S. Arai, M. Asada, Y. Miyamoto, and Y. Suematsu, IEEE J. Quantum Electron. QE-29 (1993) 2123.

Preliminary Results for GaInAs Channel MISFETs Fabricated on LEC-Grown Ternary GaInAs Substrates

R. Poirier, G. Soerensen, C. R. Bolognesi[1]
Compound Semiconductor Device Laboratory – CSDL
School of Engineering Science, and Department of Physics
Simon Fraser University, Burnaby, British Columbia, Canada, V5A 1S6

X. Xu, S. P. Watkins
Department of Physics
Simon Fraser University, Burnaby, British Columbia, Canada, V5A 1S6

B. Lent, D. Reid
Crystar Research Ltd.
721 VanAlman Avenue, Victoria, British Columbia, Canada, V8Z 3B6

We report the first demonstration of GaInAs channel functional heterostructure field-effect transistors grown on Cr-doped LEC $Ga_{1-x}In_xAs$ ternary substrates. The devices feature Schottky gates with high reverse breakdown voltages and a forward ideality factor of $n \sim 1.2$-1.3, good channel modulation and a complete pinch-off. The drain *I-V* curves display some non-idealities associated with compositional non-uniformities in the substrates. Strained channel devices with a 15% Indium composition in the active layer are also demonstrated on ternary substrates. It is expected that continued progress in GaInAs ternary substrate development will permit the fabrication of high-performance transistors relying on an Indium rich channel.

A. Introduction

The frequency performance of GaInAs-based strained-channel heterostructure field-effect transistors (HFETs) improves with increasing channel Indium content because of the reduced electron effective mass and higher peak velocities in Indium rich alloys. However, the lattice mismatch between GaAs and GaInAs limits the channel thickness to the critical thickness at which the mismatch strain is relieved by the creation of misfit dislocations [1]. The availability of ternary $Ga_{1-x}In_xAs$ substrates should in principle enable higher channel Indium concentrations for a given channel layer thickness. Some semiconductor laser structures have been reported on ternary $Ga_{1-x}In_xAs$ substrates [2-6], but there are still no published reports of HFETs implemented on such substrates, and the very feasibility of transistors on ternary GaInAs substrates still remains to be proven before performance improvements by growth on ternary substrates can be pursued.

1.Email: *colombo@sfu.ca*

In the present work, we demonstrate the feasibility of HFET fabrication on $Ga_{1-x}In_xAs$ ternary substrates and report MOCVD-grown MISFET transistors lattice matched to Cr-doped $Ga_{1-x}In_xAs$ substrates ($x \approx 0.06$) with a surface area of ~ 2.5 cm^2. The same epitaxial layers were also grown simultaneously on a high-quality "Epiready" vertical gradient freeze (VGF) GaAs substrate placed side-by-side with the $Ga_{0.94}In_{0.06}As$ wafer to provide a comparison benchmark with a mature substrate technology. Strained channel devices with a higher Indium composition (15%) were also fabricated on both substrates to investigate the feasibility of pseudomorphic devices on ternary substrates.

B. Characterization of Ternary Substrates, Epitaxy, Device Fabrication

The Cr-doped $Ga_{0.94}In_{0.06}As$ substrates used in this study were grown by continuous melt replenishment [7] Liquid Encapsulated Czochralski (LEC) growth at *Crystar Research Ltd.* Triple Axis X-ray Diffraction (TAXRD) measurements were first used to determine the absolute lattice parameter and composition of the nominally (001)-oriented GaInAs substrates to an accuracy better than 0.01% [8]. The substrate Indium mole fractions are calculated from the lattice parameter data assuming Vegard's law. The substrates used for our studies have an average Indium mole fraction of $x \sim 6\%$. Fig. 1 shows that the $Ga_{0.94}In_{0.06}As$ substrates feature broad shoulders around the main Bragg diffraction peak when compared to the Vertical Gradient Freeze (VGF) GaAs substrates. The X-ray spectra reveal a FWHM of 27 and 12 arcsec for the $Ga_{0.94}In_{0.06}As$ and GaAs substrates, respectively. The shoulders around $Ga_{0.94}In_{0.06}As$ Bragg peak are attributed to compositional fluctuations of the order of 0.5% in the substrate.

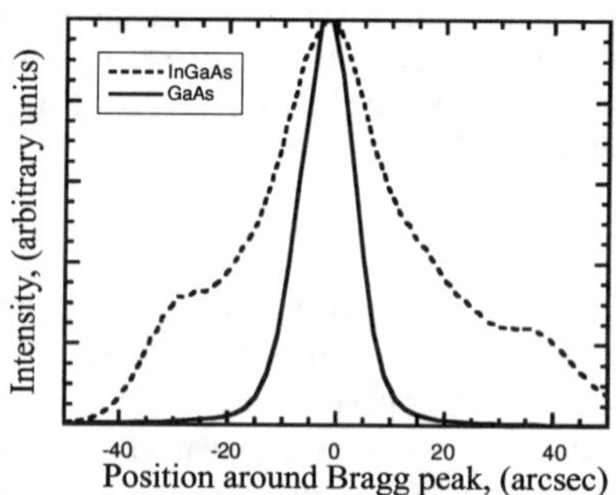

Fig. 1: Triple Axis X-ray Diffraction spectra of the GaAs and $Ga_{1-x}In_xAs$ substrates. The X-ray source was the 1.54 nm Cu k_α line. The GaAs FWHM is 12 arcsec, and the GaInAs FWHM is 27 arcsec.

Our planar MISFET structures were grown at 580°C in a Thomas-Swan vertical MOCVD reactor using a hydrogen standard carrier flow of 2.3 l/min at a pressure of 50 Torr. The precursors were triethylgallium (TEG), tertiarybutylarsine (TBA), and trimethylindium (TMIn). Doping was carried out with 100 ppm disilane diluted with high-purity hydrogen and liquid diethylzinc (DEZn). A 2000 Å GaInAs buffer lattice-matched to the GaInAs substrate was first grown to prevent channel compensation problems due to a potential Cr out-diffusion from the substrate. The buffer was p-doped to 1×10^{15} cm^{-3} with DEZn to help confine electrons in the channel.

324

This was followed by a 200 Å GaAs spacer intended to reduce hot electron injection in the buffer layer and minimize short-channel effects due to substrate currents. The spacer was followed by a 1000 Å $Ga_{1-x}In_xAs$ channel with $x = 5\%$, or 500 Å with $x = 15\%$ doped at 2×10^{17} cm^{-3} with silicon. The growth was completed by a 500 Å undoped GaAs cap required to increase the Schottky barrier height because GaInAs alloys display reduced barrier heights with increasing Indium content. No attempts were made to optimize device performance with n$^+$ caps for gate recessing in this feasibility study. Hall effect measurements on the sample grown on GaAs revealed a sheet density of $N_s = 3.3 \times 10^{12}$ cm^{-2}, and a mobility of $\mu = 2500$ cm^2/Vs at 300 K for the 5% Indium channel devices and $N_s = 1.6 \times 10^{12}$ cm^{-2} with similar mobilities for the 15% Indium channel devices. Ge/Au/Ni/Au Ohmic contacts were formed by first removing the undoped GaAs cap layer by wet etching in a phosphoric acid solution, and by annealing at 350°C in forming gas. Following mesa isolation, Ti/Au (100 nm / 300 nm) Schottky gates were defined by conventional hard contact optical lithography and lift-off. The gate finger was centered in a 5 μm source-drain space. The measured gate length of our devices is 1 μm.

C. Results

Devices fabricated on both the $Ga_{0.94}In_{0.06}As$ and GaAs substrates feature very good Schottky gate contacts with a diode ideality factor n of 1.2 to 1.3. The gate diode I-V curves for MISFETs fabricated on $Ga_{0.94}In_{0.06}As$ and GaAs feature high reverse Schottky contact breakdown voltages V in all cases, despite the compositional non-uniformity of the ternary substrate. Fig. 2 shows drain characteristics for 1 μm gate devices grown on $Ga_{0.94}In_{0.06}As$ and GaAs. Devices grown on $Ga_{0.94}In_{0.06}As$ show a lower saturation current that may be due to a small difference in miscut orientation between the two substrates (which would affect dopant incorporation in the epilayers). Avalanche breakdown near the drain occurs earlier in the devices grown on ternary substrates than in those grown on conventional substrates. The $Ga_{0.94}In_{0.06}As$ substrate non-uniformities resulted for the 5% Indium channel devices in a large spread in open channel current I_{DSS} values as shown in Table 1. Microwave measurements on 1 μm gate devices were made to measure unity current gain cut-off frequency f_T and maximum oscillation frequency f_{MAX}. Data were taken at $V_{DS} = 3$-4 V, where the source-to-drain current in the device is saturated with a low output conductance. Additional time delays due to parasitic contact resistance of the pads (C_{pad}) and gate fringe (C_{fringe}) were taken into account in the determination of f_T [10].

Table 1: DC and μ-wave Characteristics of 1 μm devices

Device type	I_{DSS} (mA)	I_{DSS} spread[a] (mA)	f_T (GHz)	f_{MAX} (GHz)
5% In on GaAs	5.53	1.48	17	31
5% In On GaInAs	2.43	2.23	14	36
15% In on GaAs	5.27	0.85	9	55
15% In on GaInAs	1.32	0.78	11	50

a.Standard deviation

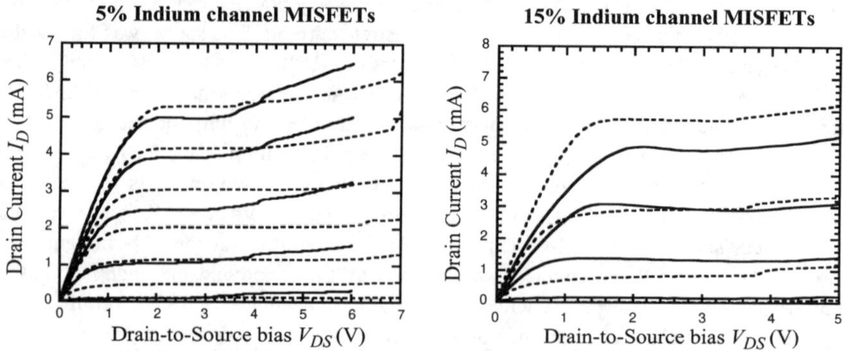

Fig. 2: Typical room temperature drain characteristics for **5% and 15% Indium channel MISFETs** for the conventional GaAs substrate (dashed lines), and for the (Ga,In)As substrate (solid lines). The gate voltage is stepped from $V_{GS} = 0.5$ Volts in increments of -0.25 Volts.

D. Conclusion

In conclusion, we have demonstrated the feasibility of lattice matched and pseudomorphic channel HFET fabrication on ternary GaInAs substrates. The devices are comparable to devices simultaneously grown on GaAs, but they exhibit performance fluctuations associated with GaInAs substrate non-uniformities. Future work will focus on substrate and processing optimization for an improved device performance and higher Indium concentration in the active channel layer. This work was jointly supported by NSERC and BC Advanced Systems Institute.

References

[1] J. W. Matthews, A. E. Blakeslee, J. Cryst. Growth., **27**, (1974) 118.

[2] H. Shoji, T. Uchida, T. Kusonoki, M. Matsuda, H. Kurakake, S. Yamazaki, K. Nakajima, H. Ishikawa, IEEE Photonics Technol. Lett., **6** (1994) 1170.

[3] H. Shoji, K. Otsubo, T. Kusunoki, T. Suzuki, T. Ushida, H. Ishikawa, Japn. J. of Appl. Phys., Part 2, l. **35** (1996) L778.

[4] A. H. Moore, B. Lent, W.A. Bonner, Electron. Letters, **32** (1996) 2018.

[5] L. Jedral, C. Edirisinghe, H. Ruda, A. Moore, B. Lent, J. of Appl. Phys., **82** (1997) 375.

[6] A.M. Jones, B. Lent, J.F. Kluender, S.D. Roh, W.A. Bonner, J.J. Coleman, IEEE Photonics Technol. Lett., **9** (1997) 1319.

[7] D. Reid, B. Lent, T. Bryskiewicz, P. Singer, E. Mortimer, W.A. Bonner, J. Cryst. Growth, **174** (1997) 250.

[8] W. L. Bond, Acta Cryst., **13** (1960) 814

[9] T. G. Andersson, Z. G. Chen, V. D. Kulakouskii, A. Uddin, J. T. Vallin, Appl. Phys. Lett. **51** (1987) 752.

[10] L. D. Nguyen, L. E. Larson, U. K. Mishra. Proc. of the IEEE, **80** (1992) 494.

Analysis of Substrate Deep-Trap Effects on the Turn-on Characteristics in GaAs MESFETs

K. Horio, A. Wakabayashi, S. Otsuka and T. Yamada

Faculty of Systems Engineering, Shibaura Institute of Technology
307 Fukasaku, Omiya 330, Japan

Abstract --- Two-dimensional analysis of the turn-on characteristics of GaAs MESFETs is made in which deep donors "EL2" in the semi-insulating substrate and surface states are considered. It is found that abnormal current overshoot due to EL2 can be seen when the off-state gate voltage is deeply negative. This is because in such a case electrons are depleted also in the semi-insulating substrate in the OFF state, and the corresponding ionized EL2 density N_{EL2}^{+} around the channel-substrate interface can be much higher than in the ON state.

A. Introduction

In GaAs MESFETs, slow current transients were often observed experimentally even if the gate voltage was changed abruptly. This phenomenon is called "gate-lag" and is a serious problem in both digital and analog GaAs ICs [1, 2], but its mechanism is not well understood. Surface states were thought to be the main cause of this phenomenon, and in fact, recent studies by 2-D simulation indicated that the gate-lag could occur due to surface-state effects [3-5]. In this work, we have also simulated the substrate deep-trap effects on the turn-on characteristics and found that abnormal current overshoot and the subsequent slow transient could occur due to substrate traps when the off-state gate voltage is deeply negative.

B. Physical Model

Fig.1 shows the device structure analyzed here. The gate length L_G is 0.3 μm. For a substrate we consider undoped semi-insulating LEC (liquid-encapsulated Czochralski) GaAs where deep donors "EL2" (N_{EL2}) compensate shallow acceptors (N_{Ai}) [6]. Here we assume $E_C - E_{EL2} = 0.69$ eV and $N_{Ai}/N_{EL2} = 0.1$ [7], where E_{EL2} is the energy level of EL2. For the surface-state model, we adopt Spicer's unified defect model [8], and assume that the surface states consist of a pair of a deep donor and a deep acceptor. The following two cases based on experiments are considered.
  a) Sample 1: $E_{SD} = 0.925$ eV, $E_{SA} = 0.8$ eV [8], b) Sample 2: $E_{SD} = 0.87$ eV, $E_{SA} = 0.7$ eV [9]
where E_{SD} is the energy difference between the bottom of the conduction band and the deep donor's energy level, and E_{SA} is the energy difference between the deep acceptor's energy level and the top of the valence band. The surface states are assumed to be distributed uniformly within 5 Å from the surface and their density N_S is typically set to 10^{13} cm^{-2} [5]. The basic equations are Poisson's equation including ionized deep-level terms, continuity equations for electrons and holes, and three rate equations for the deep levels. These equations are solved numerically in two dimension. We have calculated the turn-on characteristics when the gate voltage is changed abruptly.

C. Surface-State Effects on the Gate-Lag

First, we describe cases with surface states only, where the substrate is assumed perfectly insulating and has no traps. Fig.2 shows the calculated turn-on characteristics of GaAs MESFETs with different types of surface states (Sample 1 and Sample 2). For the Sample 1 case, the drain current shows a fast response. But, for the Sample 2 case, the drain current remains a low value for some period and begins to increase slowly around t = 10^{-2} s, showing the large gate-lag. This difference originates from the fact that the deep-acceptor surface state, which determines the Fermi level at the surface, acts as an electron trap for Sample 1, and it acts as a hole trap for Sample 2 [4].

Fig.3 shows potential profiles for the Sample 1 case. The potentials at the surface are almost unchanged between the OFF and ON states. The channel is essentially shut down under the gate and hence the drain current shows the fast response. Fig.4 shows potential profiles for the Sample 2 case. In this case, as seen in Fig.4(a), the channel is depleted entirely from source to drain. So, even if the gate voltage is changed, the drain current remains low until the deep acceptors begin to capture or emit carriers to change their ionized density. As can be understood from the profile Fig.4(b), holes are supplied from the Schottky contact and captured by the deep acceptors, and hence the ionized deep-acceptor density N_{SA}^{-} begins to decrease. Thus the width of the depletion region under the surface-state layer begins to decrease, leading to the slow increase in the drain current.

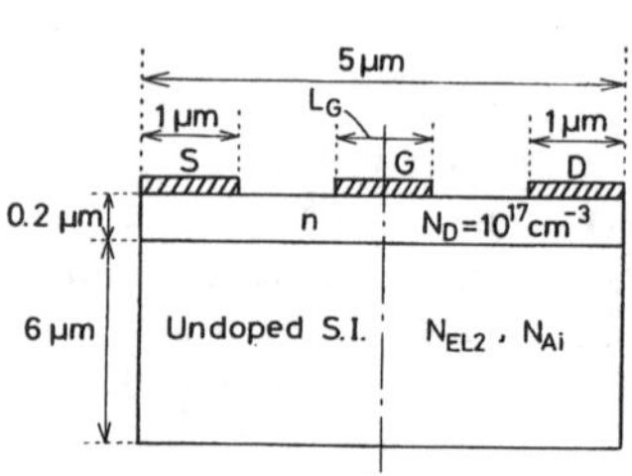

Fig.1 Device structure analyzed in this study.

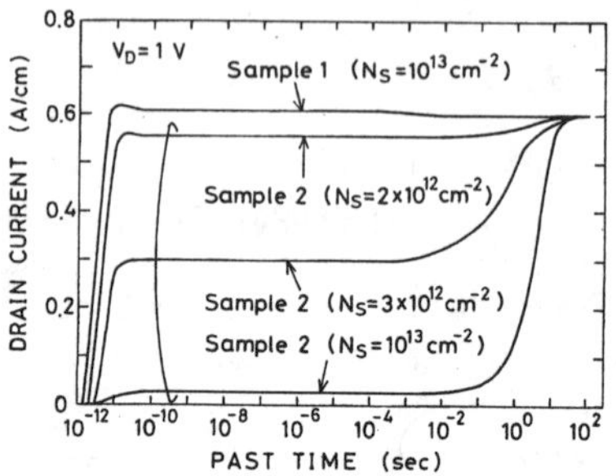

Fig.2 Calculated turn-on characteristics of GaAs MESFETs with different kinds of surface states. Perfectly insulating substrate is assumed.

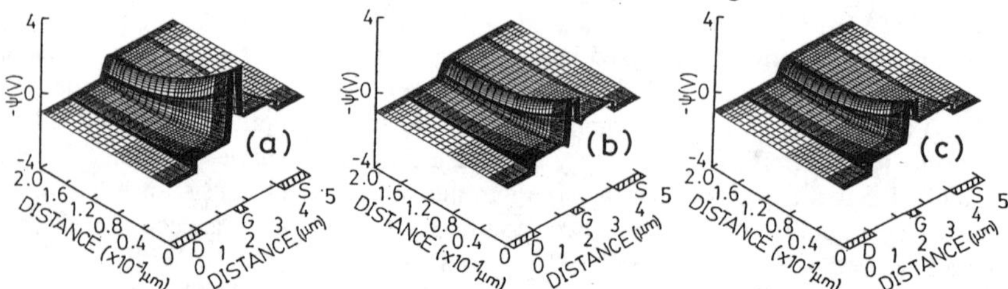

Fig.3 Potential profiles for the Sample 1 case, corresponding to Fig.2. (a) OFF, (b) $t = 10^{-6}$ s, (c) $t = 10^2$ s (ON).

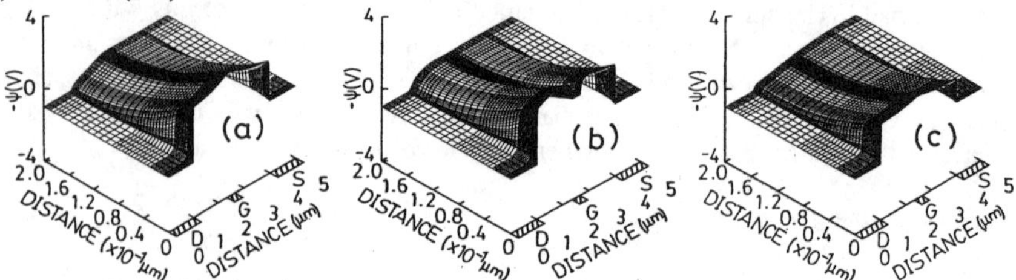

Fig.4 Potential profiles for the Sample 2 case, corresponding to Fig.2 ($N_S = 10^{13}$ cm^{-2}). (a) OFF, (b) $t = 10^{-6}$ s, (c) $t = 10^2$ s (ON).

328

D. Substrate Effects on the Turn-on Characteristics

Next, we discuss the case with substrate deep traps only, where surface states are not included. Fig.5 shows the calculated turn-on characteristics of the GaAs MESFET with $N_{Ai} = 10^{16}$ cm^{-3} as a parameter of the off-state gate voltage V_{Goff}. When V_{Goff} is equal to the threshold (pinch-off) voltage V_{th} (= − 2.3 V), the characteristics show a fast response with little gate-lag. However, when V_{Goff} is much more negative than V_{th}, abnormal current overshoot and the subsequent slow transient are observed. We will discuss below why these happen.

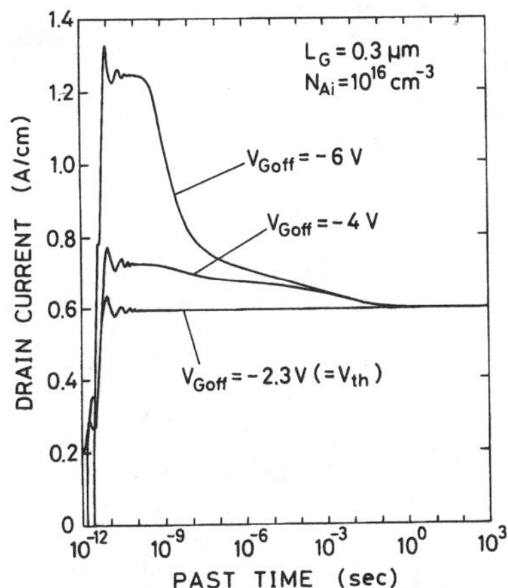

Fig.5 Calculated turn-on characteristics of GaAs MESFET made on an undoped semi-insulating substrate as a parameter of off-state gate voltage V_{Goff}. Surface states are not included.

Fig.6 shows electron density profiles for the case of $V_{Goff} = -4$ V and the corresponding profiles of ionized EL2 density N_{EL2}^{+} in the substrate. From Fig.6(a), we see that electrons are depleted also in the semi-insulating substrate, and hence N_{EL2}^{+} around the channel-substrate interface becomes much higher for the OFF state than for the ON state (Fig.6(c)). Because of this increase in positive space-charge in the substrate, when the gate voltage is switched on, electron densities in the channel becomes higher than for the ON state, as shown in Fig.6(b). Electrons are also injected into the semi-insulating substrate. These are the reasons why the current overshoot arises. It is interpreted that the drain current begins to decrease gradually as the deep donors "EL2" begin to capture electrons that are injected into the substrate. This leads to the decrease in N_{EL2}^{+}, reaching the steady-state profiles and the steady-state current.

Finally, we describe cases with both surface states and substrate deep traps. Fig.7 shows examples of the calculated turn-on characteristics. When V_{Goff} is equal to the threshold (pinch-off) voltage (V_{th1} or V_{th2}), the characteristics show fast response for the Sample 1 case and show the gate-lag behavior for the Sample 2 case, as in the cases without substrate traps (cf. Fig.2). When V_{Goff} is deeply negative (− 4 V), abnormal current overshoot and the subsequent slow transient due to EL2 in the substrate are observed for the Sample 1 case. For the Sample 2 case, the drain current becomes lower than that for $V_{Goff} = V_{th2}$ due to the surface-state effect [3], but shows slight overshoot (due to substrate traps) in relatively early periods, which was also seen qualitatively in an experiment [2].

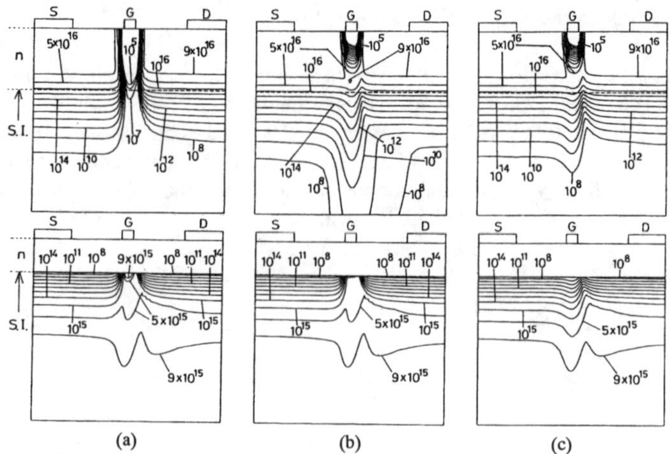

Fig.6 Electron density profiles (upper) and ionized EL2 density N_{EL2}^{+} profiles (lower) for the case of $V_{Goff} = -4$ V, corresponding to Fig.5. (a) OFF, (b) $t = 10^{-8}$ s, (c) $t = 10^{2}$ s (ON).

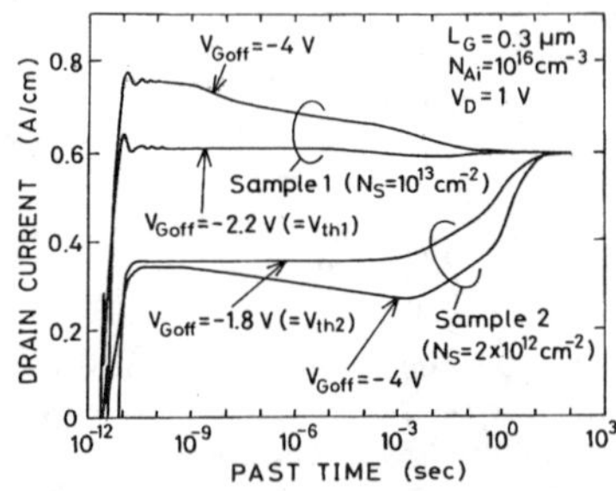

Fig.7 Calculated turn-on characteristics of GaAs MESFETs with both surface states and substrate deep traps (EL2).

E. Conclusion

Two-dimensional analysis of the turn-on characteristics in GaAs MESFETs is made in which deep donors "EL2" in the semi-insulating substrate and surface states are considered. It has been shown that abnormal current overshoot and the subsequent slow transient can be seen due to deep traps in the substrate when the off-state gate voltage is deeply negative. Both surface states and substrate deep traps should be taken into account when considering the gate-lag phenomenon.

References

[1] R. Yeats *et al., IEDM Tech. Dig.,* pp.842-845, 1988.

[2] Y. Kohno *et al., Proceedings of GaAs IC Symposium,* pp.263-266, 1994.

[3] S. H. Lo and C. P. Lee, *IEEE Trans. Electron Devices,* vol.41, pp.1504-1512, 1994.

[4] K. Horio, K. Satoh and T. Yamada, *Proceedings of SISDEP'95,* pp.78-81, 1995.

[5] K. Horio and T. Yamada, *Proceedings of GaAs IC Symposium,* pp.175-178, 1996.

[6] D. E. Holmes *et al., IEEE Trans. Electron Devices,* vol.ED-29, pp.1045-1051, 1982.

[7] K. Horio, H. Yanai and T. Ikoma, IEEE Trans. Electron Devices, vol.35, pp.1778-1785, 1988.

[8] W. E. Spicer *et al., J. Vac. Sci. Technol.,* vol.16, pp.1422-1433, 1979.

[9] H. H. Wieder, *Surface Sci.,* vol.132, pp.390-405, 1983.

A Study of Crystal Defects in Radiation Detector Grade Semi-Insulating GaAs

D. Korytár[1], C. Ferrari[2], S. Strzelecka[3], A. Šatka[4], J. Darmo[1], F. Dubecký[1], and A. Hruban[3]

[1]*Institute of Electrical Engineering, Slovak Academy of Sciences, Dúbravská cesta 9, SK-842 39 Bratislava, Slovakia*
[2]*MASPEC, CNR, Via Chiavari 18/A, I-43 100 Parma, Italy*
[3]*Institute of Electronic Materials Technology, Wolczynska 133, PL- 01 919 Warsaw, Poland*
[4]*Faculty of Electrical Engineering and Information Technolog, Slovak Technical University, Ilkovičova 3, SK-812 19 Bratislava, Slovakia*

Abstract

Crystal defects in semiconducting materials can play a crucial role in electrical parameters and performance of devices. In this work, detector grade bulk SI GaAs wafers of various producers have been examined by several characterisation techniques with the aim to compare their crystal perfection and to correlate the observed structural properties with the detecting properties of fabricated detectors. Three types of dislocation distribution by X-ray topography and etching have been identified in bulk SI GaAs materials: i) dislocation-free, ii) slip-like, and iii) cellular structure. (004) rocking curves half-width have been measured to compare the overall crystal perfection of wafers, too. Microprecipitates were studied by infrared light scattering tomography. Collection efficiency and energy resolution of radiation detectors manufactured from the bulk SI GaAs wafers showed strong dependence on substrate quality[1,2]. .Homogeneity of the detector charge collection was investigated by the scanning EBIC.

Introduction

Recently, bulk semi-insulating GaAs (SI GaAs) has found the application as a promising base material for detectors of ionising radiation (mainly X- and γ-rays), operated at room temperature (see e.g. [3] and references therein). In addition to impurity concentration and basic physical/electrical parameters, the crystal defects are especially important because the active volume of a detector is represented by the bulk of semiconductor substrate itself. X-ray diffraction, optical light scattering and other techniques can be effectively used to characterize these crystal defects and using the results of electrical testing also to study correlation between crystal quality and performance of detectors.

Experimental

Two-inch or three-inch diameter SI GaAs LEC or VGF wafers of several suppliers (A-N) have been studied by double crystal X-ray topography (DCT), high resolution X-ray diffractometry (HRXD), light scattering tomography (LST), electron beam induced current (EBIC), and detector testing techniques.

     331

Sample	FWHM	DCT	LST
A1 LEC	-	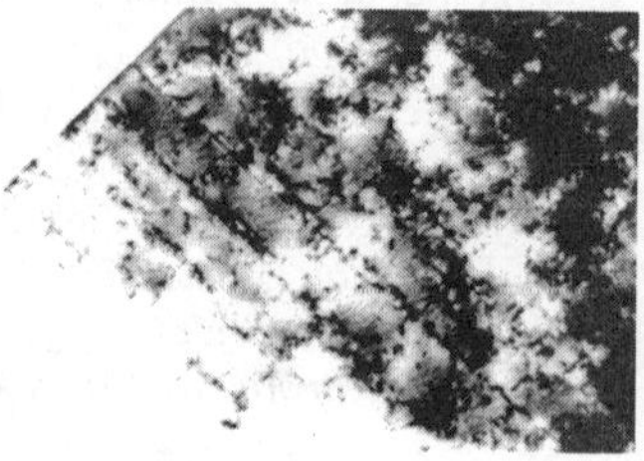	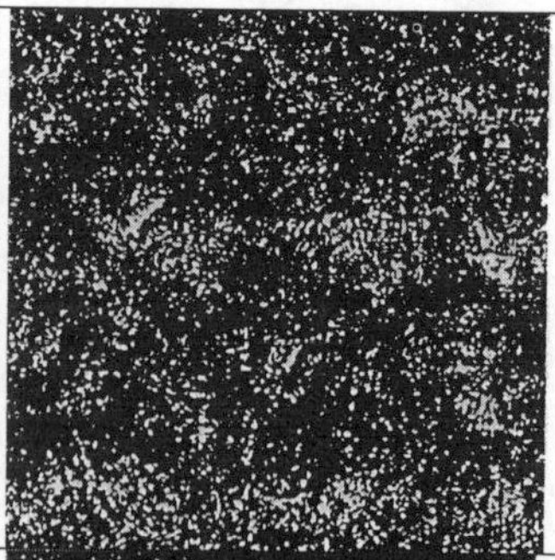
B1 LEC	7.2	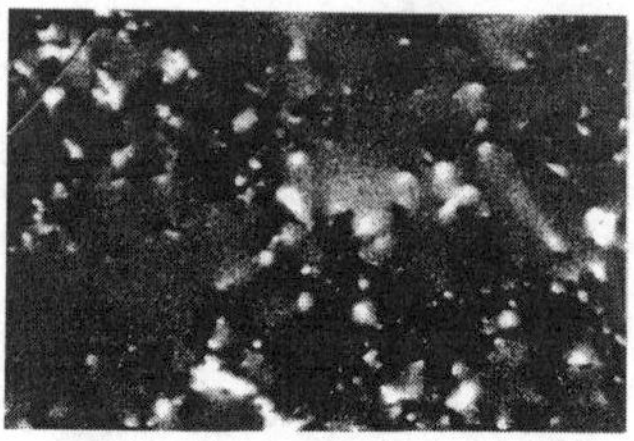	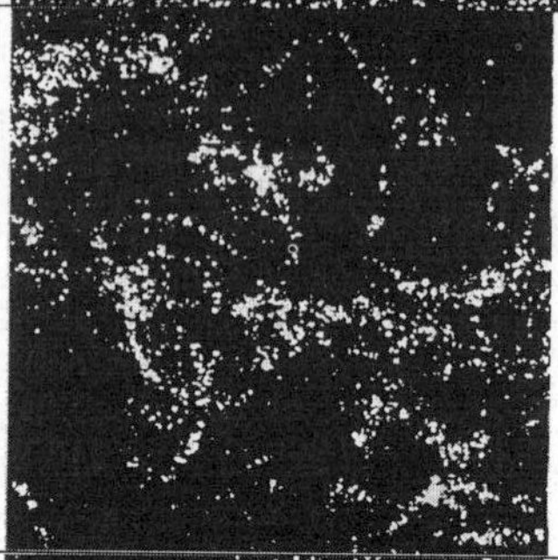
C LEC	17.4	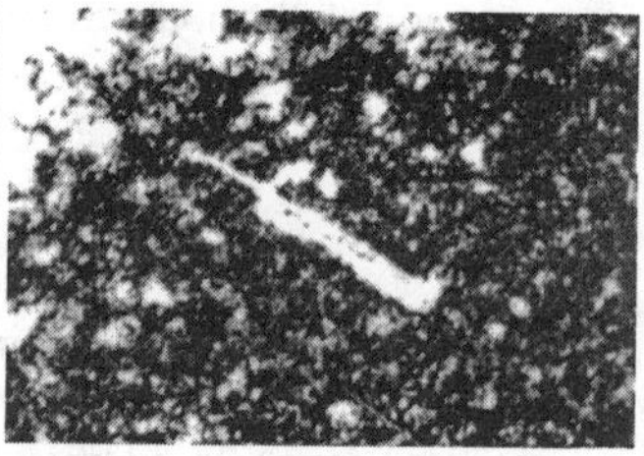	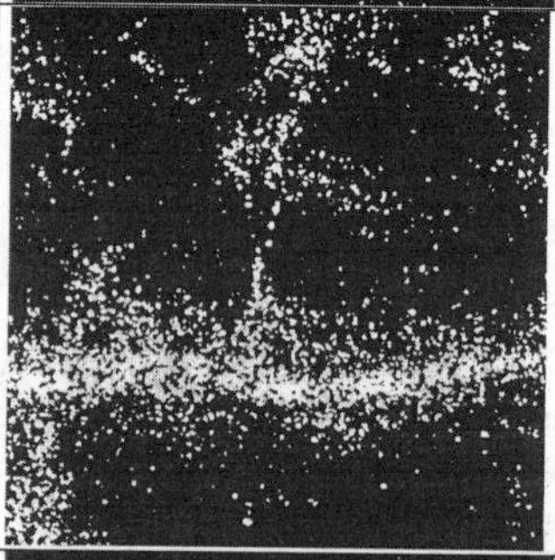
D1 VGF	6.2	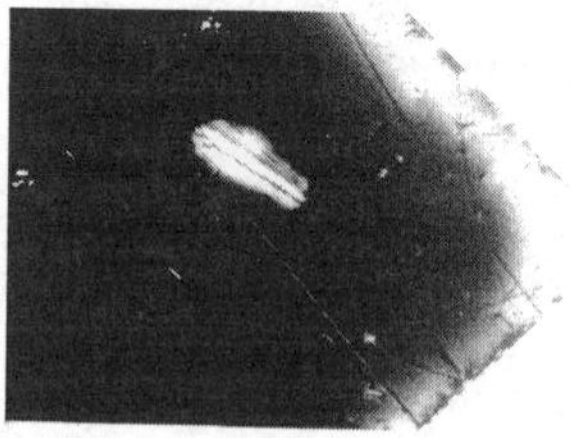	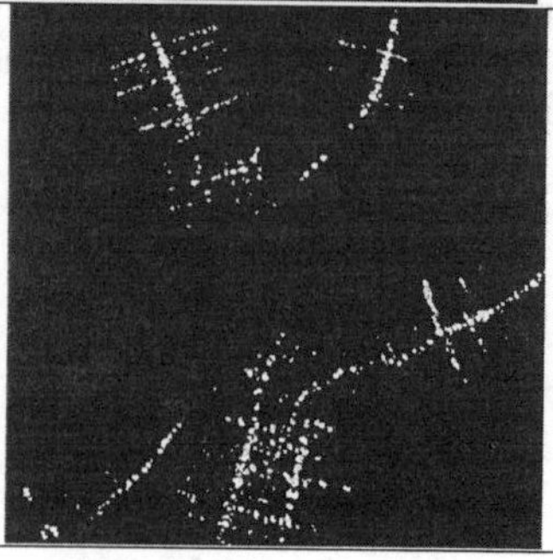
Sample	FWHM	DCT	LST

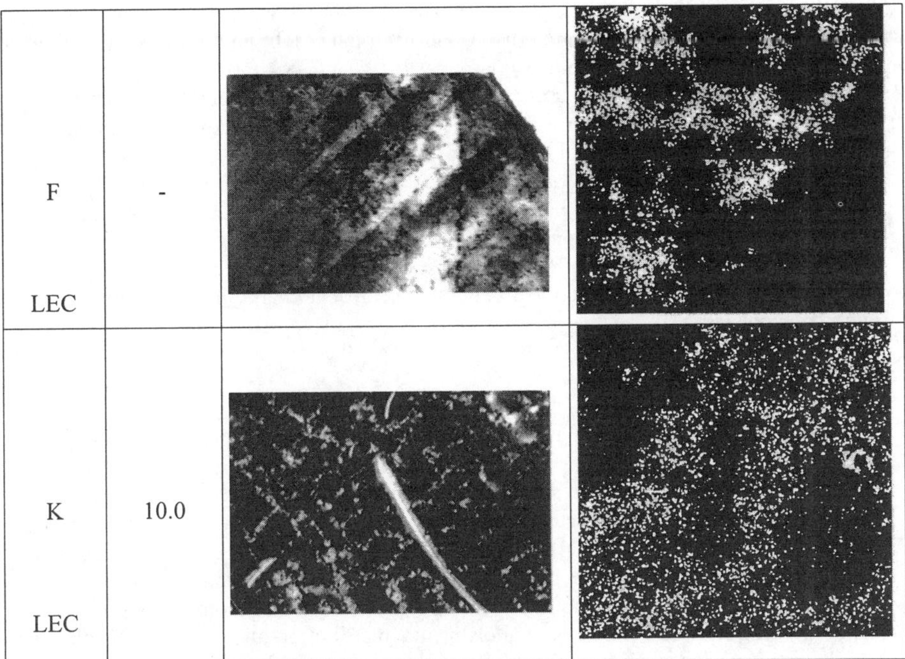

F LEC	-		
K LEC	10.0		

Figure1. DCT and LST images of SI GaAs samples of various producers. See text for the details.

a)   b)

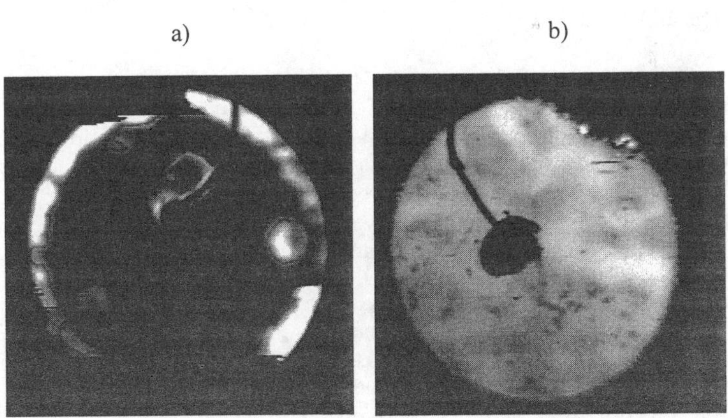

Figure 2. EBIC pictures of inhomogeneous sample K (a) and annealed, more homogeneous sample A1(b). Accelerating voltage 19 kV, high injection,, bias voltage 84 V.

Symmetrical (004) rocking curves halfwidths giving an overall information on crystal perfection have been measured by a Philips high-resolution X-ray diffractometer equipped with a four crystal Ge monochromator set at (220) $CuK_{\alpha1}$ reflection. geometry with a horizontal beam divergence of 12 arcsecs. A double crystal X-ray topographic camera [4] with Ge collimating crystal set to an asymmetric (620) $CuK_{\alpha1}$ diffraction radiation has been applied to take (620) asymmetric double crystal topographs from GaAs samples.

For Light Scattering Tomography (LST) an Nd:YAG laser emitting monomode 1.06 μm light and an optical microscope with infrared CCD camera, oriented perpendicular to the laser beam, have been used in a computerized equipment.

When applying the Electron Beam Induced Current (EBIC) technique, GaAs detector Schottky junction has been used to collect generated electrical charge in a standard scanning electron microscope (SEM).

Results

Fig.1 shows DCT and LST images of SI GaAs samples denoted by capital letters. DCT images come from about 5μm near surface region while LST images come from the central parts of the samples. Vertical sides of DCT images represent 2.7 mm except B1-case where it is 1.0 mm. Sides of LST images are 257 μm long. Full widths at half-maxima of the rocking curves measured by HRXD are given in arcsecs. Estimated dislocation density is more than 10^5 cm^{-2} only in C-sample, in other samples is less, D1 sample is dislocation-free.

Apart from scratches, pronounced cellular structure is observed in A1, B1 and K samples, both in DCT and LST images, while slip-like distribution of dislocations is observed in F-sample, in accordance with [5,6]. While the lowest microdefects concentration is in dislocation-free VGF sample D1, their most homogeneous distribution is in annealed [7] sample A1. These two wafers gave the best detectors as for detecting efficiency and peak to valley ratio [1,2] measured by means of 122 keV photons. Fig.2 shows EBIC contrast of relatively inhomogeneous sample K and more homogeneous sample A1.

Conclusions

X-ray diffraction and light scattering techniques have been used to characterize crystal perfection of radiation detector grade SI GaAs wafers of various producers. In connection with [1,2] it can be concluded that dislocation-free wafers with low concentration of microprecipitates and annealed wafers with homogeneous concentration of microprecipitates give radiation detectors of the best parameters.

References

[1]   F. Dubecký and J. Darmo, presented at this conference
[2]   F. Dubecký et al., presented at this conference
[3]   T. E. Schlesinger and R. B. James (vol. Eds.), Semiconductors for Room Temperature Nuclear Detector Applications, Semiconductors and Semimetals, **43**, treatise eds. R.K. Willardson, Albert C. Beer, and Eicke R. Weber, (Academic Press, San Diego 1995)
[4]   B. Jenichen, R., Köhler, and W. Möhling, J. Phys. E: Sci. Instrum. **21** (1988) 1062
[5]   B. Zielinska-Rohozinska, J. Cryst. Growth 87(1988) 154
[6]   D. Korytár J. Cryst. Growth 126(1993) 30
[7]   B. Hoffmann, et al., in the Proceedings of the SIMC'9, ed. Ch. Fontaine, p.63

Electrical and detection performances of particle detectors based on bulk semi-insulating InP

F. Dubecký[1], R. Fornari[2], M. Krempaský[1], J. Darmo[1], E. Gombia[2],
M. Pikna[3], P.G. Pelfer[4], M. Sekáčová[1], and M. Ruček[1]

[1]*Institute of Electrical Engineering, Slovak Academy of Sciences, Dúbravská cesta 9
SK-84239 Bratislava, Slovakia; e-mail: elekfdub@savba.sk*
[2]*MASPEC CNR, Parma, Via Chiavari 18/A, I-43100 Parma, Italy*
[3]*Faculty of Mathematics and Physics, Comenius University, Mlynská dolina
SK-842 33 Bratislava, Slovakia*
[4]*Department of Physics of the University of Florence and INFN, Largo E.Fermi 2
I-50125 Firenze, Italy*

Abstract – **A study of electrical transport properties and detection performances of the semi-insulating InP-based detectors of charged particles and ionizing radiation is presented. Detectors with a top P^+ layer and a Schottky back contact give a charge collection efficiency over 90 % and energy resolution about 3% (FWHM) for 5.48 MeV α-particles at 220 K. Detection of 60, 122 and 662 keV γ-rays in a temperature range 215÷250 K is demonstrated.**

A. Introduction

InP is considered as a potential candidate material for room temperature (RT) operating detectors of charged particles and γ-rays due to the large value of Z of In (49), and the wide band-gap (1.34 eV at RT) [1]. The first interest has come from the InP detector as a real time solar neutrino detector with the ability to measure the incident neutrino energy [2,3]. InP detectors based on conductive LPE material with a base thickness of 6÷25 μm were also fabricated and tested [4] for detection of α-particles and photons (60 and 122 keV) at RT. Detectors based on bulk SI InP:Fe were studied at RT using 5.48 MeV α-particles [5-7].

In this paper a study of the SI InP:Fe based radiation detectors with a top P^+ contact grown by the MOCVD and a Schottky back contact is presented. The I-V characteristics at different temperatures and the detection performances for α-particles (5.48 MeV) observed at 220 and 243 K are shown. Detection of 60, 122 and 662 keV γ-irradiation was possible observed at operating temperatures of 215÷250 K obtained using a Peltier cooler.

B. Experiment and discussion

Bulk LEC (liquid encapsulated Czochralski) SI InP containing Fe with RT Hall mobility of 2200 cm^2/Vs and resistivity of 1.1×10^7 Ωcm was used as base material. Detectors were fabricated from (100) oriented wafer with Fe concentration of about 1.5×10^{16} cm^{-3}. The wafer was polished from both sides down to 180 μm afterwards one side (the top) a P$^+$ layer (60 nm, Zn doping with the hole concentration of 1.8×10^{18} cm^{-3}) was deposited by MOCVD. Circular Ti/Au contacts (20/40 nm) 2 mm in diameter were formed using photolithography and MESA etching while the rest part of the top side was passivated by sputtered silicon nitride layer (100 nm). Using the same Ti/Au metallisation a full area Schottky contact was evaporated onto the back side of the wafer. Detector chips were mounted on a capton foil and contacted by a silver paste.

The "reverse" current voltage (I-V) characteristics (with the P$^+$ contact negatively biased) measured at different temperatures are reported in Fig. 1. They show a linear shape up to about 10 V followed by a weak sublinear part and a sharp current increase at a voltage over 200 V. The resistivity calculated from the linear part of the I-V characteristics is about 3 times higher then the value obtained from the galvanomagnetic measurement. This effect could be explained by a weak blocking behaviour, or free carrier extraction from detector volume near its electrodes.

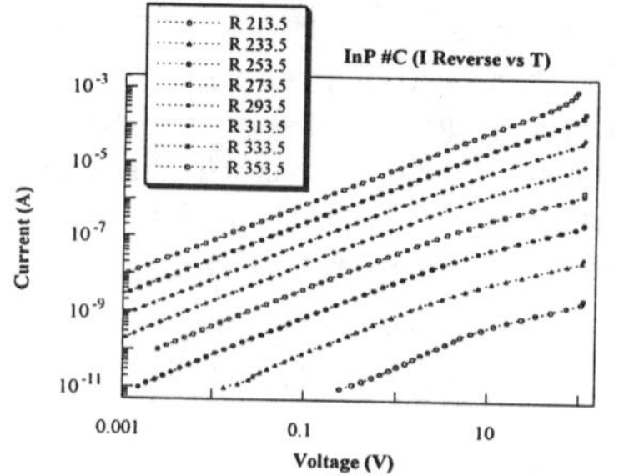

Fig. 1. I-V characteristics of SI InP:Fe detector measured at various temperatures (top bias negative).

During the test of the detector the holder was mounted on a Peltier cooler using a silicone paste. The signal produced by the ionizing radiation was detected using a standard front-end (F-E) electronic chain with a shaping time of 0.5 μs for negative and 1 μs for the positive top bias, followed by a multichannel analyzer. Pulse height spectra were acquired and stored for detectors tested at different bias voltages and temperatu-res. The F-E chain was callibrated using a stan-dard Si detector. Energy of 4.2 eV for the e-h pair production in InP [1] was used in this calculation. The spectra of α-particles from ^{241}Am source (5.48 MeV) were measured at lowered temperature with different bias voltages. Measured sample was placed in vacuum together with the radiation source. The CCE (charge collection efficiency) and the energy resolution in FWHM (full width at half maximum) observed at 243 K were calculated from the detected spectra and plotted as a function of the bias voltage (Fig. 2). The observed CCE increases until the breakdown voltage is reached, as recently published [5-7]. This result indicates that the detector breakdown occurs before than saturation of the CCE is reached. The best CCE and energy resolu-

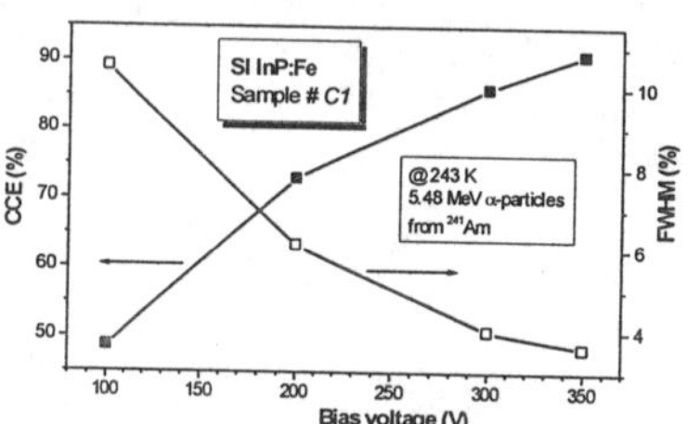

Fig. 2. Calculated CCE and FWHM vs. voltage for 5.48 MeV α-particles detected by SI InP detector.

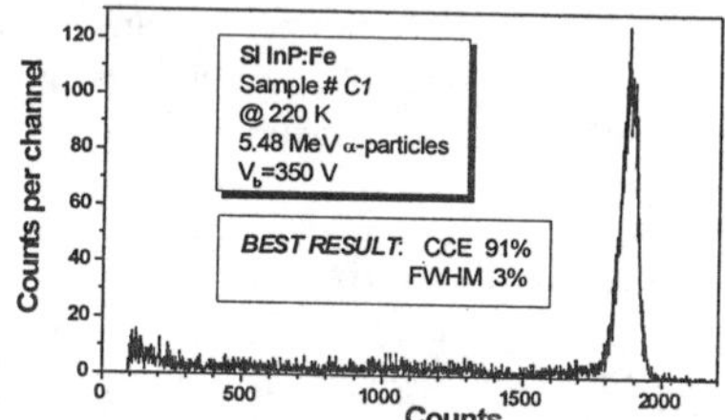

Fig. 3. Spectrum of 5.48 MeV α-particles detected by SI InP detectors at a temperature of 220 K.

tion (in FWHM) values of 91% and 3% respectively, were observed at a temperature of 220 K and a bias of 350 V. This is a substantial improvement comparing to the values obtained at RT with the same detectors (60% and 7.5% [6,7]). The values observed at 220 K are not far from the best results reported for SI GaAs detectors operating at RT (96% and 2.4% for the CCE and FWHM, respectively [8]).

The detected pulse height spectra of detectors using 60 keV X-rays ([241]Am source) are presented in Fig. 4, where the influence of the temperature at a bias voltage of 150 V (top polarity negative) is shown in Fig. 4a. As it can be see the signal due to X-rays was detected by cooling the detector at temperatures lower than –25 °C only due to a large shot noise of the detector. Two peaks are clearly revealed at temperatures <240 K; their position did not change with temperature, while the energy resolution improved by decreasing the temperature. Considering the precise energy callibration, the peak positioned at the higher channel number corresponds to the 60 keV photopeak. The peak corresponding to the lower energy has been related to the Compton effect. A higher amplitude of this peak indicates a larger content of defects in detector volume. The influ-

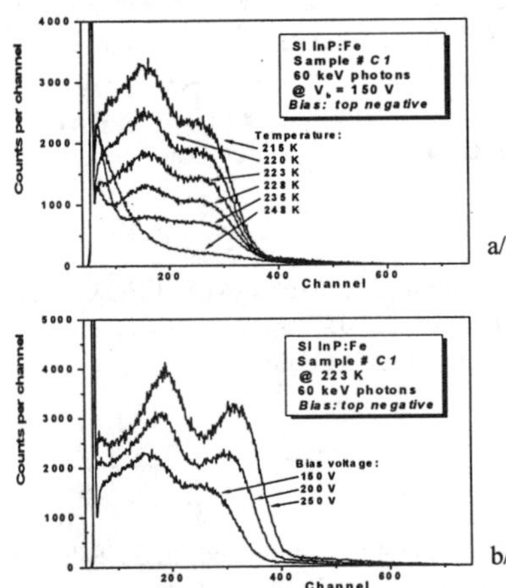

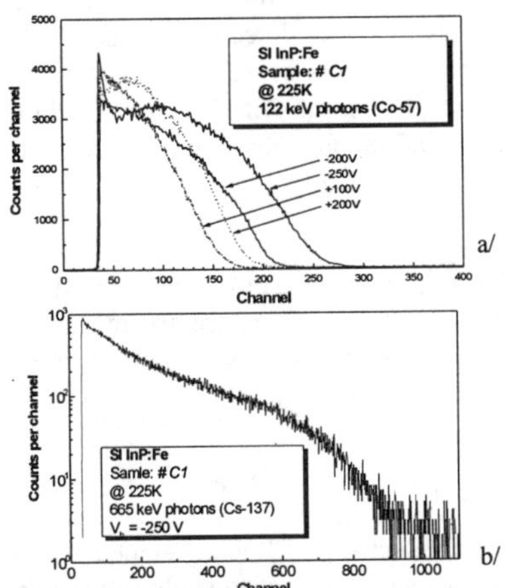

Fig. 4. 60 keV photon spectra detected by SI InP:Fe detector at various temperatures (a) and bias (b).

Fig. 5. 122 keV (a) and 662 keV (b) photon spectra detected by SI InP:Fe detector at 225 K (-250 V).

ence of the bias voltage on the measured pulse height spectra at a temperature of 223 K is illustrated in Fig. 4b. The energy resolution and the CCE improve with increasing bias. Detector operates at both bias polarities (also in the case of detection of alpha particles). It confirms that electrical field is well distributed within almost the all detector volume. However, the best detector performances (CCE, FWHM) for X-rays were obtained with the negative top bias at a voltage of 250 V and at the lowest temperature used in investigation (215 K). On the other hand, the use of positive bias gives the lower noise allowes the detection of X-rays at slightly higher temperature, at about –20 °C. Pulse height spectra obtained using gamma sources of 122 keV ([57]Co) and 662 keV (^{137}Cs) at a temperature of 225 K and bias –250 V are presented in Fig. 5a and Fig. 5b, respectively. As can be seen, no photopeak is observed in the spectra. This is explanable as a result of bad hole collection as well as a large Compton scattering of photons related to material impurities, defects and inhomogeneities.

C. Conclusions

In conclusions, radiation detectors fabricated from bulk LEC SI InP:Fe were investigated for detection of 5.48 MeV α-paricles, 60, 122 and 662 keV γ-rays at temperatures 215$\div$250 K mainted by a solid-state (Peltier) cooler. Detectors with a P^+ top MOCVD layer and back Schottky contact when cooled offer an interesting possibility for detection of γ-rays and charged particles. It is worth noting that no blocking behavior was observed in the current transport when top P^+ contacts were used. Obtained values of the CCE (over 90%) and energy resolution (about 3% FWHM) at a temperature of 220 K and a bias of 350 V present the best reported values for an InP detector. These values are not very far from the best results obtained with SI GaAs-based detectors at RT (96%, 2.4%, respectively) with thinner detector (100 μm) and at a higher bias voltage (500 V). However detection performances of the SI InP detectors for γ-rays are still far from the values reported for SI GaAs detectors (CCE>85%, FWHM<15%). In the measured spectra of X-rays two peaks were revealed. Present explanation is that the peak positioned at lower energy is due to the interaction of photons with InP matrix via Compton effect. To improve detection performances of SI InP radiation detectors bulk material with higher purity and better homogeneity should be developed.

Acknowledgements - Authors are grateful to G. Mignoni and A. Motta (MASPEC C.N.R., Parma) for the assistance in wafer treatment, and to R. Senderák and J. Ivanco for assistance in detector technology. This work was supported, in a part, by the Slovak Grant Agency through grants No. 2/5166/98 and 95/5305/509.

References

[1] Semiconductors for Room Temperature Nuclear Detector Applications, eds. T.E. Schlessinger and R.B. James in Semiconductors and Semimetals, treatise eds.: R.K. Willardson, A.C. Beer and E.R. Weber, vol. **43**, (Academic Press, San Diego, 1995).

[2] J.C. Lund, F. Olschner, F. Sinclair, and M.R. Squllante, Nucl. Instr. Meth. **A272,** (1988) 885.

[3] F. Olschner, J.C. Lund, M.R. Squillante, and D. L. Kelly, IEEE Trans. on Nucl. Sci. **NS-36,** (1989) 210.

[4] Y. Suzuki, Y. Fukuda, and Y. Nagashima, Nucl. Instr. Meth. **A275** (1989) 142.

[5] A. Valentini, A. Cola, G. Maggi, V. Paticcho, F. Quaranta, L. Vasanelli, Nucl. Instr. Meth. **A373,** (1996) 47.

[6] F. Dubecký, M. Krempaský, J. Darmo, M. Sekáčová, M. Pikna, R. Fornari, E. Gombia, P.G. Pelfer, and P. Hudek, in Proc. of the Int. Conf. ASDAM '96, Smolenice 96 eds. T. Lalinský, F. Dubecký, J. Osvald and Š. Haščík (IEE SAS, Bratislava 1996), pp 157-160.

[7] F. Dubecký, R. Fornari, J. Darmo, M. Pikna, E. Gombia, M. Krempaský, P.G. Pelfer, M. Sekáčová, P. Hudek, and M. Ruček, Nucl. Instr. Meth. A (1998), in print.

[8] M. Alietti, C. Canalli, A. Castadliny, A. Cavallini, A. Cenronio, C. Chiossi, S. D'Auria, C. del Papa, C. Lanzieri, F. Nava, P. Vanni, Nucl. Instr. Meth. **A362,** (1995) 344.

Influence of Metallic Spikes and Non-Uniform Density of Two-Dimensional Electron Gas (2DEG) on the Contact Resistance to AlGaAs/GaAs Heterostructures

Zbigniew Tkaczyk and Andrezj Wolkenberg
Insitute of Electron Technology, al. Lotnkików 32/46, 02-668 Warsaw, Poland
wolken@ite.waw.pl

An expression for the series resistance of an HEMT has been derived, taking into account for the first time the nonuniformity of the ohmic contact interface. The penetration of the AuGeNi metallization into the 2DEG in the subcontact area (in the form deep penetrating "spikes") has been investigated and the model has been elaborated in order to determine the correct value of the contact resistance depending on the fraction of the area of penetration. Simulation of the growth of the spikes depending on the thermal processes has been made. The prediction of the shape "spikes" and their density makes it possible to calculate the value of the area of penetration. Therefore, a correlation between the dependence has been experimentally verified and it gave good agreement with the model. It has established the optimum parameters of thermal processing to obtain the minimum value of series resistance. The model of nonuniformity of interface may also be used to solve the problem of the influence of a periodic electrical potential on ballistic transport in mesoscopic devices.

The extremely high frequency and high-speed switching properties achieved in the High-Electron-Mobility Transistor (HEMT) result from the superior electron dynamics at its heterointerface, where the two-dimensional electron gas (2DEG) arises. The improvement of its performance is limited by its series resistance - R_s. The series resistance of the source is especially important because it causes negative feedback (on the input of the transistor) and therefore, the effective transconductance (the main electrical parameter) is limited. Because of these undesirable effects, the magnitude of R_s must be reduced to a minimum. It may be achieved by appropriate design of the HEMT structure. The reduction of the dimensions of the active regions of a transistor makes it possible to reduce the value of series resistance. But one component of the series resistance, contact resistance - R_c does not depend directly on the dimensions of a transistor. The value of the contact resistance depends rather on the parameters of subcontact layers and on the technology of ohmic contacts (on thermal annealing). Therefore, especially in the submicrometer gate-length transistors, it is essential to consider the influence of contact resistance on the series resistance, where the parameters of the heterostructure (AlGaAs/GaAs) from which the HEMT is made should be taken into account.

For the determination of the source resistance and the contribution of different heterostructure resistive layers and related contact resistances, a two-layer transmission line model has been commonly used [1-3]. In order to describe the influence of the contact resistance on the total value of the series resistance, different models have been developed depending on the ohmic contact penetration of 2DEG into the subcontract area. In the case of deep alloying contacts, the contact resistance elements have been treated as lumped elements R_{c1} and R_{c2}, representing contacts to the respective active layers (Feuer's model, Fig. 1). Whereas in the case of shallow alloying contacts, the whole area of contact is treated as a multi-layer transmission line using the distributed contact resistance and the sheet resistivity data of the associated semiconductor layer (Look's model, Fig. 2). It appears that neither of these approaches accurately represents the real situation in the contact. The main disadvantage of the commonly used AuGeNi metallization to the n-type heterostructure is non-uniform penetration of the contact interface during annealing (Fig. 3) [4].

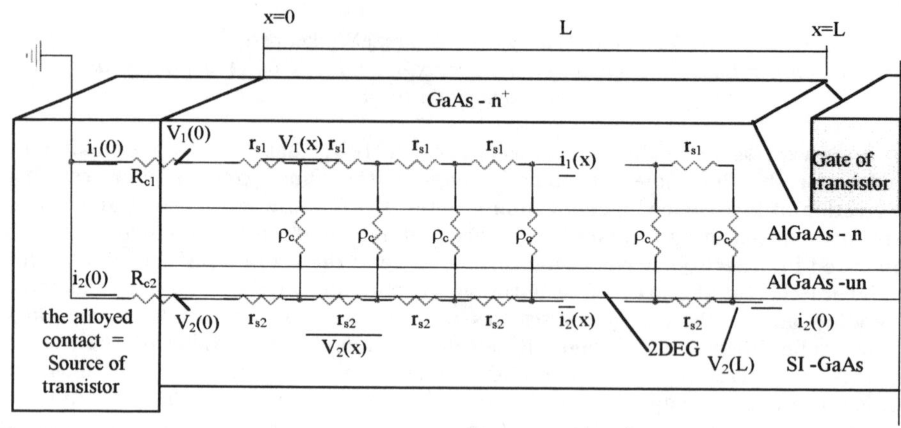

Fig. 1. The lumped elements R$_{c1}$ and Rc$_2$ representing contact resistance according to Feuer's model.

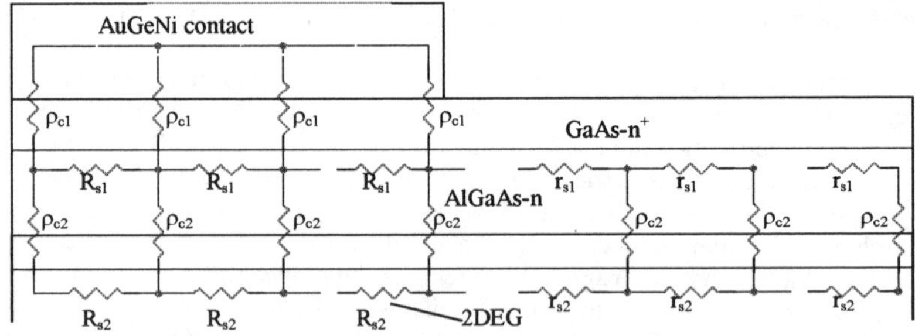

Fig. 2. Representation of the multi-layer transmission line of the contact area according to Look.

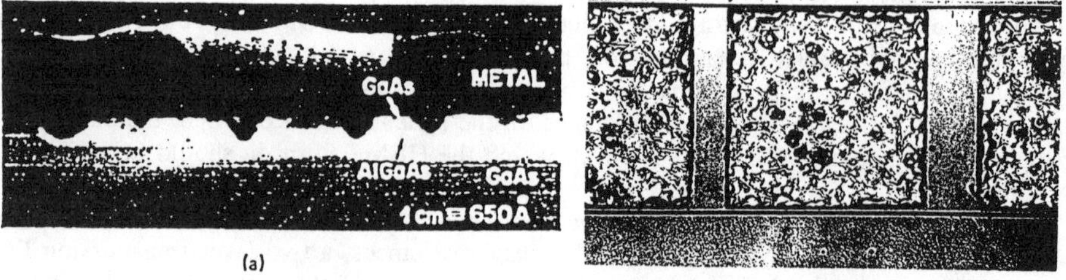

Fig. 3. a) Cross-sectional TEM micrography AuGeNi alloyed ohmic contact (420°C, 20s) [4],
     b) Optical micrograph of AuGeNi contact after metal removal in our TLM structures.

Experiments [5] revealed that the minimum resistance of an ohmic contact is achieved when two regions co-exist in the subcontact area, namely, a penetration region with deep, metallic "spikes" (corresponding to the alloying contacts) and a non-penetration region of 2DEG (corresponding to the non-alloying contacts). The deep "spikes" change the sheet

340

density of the 2DEG, making the resistance of these regions different from that of the regions without penetration. Therefore, the contact resistance depends on the density of the "spikes."

In this paper we propose for the first time a model to calculate the ohmic contact resistance, taking into account different components of the contact area and treating the total resistance as a parallel connection of the respective regions (penetrated and non-penetrated) (Fig. 4). The expression for the series resistance of an HEMT transistor, including a non-uniform interface of contact resistance R_c, as a function of the fraction penetrated area over the whole area of contact - x_f, has developed.

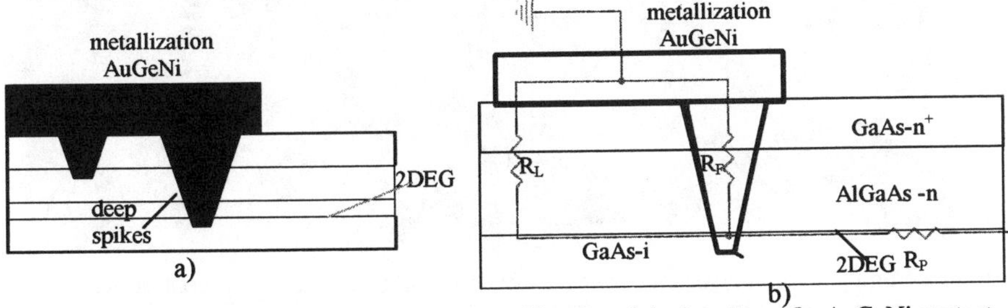

Fig. 4. a) The schematic presentation of the nonuniformity of the interface of a AuGeNi contact, b) The parallel connection of the suitable components representing the different area of contact.

In our model the effective contact resistance changes from the value obtained from Look's model (corresponding to the shallow alloyed, non-perforated region) to the value of Feuer's model (corresponding to the entirely perforated region) with an increase in the percentage of "spiking area" - x_f. A minimum has appeared in this dependence which is illustrated in Fig. 5 (curve - T).

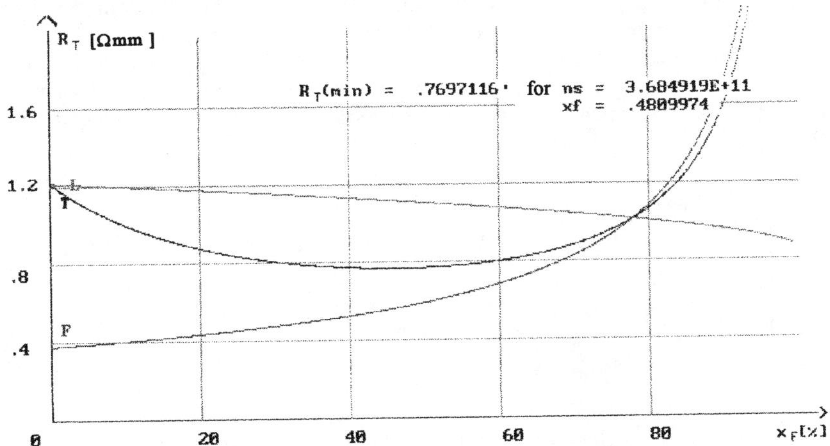

Fig. 5. The series resistance versus the fraction (%) of the penetrating area over the whole area of contact.

As the size and density of the "spikes" depend on thermal treatment of the contacts, the minimum contact resistance may be reached by appropriate adjustment of the thermal cycle in making the contacts. A model has been elaborated to stimulate the growth of the "spikes" depending on the parameters of the process of annealing of contacts (Figs. 6, 7).

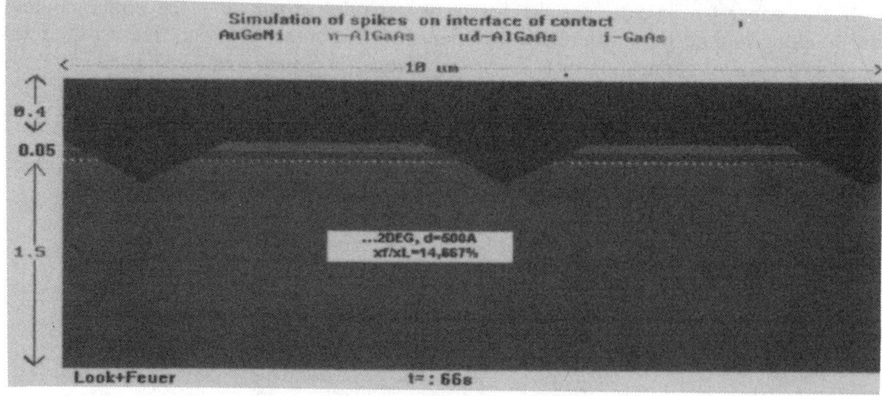

Fig. 6. Vertical simulation of growth of the "spikes" on the interface of the contact over time.

This model makes it possible to determine the value of the penetrating area and therefore, to predict the dependence of the contact resistance on the temperature of the annealing of contacts. The expression of the series resistance of the HEMT enables its value to be calculated depending on the alloying conditions of the contacts. To confirm our model, we have derived the experimental correlation between contact resistance and the value of the penetration area. The value of the contact resistance has been measured using a test structure (with TLM configuration) and the penetration area was determined from microscopic observation of the subcontract area revealed after removing the outer, non-alloying metallic layer (the size and density of the "spikes" was calculated). The dependence obtained from these experiments agrees well with the correlation predicted by our model. Experimental dependence between the series resistance and the temperature of the annealing of contacts established by other researchers also confirms our model [6,7].

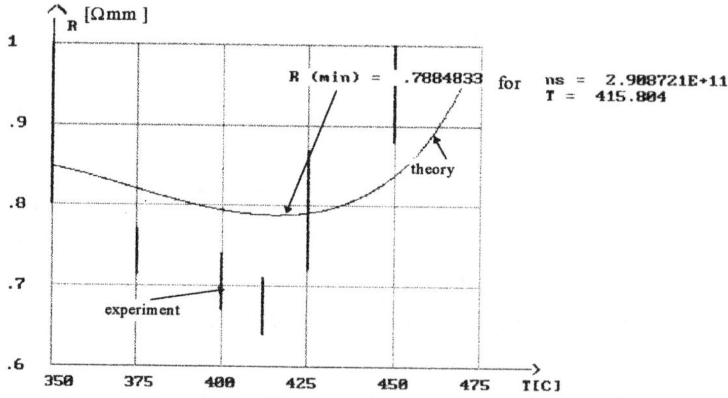

Fig. 7. Comparison of the theoretical calculation of series resistance dependent on the temperature with experimental results.

The results of this paper will be used in the manufacturing of low resistive, reliable ohmic contacts to HEMT transistors and may also be used to determine the contact resistance in mesoscopic devices, where the influence of periodic electric potential upon ballistic transport should be considered.

REFERENCES

[1]   M.D. Feuer, IEEE Trans. Electron. Dev. 32 (1985) 7.
[2]   D.C. Look, IEEE Trans. Electron. Dev. 35 (1988) 133.
[3]   S.J. Hawksworth, J.M. Chamberlain, T.S. Cheng, M. Henini, M. Heath, M. Davies and A. J. Page, Semicond. Sci. Technol. 7 (1992) 1085.
[4]   F. Ren, A.B. Emerson, S.J. Pearton, T.R. Fullowan and J. M. Brown, Appl. Phys. Lett. 58 (1991) 1030.
[5]   R.P. Taylor, P.T. Coleridge, M. Davies, Y. Feng, J.P. McCaffrey and P.A. Marshall, J. Appl. Phys. 76 (1994) 7966.
[6]   A. Ketterson, F. Ponse, T. Henderson, J. Klem and H. Morkoç, J. Appl. Phys. 57 (1985) 2305.
[7]   A. Christou and N. Papanicolaou, Solid State Electron. 29 (1986) 189.
[8]   V. Chabasseur-Molyneux, J.E.F. Frost, M.Y. Simmons, D.A. Ritchie and M. Pepper, Appl. Phys. Lett. 68 (1996) 3434.